Wave-Particle Duality

Edited by Paul F. Kisak

Contents

i

Chapter 1

Wave–particle duality

Wave–particle duality is the fact that every elementary particle or quantic entity exhibits the properties of not only particles, but also waves. It addresses the inability of the classical concepts "particle" or "wave" to fully describe the behavior of quantum-scale objects. As Einstein wrote: *"It seems as though we must use sometimes the one theory and sometimes the other, while at times we may use either. We are faced with a new kind of difficulty. We have two contradictory pictures of reality; separately neither of them fully explains the phenomena of light, but together they do"*.[1]

Through the work of Max Planck, Albert Einstein, Louis de Broglie, Arthur Compton, Niels Bohr, and many others, current scientific theory holds that *all* particles *also* have a wave nature (and vice versa).[2] This phenomenon has been verified not only for elementary particles, but also for compound particles like atoms and even molecules. For macroscopic particles, because of their extremely short wavelengths, wave properties usually cannot be detected.[3]

Although the use of the wave-particle duality has worked well in physics, the *meaning* or *interpretation* has not been satisfactorily resolved; see Interpretations of quantum mechanics.

Niels Bohr regarded the "duality paradox" as a fundamental or metaphysical fact of nature. A given kind of quantum object, will exhibit sometimes wave, sometimes particle, character, in respectively different physical settings. He saw such duality as one aspect of the concept of complementarity.[4] Bohr regarded renunciation of the cause-effect relation, or complementarily, of the space-time picture, as essential to the quantum mechanical account.[5]

Werner Heisenberg considered the question further. He saw the duality as present for all quantic entities, but not quite in the usual quantum mechanical account considered by Bohr. He saw it in what is called second quantization, which generates an entirely new concept of fields which exist in ordinary space-time, causality still being visualizable. Classical field values (e.g. the electric and magnetic field strengths of Maxwell) are replaced by an entirely new kind of field value, as considered in quantum field theory. Turning the reasoning around, ordinary quantum mechanics can be deduced as a specialized consequence of quantum field theory.[6][7]

1.1 Brief history of wave and particle viewpoints

Aristotle was one of the first to publicly hypothesize about the nature of light, proposing that light is a disturbance in the element aether (that is, it is a wave-like phenomenon). On the other hand, Democritus—the original *atomist*—argued that all things in the universe, including light, are composed of indivisible sub-components (light being some form of solar atom).[8] At the beginning of the 11th Century, the Arabic scientist Alhazen wrote the first comprehensive treatise on optics; describing refraction, reflection, and the operation of a pinhole lens via rays of light traveling from the point of emission to the eye. He asserted that these rays were composed of particles of light. In 1630, René Descartes popularized and accredited the opposing wave description in his treatise on light, showing that the behavior of light could be re-created by modeling wave-like disturbances in a universal medium ("plenum"). Beginning in 1670 and progressing over three decades, Isaac Newton developed and championed his corpuscular hypothesis, arguing that the perfectly straight lines of reflection demonstrated light's particle nature; only particles could travel in such straight lines. He explained refraction by positing that particles of light accelerated laterally upon entering a denser medium. Around the same time, Newton's contemporaries Robert Hooke and Christiaan Huygens—and later Augustin-Jean Fresnel—mathematically refined the wave viewpoint, showing that if light traveled at different speeds in different media (such as water and air), refraction could be easily explained as the medium-dependent propagation of light waves. The resulting Huygens–Fresnel principle was extremely successful at reproducing light's behavior and was subsequently supported by Thomas Young's 1803 discovery of double-slit interference.[9][10] The wave view did not immediately displace the ray and particle view, but began to dominate scientific thinking about light in the mid 19th century, since it could explain polarization phenomena that the alternatives could not.[11]

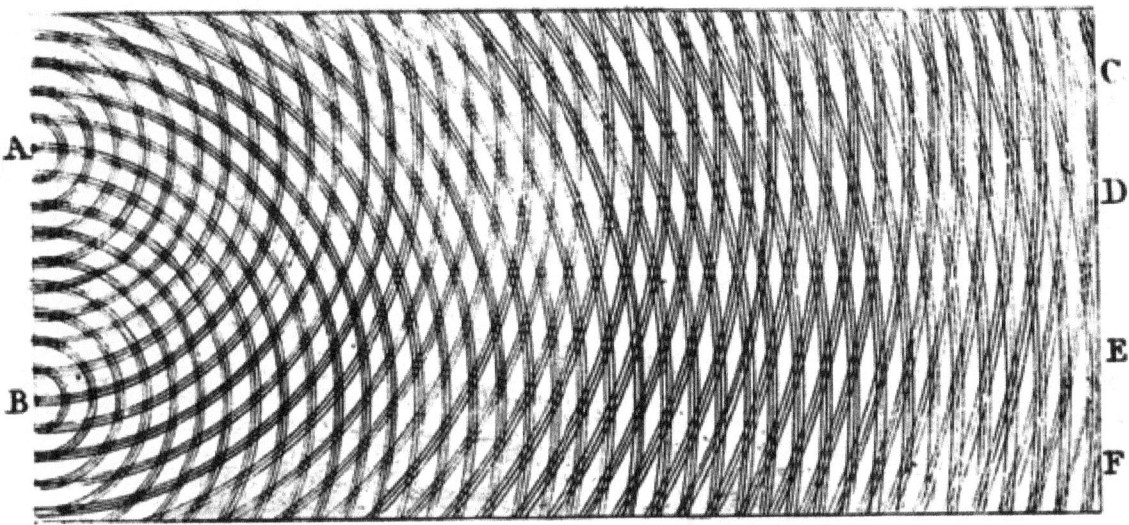

Thomas Young's sketch of two-slit diffraction of waves, 1803

James Clerk Maxwell discovered that he could combine four simple equations, which had been previously discovered, along with a slight modification to describe self-propagating waves of oscillating electric and magnetic fields. When the propagation speed of these electromagnetic waves was calculated, the speed of light fell out. It quickly became apparent that visible light, ultraviolet light, and infrared light (phenomena thought previously to be unrelated) were all electromagnetic waves of differing frequency. The wave theory had prevailed—or at least it seemed to.

While the 19th century had seen the success of the wave theory at describing light, it had also witnessed the rise of the atomic theory at describing matter. Antoine Lavoisier deduced the law of conservation of mass and categorized many new chemical elements and compounds; and Joseph Louis Proust advanced chemistry towards the atom by showing that elements combined in definite proportions. This led John Dalton to propose that elements were invisible sub components; Amedeo Avogadro discovered diatomic gases and completed the basic atomic theory, allowing the correct molecular formulae of most known compounds—as well as the correct weights of atoms—to be deduced and categorized in a consistent manner. Dimitri Mendeleev saw an order in recurring chemical properties, and created a table presenting the elements in unprecedented order and symmetry.

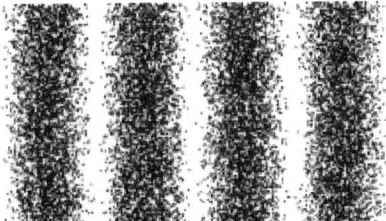

Particle impacts make visible the interference pattern of waves.

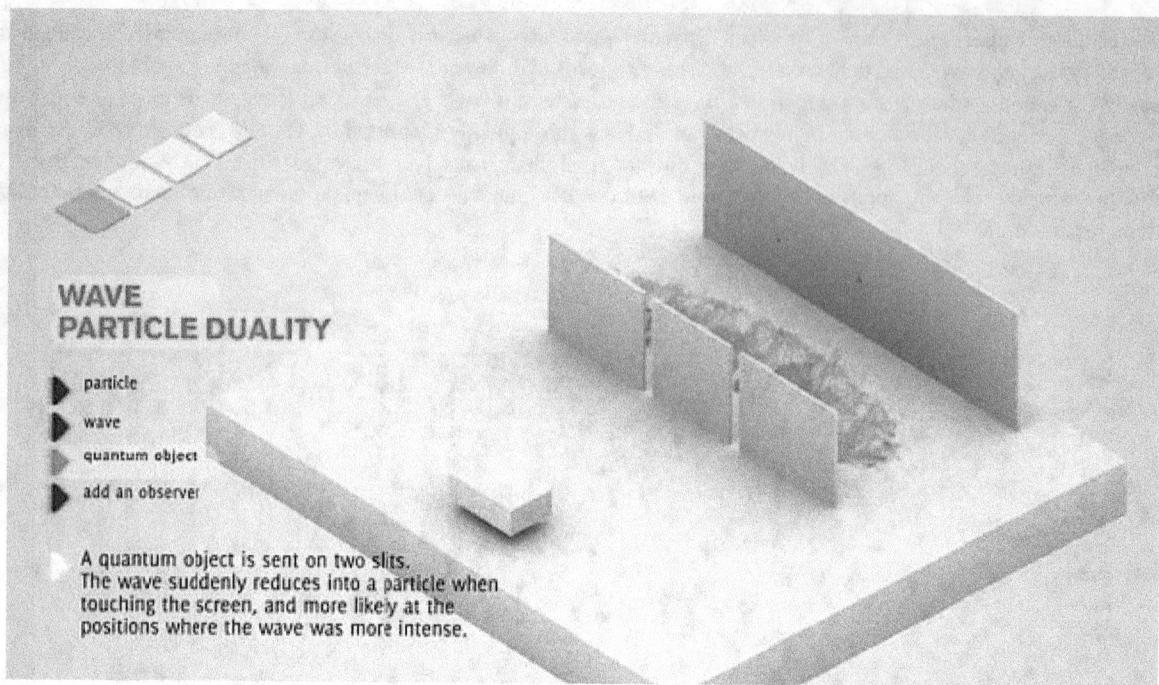

Animation showing the wave-particle duality with a double slit experiment and effect of an observer. Increase size to see explanations in the video itself. See also quiz based on this animation.

A quantum particle is represented by a wave packet.

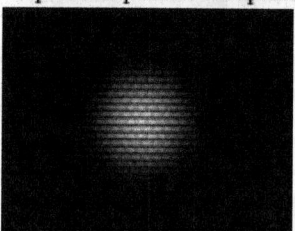

Interference of a quantum particle with itself.
Click images for animations.

1.2   Turn of the 20th century and the paradigm shift

1.2.1   Particles of electricity

At the close of the 19th century, the reductionism of atomic theory began to advance into the atom itself; determining, through physics, the nature of the atom and the operation of chemical reactions. Electricity, first thought to be a fluid, was now understood to consist of particles called electrons. This was first demonstrated by J. J. Thomson in 1897 when, using a cathode ray tube, he found that an electrical charge would travel across a vacuum (which would possess infinite resistance in classical theory). Since the vacuum offered no medium for an electric fluid to travel, this discovery could only be explained via a particle carrying a negative charge and moving through the vacuum. This *electron* flew in the face of classical electrodynamics, which had successfully treated electricity as a fluid for many years (leading to the invention of batteries, electric motors, dynamos, and arc lamps). More importantly, the intimate relation between electric charge and electromagnetism had been well documented following the discoveries of Michael Faraday and James Clerk Maxwell. Since electromagnetism was *known* to be a wave generated by a changing electric or magnetic *field* (a continuous, wave-like entity itself) an atomic/particle description of electricity and charge was a non sequitur. Furthermore, classical electrodynamics was not the only classical theory rendered incomplete.

1.2.2   Radiation quantization

Main article: Planck's law

In 1901, Max Planck published an analysis that succeeded in reproducing the observed spectrum of light emitted by a glowing object. To accomplish this, Planck had to make an ad hoc mathematical assumption of quantized energy of the oscillators (atoms of the black body) that emit radiation. It was Einstein who later proposed that it is the electromagnetic radiation itself that is quantized, and not the energy of radiating atoms.

Black-body radiation, the emission of electromagnetic energy due to an object's heat, could not be explained from classical arguments alone. The equipartition theorem of classical mechanics, the basis of all classical thermodynamic theories, stated that an object's energy is partitioned equally among the object's vibrational modes. But applying the same reasoning to the electromagnetic emission of such a thermal object was not so successful. It had been long known that thermal objects emit light. Since light was known to be waves of electromagnetism, physicists hoped to describe this emission via classical laws. This became known as the black body problem. Since the equipartition theorem worked so well in describing the vibrational modes of the thermal object itself, it was natural to assume that it would perform equally well in describing the radiative emission of such objects. But a problem quickly arose: if each mode received an equal partition of energy, the short wavelength modes would consume all the energy. This became clear when plotting the Rayleigh–Jeans law which, while correctly predicting the intensity of long wavelength emissions, predicted infinite total energy as the intensity diverges to infinity for short wavelengths. This became known as the ultraviolet catastrophe.

In 1900 when Max Planck hypothesized that the frequency of light emitted by the black body depended on the frequency of the *oscillator* that emitted it, and the energy of these oscillators increased linearly with frequency (according to his constant h, where $E = h\nu$). This was not an unsound proposal considering that macroscopic oscillators operate similarly: when studying five simple harmonic oscillators of equal amplitude but different frequency, the oscillator with the highest frequency possesses the highest energy (though this relationship is not linear like Planck's). By demanding that high-frequency light must be emitted by an oscillator of equal frequency, and further requiring that this oscillator occupy higher energy than one of a lesser frequency, Planck avoided any catastrophe; giving an equal partition to high-frequency oscillators produced successively fewer oscillators and less emitted light. And as in the Maxwell–Boltzmann distribution, the low-frequency, low-energy oscillators were suppressed by the onslaught of thermal jiggling from higher energy oscillators, which necessarily increased their energy and frequency.

The most revolutionary aspect of Planck's treatment of the black body is that it inherently relies on an integer number of oscillators in thermal equilibrium with the electromagnetic field. These oscillators *give* their entire energy to the electromagnetic field, creating a quantum of light, as often as they are *excited* by the electromagnetic field, absorbing a quantum of light and beginning to oscillate at the corresponding frequency. Planck had intentionally created an atomic theory of the black body, but had unintentionally generated an atomic theory of light, where the black body never generates quanta of light at a given frequency with an energy less than $h\nu$. However, once realizing that he had quantized the electromagnetic field, he denounced particles of light as a limitation of his approximation, not a property of reality.

1.2.3   Photoelectric effect illuminated

While Planck had solved the ultraviolet catastrophe by using atoms and a quantized electromagnetic field, most contemporary physicists agreed that Planck's "light quanta" represented only flaws in his model. A more-complete derivation of black body radiation would yield a fully continuous and 'wave-like' electromagnetic field with no quantization. However, in 1905 Albert Einstein took Planck's black body model to produce his solution to another outstanding problem of the day: the photoelectric effect, wherein electrons are emitted from atoms when they absorb energy from light. Since their discovery eight years previously, electrons had been *the* thing to study in physics laboratories worldwide.

In 1902 Philipp Lenard discovered that the energy of these ejected electrons did *not* depend on the intensity of the incoming light, but instead on its *frequency*. So if one shines a little low-frequency light upon a metal, a few low energy electrons are ejected. If one now shines a very intense beam of low-frequency light upon the same metal, a whole slew of electrons are ejected; however they possess the same low energy, there are merely *more of them*. The more light there is, the more electrons are ejected. Whereas in order to get high energy electrons, one must illuminate the metal with high-frequency light. Like blackbody radiation, this was at odds with a theory invoking continuous transfer of energy between radiation and matter. However, it can still be explained using a fully classical description of light, as long as matter is quantum mechanical in nature.[12]

If one used Planck's energy quanta, and demanded that electromagnetic radiation at a given frequency could only transfer energy to matter in integer multiples of an energy quantum **h**v, then the photoelectric effect could be explained very simply. Low-frequency light only ejects low-energy electrons because each electron is excited by the absorption of a single photon. Increasing the intensity of the low-frequency light (increasing the number of photons) only increases the number of excited electrons, not their energy, because the energy of each photon remains low. Only by increasing the frequency of the light, and thus increasing the energy of the photons, can one eject electrons with higher energy. Thus, using Planck's constant h to determine the energy of the photons based upon their frequency, the energy of ejected electrons should also increase linearly with frequency; the gradient of the line being Planck's constant. These results were not confirmed until 1915, when Robert Andrews Millikan, who had previously determined the charge of the electron, produced experimental results in perfect accord with Einstein's predictions. While the energy of ejected electrons reflected Planck's constant, the existence of photons was not explicitly proven until the discovery of the photon antibunching effect, of which a modern experiment can be performed in undergraduate-level labs.[13] This phenomenon could only be explained via photons, and not through any semi-classical theory (which could alternatively explain the photoelectric effect). When Einstein received his Nobel Prize in 1921, it was not for his more difficult and mathematically laborious special and general relativity, but for the simple, yet totally revolutionary, suggestion of quantized light. Einstein's "light quanta" would not be called photons until 1925, but even in 1905 they represented the quintessential example of wave-particle duality. Electromagnetic radiation propagates following linear wave equations, but can only be emitted or absorbed as discrete elements, thus acting as a wave and a particle simultaneously.

1.2.4 Einstein's explanation of the photoelectric effect

Main article: Photoelectric effect

In 1905, Albert Einstein provided an explanation of the photoelectric effect, a hitherto troubling experiment that the wave theory of light seemed incapable of explaining. He did so by postulating the existence of photons, quanta of light energy with particulate qualities.

In the photoelectric effect, it was observed that shining a light on certain metals would lead to an electric current in a circuit. Presumably, the light was knocking electrons out of the metal, causing current to flow. However, using the case of potassium as an example, it was also observed that while a dim blue light was enough to cause a current, even the strongest, brightest red light available with the technology of the time caused no current at all. According to the classical theory of light and matter, the strength or amplitude of a light wave was in proportion to its brightness: a bright light should have been easily strong enough to create a large current. Yet, oddly, this was not so.

Einstein explained this conundrum by postulating that the electrons can receive energy from electromagnetic field only in discrete portions (quanta that were called photons): an amount of energy E that was related to the frequency f of the light by

$$E = hf$$

where h is Planck's constant (6.626×10^{-34} J seconds). Only photons of a high enough frequency (above a certain *threshold* value) could knock an electron free. For example, photons of blue light had sufficient energy to free an electron from the metal, but photons of red light did not. More intense light above the threshold frequency could release more electrons,

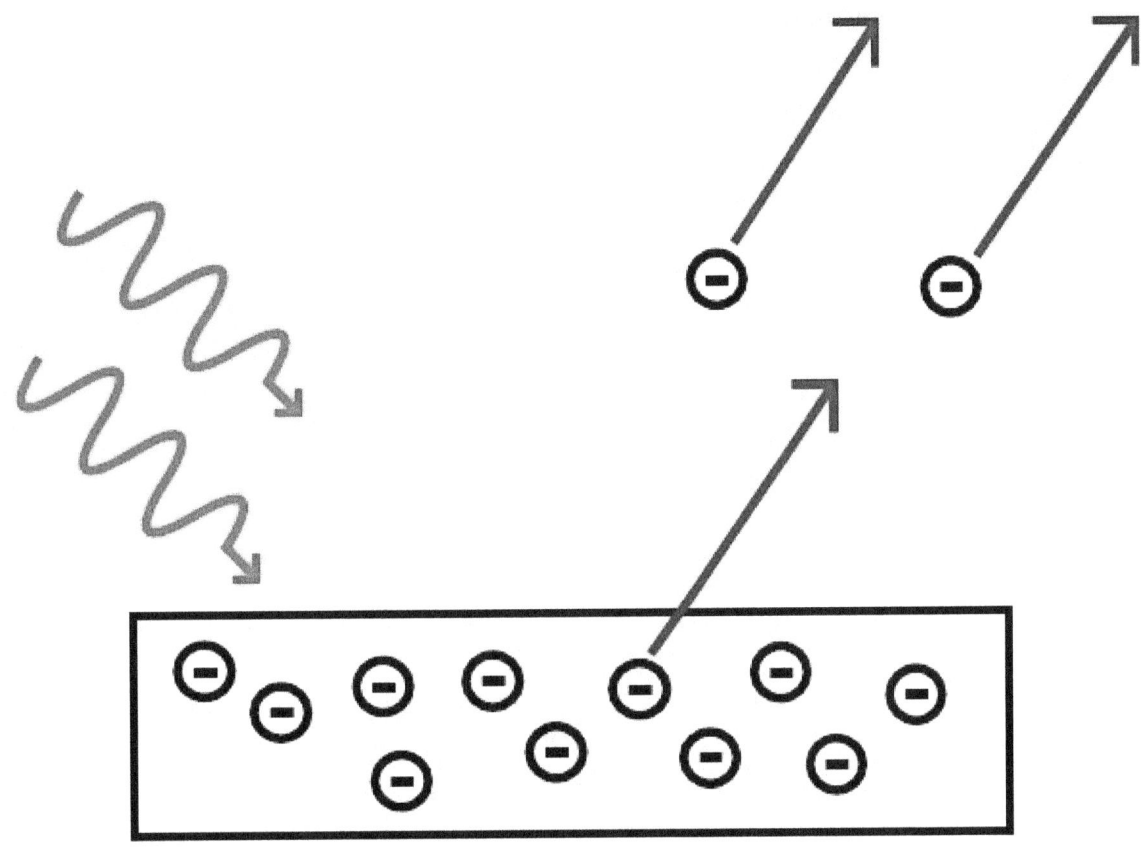

The photoelectric effect. Incoming photons on the left strike a metal plate (bottom), and eject electrons, depicted as flying off to the right.

but no amount of light (using technology available at the time) below the threshold frequency could release an electron. To "violate" this law would require extremely high intensity lasers which had not yet been invented. Intensity-dependent phenomena have now been studied in detail with such lasers.[14]

Einstein was awarded the Nobel Prize in Physics in 1921 for his discovery of the law of the photoelectric effect.

1.2.5   De Broglie's wavelength

Main article: Matter wave
In 1924, Louis-Victor de Broglie formulated the de Broglie hypothesis, claiming that *all* matter,[15][16] not just light, has a wave-like nature; he related wavelength (denoted as λ), and momentum (denoted as p):

$$\lambda = \frac{h}{p}$$

This is a generalization of Einstein's equation above, since the momentum of a photon is given by $p = \frac{E}{c}$ and the wavelength (in a vacuum) by $\lambda = \frac{c}{f}$, where c is the speed of light in vacuum.

De Broglie's formula was confirmed three years later for electrons (which differ from photons in having a rest mass) with the observation of electron diffraction in two independent experiments. At the University of Aberdeen, George Paget Thomson passed a beam of electrons through a thin metal film and observed the predicted interference patterns. At Bell Labs Clinton Joseph Davisson and Lester Halbert Germer guided their beam through a crystalline grid.

De Broglie was awarded the Nobel Prize for Physics in 1929 for his hypothesis. Thomson and Davisson shared the Nobel Prize for Physics in 1937 for their experimental work.

1.2.6 Heisenberg's uncertainty principle

Main article: Heisenberg uncertainty principle

In his work on formulating quantum mechanics, Werner Heisenberg postulated his uncertainty principle, which states:

$$\Delta x \Delta p \geq \frac{\hbar}{2}$$

where

   Δ here indicates standard deviation, a measure of spread or uncertainty;

   x and p are a particle's position and linear momentum respectively.

   $\hbar$ is the reduced Planck's constant (Planck's constant divided by 2π).

Heisenberg originally explained this as a consequence of the process of measuring: Measuring position accurately would disturb momentum and vice versa, offering an example (the "gamma-ray microscope") that depended crucially on the de Broglie hypothesis. It is now thought, however, that this only partly explains the phenomenon, but that the uncertainty also exists in the particle itself, even before the measurement is made.

In fact, the modern explanation of the uncertainty principle, extending the Copenhagen interpretation first put forward by Bohr and Heisenberg, depends even more centrally on the wave nature of a particle: Just as it is nonsensical to discuss the precise location of a wave on a string, particles do not have perfectly precise positions; likewise, just as it is nonsensical to discuss the wavelength of a "pulse" wave traveling down a string, particles do not have perfectly precise momenta (which corresponds to the inverse of wavelength). Moreover, when position is relatively well defined, the wave is pulse-like and has a very ill-defined wavelength (and thus momentum). And conversely, when momentum (and thus wavelength) is relatively well defined, the wave looks long and sinusoidal, and therefore it has a very ill-defined position.

1.2.7 de Broglie–Bohm theory

De Broglie himself had proposed a pilot wave construct to explain the observed wave-particle duality. In this view, each particle has a well-defined position and momentum, but is guided by a wave function derived from Schrödinger's equation. The pilot wave theory was initially rejected because it generated non-local effects when applied to systems involving more than one particle. Non-locality, however, soon became established as an integral feature of quantum theory (see EPR paradox), and David Bohm extended de Broglie's model to explicitly include it.

In the resulting representation, also called the de Broglie–Bohm theory or Bohmian mechanics,[18] the wave-particle duality vanishes, and explains the wave behaviour as a scattering with wave appearance, because the particle's motion is subject to a guiding equation or quantum potential. *"This idea seems to me so natural and simple, to resolve the wave-particle dilemma in such a clear and ordinary way, that it is a great mystery to me that it was so generally ignored"*,[19] J.S.Bell.

The best illustration of the *pilot-wave model* was given by Couder's 2010 "walking droplets" experiments,[20] demonstrating the pilot-wave behaviour in a macroscopic mechanical analog.[17]

1.3 Wave behavior of large objects

Since the demonstrations of wave-like properties in photons and electrons, similar experiments have been conducted with neutrons and protons. Among the most famous experiments are those of Estermann and Otto Stern in 1929.[21] Authors of similar recent experiments with atoms and molecules, described below, claim that these larger particles also act like waves. A wave is basically a group of particles which moves in a particular form of motion i.e. to and fro, if we break that flow by an object it will convert into radiants.

A dramatic series of experiments emphasizing the action of gravity in relation to wave–particle duality was conducted in the 1970s using the neutron interferometer.[22] Neutrons, one of the components of the atomic nucleus, provide much of the mass of a nucleus and thus of ordinary matter. In the neutron interferometer, they act as quantum-mechanical waves directly subject to the force of gravity. While the results were not surprising since gravity was known to act on everything, including light (see tests of general relativity and the Pound–Rebka falling photon experiment), the self-interference of the quantum mechanical wave of a massive fermion in a gravitational field had never been experimentally confirmed before.

In 1999, the diffraction of C_{60} fullerenes by researchers from the University of Vienna was reported.[23] Fullerenes are comparatively large and massive objects, having an atomic mass of about 720 u. The de Broglie wavelength is 2.5 pm, whereas the diameter of the molecule is about 1 nm, about 400 times larger. In 2012, these far-field diffraction experiments could be extended to phthalocyanine molecules and their heavier derivatives, which are composed of 58 and 114 atoms respectively. In these experiments the build-up of such interference patterns could be recorded in real time and with single molecule sensitivity.[24][25]

In 2003, the Vienna group also demonstrated the wave nature of tetraphenylporphyrin[26]—a flat biodye with an extension of about 2 nm and a mass of 614 u. For this demonstration they employed a near-field Talbot Lau interferometer.[27][28] In the same interferometer they also found interference fringes for $C_{60}F_{48}$., a fluorinated buckyball with a mass of about 1600 u, composed of 108 atoms.[26] Large molecules are already so complex that they give experimental access to some aspects of the quantum-classical interface, i.e., to certain decoherence mechanisms.[29][30] In 2011, the interference of molecules as heavy as 6910 u could be demonstrated in a Kapitza–Dirac–Talbot–Lau interferometer.[31] In 2013, the interference of molecules beyond 10,000 u has been demonstrated.[32]

Whether objects heavier than the Planck mass (about the weight of a large bacterium) have a de Broglie wavelength is theoretically unclear and experimentally unreachable; above the Planck mass a particle's Compton wavelength would be smaller than the Planck length and its own Schwarzschild radius, a scale at which current theories of physics may break down or need to be replaced by more general ones.[33]

Recently Couder, Fort, *et al.* showed[34] that we can use macroscopic oil droplets on a vibrating surface as a model of wave–particle duality—localized droplet creates periodical waves around and interaction with them leads to quantum-like phenomena: interference in double-slit experiment,[35] unpredictable tunneling[36] (depending in complicated way on practically hidden state of field), orbit quantization[37] (that particle has to 'find a resonance' with field perturbations it creates—after one orbit, its internal phase has to return to the initial state) and Zeeman effect.[38]

1.4   Treatment in modern quantum mechanics

Wave–particle duality is deeply embedded into the foundations of quantum mechanics. In the formalism of the theory, all the information about a particle is encoded in its *wave function*, a complex-valued function roughly analogous to the amplitude of a wave at each point in space. This function evolves according to a differential equation (generically called the Schrödinger equation). For particles with mass this equation has solutions that follow the form of the wave equation. Propagation of such waves leads to wave-like phenomena such as interference and diffraction. Particles without mass, like photons, have no solutions of the Schrödinger equation so have another wave.

The particle-like behavior is most evident due to phenomena associated with measurement in quantum mechanics. Upon measuring the location of the particle, the particle will be forced into a more localized state as given by the uncertainty principle. When viewed through this formalism, the measurement of the wave function will randomly "collapse", or rather "decohere", to a sharply peaked function at some location. For particles with mass the likelihood of detecting the particle at any particular location is equal to the squared amplitude of the wave function there. The measurement will return a well-defined position, (subject to uncertainty), a property traditionally associated with particles. It is important to note that a measurement is only a particular type of interaction where some data is recorded and the measured quantity is forced into a particular eigenstate. The act of measurement is therefore not fundamentally different from any other interaction.

Following the development of quantum field theory the ambiguity disappeared. The field permits solutions that follow the wave equation, which are referred to as the wave functions. The term particle is used to label the irreducible representations of the Lorentz group that are permitted by the field. An interaction as in a Feynman diagram is accepted as a calculationally convenient approximation where the outgoing legs are known to be simplifications of the propagation and the internal lines are for some order in an expansion of the field interaction. Since the field is non-local and quantized, the phenomena which previously were thought of as paradoxes are explained. Within the limits of the wave-particle duality the quantum field theory gives the same results.

1.4.1 Visualization

There are two ways to visualize the wave-particle behaviour: by the "standard model", described below; and by the Broglie–Bohm model, where no duality is perceived.

Below is an illustration of wave–particle duality as it relates to De Broglie's hypothesis and Heisenberg's uncertainty principle (above), in terms of the position and momentum space wavefunctions for one spinless particle with mass in one dimension. These wavefunctions are Fourier transforms of each other.

The more localized the position-space wavefunction, the more likely the particle is to be found with the position coordinates in that region, and correspondingly the momentum-space wavefunction is less localized so the possible momentum components the particle could have are more widespread.

Conversely the more localized the momentum-space wavefunction, the more likely the particle is to be found with those values of momentum components in that region, and correspondingly the less localized the position-space wavefunction, so the position coordinates the particle could occupy are more widespread.

1.5 Alternative views

Wave–particle duality is an ongoing conundrum in modern physics. Most physicists accept wave-particle duality as the best explanation for a broad range of observed phenomena; however, it is not without controversy. Alternative views are also presented here. These views are not generally accepted by mainstream physics, but serve as a basis for valuable discussion within the community.

1.5.1 Both-particle-and-wave view

The pilot wave model, originally developed by Louis de Broglie and further developed by David Bohm into the hidden variable theory proposes that there is no duality, but rather a system exhibits both particle properties and wave properties simultaneously, and particles are guided, in a deterministic fashion, by the pilot wave (or its "quantum potential") which will direct them to areas of constructive interference in preference to areas of destructive interference. This idea is held by a significant minority within the physics community.[39]

At least one physicist considers the "wave-duality" as not being an incomprehensible mystery. L.E. Ballentine, *Quantum Mechanics, A Modern Development*, p. 4, explains:

> When first discovered, particle diffraction was a source of great puzzlement. Are "particles" really "waves?" In the early experiments, the diffraction patterns were detected holistically by means of a photographic plate, which could not detect individual particles. As a result, the notion grew that particle and wave properties were mutually incompatible, or complementary, in the sense that different measurement apparatuses would be required to observe them. That idea, however, was only an unfortunate generalization from a technological limitation. Today it is possible to detect the arrival of individual electrons, and to see the diffraction pattern emerge as a statistical pattern made up of many small spots (Tonomura et al., 1989). Evidently, quantum particles are indeed particles, but whose behaviour is very different from classical physics would have us to expect.

It has been claimed that the Afshar experiment[40] (2007) shows that it is possible to simultaneously observe both wave and particle properties of photons. This claim is, however, rejected by other scientists.

1.5.2   Wave-only view

At least one scientist proposes that the duality can be replaced by a "wave-only" view. In his book *Collective Electrodynamics: Quantum Foundations of Electromagnetism* (2000), Carver Mead purports to analyze the behavior of electrons and photons purely in terms of electron wave functions, and attributes the apparent particle-like behavior to quantization effects and eigenstates. According to reviewer David Haddon:[41]

> Mead has cut the Gordian knot of quantum complementarity. He claims that atoms, with their neutrons, protons, and electrons, are not particles at all but pure waves of matter. Mead cites as the gross evidence of the exclusively wave nature of both light and matter the discovery between 1933 and 1996 of ten examples of pure wave phenomena, including the ubiquitous laser of CD players, the self-propagating electrical currents of superconductors, and the Bose–Einstein condensate of atoms.

Albert Einstein, who, in his search for a Unified Field Theory, did not accept wave-particle duality, wrote:[42]

> This double nature of radiation (and of material corpuscles)...has been interpreted by quantum-mechanics in an ingenious and amazingly successful fashion. This interpretation...appears to me as only a temporary way out...

The many-worlds interpretation (MWI) is sometimes presented as a waves-only theory, including by its originator, Hugh Everett who referred to MWI as "the wave interpretation".[43]

The *Three Wave Hypothesis* of R. Horodecki relates the particle to wave.[44][45] The hypothesis implies that a massive particle is an intrinsically spatially as well as temporally extended wave phenomenon by a nonlinear law.

1.5.3   Particle-only view

Still in the days of the old quantum theory, a pre-quantum-mechanical version of wave–particle duality was pioneered by William Duane,[46] and developed by others including Alfred Landé.[47] Duane explained diffraction of x-rays by a crystal in terms solely of their particle aspect. The deflection of the trajectory of each diffracted photon was due to quantal translative momentum transfer from the spatially regular structure of the diffracting crystal.[48] Fourier analysis reveals the wave–particle duality as a simply mathematical equivalence, always present, and universal for all quanta. The same reasoning applies for example to diffraction of electrons by a crystal.

1.5.4   Neither-wave-nor-particle view

It has been argued that there are never exact particles or waves, but only some compromise or intermediate between them. For this reason, in 1928 Arthur Eddington[49] coined the name "*wavicle*" to describe the objects although it is not regularly used today. One consideration is that zero-dimensional mathematical points cannot be observed. Another is that the formal representation of such points, the Dirac delta function is unphysical, because it cannot be normalized. Parallel arguments apply to pure wave states. Roger Penrose states:[50]

> "Such 'position states' are idealized wavefunctions in the opposite sense from the momentum states. Whereas the momentum states are infinitely spread out, the position states are infinitely concentrated. Neither is normalizable [...]."

1.5.5   Relational approach to wave–particle duality

Relational quantum mechanics is developed which regards the detection event as establishing a relationship between the quantized field and the detector. The inherent ambiguity associated with applying Heisenberg's uncertainty principle and thus wave–particle duality is subsequently avoided.[51]

1.6 Applications

Although it is difficult to draw a line separating wave–particle duality from the rest of quantum mechanics, it is nevertheless possible to list some applications of this basic idea.

- Wave–particle duality is exploited in electron microscopy, where the small wavelengths associated with the electron can be used to view objects much smaller than what is visible using visible light.

- Similarly, neutron diffraction uses neutrons with a wavelength of about 0.1 nm, the typical spacing of atoms in a solid, to determine the structure of solids.

1.7 See also

- Arago spot

- Afshar experiment

- Basic concepts of quantum mechanics

- Complementarity (physics)

- Englert–Greenberger–Yasin duality relation

- Kapitsa–Dirac effect

- Electron wave-packet interference

- Faraday wave

- Hanbury Brown and Twiss effect

- Photon polarization

- Scattering theory

- Wavelet

- Wheeler's delayed choice experiment

1.8 Notes and references

[1] Harrison, David (2002). "Complementarity and the Copenhagen Interpretation of Quantum Mechanics". *UPSCALE*. Dept. of Physics, U. of Toronto. Retrieved 2008-06-21.

[2] Walter Greiner (2001). *Quantum Mechanics: An Introduction*. Springer. ISBN 3-540-67458-6.

[3] R. Eisberg and R. Resnick (1985). *Quantum Physics of Atoms, Molecules, Solids, Nuclei, and Particles* (2nd ed.). John Wiley & Sons. pp. 59–60. ISBN 047187373X. For both large and small wavelengths, both matter and radiation have both particle and wave aspects.... But the wave aspects of their motion become more difficult to observe as their wavelengths become shorter.... For ordinary macroscopic particles the mass is so large that the momentum is always sufficiently large to make the de Broglie wavelength small enough to be beyond the range of experimental detection, and classical mechanics reigns supreme.

[4] Kumar, Manjit (2011). *Quantum: Einstein, Bohr, and the Great Debate about the Nature of Reality* (Reprint ed.). W. W. Norton & Company. p. 242, 375-376. ISBN 978-0393339888.

[5] Bohr, N. (1927/1928). The quantum postulate and the recent development of atomic theory, *Nature* Supplement April 14 1928, 121: 580–590.

[6] Camilleri, K. (2009). *Heisenberg and the Interpretation of Quantum Mechanics: the Physicist as Philosopher*, Cambridge University Press, Cambridge UK, ISBN 978-0-521-88484-6.

[7] Preparata, G. (2002). *An Introduction to a Realistic Quantum Physics*, World Scientific, River Edge NJ, ISBN 978-981-238-176-7.

[8] Nathaniel Page Stites, M.A./M.S. "Light I: Particle or Wave?," Visionlearning Vol. PHY-1 (3), 2005. http://www.visionlearning.com/library/module_viewer.php?mid=132

[9] Young, Thomas (1804). "Bakerian Lecture: Experiments and calculations relative to physical optics". *Philosophical Transactions of the Royal Society* **94**: 1–16. Bibcode:1804RSPT...94....1Y. doi:10.1098/rstl.1804.0001.

[10] Thomas Young: The Double Slit Experiment

[11] Buchwald, Jed (1989). *The Rise of the Wave Theory of Light: Optical Theory and Experiment in the Early Nineteenth Century*. Chicago: University of Chicago Press. ISBN 0-226-07886-8. OCLC 18069573 59210058.

[12] Lamb, Willis E.; Scully, Marlan O. (1968). "The photoelectric effect without photons" (PDF).

[13] "Observing the quantum behavior of light in an undergraduate laboratory". *American Journal o f Physics* **72**: 1210 doi:10.1119/1.1737397.

[14] Zhang, Q (1996). "Intensity dependence of the photoelectric effect induced by a circularly polarized laser beam". *Physics Letters A* **216** (1-5): 125–128. Bibcode:1996PhLA..216..125Z. doi:10.1016/0375-9601(96)00259-9.

[15] Donald H Menzel, "*Fundamental formulas of Physics*", volume 1, page 153; Gives the de Broglie wavelengths for composite particles such as protons and neutrons.

[16] Brian Greene, The Elegant Universe, page 104 "all matter has a wave-like character"

[17] See this Science Channel production (Season II, Episode VI "How Does The Universe Work?"), presented by Morgan Freeman. https://www.youtube.com/watch?v=W9yWv5dqSKk

[18] Bohmian Mechanics, *Stanford Encyclopedia of Philosophy*.

[19] Bell, J. S., "Speakable and Unspeakable in Quantum Mechanics", Cambridge: Cambridge University Press, 1987.

[20] Y. Couder, A. Boudaoud, S. Protière, Julien Moukhtar, E. Fort: *Walking droplets: a form of wave-particle duality at macroscopic level?* , doi:10.1051/epn/2010101, (PDF)

[21] Estermann, I.; Stern O. (1930). "Beugung von Molekularstrahlen". *Zeitschrif t für Physik* **61** (1-2): 95–125. Bibcode: doi:10.1007/BF01340293.

[22] R. Colella, A. W. Overhauser and S. A. Werner, Observation of Gravitationally Induced Quantum Interference, *Phys. Rev. Lett.* **34**, 1472–1474 (1975).

[23] Arndt, Markus; O. Nairz; J. Voss-Andreae, C. Keller, G. van der Zouw, A. Zeilinger (14 October 1999). "Wave–particle duality of C_{60}". *Nature* **401** (6754): 680–682. Bibcode:1999Natur.401..680A. doi:10.1038/44348. PMID 18494170.

[24] Juffmann, Thomas; et al. (25 March 2012). "Real-time single-molecule imaging of quantum interference". Nature Nanotechnology. Retrieved 27 March 2012.

[25] Quantumnanovienna. "Single molecules in a quantum interference movie". Retrieved 2012-04-21.

[26] Hackermüller, Lucia; Stefan Uttenthaler; Klaus Hornberger; Elisabeth Reiger; Björn Brezger; Anton Zeilinger; Markus Arndt (2003). "The wave nature of biomolecules and fluorofullerenes". *Phys. Rev. Lett.* **91** (9): 090408. arXiv:quant-ph/0309016. Bibcode:2003PhRvL..91i0408H. doi:10.1103/PhysRevLett.91.090408. PMID 14525169.

[27] Clauser, John F.; S. Li (1994). "Talbot von Lau interefometry with cold slow potassium atoms.". *Phys. Rev. A* **49** (4): R2213–17. Bibcode:1994PhRvA..49.2213C. doi:10.1103/PhysRevA.49.R2213. PMID 9910609.

[28] Brezger, Björn; Lucia Hackermüller; Stefan Uttenthaler; Julia Petschinka; Markus Arndt; Anton Zeilinger (2002). "Matter-wave interferometer for large molecules". *Phys. Rev. Lett.* **88** (10): 100404. arXiv:quant-ph/0202158. Bibcode:2002PhRvL..88j0404B.doi:10.1103/PhysRevLett.88.100404. PMID 11909334.

[29] Hornberger, Klaus; Stefan Uttenthaler; Björn Brezger; Lucia Hackermüller; Markus Arndt; Anton Zeilinger (2003). "Observation of Collisional Decoherence in Interferometry". *Phys. Rev. Lett.* **90** (16): 160401. arXiv:quant-ph/0303093. Bibcode:2003PhRv L..90p0401H.doi:10.1103/PhysRevLett.90.160401. PMID 12731960.

[30] Hackermüller, Lucia; Klaus Hornberger; Björn Brezger; Anton Zeilinger; Markus Arndt (2004). "Decoherence of matter waves by thermal emission of radiation". *Nature* **427** (6976): 711–714. arXiv:quant-ph/0402146. Bibcode:2004Natur.427..711H. doi:10.1038/nature02276. PMID 14973478.

[31] Gerlich, Stefan; et al. (2011). "Quantum interference of large organi c molecules". *Nature Communications* **2** (263). Bibcode doi:10.1038/ncomms1263. PMC 3104521. PMID 21468015.

[32] Eibenberger, S.; Gerlich, S.; Arndt, M.; Mayor, M.; Tüxen, J. (2013). "Matter–wave interference of particles selected from a molecular library with masses exceeding 10 000 amu". *Physical Chemistry Chemical Physics* **15** (35): 14696–14700. doi:10.103 9/c3cp51500a.PMID 23900710.

[33] Peter Gabriel Bergmann, *The Riddle of Gravitation*, Courier Dover Publications, 1993 ISBN 0-486-27378-4 online

[34] http://www.youtube.com/watch?v=W9yWv5dqSKk - You Tube video - Yves Couder Explains Wave/Particle Duality via Silicon Droplets

[35] Y. Couder, E. Fort, *Single-Particle Diffraction and Interference at a Macroscopic Scale*, PRL 97, 154101 (2006) online

[36] A. Eddi, E. Fort, F. Moisy, Y. Couder, *Unpredictable Tunneling of a Classical Wave–Particle Association*, PRL 102, 240401 (2009)

[37] Fort, E.; Eddi, A.; Boudaoud, A.; Moukhtar, J.; Couder, Y. (2010). "Path-memory induced quantization of classical orbits". *PNAS* **107** (41): 17515–17520. doi:10.1073/pnas.1007386107.

[38] http://prl.aps.org/abstract/PRL/v108/i26/e264503 - Level Splitting at Macroscopic Scale

[39] (Buchanan pp. 29–31)

[40] Afshar S.S. et al: Paradox in Wave Particle Duality. Found. Phys. 37, 295 (2007) http://arxiv.org/abs/quant-ph/0702188 arXiv:quant-ph/0702188

[41] David Haddon. "Recovering Rational Science". *Touchstone*. Retrieved 2007-09-12.

[42] Paul Arthur Schilpp, ed, *Albert Einstein: Philosopher-Scientist*, Open Court (1949), ISBN 0-87548-133-7, p 51.

[43] See section VI(e) of Everett's thesis: *The Theory of the Universal Wave Function*, in Bryce Seligman DeWitt, R. Neill Graham, eds, *The Many-Worlds Interpretation of Quantum Mechanics*, Princeton Series in Physics, Princeton University Press (1973), ISBN 0-691-08131-X, pp 3–140.

[44] Horodecki, R. (1981). "De broglie wave and its dual wave". *Phys. Lett.* A **87** (3): 95–97. Bibcode:1981PhLA...87...95H. doi:10.1016/0375-9601(81)90571-5.

[45] Horodecki, R. (1983). "Superluminal singular dual wave". *Lett. Novo Cimento* **38**: 509–511.

[46] Duane, W. (1923). The transfer in quanta of radiation momentum to matter, *Proc. Natl. Acad. Sci.* **9**(5): 158–164.

[47] Landé, A. (1951). *Quantum Mechanics*, Sir Isaac Pitman and Sons, London, pp. 19–22.

[48] Heisenberg, W. (1930). *The Physical Principles of the Quantum Theory*, translated by C. Eckart and F.C. Hoyt, University of Chicago Press, Chicago, pp. 77–78.

[49] Eddington, Arthur Stanley (1928). *The Nature of the Physical World*. Cambridge, UK.: MacMillan. p. 201.

[50] Penrose, Roger (2007). *The Road to Reality: A Complete Guide to the Laws of the Universe*. Vintage. p. 521, §21.10. ISBN 978-0-679-77631-4.

[51] http://www.quantum-relativity.org/Quantum-Relativity.pdf. See Q. Zheng and T. Kobayashi, *Quantum Optics as a Relativistic Theory of Light*; Physics Essays 9 (1996) 447. Annual Report, Department of Physics, School of Science, University of Tokyo (1992) 240.

1.9 External links

- Animation, applications and research linked to the wave-particle duality and other basic quantum phenomena (Université Paris Sud)

- H. Nikolic. "Quantum mechanics: Myths and facts". arXiv:quant-ph/0609163.

- Young & Geller. "College Physics".

- B. Crowell. "Light as a Particle" (Web page). Retrieved December 10, 2006.

- E.H. Carlson, *Wave–Particle Duality: Light* on Project PHYSNET

- R. Nave. "Wave–Particle Duality" (Web page). *HyperPhysics*. Georgia State University, Department of Physics and Astronomy. Retrieved December 12, 2005.

- Juffmann, Thomas; et al. (25 March 2012). "Real-time single-molecule imaging of quantum interference". Nature Nanotechnology. Retrieved 21 January 2014.

$$\Psi = Ae^{i(px - \omega t)}$$

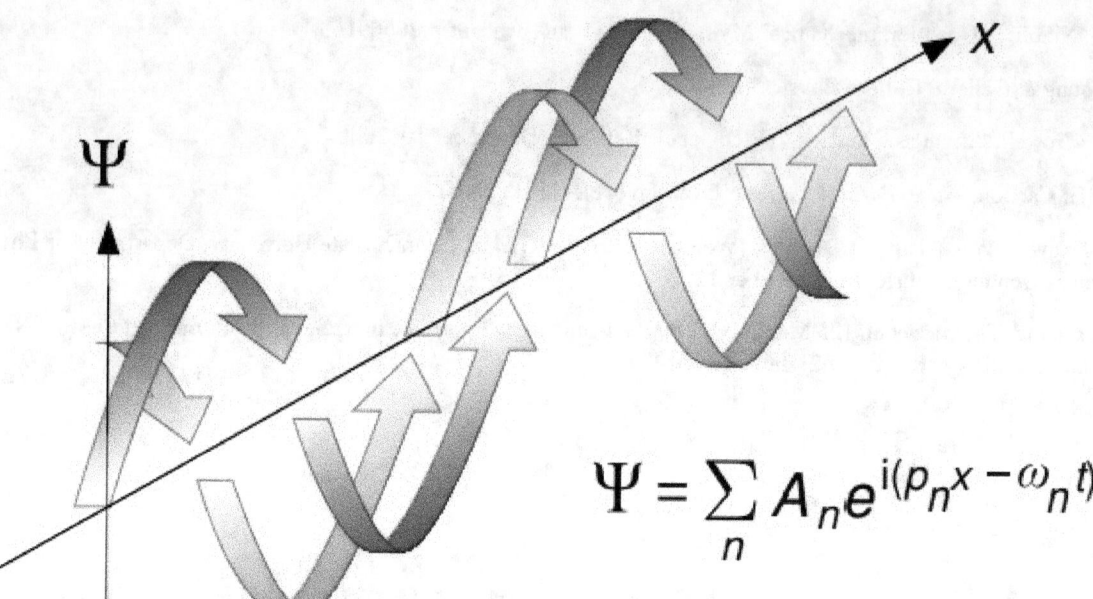

$$\Psi = \sum_n A_n e^{i(p_n x - \omega_n t)}$$

Propagation of de Broglie waves in 1d—real part of the complex amplitude is blue, imaginary part is green. The probability (shown as the colour opacity) of finding the particle at a given point x is spread out like a waveform; there is no definite position of the particle. As the amplitude increases above zero the curvature decreases, so the amplitude decreases again, and vice versa—the result is an alternating amplitude: a wave. Top: Plane wave. Bottom: Wave packet.

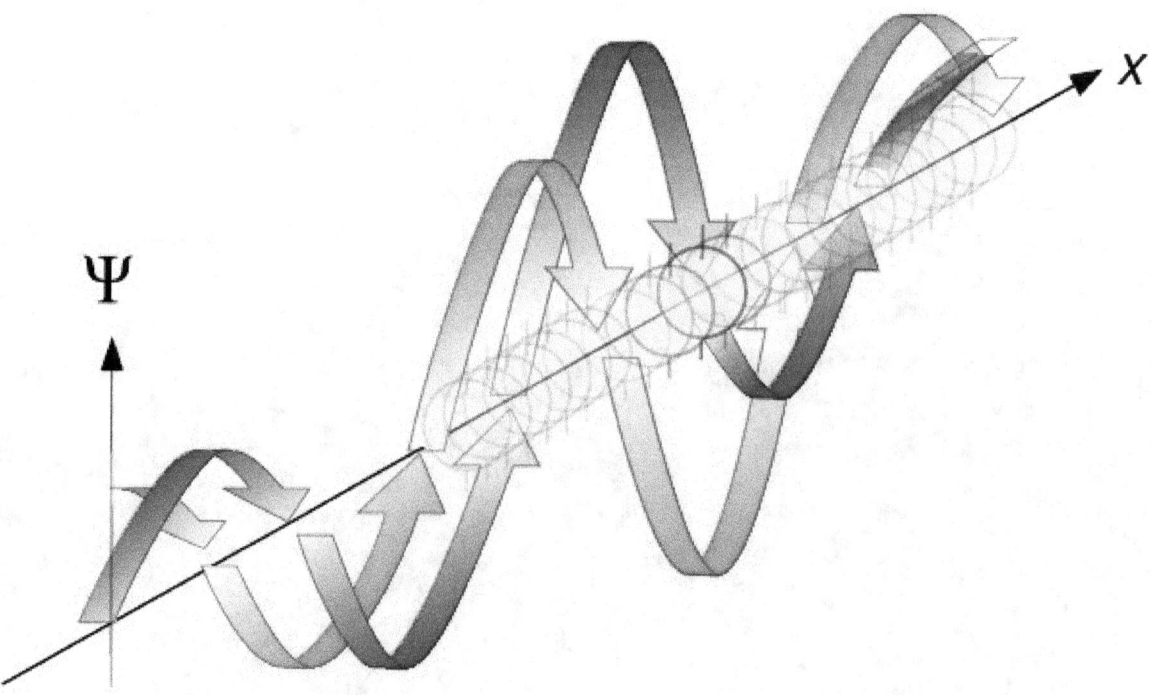

Couder experiments,[17] "materializing" the pilot wave model.

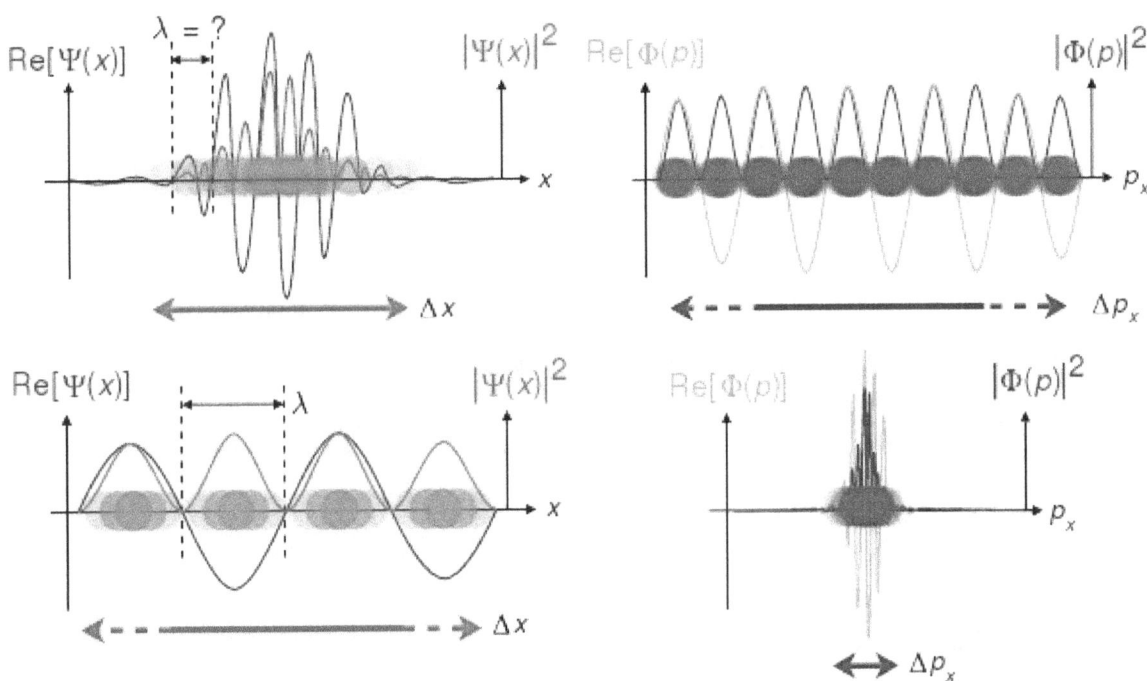

Position x and momentum p wavefunctions corresponding to quantum particles. The colour opacity (%) of the particles corresponds to the probability density of finding the particle with position x or momentum component p.

Top: *If wavelength λ is unknown, so are momentum p, wave-vector k and energy E (de Broglie relations). As the particle is more localized in position space, Δx is smaller than for Δp$_x$.*

Bottom: *If λ is known, so are p, k, and E. As the particle is more localized in momentum space, Δp is smaller than for Δx.*

Chapter 2

Elementary particle

This article is about the physics concept. For the novel, see The Elementary Particles.

In particle physics, an **elementary particle** or **fundamental particle** is a particle whose substructure is unknown, thus

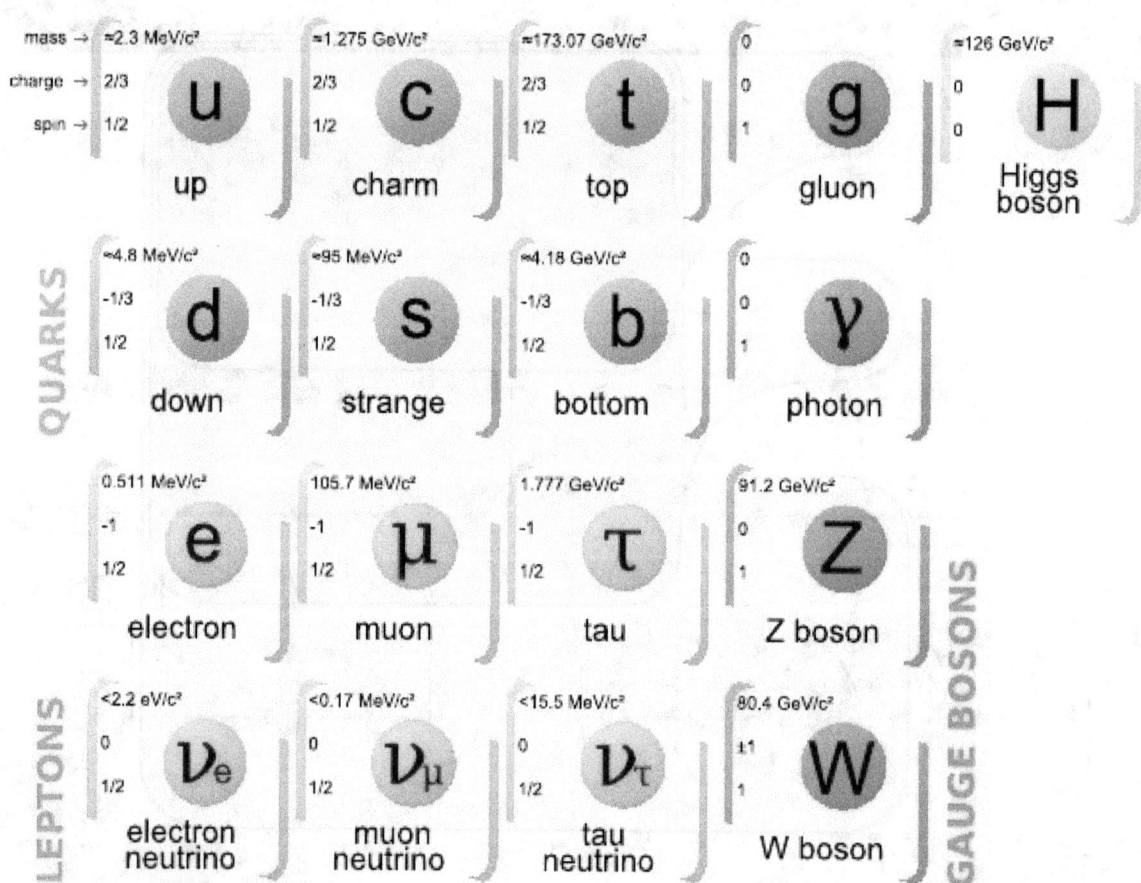

Elementary particles included in the Standard Model

it is unknown whether it is composed of other particles.[1] Known elementary particles include the fundamental fermions (quarks, leptons, antiquarks, and antileptons), which generally are "matter particles" and "antimatter particles", as well as the fundamental bosons (gauge bosons and the Higgs boson), which generally are "force particles" that mediate interactions among fermions.[1] A particle containing two or more elementary particles is a *composite particle*.

Everyday matter is composed of atoms, once presumed to be matter's elementary particles—*atom* meaning "indivisible" in Greek—although the atom's existence remained controversial until about 1910, as some leading physicists regarded molecules as mathematical illusions, and matter as ultimately composed of energy.[1][2] Soon, subatomic constituents of the atom were identified. As the 1930s opened, the electron and the proton had been observed, along with the photon, the particle of electromagnetic radiation.[1] At that time, the recent advent of quantum mechanics was radically altering the conception of particles, as a single particle could seemingly span a field as would a wave, a paradox still eluding satisfactory explanation.[3][4][5]

Via quantum theory, protons and neutrons were found to contain quarks—up quarks and down quarks—now considered elementary particles.[1] And within a molecule, the electron's three degrees of freedom (charge, spin, orbital) can separate via wavefunction into three quasiparticles (holon, spinon, orbiton).[6] Yet a free electron—which, not orbiting an atomic nucleus, lacks orbital motion—appears unsplittable and remains regarded as an elementary particle.[6]

Around 1980, an elementary particle's status as indeed elementary—an *ultimate constituent* of substance—was mostly discarded for a more practical outlook,[1] embodied in particle physics' Standard Model, science's most experimentally successful theory.[5][7] Many elaborations upon and theories beyond the Standard Model, including the extremely popular supersymmetry, double the number of elementary particles by hypothesizing that each known particle associates with a "shadow" partner far more massive,[8][9] although all such superpartners remain undiscovered.[7][10] Meanwhile, an elementary boson mediating gravitation—the graviton—remains hypothetical.[1]

2.1   Overview

Main article: Standard Model
See also: Physics beyond the Standard Model
All elementary particles are—depending on their *spin*—either bosons or fermions. These are differentiated via the spin–

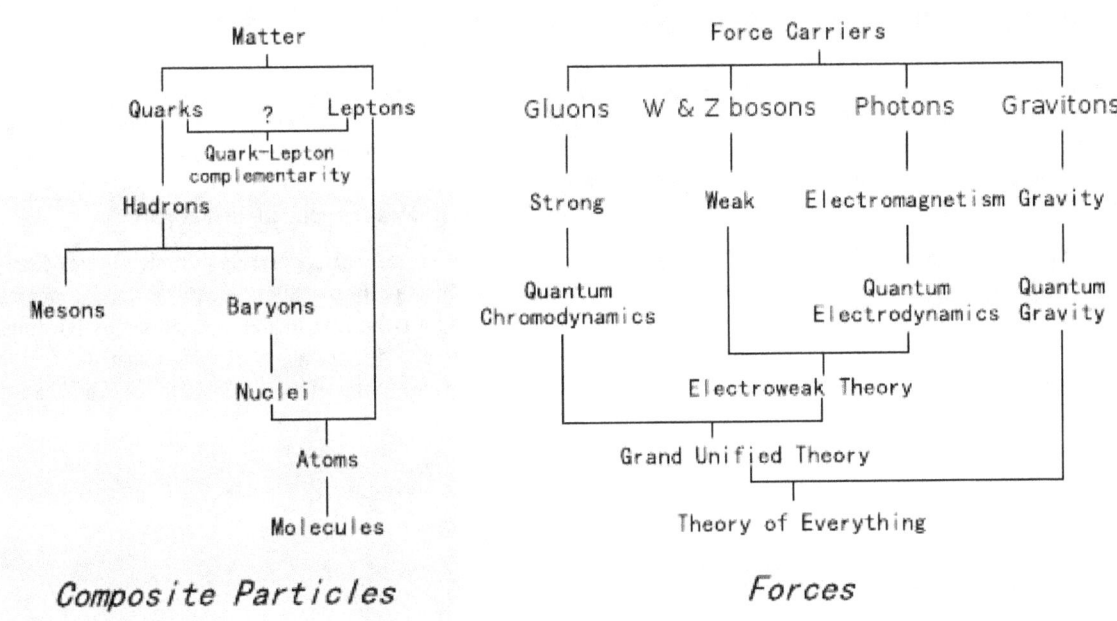

An overview of the various families of elementary and composite particles, and the theories describing their interactions

statistics theorem of quantum statistics. Particles of *half-integer* spin exhibit Fermi–Dirac statistics and are fermions.[1] Particles of *integer* spin, in other words full-integer, exhibit Bose–Einstein statistics and are bosons.[1]

Elementary fermions:

- Matter particles
    - Quarks:
        - up, down
        - charm, strange
        - top, bottom
    - Leptons:
        - electron, electron neutrino (a.k.a., "neutrino")
        - muon, muon neutrino
        - tau, tau neutrino
- Antimatter particles
    - Antiquarks
    - Antileptons

Elementary bosons:

- Force particles (gauge bosons):
    - photon
    - gluon (numbering eight)[1]
    - W^+, W^-, and Z^0 bosons
    - graviton (hypothetical)[1]
- Scalar boson
    - Higgs boson

A particle's mass is quantified in units of energy versus the electron's (electronvolts). Through conversion of energy into mass, any particle can be produced through collision of other particles at high energy,[1][11] although the output particle might not contain the input particles, for instance matter creation from colliding photons. Likewise, the composite fermions protons were collided at nearly light speed to produce the relatively more massive Higgs boson.[11] The most massive elementary particle, the top quark, rapidly decays, but apparently does not contain, lighter particles.

When probed at energies available in experiments, particles exhibit spherical sizes. In operating particle physics' Standard Model, elementary particles are usually represented for predictive utility as point particles, which, as zero-dimensional, lack spatial extension. Though extremely successful, the Standard Model is limited to the microcosm by its omission of gravitation, and has some parameters arbitrarily added but unexplained.[12] Seeking to resolve those shortcomings, string theory posits that elementary particles are ultimately composed of one-dimensional energy strings whose absolute minimum size is the Planck length.

2.2 Common elementary particles

Main article: cosmic abundance of elements

According to the current models of big bang nucleosynthesis, the primordial composition of visible matter of the universe should be about 75% hydrogen and 25% helium-4 (in mass). Neutrons are made up of one up and two down quark, while protons are made of two up and one down quark. Since the other common elementary particles (such as electrons, neutrinos, or weak bosons) are so light or so rare when compared to atomic nuclei, we can neglect their mass contribution to the observable universe's total mass. Therefore, one can conclude that most of the visible mass of the universe consists of protons and neutrons, which, like all baryons, in turn consist of up quarks and down quarks.

Some estimates imply that there are roughly 10^{80} baryons (almost entirely protons and neutrons) in the observable universe.[13][14][15]

The number of protons in the observable universe is called the Eddington number.

In terms of number of particles, some estimates imply that nearly all the matter, excluding dark matter, occurs in neutrinos, and that roughly 10^{86} elementary particles of matter exist in the visible universe, mostly neutrinos.[15] Other estimates imply that roughly 10^{97} elementary particles exist in the visible universe (not including dark matter), mostly photons, gravitons, and other massless force carriers.[15]

2.3 Standard Model

Main article: Standard Model
 The Standard Model of particle physics contains 12 flavors of elementary fermions, plus their corresponding antiparticles,

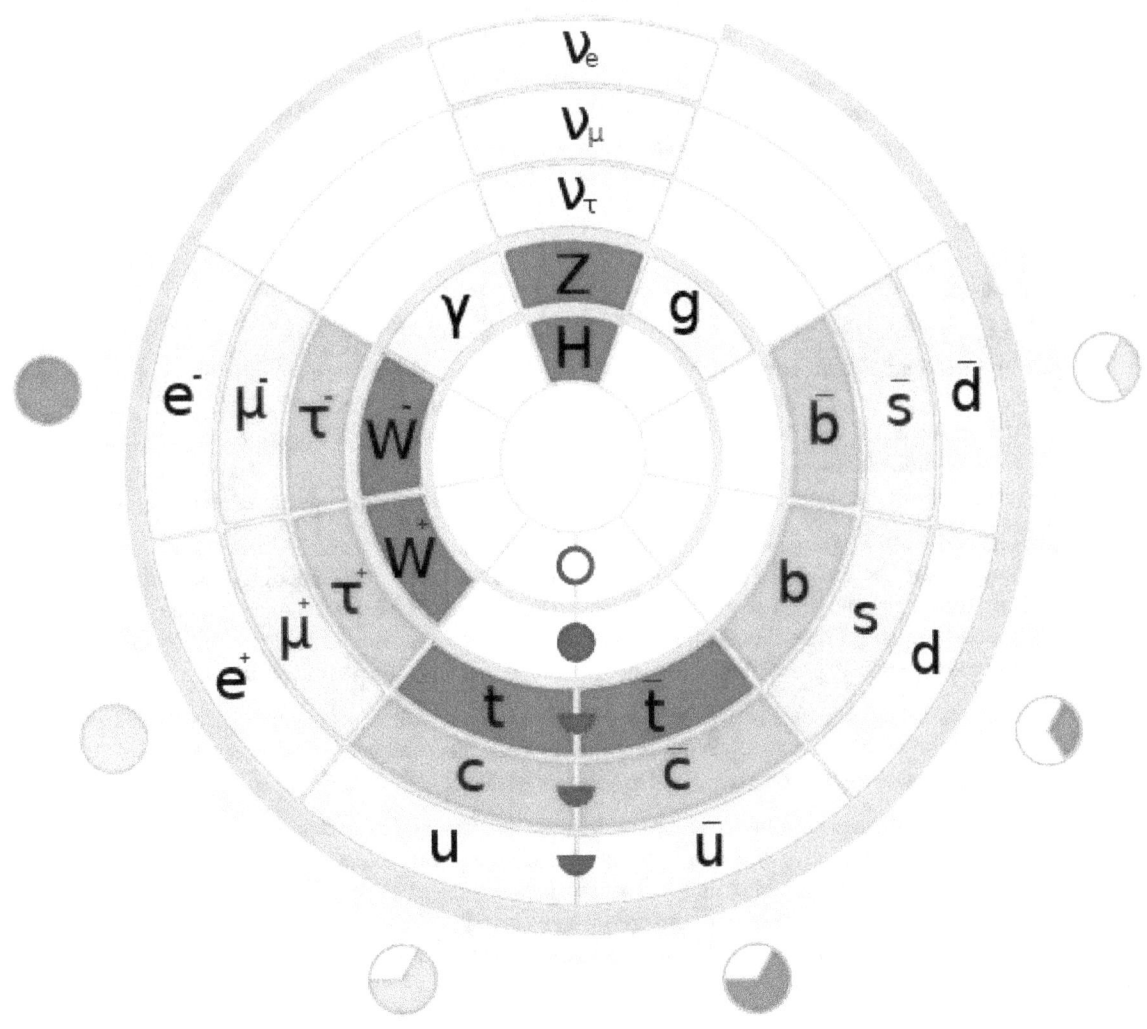

Graphic representation of the standard model. Spin, charge, mass and participation in different force interactions are shown. Click on the image to see the full description

as well as elementary bosons that mediate the forces and the Higgs boson, which was reported on July 4, 2012, as having been likely detected by the two main experiments at the LHC (ATLAS and CMS). However, the Standard Model is widely considered to be a provisional theory rather than a truly fundamental one, since it is not known if it is compatible with Einstein's general relativity. There may be hypothetical elementary particles not described by the Standard Model, such as the graviton, the particle that would carry the gravitational force, and sparticles, supersymmetric partners of the ordinary particles.

2.3.1 Fundamental fermions

Main article: Fermion

The 12 fundamental fermionic flavours are divided into three generations of four particles each. Six of the particles are quarks. The remaining six are leptons, three of which are neutrinos, and the remaining three of which have an electric charge of −1: the electron and its two cousins, the muon and the tau.

Antiparticles

Main article: Antimatter

There are also 12 fundamental fermionic antiparticles that correspond to these 12 particles. For example, the antielectron (positron) $e+$ is the electron's antiparticle and has an electric charge of +1.

Quarks

Main article: Quark

Isolated quarks and antiquarks have never been detected, a fact explained by confinement. Every quark carries one of three color charges of the strong interaction; antiquarks similarly carry anticolor. Color-charged particles interact via gluon exchange in the same way that charged particles interact via photon exchange. However, gluons are themselves color-charged, resulting in an amplification of the strong force as color-charged particles are separated. Unlike the electromagnetic force, which diminishes as charged particles separate, color-charged particles feel increasing force.

However, color-charged particles may combine to form color neutral composite particles called hadrons. A quark may pair up with an antiquark: the quark has a color and the antiquark has the corresponding anticolor. The color and anticolor cancel out, forming a color neutral meson. Alternatively, three quarks can exist together, one quark being "red", another "blue", another "green". These three colored quarks together form a color-neutral baryon. Symmetrically, three antiquarks with the colors "antired", "antiblue" and "antigreen" can form a color-neutral antibaryon.

Quarks also carry fractional electric charges, but, since they are confined within hadrons whose charges are all integral, fractional charges have never been isolated. Note that quarks have electric charges of either +2/3 or −1/3, whereas antiquarks have corresponding electric charges of either −2/3 or +1/3.

Evidence for the existence of quarks comes from deep inelastic scattering: firing electrons at nuclei to determine the distribution of charge within nucleons (which are baryons). If the charge is uniform, the electric field around the proton should be uniform and the electron should scatter elastically. Low-energy electrons do scatter in this way, but, above a particular energy, the protons deflect some electrons through large angles. The recoiling electron has much less energy and a jet of particles is emitted. This inelastic scattering suggests that the charge in the proton is not uniform but split among smaller charged particles: quarks.

2.3.2 Fundamental bosons

Main article: Boson

In the Standard Model, vector (spin−1) bosons (gluons, photons, and the W and Z bosons) mediate forces, whereas the Higgs boson (spin-0) is responsible for the intrinsic mass of particles. Bosons differ from fermions in the fact that multiple bosons can occupy the same quantum state (Pauli exclusion principle). Also, bosons can be either elementary, like photons, or a combination, like mesons. The spin of bosons are integers instead of half integers.

Gluons

Main article: Gluon

Gluons mediate the strong interaction, which join quarks and thereby form hadrons, which are either baryons (three quarks) or mesons (one quark and one antiquark). Protons and neutrons are baryons, joined by gluons to form the atomic nucleus. Like quarks, gluons exhibit colour and anticolour—unrelated to the concept of visual color—sometimes in combinations, altogether eight variations of gluons.

Electroweak bosons

Main articles: W and Z bosons and Photon

There are three weak gauge bosons: W^+, W^-, and Z^0; these mediate the weak interaction. The W bosons are known for their mediation in nuclear decay. The W^- converts a neutron into a proton then decay into an electron and electron antineutrino pair. The Z^0 does not convert charge but rather changes momentum and is the only mechanism for elastically scattering neutrinos. The weak gauge bosons were discovered due to momentum change in electrons from neutrino-Z exchange. The massless photon mediates the electromagnetic interaction. These four gauge bosons form the electroweak interaction among elementary particles.

Higgs boson

Main article: Higgs boson

Although the weak and electromagnetic forces appear quite different to us at everyday energies, the two forces are theorized to unify as a single electroweak force at high energies. This prediction was clearly confirmed by measurements of cross-sections for high-energy electron-proton scattering at the HERA collider at DESY. The differences at low energies is a consequence of the high masses of the W and Z bosons, which in turn are a consequence of the Higgs mechanism. Through the process of spontaneous symmetry breaking, the Higgs selects a special direction in electroweak space that causes three electroweak particles to become very heavy (the weak bosons) and one to remain massless (the photon). On 4 July 2012, after many years of experimentally searching for evidence of its existence, the Higgs boson was announced to have been observed at CERN's Large Hadron Collider. Peter Higgs who first posited the existence of the Higgs boson was present at the announcement.[16] The Higgs boson is believed to have a mass of approximately 125 GeV.[17] The statistical significance of this discovery was reported as 5-sigma, which implies a certainty of roughly 99.99994%. In particle physics, this is the level of significance required to officially label experimental observations as a discovery. Research into the properties of the newly discovered particle continues.

Graviton

Main article: Graviton

The graviton is hypothesized to mediate gravitation, but remains undiscovered and yet is sometimes included in tables of elementary particles.[1] Its spin would be two—thus a boson—and it would lack charge or mass. Besides mediating an extremely feeble force, the graviton would have its own antiparticle and rapidly annihilate, rendering its detection extremely difficult even if it exists.

2.4   Beyond the Standard Model

Although experimental evidence overwhelmingly confirms the predictions derived from the Standard Model, some of its parameters were added arbitrarily, not determined by a particular explanation, which remain mysteries, for instance the hierarchy problem. Theories beyond the Standard Model attempt to resolve these shortcomings.

2.4.1   Grand unification

Main article: Grand Unified Theory

One extension of the Standard Model attempts to combine the electroweak interaction with the strong interaction into a single 'grand unified theory' (GUT). Such a force would be spontaneously broken into the three forces by a Higgs-like mechanism. The most dramatic prediction of grand unification is the existence of X and Y bosons, which cause proton decay. However, the non-observation of proton decay at the Super-Kamiokande neutrino observatory rules out the simplest GUTs, including SU(5) and SO(10).

2.4.2   Supersymmetry

Main article: Supersymmetry

Supersymmetry extends the Standard Model by adding another class of symmetries to the Lagrangian. These symmetries exchange fermionic particles with bosonic ones. Such a symmetry predicts the existence of supersymmetric particles, abbreviated as *sparticles*, which include the sleptons, squarks, neutralinos, and charginos. Each particle in the Standard Model would have a superpartner whose spin differs by 1/2 from the ordinary particle. Due to the breaking of supersymmetry, the sparticles are much heavier than their ordinary counterparts; they are so heavy that existing particle colliders would not be powerful enough to produce them. However, some physicists believe that sparticles will be detected by the Large Hadron Collider at CERN.

2.4.3   String theory

Main article: String theory

String theory is a model of physics where all "particles" that make up matter are composed of strings (measuring at the Planck length) that exist in an 11-dimensional (according to M-theory, the leading version) universe. These strings vibrate at different frequencies that determine mass, electric charge, color charge, and spin. A string can be open (a line) or closed in a loop (a one-dimensional sphere, like a circle). As a string moves through space it sweeps out something called a *world sheet*. String theory predicts 1- to 10-branes (a 1-brane being a string and a 10-brane being a 10-dimensional object) that prevent tears in the "fabric" of space using the uncertainty principle (E.g., the electron orbiting a hydrogen atom has the probability, albeit small, that it could be anywhere else in the universe at any given moment).

String theory proposes that our universe is merely a 4-brane, inside which exist the 3 space dimensions and the 1 time dimension that we observe. The remaining 6 theoretical dimensions either are very tiny and curled up (and too small to be macroscopically accessible) or simply do not/cannot exist in our universe (because they exist in a grander scheme called the "multiverse" outside our known universe).

Some predictions of the string theory include existence of extremely massive counterparts of ordinary particles due to vibrational excitations of the fundamental string and existence of a massless spin-2 particle behaving like the graviton.

2.4.4   Technicolor

Main article: Technicolor (physics)

Technicolor theories try to modify the Standard Model in a minimal way by introducing a new QCD-like interaction. This means one adds a new theory of so-called Techniquarks, interacting via so called Technigluons. The main idea is that the Higgs-Boson is not an elementary particle but a bound state of these objects.

2.4.5   Preon theory

Main article: Preon

According to preon theory there are one or more orders of particles more fundamental than those (or most of those) found in the Standard Model. The most fundamental of these are normally called preons, which is derived from "pre-quarks". In essence, preon theory tries to do for the Standard Model what the Standard Model did for the particle zoo that came before it. Most models assume that almost everything in the Standard Model can be explained in terms of three to half a dozen more fundamental particles and the rules that govern their interactions. Interest in preons has waned since the simplest models were experimentally ruled out in the 1980s.

2.4.6   Acceleron theory

Accelerons are the hypothetical subatomic particles that integrally link the newfound mass of the neutrino and to the dark energy conjectured to be accelerating the expansion of the universe.[18]

In theory, neutrinos are influenced by a new force resulting from their interactions with accelerons. Dark energy results as the universe tries to pull neutrinos apart.[18]

2.5   See also

- Asymptotic freedom

- List of particles

- Physical ontology

- Quantum field theory

- Quantum gravity

- Quantum triviality

- UV fixed point

2.6   Notes

[1] Sylvie Braibant; Giorgio Giacomelli; Maurizio Spurio (2012). *Particles and Fundamental Interactions: An Introduction to Particle Physics* (2nd ed.). Springer. pp. 1–3. ISBN 978-94-007-2463-1.

[2] Ronald Newburgh; Joseph Peidle; Wolfgang Rueckner (2006). "Einstein, Perrin, and the reality of atoms: 1905 revisited" (PDF). *American Journal of Physics.* 74 (6): 478–481. Bibcode:2006AmJPh..74..478N. doi:10.1119/1.2188962.

[3] Friedel Weinert (2004). *The Scientist as Philosopher: Philosophical Consequences of Great Scientific Discoveries.* Springer. p. 43. ISBN 978-3-540-20580-7.

[4] Friedel Weinert (2004). *The Scientist as Philosopher: Philosophical Consequences of Great Scientific Discoveries*. Springer. pp. 57–59. ISBN 978-3-540-20580-7.

[5] Meinard Kuhlmann (24 Jul 2013). "Physicists debate whether the world is made of particles or fields—or something else entirely". *Scientific American*.

[6] Zeeya Merali (18 Apr 2012). "Not-quite-so elementary, my dear electron: Fundamental particle 'splits' into quasiparticles, including the new 'orbiton'". *Nature*. doi:10.1038/nature.2012.10471.

[7] Ian O'Neill (24 Jul 2013). "LHC discovery maims supersymmetry, again". *Discovery News*. Retrieved 2013-08-28.

[8] Particle Data Group. "Unsolved mysteries—supersymmetry". *The Particle Adventure*. Berkeley Lab. Retrieved 2013-08-28.

[9] National Research Council (2006). *Revealing the Hidden Nature of Space and Time: Charting the Course for Elementary Particle Physics*. National Academies Press. p. 68. ISBN 978-0-309-66039-6.

[10] "CERN latest data shows no sign of supersymmetry—yet". *Phys.Org*. 25 Jul 2013. Retrieved 2013-08-28.

[11] Ryan Avent (19 Jul 2012). "The Q&A: Brian Greene—Life after the Higgs". *The Economist*. Retrieved 2013-08-28.

[12] Sylvie Braibant; Giorgio Giacomelli; Maurizio Spurio (2012). *Particles and Fundamental Interactions: An Introduction to Particle Physics* (2nd ed.). Springer. p. 384. ISBN 978-94-007-2463-1.

[13] Frank Heile. "Is the Total Number of Particles in the Universe Stable Over Long Periods of Time?". 2014.

[14] Jared Brooks. "Galaxies and Cosmology". 2014. p. 4, equation 16.

[15] Robert Munafo (24 Jul 2013). "Notable Properties of Specific Numbers". Retrieved 2013-08-28.

[16] Lizzy Davies (4 July 2014). "Higgs boson announcement live: CERN scientists discover subatomic particle". *The Guardian*. Retrieved 2012-07-06.

[17] Lucas Taylor (4 Jul 2014). "Observation of a new particle with a mass of 125 GeV". CMS. Retrieved 2012-07-06.

[18] "New theory links neutrino's slight mass to accelerating Universe expansion". *ScienceDaily*. 28 Jul 2004. Retrieved 2008-06-05.

2.7 Further reading

2.7.1 General readers

- Feynman, R.P. & Weinberg, S. (1987) *Elementary Particles and the Laws of Physics: The 1986 Dirac Memorial Lectures*. Cambridge Univ. Press.

- Ford, Kenneth W. (2005) *The Quantum World*. Harvard Univ. Press.

- Brian Greene (1999). *The Elegant Universe*. W.W.Norton & Company. ISBN 0-393-05858-1.

- John Gribbin (2000) *Q is for Quantum – An Encyclopedia of Particle Physics*. Simon & Schuster. ISBN 0-684-85578-X.

- Oerter, Robert (2006) *The Theory of Almost Everything: The Standard Model, the Unsung Triumph of Modern Physics*. Plume.

- Schumm, Bruce A. (2004) *Deep Down Things: The Breathtaking Beauty of Particle Physics*. Johns Hopkins University Press. ISBN 0-8018-7971-X.

- Martinus Veltman (2003). *Facts and Mysteries in Elementary Particle Physics*. World Scientific. ISBN 981-238-149-X.

- Frank Close (2004). *Particle Physics: A Very Short Introduction*. Oxford: Oxford University Press. ISBN 0-19-280434-0.

- Seiden, Abraham (2005). *Particle Physics – A Comprehensive Introduction*. Addison Wesley. ISBN 0-8053-8736-6.

2.7.2   Textbooks

- Bettini, Alessandro (2008) *Introduction to Elementary Particle Physics.* Cambridge Univ. Press. ISBN 978-0-521-88021-3

- Coughlan, G. D., J. E. Dodd, and B. M. Gripaios (2006) *The Ideas of Particle Physics: An Introduction for Scientists,* 3rd ed. Cambridge Univ. Press. An undergraduate text for those not majoring in physics.

- Griffiths, David J. (1987) *Introduction to Elementary Particles.* John Wiley & Sons. ISBN 0-471-60386-4.

- Kane, Gordon L. (1987). *Modern Elementary Particle Physics.* Perseus Books. ISBN 0-201-11749-5.

- Perkins, Donald H. (2000) *Introduction to High Energy Physics,* 4th ed. Cambridge Univ. Press.

2.8   External links

The most important address about the current experimental and theoretical knowledge about elementary particle physics is the Particle Data Group, where different international institutions collect all experimental data and give short reviews over the contemporary theoretical understanding.

- Particle Data Group

other pages are:

- Greene, Brian, "*Elementary particles*", The Elegant Universe, NOVA (PBS)

- particleadventure.org, a well-made introduction also for non physicists

- CERNCourier: Season of Higgs and melodrama

- Pentaquark information page

- Interactions.org, particle physics news

- Symmetry Magazine, a joint Fermilab/SLAC publication

- "Sized Matter: perception of the extreme unseen", Michigan University project for artistic visualisation of subatomic particles

- Elementary Particles made thinkable, an interactive visualisation allowing physical properties to be compared

Chapter 3

Quantum

For other uses, see Quantum (disambiguation).

In physics, a **quantum** (plural: **quanta**) is the minimum amount of any physical entity involved in an interaction. Behind this, one finds the fundamental notion that a physical property may be "quantized," referred to as "the hypothesis of quantization".[1] This means that the magnitude can take on only certain discrete values.

A photon is a single quantum of (visible) light as well as all other forms of electromagnetic radiation and can be referred to as a "light quantum". The energy of an electron bound to an atom is quantized, which results in the stability of atoms, and hence of matter in general.

As incorporated into the theory of quantum mechanics, this is regarded by physicists as part of the fundamental framework for understanding and describing nature.

3.1 Etymology and discovery

The word "quantum" comes from the Latin "quantus", meaning "how much". "Quanta", short for "quanta of electricity" (electrons) was used in a 1902 article on the photoelectric effect by Philipp Lenard, who credited Hermann von Helmholtz for using the word in the area of electricity. However, the word quantum in general was well known before 1900.[2] It was often used by physicians, such as in the term quantum satis. Both Helmholtz and Julius von Mayer were physicians as well as physicists. Helmholtz used "quantum" with reference to heat in his article[3] on Mayer's work, and indeed, the word "quantum" can be found in the formulation of the first law of thermodynamics by Mayer in his letter[4] dated July 24, 1841. Max Planck used "quanta" to mean "quanta of matter and electricity",[5] gas, and heat.[6] In 1905, in response to Planck's work and the experimental work of Lenard (who explained his results by using the term "quanta of electricity"), Albert Einstein suggested that radiation existed in spatially localized packets which he called "quanta of light" ("Lichtquanta").[7]

The concept of quantization of radiation was discovered in 1900 by Max Planck, who had been trying to understand the emission of radiation from heated objects, known as black-body radiation. By assuming that energy can only be absorbed or released in tiny, differential, discrete packets he called "bundles" or "energy elements",[8] Planck accounted for the fact that certain objects change colour when heated.[9] On December 14, 1900, Planck reported his revolutionary findings to the German Physical Society, and introduced the idea of quantization for the first time as a part of his research on black-body radiation.[10] As a result of his experiments, Planck deduced the numerical value of h, known as the Planck constant, and could also report a more precise value for the Avogadro–Loschmidt number, the number of real molecules in a mole and the unit of electrical charge, to the German Physical Society. After his theory was validated, Planck was awarded the Nobel Prize in Physics in 1918 for his discovery.

3.2   Beyond electromagnetic radiation

While quantization was first discovered in electromagnetic radiation, it describes a fundamental aspect of energy not just restricted to photons.[11] In the attempt to bring experiment into agreement with theory, Max Planck postulated that electromagnetic energy is absorbed or emitted in discrete packets, or quanta.[12]

3.3   See also

- Elementary particle

- Graviton

- Introduction to quantum mechanics

- Magnetic flux quantum

- Photon

- Photon polarization

- Quantal analysis

- Quantization (physics)

- Quantum cellular automata

- Quantum channel

- Quantum coherence

- Quantum chromodynamics

- Quantum computer

- Quantum cryptography

- Quantum dot

- Quantum electronics

- Quantum entanglement

- Quantum immortality

- Quantum lithography

- Quantum mechanics

- Quantum number

- Quantum sensor

- Quantum state

- Subatomic particle

3.4 References

[1] Wiener, N. (1966). *Differential Space, Quantum Systems, and Prediction*. Cambridge: The Massachusetts Institute of Technology Press

[2] E. Cobham Brewer 1810–1897. Dictionary of Phrase and Fable. 1898.

[3] E. Helmholtz, Robert Mayer's Priorität (German)

[4] Herrmann,A. Weltreich der Physik, GNT-Verlag (1991) (German)

[5] Planck, M. (1901). "Ueber die Elementarquanta der Materie und der Elektricität". *Annalen der Physik* (in German) 309 (3): 564–566. Bibcode:1901AnP...309..564P. doi:10.1002/andp.19013090311.

[6] Planck, Max (1883). "Ueber das thermodynamische Gleichgewicht von Gasgemengen". *Annalen der Physik* (in German) 255 (6): 358. Bibcode:1883AnP...255..358P. doi:10.1002/andp.18832550612.

[7] Einstein, A. (1905). "Über einen die Erzeugung und Verwandlung des Lichtes betreffenden heuristischen Gesichtspunkt" (PDF). *Annalen der Physik* (in German) 17 (6): 132–148. Bibcode:1905AnP...322..132E. doi:10.1002/andp.19053220607.. A partial English translation is available from Wikisource.

[8] Max Planck (1901). "Ueber das Gesetz der Energieverteilung im Normalspectrum (On the Law of Distribution of Energy in the Normal Spectrum)". *Annalen der Physik* 309 (3): 553. Bibcode:1901AnP...309..553P. doi:10.1002/andp.19013090310. Archived from the original on 2008-04-18.

[9] Brown, T., LeMay, H., Bursten, B. (2008). *Chemistry: The Central Science* Upper Saddle River, NJ: Pearson Education ISBN 0-13-600617-5

[10] Klein, Martin J. (1961). "Max Planck and the beginnings of the quantum theory". *Archive for History of Exact Sciences* 1 (5): 459. doi:10.1007/BF00327765.

[11] Melville, K. (2005, February 11). Real-World Quantum Effects Demonstrated

[12] Modern Applied Physics-Tippens third edition; McGraw-Hill.

3.5 Further reading

- B. Hoffmann, *The Strange Story of the Quantum*, Pelican 1963.

- Lucretius, *On the Nature of the Universe*, transl. from the Latin by R.E. Latham, Penguin Books Ltd., Harmondsworth 1951. There are, of course, many translations, and the translation's title varies. Some put emphasis on how things work, others on what things are found in nature.

- J. Mehra and H. Rechenberg, *The Historical Development of Quantum Theory*, Vol.1, Part 1, Springer-Verlag New York Inc., New York 1982.

- M. Planck, *A Survey of Physical Theory*, transl. by R. Jones and D.H. Williams, Methuen & Co., Ltd., London 1925 (Dover editions 1960 and 1993) including the Nobel lecture.

- Rodney, Brooks (2011) *Fields of Color: The theory that escaped Einstein*. Allegra Print & Imaging.

Chapter 4

Interpretations of quantum mechanics

An **interpretation of quantum mechanics** is a set of statements which attempt to explain quantum mechanics. Although quantum mechanics has held up to rigorous and thorough experimental testing, many of these experiments are open to different interpretations. There exist a number of contending schools of thought, differing over whether quantum mechanics can be understood to be deterministic, which elements of quantum mechanics can be considered "real", and other matters.

This question is of special interest to philosophers of physics, as physicists continue to show a strong interest in the subject. They usually consider an interpretation of quantum mechanics as an interpretation of the mathematical formalism of quantum mechanics, specifying the physical meaning of the mathematical entities of the theory.

4.1 History of interpretations

Main quantum mechanics interpreters

Schrödinger

Born

Everett

The definition of quantum theorists' terms, such as *wavefunctions* and *matrix mechanics*, progressed through many stages. For instance, Erwin Schrödinger originally viewed the electron's wavefunction as its charge density smeared across the field, whereas Max Born reinterpreted it as the electron's probability density distributed across the field. There was detailed and vigorous debate about this and many other related questions at the fifth Solvay Conference in 1927.[1] Debate has continued right to present times.[2][3]

An early interpretation has acquired the label Copenhagen interpretation, and is often used. Amongst more recent interpretational concepts are quantum decoherence[4][5][6][7] and many worlds.[8][9][10]

During most of the 20th century, collapse theories were clearly the mainstream view, and the question of "interpretation" of quantum mechanics mostly revolved around how to interpret "collapse". Proponents of either "pilot-wave" (de Broglie-Bohm-like) or "many-worlds" (Everettian) interpretations tend to emphasize how their respective camps were intellectually marginalized throughout 1950s to 1980s. In this sense, all non-collapse theories are (historically) "minority" interpretations.

However, since the 1990s, there has been a resurgence of interest in non-collapse theories. The *Stanford Encyclopedia* as of 2015 groups interpretations of quantum mechanics into "Bohmian mechanics" (pilot-wave theories),[11] "collapse theories",[12] "many-worlds interpretations",[13] "modal interpretation"[14] and "relational interpretations"[15] as classes of into which most suggestions may be grouped.

As a rough guide development of the mainstream view during the 1990s to 2000s, consider the "snapshot" of opinions collected in a poll by Schlosshauer et al. at the 2011 "Quantum Physics and the Nature of Reality" conference of July 2011.[16] The authors reference a similarly informal poll carried out by Max Tegmark at the "Fundamental Problems in Quantum Theory" conference in August 1997. The main conclusion of the authors is that "the Copenhagen interpretation still reigns supreme", receiving the most votes in their poll (42%), besides the rise to mainstream notability of the many-worlds interpretations:

> "The Copenhagen interpretation still reigns supreme here, especially if we lump it together with intellectual offsprings such as information-based interpretations and the Quantum Bayesian interpretation. In Tegmark's poll, the Everett interpretation received 17% of the vote, which is similar to the number of votes (18%) in our poll."

4.2   Nature of interpretation

An interpretation of quantum mechanics is a conceptual or argumentative way of relating between:

- the *formalism* of quantum mechanics—mathematical objects, relations, and conceptual principles that are intended to be interpreted as representing quantum physical objects and processes of interest, and

- the *phenomenology* of quantum physics—observations made in empirical investigations of those quantum physical objects and processes, and

- the physical meaning of the phenomena, in terms of ordinary understanding.

Two qualities vary among interpretations:

1. Ontology—claims about what things, such as categories and entities, *exist* in the world, and what theoretical objects are related to those real existents

2. Epistemology—claims about the possibility, scope, and means toward relevant *knowledge* of the world

In philosophy of science, the distinction of knowledge versus reality is termed *epistemic* versus *ontic*. A general law is a *regularity* of outcomes (epistemic), whereas a causal mechanism may *regulate* the outcomes (ontic). A phenomenon can receive interpretation either ontic or epistemic. For instance, indeterminism may be attributed to limitations of human observation and perception (epistemic), or may be explained as a real existing *maybe* encoded in the universe (ontic). Confusing the epistemic with the ontic, like if one were to presume that a general law actually "governs" outcomes—and that the statement of a regularity has the role of a causal mechanism—is a category mistake.

In a broad sense, scientific theory can be viewed as offering scientific realism—approximately true description or explanation of the natural world—or might be perceived with antirealism. A realist stance seeks the epistemic and the ontic, whereas an antirealist stance seeks epistemic but not the ontic. In the 20th century's first half, antirealism was mainly logical positivism, which sought to exclude unobservable aspects of reality from scientific theory.

Since the 1950s, antirealism is more modest, usually instrumentalism, permitting talk of unobservable aspects, but ultimately discarding the very question of realism and posing scientific theory as a tool to help humans make predictions, not to attain metaphysical understanding of the world. The instrumentalist view is carried by the famous quote of David Mermin, "Shut up and calculate".[17]

Other approaches to resolve conceptual problems introduce new mathematical formalism, and so propose additional theories with their interpretations. An example is Bohmian mechanics, which is empirically equivalent with the standard formalisms, but requires extra equations to describe the precise trajectory through state space taken by the actual world. This extra ontological cost provides the explanatory benefit of explaining how the probabilities observed in measurements can arise somewhat naturally from a deterministic process.

4.3   Challenges for interpretations

Interpretations of quantum mechanics attempt to provide a conceptual framework for understanding the many aspects of quantum mechanics which are not easily handled by the conceptual framework used for classical physics:

1. Abstract, mathematical nature of quantum field theories

2. Existence of apparently indeterministic and yet reversible processes

3. Role of the observer in determining outcomes

4. Distinction between preparation and measurement

5. Correlations between remote objects

6. Complementarity of offered descriptions

7. Rapidly rising intricacy, far exceeding humans' present calculational capacity, as a system's size increases

The mathematical structure of quantum mechanics is based on rather abstract mathematics, like Hilbert spaces. In classical field theory, a physical property at a given location in the field is readily derived. In Heisenberg's formalism, on the other hand, to derive physical information about a location in the field, one must apply a quantum operation to a quantum state, an elaborate mathematical process.[18]

Schrödinger's formalism describes a waveform governing the probability of outcomes across a field. Yet how do we find in a specific location a particle whose wavefunction, a mere probability distribution of existence, spans a vast region of space?

The act of measurement can interact with the system state in peculiar ways, as found in double-slit experiments. The Copenhagen interpretation holds that the myriad probabilities across a quantum field are unreal, yet that the act of observation/measurement collapses the wavefunction and sets a single possibility to become real. Yet quantum decoherence grants that all the possibilities can be real, and that the act of observation/measurement sets up new subsystems.[19]

A key interpretational question is posed by Dirac's famous sentences about quantum interference: "Each photon then interferes only with itself. Interference between two different photons never occurs." Dirac stops short of repeating this statement for objects other than photons, such as electrons, contenting himself with saying "... for particles even as light as electrons the associated wave frequency is so high that it is not easy to demonstrate interference."[20] Dirac was of course well familiar with the concept of electron diffraction by crystals, which is usually regarded as an interference phenomenon. The comments of Dirac surrounding these sentences indicate that he considers them to be interpretive. No experiment can directly test them, an actual particular photon being detectable only once.

Quantum entanglement, as illustrated in the EPR paradox, seemingly violates principles of local causality.[21]

Complementarity holds that no set of classical physical concepts can simultaneously refer to all properties of a quantum system. For instance, wave description A and particulate description B can each describe a quantum system S, but not simultaneously. Still, complementarity does not usually imply that classical logic is at fault (although Hilary Putnam took such view in "Is logic empirical?"); rather, the composition of physical properties of S does not obey the rules of classical propositional logic when using propositional connectives (see "Quantum logic"). As now well known, the "origin of complementarity lies in the non-commutativity of operators" that describe quantum objects (Omnès 1999).

4.3.1 Instrumentalist description

Further information: Instrumentalism

Any modern scientific theory requires at the very least a description that relates the mathematical formalism to experimental practice and prediction. In the case of quantum mechanics, the most common description is an assertion of statistical regularity between state preparation processes and measurement processes. That is, if a measurement of a real-value quantity is performed many times, each time starting with the same initial conditions, the outcome is a well-defined probability distribution agreeing with the real numbers; moreover, quantum mechanics provides a computational instrument to determine statistical properties of this distribution, such as its expectation value.

Calculations for measurements performed on a system S postulate a Hilbert space H over the complex numbers. When the system S is prepared in a pure state, it is associated with a vector in H. Measurable quantities are associated with Hermitian operators acting on H: these are referred to as observables.

Repeated measurement of an observable A where S is prepared in state ψ yields a distribution of values. The expectation value of this distribution is given by the expression

$$\langle\psi|A|\psi\rangle.$$

This mathematical machinery gives a simple, direct way to compute a statistical property of the outcome of an experiment, once it is understood how to associate the initial state with a Hilbert space vector, and the measured quantity with an observable (that is, a specific Hermitian operator).[22]

As an example of such a computation, the probability of finding the system in a given state $|\phi\rangle$ is given by computing the expectation value of a (rank-1) projection operator.

$$\Pi = |\phi\rangle\langle\phi|.$$

The probability is then the non-negative real number given by

$$P = \langle\psi|\Pi|\psi\rangle = |\langle\phi|\psi\rangle|^2.$$

In the context of quantum mechanics, the "instrumentalist interpretation" refers to the position that there can be no meaningful "interpretation" beyond the mere instrumentalist description, a position often equated with eschewing all interpretation. It is summarized by the sentence "Shut up and calculate!".[23] [24] Depending on the understanding of the term instrumentalism, the "instrumentalist" or "shut up and calculate"-interpretation may either express pragmatism (the formalism of quantum mechanics has no application outside of the description of predicting experimental outcomes) or radical empiricism or phenomenalism (there is no valid concept of reality outside of perceptual phenomena).

Paul Dirac was remarkably silent on issues of interpretation although he is classified as having an instrumentalist attitude.[25] and one of his very few statements on this issue is quoted as "I was not very much interested. I was more interested in getting the correct equations.[25] A similar silent attitude was assumed by John Ward[26] while Willis Lamb criticized many of the interpretational trends.[27]

4.3.2   Concerns of Einstein

An interpretation of the mathematical formalism of quantum mechanics can be characterized by its treatment of some physical or micro-cosmological problems that Einstein saw in Copenhagenism, such as:

- Realism
- Completeness
- Local realism
- Determinism
- Causality

To explain these problems, we need to be more explicit about the kind of picture an interpretation provides. To that end we will regard an interpretation as a correspondence between the elements of the mathematical formalism **M** and the elements of an interpreting structure **I**, where:

- The *mathematical formalism* **M** consists of the Hilbert space machinery of ket-vectors, self-adjoint operators acting on the space of ket-vectors, unitary time dependence of the ket-vectors, and measurement operations. In this context a measurement operation is a transformation which turns a ket-vector into a probability distribution (for a formalization of this concept see quantum operations).

- The *interpreting structure* **I** includes states, transitions between states, measurement operations, and possibly information about spatial extension of these elements. A measurement operation refers to an operation which returns a value and might result in a system state change. Spatial information would be exhibited by states represented as functions on configuration space. The transitions may be non-deterministic or probabilistic or there may be infinitely many states.

One way of assessing an interpretation is whether the elements of **I** are regarded as physically real. Hence the bare instrumentalist view of quantum mechanics outlined in the previous section is not an interpretation at all, for it makes no claims about elements of physical reality.

The current usage of realism and completeness originated in the 1935 paper in which Einstein and others proposed the EPR paradox.[28] In that paper the authors proposed the concepts of *element of reality* and of *completeness of a physical theory*. They characterised element of reality as a quantity whose value can be predicted with certainty before measuring or otherwise disturbing it, and defined a complete physical theory as one in which every element of physical reality is accounted for by the theory. The paper proposed that an interpretation is complete if every element of the interpreting structure is present in the mathematics. Realism is also a property of each of the elements of the maths; an element is real if it corresponds to something physical in the interpreting structure. For example, in some interpretations of quantum mechanics (such as the many-worlds interpretation) the ket vector associated to the system state is said to correspond to an element of physical reality, while in other interpretations it is not. Einstein was not the active author of the EPR paper, and it did not quite focus on his principal concern, which was about causality.

Determinism is a property characterizing state changes due to the passage of time, namely that the state at a future instant is a uniquely defined mathematical function of the state in the present (see time evolution). It may not always be clear whether a particular interpretation is deterministic or not, as there may not be a clear choice of a time parameter. Moreover, a given theory may have two interpretations, one of which is deterministic and the other not.

Local realism is an attempt to formulate in relevant mathematical terms the subtle physical, micro-cosmological, or meta-physical concept of causality. It has two aspects:

- The value returned by a measurement corresponds to the value of some function in the state space. In other words, that value is an element of reality;

- The effects of measurement have a propagation speed not exceeding some universal limit (e.g. the speed of light). In order for this to make sense, measurement operations in the interpreting structure must be localized.

A precise formulation of local realism in terms of a local hidden variable theory was proposed by John Bell. Bell's theorem, combined with experimental testing, restricts the kinds of properties a quantum theory can have, the primary implication being that quantum mechanics cannot satisfy both the principle of locality and counterfactual definiteness.

4.4 Summary of interpretations

4.4.1 Collapse theories

Further information: Wave function collapse

The Copenhagen interpretation

Main article: Copenhagen interpretation

The Copenhagen interpretation is the interpretation of quantum mechanics formulated by Niels Bohr and Werner Heisenberg while collaborating in Copenhagen around 1927. For most of the twentieth century it was regarded as the "standard" interpretation. Bohr and Heisenberg extended the probabilistic interpretation of the wavefunction proposed originally by Max Born. The Copenhagen interpretation rejects questions like "where was the particle before I measured its position?" as meaningless. The measurement process picks out exactly one of the many possibilities allowed for by the state's wave function in a manner consistent with the well-defined probabilities that are assigned to each possible state. According to the interpretation, the interaction of an observer or apparatus that is external to the quantum system is the cause of wave function collapse, thus according to Paul Davies, "reality is in the observations, not in the electron".[29]

Consciousness causes collapse

Main article: Von Neumann–Wigner interpretation

In his treatise *The Mathematical Foundations of Quantum Mechanics*, John von Neumann deeply analyzed the so-called measurement problem. He concluded that the entire physical universe could be made subject to the Schrödinger equation (the universal wave function). He also described how measurement could cause a collapse of the wave function.[30] This point of view was prominently expanded on by Eugene Wigner, who argued that human experimenter consciousness (or maybe even dog consciousness) was critical for the collapse, but he later abandoned this interpretation.[31][32]

Variations of the von Neumann interpretation include:

Subjective reduction research

This principle, that consciousness causes the collapse, is the point of intersection between quantum mechanics and the mind/body problem; and researchers are working to detect conscious events correlated with physical events that, according to quantum theory, should involve a wave function collapse; but, thus far, results are inconclusive.[33][34][35]

Participatory anthropic principle (PAP)

Main article: Anthropic principle

John Archibald Wheeler's participatory anthropic principle says that consciousness plays some role in bringing the universe into existence.[36]

Other physicists have elaborated their own variations of the von Neumann interpretation; including:

- Henry P. Stapp (*Mindful Universe: Quantum Mechanics and the Participating Observer*)

- Bruce Rosenblum and Fred Kuttner (*Quantum Enigma: Physics Encounters Consciousness*)

Objective collapse theories

Main article: Objective collapse theory

Objective collapse theories differ from the Copenhagen interpretation in regarding both the wavefunction and the process of collapse as ontologically objective. In objective theories, collapse occurs randomly ("spontaneous localization"), or when some physical threshold is reached, with observers having no special role. Thus, they are realistic, indeterministic, no-hidden-variables theories. The mechanism of collapse is not specified by standard quantum mechanics, which needs to be extended if this approach is correct, meaning that Objective Collapse is more of a theory than an interpretation. Examples include the Ghirardi-Rimini-Weber theory[37] and the Penrose interpretation.[38]

4.4.2   Many worlds theories

Main article: Many-worlds interpretation

The many-worlds interpretation is an interpretation of quantum mechanics in which a universal wavefunction obeys the same deterministic, reversible laws at all times; in particular there is no (indeterministic and irreversible) wavefunction collapse associated with measurement. The phenomena associated with measurement are claimed to be explained by decoherence, which occurs when states interact with the environment producing entanglement, repeatedly splitting the universe into mutually unobservable alternate histories—distinct universes within a greater multiverse. In this interpretation the wavefunction has objective reality.

Many minds

Main article: Many-minds interpretation

The many-minds interpretation of quantum mechanics extends the many-worlds interpretation by proposing that the distinction between worlds should be made at the level of the mind of an individual observer.

4.4.3 Hidden variables

Main article: Hidden variable theory
Further information: Bell's theorem, Bell test experiments, Local hidden variable and Spekkens Toy Model

Pilot-wave theories

Main articles: de Broglie–Bohm theory and Pilot wave

The de Broglie–Bohm or "pilot wave" theories form a class of interpretations of quantum mechanics based on a theory of Louis de Broglie later extended by David Bohm. Particles, which always have positions, are guided by the wavefunction. The wavefunction evolves according to the Schrödinger wave equation, and the wavefunction never collapses. The theory takes place in a single space-time, is non-local, and is deterministic. The simultaneous determination of a particle's position and velocity is subject to the usual uncertainty principle constraint. The theory is considered to be a hidden variable theory, and by embracing non-locality it satisfies Bell's inequality. The measurement problem is resolved, since the particles have definite positions at all times.[39] The appearance of collapse is explained as phenomenological.[40]

Time-symmetric theories

Several theories have been proposed which modify the equations of quantum mechanics to be symmetric with respect to time reversal.[41][42][43][44][45][46] (E.g. see Wheeler–Feynman time-symmetric theory). This creates retrocausality: events in the future can affect ones in the past, exactly as events in the past can affect ones in the future. In these theories, a single measurement cannot fully determine the state of a system (making them a type of hidden variables theory), but given two measurements performed at different times, it is possible to calculate the exact state of the system at all intermediate times. The collapse of the wavefunction is therefore not a physical change to the system, just a change in our knowledge of it due to the second measurement. Similarly, they explain entanglement as not being a true physical state but just an illusion created by ignoring retrocausality. The point where two particles appear to "become entangled" is simply a point where each particle is being influenced by events that occur to the other particle in the future.

Not all advocates of time-symmetric causality favour modifying the unitary dynamics of standard quantum mechanics. Thus a leading exponent of the two-state vector formalism, Lev Vaidman, highlights how well the two-state vector formalism dovetails with Hugh Everett's many-worlds interpretation.[47]

Transactional Interpretation   Main article: Transactional interpretation

The transactional interpretation of quantum mechanics (TIQM) by John G. Cramer is an interpretation of quantum mechanics inspired by the Wheeler–Feynman absorber theory.[48] It describes a quantum interaction in terms of a standing wave formed by the sum of a retarded (forward-in-time) and an advanced (backward-in-time) wave. The author argues that it avoids the philosophical problems with the Copenhagen interpretation and the role of the observer, and resolves various quantum paradoxes.

Stochastic mechanics

Main articles: Stochastic interpretation and Quantum foam

An entirely classical derivation and interpretation of Schrödinger's wave equation by analogy with Brownian motion was suggested by Princeton University professor Edward Nelson in 1966.[49] Similar considerations had previously been published, for example by R. Fürth (1933), I. Fényes (1952), and Walter Weizel (1953), and are referenced in Nelson's paper. More recent work on the stochastic interpretation has been done by M. Pavon.[50] An alternative stochastic interpretation was developed by Roumen Tsekov.[51]

Scale relativity

Main article: Scale relativity

An approach closely related to stochastic mechanics is scale relativity developed by Laurent Nottale. The main difference with stochastic mechanics is that the stochastic fluctuations which transform classical mechanics into quantum mechanics are the consequence of the scale relativistic nature of the law of motion. This is a generalisation of Einstein's theory of relativity to include scale transformations. It is scale covariance that makes spacetime fractal and hence particle paths become non-differentiable fractal paths just like quantum paths.

Scale relativity is also more general than Nelson stochastic mechanics since it not only derives the Schrödinger equation from quantum mechanics but also the equations of Quantum field theory. Depending of the form of the scale covariance law one gets different theories ranging from standard non-relativistic quantum mechanics, over non-linear non-relativistic Schrödinger equations, to relativistic quantum mechanics (Klein-Gordon & Dirac equation) and various quantum field theories.

Popper's experiment

Main article: Popper's experiment

Karl Popper took part in the "EPR controversy", by exchanging letters with Einstein, Bell etc. about the issue, and by proposing his own Bell test experiment. Popper first proposed an experiment that would test indeterminacy in Quantum Mechanics in two works of 1934.[52][53] However, Einstein wrote a letter to Popper about the experiment in which he raised some crucial objections, causing Popper to admit that his initial idea was "based on a mistake".[54] In the 1950s he returned to the subject and formulated this later experiment, which was finally published in 1982.[55][56]

Since Popper holds both counterfactual definiteness and locality to be true, it is under dispute whether his view is an interpretation (which is what he claimed) or a modification of Quantum Mechanics (which is what many Physicists claim), and, in case of the latter, if it has been empirically refuted or not by Bell test experiments.

Other than that, Popper's theory is a variant of the de Broglie–Bohm theory that interprets the probabilities as a stochastic element in the particle movement instead of uncertainties in their initial position. In this sense it is a position between de Broglie–Bohm and stochastic mechanics, accepting the reality of the wave function like the former (in Popper's view, it is a propensity field), and the stochastic element like the latter (see also note 138 in Popper's autobiography *Unended Quest*, where he expresses some sympathy for Nelson's stochastic interpretation).

4.4.4   Information-based interpretations

Further information: Quantum logic and Digital physics

Quantum informational approaches have attracted growing support during the 2000s.[57] Hagar and Hemmo (2008) even refer to it (critically) as "a new orthodoxy in the foundations of quantum mechanics."[58]

J. A. Wheeler (1990) with his "It from Bit" ("It": physical entity, "Bit": unit of information) has been described as "the cheerleader of this sort of view".[59] These approaches have been described as a revival of immaterialism[60]

Relational quantum mechanics

Main article: Relational quantum mechanics

The essential idea behind relational quantum mechanics, following the precedent of special relativity, is that different observers may give different accounts of the same series of events: for example, to one observer at a given point in time, a system may be in a single, "collapsed" eigenstate, while to another observer at the same time, it may be in a superposition of two or more states. Consequently, if quantum mechanics is to be a complete theory, relational quantum mechanics argues that the notion of "state" describes not the observed system itself, but the relationship, or correlation, between the system and its observer(s). The state vector of conventional quantum mechanics becomes a description of the correlation of some *degrees of freedom* in the observer, with respect to the observed system. However, it is held by relational quantum mechanics that this applies to all physical objects, whether or not they are conscious or macroscopic. Any "measurement event" is seen simply as an ordinary physical interaction, an establishment of the sort of correlation discussed above. Thus the physical content of the theory has to do not with objects themselves, but the relations between them.[61][62]

An independent relational approach to quantum mechanics was developed in analogy with David Bohm's elucidation of special relativity,[63] in which a detection event is regarded as establishing a relationship between the quantized field and the detector. The inherent ambiguity associated with applying Heisenberg's uncertainty principle is subsequently avoided.[64]

Quantum Bayesianism

Main article: Quantum Bayesianism

Interpretations where quantum mechanics is said to describe an observer's knowledge of the world, rather than the world itself. Quantum Bayesianism ("QBism") aims at giving a "subjective Bayesian account of quantum probability",[65] to derive quantum mechanics from informational considerations.

This approach has some similarity with Bohr's thinking.[66] Collapse (also known as reduction) is often interpreted as an observer acquiring information from a measurement, rather than as an objective event. These approaches have been appraised as similar to instrumentalism.[67]

4.4.5 Other

Further information: Minority interpretations of quantum mechanics

Ensemble interpretation

Main article: Ensemble interpretation

The ensemble interpretation, or statistical interpretation can be viewed as a minimalist approach; it is a quantum mechanical interpretation that claims to make the fewest assumptions associated with the standard mathematical formalization. At its heart, it takes to the fullest extent the statistical Born rule.[68]

It does not attempt to justify, or otherwise derive, or explain quantum mechanics from any deterministic process, or make any other statement about the real nature of quantum phenomena; it is simply a statement as to the manner of wave function interpretation. The wave function in this interpretation is not a property of any individual system, it is by its nature a statistical description of a hypothetical "ensemble" of similar systems. The probabilistic nature of quantum mechanical predictions thus follow directly from the construction or scope of the theory rather than from any intrinsic property of nature.

Probably the most notable supporter of such an interpretation was Albert Einstein:

> The attempt to conceive the quantum-theoretical description as the complete description of the individual systems leads to unnatural theoretical interpretations, which become immediately unnecessary if one accepts the interpretation that the description refers to ensembles of systems and not to individual systems.
> — Albert Einstein[69]

A prominent advocate of the ensemble interpretation is Leslie E. Ballentine, Professor at Simon Fraser University, and writer of the graduate-level textbook "Quantum Mechanics, A Modern Development".[70]

Modal interpretations

Modal interpretations of quantum mechanics were first conceived of in 1972 by B. van Fraassen, in his paper "A formal approach to the philosophy of science." However, this term now is used to describe a class of models that grew out of this approach.

The *Stanford Encyclopedia of Philosophy* has an article on "Modal Interpretations of Quantum Mechanics"[71] which presents "modal interpretations" as a class contrasting with objective collapse theories, pilot-wave theories and many-worlds interpretations.

Van Fraassen's proposal distinguished a *dynamical state* from a *value state*. The dynamical state corresponds to the ordinary quantum state, which however never collapses. The value state is the feature which replaces the idea of "collapse". An observable of a system is taken to have a sharp value even if the dynamical state is not an eigenstate of that same observable. Van Fraassen's proposal is "modal" because it leads to a modal logic of quantum propositions. Since the 1980s, a number of authors have developed other "realist" proposals which can in retrospect be classed with van Fraassen's "modal" proposal.

Consistent histories

Main article: Consistent histories

The consistent histories interpretation is based on a consistency criterion that allows the history of a system to be described so that the probabilities for each history obey the additive rules of classical probability.

According to this interpretation, the purpose of a quantum-mechanical theory is to predict the relative probabilities of various alternative histories (for example, of a particle). It is claimed to be consistent with the Schrödinger equation. It attempts to provide a natural interpretation of quantum cosmology.

According to Robert E. Griffiths "It is in fact not necessary to interpret quantum mechanics in terms of measurements."[72]

Nevertheless, Griffiths also says "A quantum theory of measurements is a necessary part of any consistent way of understanding quantum theory for a fairly obvious reason." Griffiths' explanation of this is that quantum measurement theory is derived from the principles of quantum mechanics, which, however, do not themselves explicitly postulate a primary ontological category of measurement in its own right, and which can be interpreted without explicit talk of measurement. Griffiths writes "Thus quantum measurements can, at least in principle, be analyzed using quantum theory."[73] This contradicts the postulate of the orthodox interpretation, that the wave function changes in two ways, (1) according to the Schrödinger equation, which does not involve measurement, and (2) in the so-called 'collapse' or 'reduction' that occurs upon particle detection in the process of measurement.[74]

4.5   Tabular comparison

The most common interpretations are summarized in the table below. The values shown in the cells of the table are not without controversy, for the precise meanings of some of the concepts involved are unclear and, in fact, are themselves at the center of the controversy surrounding the given interpretation.

No empirical evidence exists that distinguishes among these interpretations. To that extent, the physical theory stands, and is consistent within itself and with observation and experiment; difficulties arise only when one attempts to "interpret" the theory. Nevertheless, designing experiments which would test the various interpretations is the subject of active research.

Most of these interpretations have variants. For example, it is difficult to get a precise definition of the Copenhagen interpretation as it was developed and argued about by many people.

- [1] According to Bohr, the concept of a physical state independent of the conditions of its experimental observation does not have a well-defined meaning. According to Heisenberg the wavefunction represents a probability, but not an objective reality itself in space and time.

- [2] According to the Copenhagen interpretation, the wavefunction collapses when a measurement is performed.

- [3] Both particle AND guiding wavefunction are real.

- [4] Unique particle history, but multiple wave histories.

- [5] But quantum logic is more limited in applicability than Coherent Histories.

- [6] Quantum mechanics is regarded as a way of predicting observations, or a theory of measurement.

- [7] Observers separate the universal wavefunction into orthogonal sets of experiences.

- [8] If wavefunction is real then this becomes the many-worlds interpretation. If wavefunction is less than real, but more than just information, then Zurek calls this the "existential interpretation".

- [9] In the TI the collapse of the state vector is interpreted as the completion of the transaction between emitter and absorber.

- [10] Comparing histories between systems in this interpretation has no well-defined meaning.

- [11] Any physical interaction is treated as a collapse event relative to the systems involved, not just macroscopic or conscious observers.

- [12] The state of the system is observer-dependent, i.e., the state is specific to the reference frame of the observer.

- [13] Since Popper holds both CFD and locality to be true, it is under dispute whether his view is an interpretation (which is what he claimed) or a modification of Quantum Mechanics (which is what many Physicists claim), and, in case of the latter, if it has been empirically refuted or not by Bell test experiments.

- [14] The transactional interpretation is explicitly non-local.

- [15] The assumption of intrinsic periodicity is an element of non-locality consistent with relativity as the periodicity varies in a causal way.

- [16] In the stochastic interpretation it's not possible to define velocities for particles, i.e. the paths are not smooth. Moreover, to know the motion of the particles at any moment, you have to know what the Markov process is. However, once we know the exact initial conditions and the Markov process, the theory is in fact a realistic interpretation of quantum mechanics; trajectories are continuous.

- [17] The kind of locality violated by the theory is the one assumed in deriving Bell inequalities. In particular, this kind non-locality is compatible with no signaling theorem and so with relativity.

- [18] The interpretation is compatible with the view of a deterministic world as a whole, but does not exclude indeterminism.

- [19] There are no hidden variables associated with the state of the quantum entity, but there are hidden variables associated with the measurement-interactions.

4.6 See also

4.7 Sources

- Bub, J.; Clifton, R. (1996). "A uniqueness theorem for interpretations of quantum mechanics". *Studies in History and Philosophy of Modern Physics* **27B**: 181–219.

- Rudolf Carnap, 1939, "The interpretation of physics", in *Foundations of Logic and Mathematics* of the *International Encyclopedia of Unified Science*. University of Chicago Press.

- Dickson, M., 1994, "Wavefunction tails in the modal interpretation" in Hull, D., Forbes, M., and Burian, R., eds., *Proceedings of the PSA* 1" 366–76. East Lansing, Michigan: Philosophy of Science Association.

- --------, and Clifton, R., 1998, "Lorentz-invariance in modal interpretations" in Dieks, D. and Vermaas, P., eds., *The Modal Interpretation of Quantum Mechanics*. Dordrecht: Kluwer Academic Publishers: 9–48.

- Fuchs, Christopher, 2002, "Quantum Mechanics as Quantum Information (and only a little more)." arXiv:quant-ph/0205039

- -------- and A. Peres, 2000, "Quantum theory needs no 'interpretation'", *Physics Today*.

- Herbert, N., 1985. *Quantum Reality: Beyond the New Physics*. New York: Doubleday. ISBN 0-385-23569-0.

- Hey, Anthony, and Walters, P., 2003. *The New Quantum Universe*, 2nd ed. Cambridge Univ. Press. ISBN 0-521-56457-3.

- Jackiw, Roman; Kleppner, D. (2000). "One Hundred Years of Quantum Physics". *Science* **289** (5481): 893. doi:10.1126/science.289.5481.893.

- Max Jammer, 1966. *The Conceptual Development of Quantum Mechanics*. McGraw-Hill.

- --------, 1974. *The Philosophy of Quantum Mechanics*. Wiley & Sons.

- Al-Khalili, 2003. *Quantum: A Guide for the Perplexed*. London: Weidenfeld & Nicholson.

- de Muynck, W. M., 2002. *Foundations of quantum mechanics, an empiricist approach*. Dordrecht: Kluwer Academic Publishers. ISBN 1-4020-0932-1.[77]

- Roland Omnès, 1999. *Understanding Quantum Mechanics*. Princeton Univ. Press.

- Karl Popper, 1963. *Conjectures and Refutations*. London: Routledge and Kegan Paul. The chapter "Three views Concerning Human Knowledge" addresses, among other things, instrumentalism in the physical sciences.

- Hans Reichenbach, 1944. *Philosophic Foundations of Quantum Mechanics*. Univ. of California Press.

- Tegmark, Max; Wheeler, J. A. (2001). "100 Years of Quantum Mysteries". *Scientific American* **284**: 68. 68.

- Bas van Fraassen, 1972, "A formal approach to the philosophy of science", in R. Colodny, ed., *Paradigms and Paradoxes: The Philosophical Challenge of the Quantum Domain*. Univ. of Pittsburgh Press: 303-66.

- John A. Wheeler and Wojciech Hubert Zurek (eds), *Quantum Theory and Measurement*, Princeton: Princeton University Press, ISBN 0-691-08316-9, LoC QC174.125.Q38 1983.

4.8   References

[1] *Électrons et Photons: Rapports et Discussions du Cinquième Conseil de Physique, tenu à Bruxelles du 24 au 29 Octobre 1927, sous les Auspices de l'Institut International de Physique Solvay* (1928), Gauthier-Villars, Paris.

[2] Jammer, M. (1974). *The Philosophy of Quantum Mechanics: the Interpretations of QM in Historical Perspective*, Wiley, USA, ISBN 0-471-43958-4.

[3] Bacciagaluppi, G., Valentini, A. (2009), *Quantum Theory at the Crossroads: Reconsidering the 1927 Solvay Conference*, Cambridge University Press, Cambridge UK, ISBN 978-0-521-81421-8.

[4] Roland Omnes (1994). *The Interpretation of Quantum Mechanics*, Princeton University Press, ISBN 0-691-03669-1.

[5] Giulini, D., Joos, E., Kiefer, C., Kupsch, J., Stamatescu, I.-O., Zeh, H.D, (1996). *Decoherence and the Appearance of a Classical World in Quantum Theory*, Springer, Berlin, ISBN 3-540-61394-3.

[6] Roland Omnes (1999). *Understanding Quantum Mechanics*, Princeton University Press.

[7] Roland Omnes (1994). *Quantum Philosophy: Understanding and Interpreting Contemporary Science*, Princeton University Press.

[8] Deutsch, D. (1986). Three connections between Everett's interpretation and experiment, pp. 215–225 in *Quantum Concepts of Space and Time*, edited by R. Penrose and C.J. Isham, Oxford University Press, Oxford UK, ISBN 0-19-851972-9.

[9] Vaidman, L. (2002, March 24). Many-Worlds Interpretation of Quantum Mechanics. Retrieved March 19, 2010, from Stanford Encyclopedia of Philosophy: http://plato.stanford.edu/entries/qm-manyworlds/#Teg98

[10] A controversial poll mentioned in *The Physics of Immortality* (1994) found that of 72 "leading cosmologists and other quantum field theorists", 58% including Stephen Hawking, Murray Gell-Mann, and Richard Feynman supported a many-worlds interpretation ["Who believes in many-worlds?", *Hedweb.com*, Accessed online: 24 Jan 2011].

[11] Goldstein, Sheldon, "Bohmian Mechanics", The Stanford Encyclopedia of Philosophy (Spring 2013 Edition).

[12] Ghirardi, Giancarlo, "Collapse Theories", The Stanford Encyclopedia of Philosophy (Winter 2011 Edition).

[13] Vaidman, Lev, "Many-Worlds Interpretation of Quantum Mechanics", The Stanford Encyclopedia of Philosophy (Spring 2015 Edition)

[14] Lombardi, Olimpia and Dieks, Dennis, "Modal Interpretations of Quantum Mechanics", The Stanford Encyclopedia of Philosophy (Spring 2014 Edition).

[15] Laudisa, Federico and Rovelli, Carlo, "Relational Quantum Mechanics", The Stanford Encyclopedia of Philosophy (Summer 2013 Edition)

[16] Schlosshauer, Maximilian; Kofler, Johannes; Zeilinger, Anton (2013-01-06). "A Snapshot of Foundational Attitudes Toward Quantum Mechanics". *Studies in History and Philosophy of Science Part B: Studies in History and Philosophy of Modern Physics* 44 (3): 222–230. arXiv:1301.1069. doi:10.1016/j.shpsb.2013.04.004.

[17] For a discussion of the provenance of the phrase "shut up and calculate", see

[18] Meinard Kuhlmann, "Physicists debate whether the world is made of particles or fields—or something else entirely", *Scientific American*, 24 Jul 2013.

[19] Guido Bacciagaluppi, "The role of decoherence in quantum mechanics", *The Stanford Encyclopedia of Philosophy* (Winter 2012), Edward N Zalta, ed.

[20] Dirac, P.A.M. (1930/1958). *The Principles of Quantum Mechanics*, fourth edition, Oxford University Press, Oxford UK, ISBN 0-19-852011-5, pp. 9, 10

[21] *La nouvelle cuisine*, by John S Bell, last article of Speakable and Unspeakable in Quantum Mechanics, second edition.

[22] Fuchs, C.A., Peres, A. (2000). Quantum Theory Needs No 'Interpretation', *Physics Today* 53(3): 70–71.

[23] N. David Mermin. "Could Feynman Have Said This?". *Physics Today* 57 (5). Bibcode:2004PhT....57e..10M. doi:10.1063/1.17

[24] Paul Budnik , The shut up and calculate interpretation: "This is the most popular of interpretations."

[25] H. S. Kragh, *Dirac: A Scientific Biography* (Cambridge University, Cambridge, 1990) ISBN 0-521-38089-8 p. 82

[26] F. J. Duarte, *Quantum Optics for Engineers* (CRC, New York, 2014) ISBN 978-1-4398-8853-7 p. 330

[27] W. E. Lamb, Superclassical quantum mechanics: the best interpretation of nonrelativistic quantum mechanics, *Am. J. Phys.* 69, 413-421 (2001).

[28] Einstein, A.; Podolsky, B.; Rosen, N. (1935). "Can quantum-mechanical description of physical reality be considered complete?". *Phys. Rev* 47: 777. Bibcode:1935PhRv...47..777E. doi:10.1103/physrev.47.777.

[29] http://www.naturalthinker.net/trl/texts/Heisenberg,Werner/Heisenberg,%20Werner%20-%20Physics%20and%20philosophy.pdf

[30] von Neumann, John. (1932/1955). *Mathematical Foundations of Quantum Mechanics*. Princeton: Princeton University Press. Translated by Robert T. Beyer.

[31] [Michael Esfeld, (1999), "Essay Review: Wigner's View of Physical Reality", published in *Studies in History and Philosophy of Modern Physics*, 30B, pp. 145–154, Elsevier Science Ltd.]

[32] Zvi Schreiber (1995). "The Nine Lives of Schrödinger's Cat". arXiv:quant-ph/9501014.

[33] Dick J. Bierman and Stephen Whitmarsh. (2006). *Consciousness and Quantum Physics: Empirical Research on the Subjective Reduction of the State Vector.* in Jack A. Tuszynski (Ed). **The Emerging Physics of Consciousness.** p. 27-48.

[34] Nunn, C. M. H.; et al. (1994). *"Collapse of a Quantum Field may Affect Brain Function. '".* *Journal of Consciousness Studies'* 1 (1): 127–139.

[35] Arvan, Marcus (2013). "A New Theory of Free Will". *The Philosophical Forum* 44 (1): 1–48. doi:10.1111/phil.12000.

[36] "- The anthropic universe". Abc.net.au. 2006-02-18. Retrieved 2011-01-24.

[37] "Frigg, R. GRW theory" (PDF). Retrieved 2011-01-24.

[38] "Review of Penrose's Shadows of the Mind". Thymos.com. Retrieved 2011-01-24.

[39] *Why Bohm's Theory Solves the Measurement Problem* by T. Maudlin, Philosophy of Science 62, pp. 479-483 (September, 1995).

[40] *Bohmian Mechanics as the Foundation of Quantum Mechanics* by D. Durr, N. Zanghi, and S. Goldstein in **Bohmian Mechanics and Quantum Theory: An Appraisal**, edited by J.T. Cushing, A. Fine, and S. Goldstein, Boston Studies in the Philosophy of Science 184, 21-44 (Kluwer, 1996) 1997 arXiv:quant-ph/9511016

[41] Watanabe, Satosi. "Symmetry of physical laws. Part III. Prediction and retrodiction." *Reviews of Modern Physics* 27.2 (1955): 179.

[42] Aharonov, Y. et al., "Time Symmetry in the Quantum Process of Measurement." *Phys. Rev.* 134, B1410–1416 (1964).

[43] Aharonov, Y. and Vaidman, L. "On the Two-State Vector Reformulation of Quantum Mechanics." *Physica Scripta*, Volume T76, pp. 85–92 (1998).

[44] Wharton, K. B. "Time-Symmetric Quantum Mechanics." *Foundations of Physics*, 37(1), pp. 159–168 (2007).

[45] Wharton, K. B. "A Novel Interpretation of the Klein–Gordon Equation." *Foundations of Physics*, 40(3), pp. 313–332 (2010).

[46] Heaney, M. B. "A Symmetrical Interpretation of the Klein–Gordon Equation." *Foundations of Physics* (2013): http://link.springer.com/article/10.1007%2Fs10701-013-9713-9.

[47] Yakir Aharonov, Lev Vaidman: *The Two-State Vector Formalism of Quantum Mechanics: an Updated Review.* In: Juan Gonzalo Muga, Rafael Sala Mayato, Íñigo Egusquiza (eds.): *Time in Quantum Mechanics*, Volume 1, Lecture Notes in Physics 734, pp. 399–447, 2nd ed., Springer, 2008, ISBN 978-3540734727, DOI 10.1007/978-3-540-73473-4_13, arXiv:quant-ph/0105101v2 (submitted 21 May 2001, version of 10 June 2007), p. 443

[48] "Quantum Nocality - Cramer". Npl.washington.edu. Retrieved 2011-01-24.

[49] Nelson, E (1966). "Derivation of the Schrödinger Equation from Newtonian Mechanics". *Phys. Rev.* 150: 1079–1085. Bibcode:1966PhRv..150.1079N. doi:10.1103/physrev.150.1079.

[50] M. Pavon, "Stochastic mechanics and the Feynman integral", *J. Math. Phys.* 41, 6060-6078 (2000)

[51] Roumen Tsekov (2012). "Bohmian Mechanics versus Madelung Quantum Hydrodynamics". *Ann. Univ. Sofia, Fac. Phys.* SE: 112–119. arXiv:0904.0723. Bibcode:2009arXiv0904.0723T. doi:10.13140/RG.2.1.3663.8245.

[52] Popper, K.R. Quantum Theory and the Schism in Physics, *Die Naturwissenshaften*, 22, 807 (1934)

[53] Popper, K.R.,*The Logic of Scientific Discovery*, 1934 (as *Logik der Forschung*, English translation 1959), ISBN 0-415-27844-9

[54] Popper, K.R.,*The Logic of Scientific Discovery*, (1959), p. 236 note.

[55] Hacohen, M.H., *Karl Popper: the formative years, 1902-1945 : politics and philosophy in interwar Vienna*, CUP, 2002, p. 259.

[56] William M. Shields (2012). "A Historical Survey of Sir Karl Popper's Contribution to Quantum Mechanics". *Quanta* 1 (1): 1–12. doi:10.12743/quanta.v1i1.4.

[57] Schlosshauer, Maximilian; Kofler, Johannes; Zeilinger, Anton (2013-01-06). "A Snapshot of Foundational Attitudes Toward Quantum Mechanics". *Studies in History and Philosophy of Science Part B: Studies in History and Philosophy of Modern Physics* **44** (3): 222–230. arXiv:1301.1069. doi:10.1016/j.shpsb.2013.04.004. "Evidently, there is broad enthusiasm—or at least open-mindedness—about quantum information, with three in four respondents regarding quantum information as "a breath of fresh air for quantum foundations." Indeed, it is hard to deny the impact quantum information theory has had on the field of quantum foundations over the past decade. It has inspired new ways of thinking about quantum theory and has produced information-theoretic derivations (reconstructions) of the structure of the theory. On the practical side, the quantum-information boom has helped fund numerous foundational research projects. Last but not least, quantum information has given foundational pursuits new legitimacy."

[58] Hagar, Amit; Hemmo, Meir (2006 (v2 2008)). "Explaining the Unobserved—Why Quantum Mechanics Ain't Only About Information". *Foundations of Physics* **36** (9): 1295–1324. arXiv:quant-ph/0512095v2. Bibcode:2006FoPh...36.1295H. doi:10.1007/s10701-006-9065-9. Chec k date values in: |date= (help)

[59] Information, Immaterialism, Instrumentalism: Old and New in Quantum Information. Christopher G. Timpson

[60] Timpson,Op. Cit.: "Let us call the thought that information might be the basic category from which all else flows informational immaterialism."

[61] "Relational Quantum Mechanics (Stanford Encyclopedia of Philosophy)". Plato.stanford.edu. Retrieved 2011-01-24.

[62] For more information, see Carlo Rovelli (1996). "Relational Quantum Mechanics". *International Journal of Theoretical Physics* **35** (8): 1637. arXiv:quant-ph/9609002. Bibcode:1996IJTP...35.1637R. doi:10.1007/BF02302261.

[63] David Bohm, *The Special Theory of Relativity*, Benjamin, New York, 1965

[64] . For a full account , see Q. Zheng and T. Kobayashi, 1996, "Quantum Optics as a Relativistic Theory of Light", *Physics Essays* 9: 447. Annual Report, Department of Physics, School of Science, University of Tokyo (1992) 240.

[65] Stairs, Allen (2011). "A loose and separate certainty: Caves, Fuchs and Schack on quantum probability one" (pdf). *Studies In History and Philosophy of Science Part B: Studies In History and Philosophy of Modern Physics* **42** (3): 158–166. doi:10.1016/j.shpsb.2011.02.001.Retrieved April 2012.

[66] "Physics concerns what we can say about nature". (Niels Bohr, quoted in Petersen, A. (1963). The philosophy of Niels Bohr. *Bulletin of the Atomic Scientists*, 19(7):8–14.)

[67] "The 'reduction of the wavepacket' does take place in the consciousness of the observer, not because of any unique physical process which takes place there, but only because the state is a construct of the observer and not an objective property of the physical system" Hartle, J. B. (1968). Quantum mechanics of individual systems. Am. J. Phys., 36(8):704– 712.

[68] "The statistical interpretation of quantum mechanics" (PDF). *Nobel Lecture*. December 11, 1954.

[69] Einstein: Philosopher-Scientist, ed. P.A. Schilpp (Harper & Row, New York)

[70] Leslie E. Ballentine (1998). *Quantum Mechanics: A Modern Development*. World Scientific. Chapter 9. ISBN 981-02-4105-4.

[71] Olimpia Lombardi, Dennis Dieks (2012). "Modal Interpretations of Quantum Mechanics". *Stanford Encyclopedia of Philosophy*.

[72] Griffiths, R.B. (2002). *Consistent Quantum Theory*, Cambridge University Press, Cambridge UK, p. 10.

[73] Griffiths, R.B. (2002). *Consistent Quantum Theory*, Cambridge University Press, Cambridge UK, p. 6.

[74] Wigner, E.P. (1963). "The problem of measurement". *Am. J. Phys.* **31**: 6–15. Bibcode:1963AmJPh..31....6W. doi:10.111

[75] Marie-Christine Combourieu: Karl R. Popper, 1992: About the EPR controversy. *Foundations of Physics* 22:10, 1303-1323

[76] Karl Popper: The Propensity Interpretation of the Calculus of Probability and of the Quantum Theory. *Observation and Interpretation*. Buttersworth Scientific Publications, Korner & Price (eds.) 1957. pp 65–70.

[77] de Muynck, Willem M (2002). *Foundations of quantum mechanics: an empiricist approach*. Klower Academic Publishers. ISBN 1-4020-0932-1. Retrieved 2011-01-24.

4.9   Further reading

Almost all authors below are professional physicists.

- David Z Albert, 1992. *Quantum Mechanics and Experience*. Harvard Univ. Press. ISBN 0-674-74112-9.

- John S. Bell, 1987. *Speakable and Unspeakable in Quantum Mechanics*. Cambridge Univ. Press, ISBN 0-521-36869-3. The 2004 edition (ISBN 0-521-52338-9) includes two additional papers and an introduction by Alain Aspect.

- Dmitrii Ivanovich Blokhintsev, 1968. *The Philosophy of Quantum Mechanics*. D. Reidel Publishing Company. ISBN 90-277-0105-9.

- David Bohm, 1980. *Wholeness and the Implicate Order*. London: Routledge. ISBN 0-7100-0971-2.

- Adan Cabello (15 November 2004). "Bibliographic guide to the foundations of quantum mechanics and quantum information". arXiv:quant-ph/0012089.

- John G. Cramer, *The Quantum Handshake: Entanglement, Nonlocality and Transactions*, Springer Verlag 2015, ISBN 978-3-319-24640-6.

- David Deutsch, 1997. *The Fabric of Reality*. London: Allen Lane. ISBN 0-14-027541-X; ISBN 0-7139-9061-9. Argues forcefully *against* instrumentalism. For general readers.

- Bernard d'Espagnat, 1976. *Conceptual Foundation of Quantum Mechanics*, 2nd ed. Addison Wesley. ISBN 0-8133-4087-X.

- --------, 1983. *In Search of Reality*. Springer. ISBN 0-387-11399-1.

- --------, 2003. *Veiled Reality: An Analysis of Quantum Mechanical Concepts*. Westview Press.

- --------, 2006. *On Physics and Philosophy*. Princeton Univ. Press.

- Arthur Fine, 1986. *The Shaky Game: Einstein Realism and the Quantum Theory. Science and its Conceptual Foundations*. Univ. of Chicago Press. ISBN 0-226-24948-4.

- Ghirardi, Giancarlo, 2004. *Sneaking a Look at God's Cards*. Princeton Univ. Press.

- Gregg Jaeger (2009) *Entanglement, Information, and the Interpretation of Quantum Mechanics*. Springer. ISBN 978-3-540-92127-1.

- N. David Mermin (1990) *Boojums all the way through*. Cambridge Univ. Press. ISBN 0-521-38880-5.

- Roger Penrose, 1989. *The Emperor's New Mind*. Oxford Univ. Press. ISBN 0-19-851973-7. Especially chpt. 6.

- --------, 1994. *Shadows of the Mind*. Oxford Univ. Press. ISBN 0-19-853978-9.

- --------, 2004. *The Road to Reality*. New York: Alfred A. Knopf. Argues that quantum theory is incomplete.

- Styer, Daniel F.; Balkin, Miranda S.; Becker, Kathryn M.; Burns, Matthew R.; Dudley, Christopher E.; Forth, Scott T.; Gaumer, Jeremy S.; Kramer, Mark A.; et al. (March 2002). "Nine formulations of quantum mechanics". *American Journal of Physics* **70** (3): 288–297. Bibcode:2002AmJPh..70..288S. doi:10.1119/1.1445404.

4.10   External links

- Interpretations of Quantum Mechanics at the *Internet Encyclopedia of Philosophy*.

- Stanford Encyclopedia of Philosophy:

  - "Bohmian mechanics" by Sheldon Goldstein.
  - "Collapse Theories." by Giancarlo Ghirardi.
  - "Copenhagen Interpretation of Quantum Mechanics" by Jan Faye.
  - "Everett's Relative State Formulation of Quantum Mechanics" by Jeffrey Barrett.
  - "Many-Worlds Interpretation of Quantum Mechanics" by Lev Vaidman.
  - "Modal Interpretation of Quantum Mechanics" by Michael Dickson and Dennis Dieks.
  - "Quantum Entanglement and Information" by Jeffrey Bub.
  - "Quantum mechanics" by Jenann Ismael.
  - "Relational Quantum Mechanics" by Federico Laudisa and Carlo Rovelli.
  - "The Role of Decoherence in Quantum Mechanics" by Guido Bacciagaluppi.

- Willem M. de Muynck, Broad overview of the realist vs. empiricist interpretations, against oversimplified view of the measurement process.

- Schreiber, Z., "The Nine Lives of Schrodinger's Cat." Overview of competing interpretations.

- Interpretations of quantum mechanics on arxiv.org.

- The many worlds of quantum mechanics.

- Erich Joos' Decoherence Website.

- Quantum Mechanics for Philosophers. Argues for the superiority of the Bohm interpretation.

- Hidden Variables in Quantum Theory: The Hidden Cultural Variables of their Rejection.

- Numerous Many Worlds-related Topics and Articles.

- Relational Approach to Quantum Physics.

- Theory of incomplete measurements. Deriving quantum mechanics axioms from properties of acceptable measurements.

- Alfred Neumaier's FAQ.

- Measurement in Quantum Mechanics FAQ.

- "Quantum Mechanics - the dream stuff is made of" (September 2015)

Chapter 5

Complementarity (physics)

In physics, **complementarity** is both a theoretical and an experimental result [1][2][3] of quantum mechanics, also referred as **principle of complementarity**, closely associated with the Copenhagen interpretation. It holds that objects have complementary properties which cannot be measured accurately at the same time. The more accurately one property is measured, the less accurately the complementary property is measured, according to the Heisenberg uncertainty principle. Further, a full description of a particular type of phenomenon can only be achieved through measurements made in each of the various possible bases — which are thus complementary. The complementarity principle was formulated by Niels Bohr, a leading founder of quantum mechanics.[4]

Examples of complementary properties:

- Position and momentum

- Energy and duration

- Spin on different axis

- Wave and particle

- Value of a field and its change (at a certain position)

- Entanglement and coherence [5]

Bohr's principle has only recently been formalized in *universal complementarity relations*, such as those due to Ozawa[6] and Hall.[7][8][9]

5.1 Concept

Bohr summarized the principle as follows:

> *...however far the [quantum physical] phenomena transcend the scope of classical physical explanation, the account of all evidence must be expressed in classical terms.* The argument is simply that by the word "experiment" we refer to a situation where we can tell others what we have done and what we have learned and that, therefore, the account of the experimental arrangements and of the results of the observations must be expressed in unambiguous language with suitable application of the terminology of classical physics.

> This crucial point...implies the *impossibility of any sharp separation between the behaviour of atomic objects and the interaction with the measuring instruments which serve to define the conditions under which the phenomena appear....* Consequently, evidence obtained under different experimental conditions cannot be comprehended within a single picture, but must be regarded as *complementary* in the sense that only the totality of the phenomena exhausts the possible information about the objects.[10]

For example, the particle and wave aspects of physical objects are such complementary phenomena. Both concepts are borrowed from classical mechanics, where it is impossible to be a particle and wave at the same time. Therefore it is impossible to measure the *full* properties of the wave and particle at a particular moment.[11] Moreover, Bohr implies that it is not possible to regard objects governed by quantum mechanics as having intrinsic properties independent of determination with a measuring device. The type of measurement determines which property is shown. However the single and double-slit experiment and other experiments show that *some* effects of wave and particle can be measured in one measurement.[12]

5.2   Nature

A profound aspect of complementarity is that it not only applies to measurability or knowability of some property of a physical entity, but more importantly it applies to the limitations of that physical entity's very manifestation of the property in the physical world. All properties of physical entities exist only in pairs, which Bohr described as complementary or conjugate pairs (which are also Fourier transform pairs). Physical reality is determined and defined by manifestations of properties which are limited by trade-offs between these complementary pairs. For example, an electron can manifest a greater and greater accuracy of its position only in even trade for a complementary loss in accuracy of manifesting its momentum. This means that there is a limitation on the precision with which an electron can possess (i.e., manifest) position, since an infinitely precise position would dictate that its manifested momentum would be infinitely imprecise, or undefined (i.e., non-manifest or not possessed), which is not possible. The ultimate limitations in precision of property manifestations are quantified by the Heisenberg uncertainty principle and Planck units. Complementarity and Uncertainty dictate that therefore all properties and actions in the physical world manifest themselves as non-deterministic to some degree.

Physicists F.A.M. Frescura and Basil Hiley have summarized the reasons for the introduction of the principle of complementarity in physics as follows:[13]

> "In the traditional view, it is assumed that there exists a reality in space-time and that this reality is a given thing, all of whose aspects can be viewed or articulated at any given moment. Bohr was the first to point out that quantum mechanics called this traditional outlook into question. To him the 'indivisibility of the quantum of action', which was his way of describing the uncertainty principle, implied that not all aspects of a system can be viewed simultaneously. By using one particular piece of apparatus only certain features could be made manifest at the expense of others, while with a different piece of apparatus another complementary aspect could be made manifest in such a way that the original set became non-manifest, that is, the original attributes were no longer well defined. For Bohr, this was an indication that the principle of complementarity, a principle that he had previously known to appear extensively in other intellectual disciplines but which did not appear in classical physics, should be adopted as a universal principle."

The emergence of complementarity in a system occurs when one considers the circumstances under which one attempts to measure its properties; as Bohr noted, the principle of complementarity "implies the impossibility of any sharp separation between the behaviour of atomic objects and the interaction with the measuring instruments that serve to define the conditions under which the phenomena appear."[14] It is important to distinguish, as did Bohr in his original statements, the principle of complementarity from a statement of the uncertainty principle. For a technical discussion of contemporary issues surrounding complementarity in physics see, e.g., Bandyopadhyay (2000),[15] from which parts of this discussion were drawn.

5.3   Additional considerations

In his original lecture on the topic, Bohr pointed out that just as the finitude of the speed of light implies the impossibility of a sharp separation between space and time (relativity), the finitude of the quantum of action implies the impossibility of a sharp separation between the behavior of a system and its interaction with the measuring instruments and leads to the well known difficulties with the concept of 'state' in quantum theory; the notion of complementarity is intended to symbolize this new situation in epistemology created by quantum theory. Some people consider it a philosophical adjunct to quantum mechanics, while others consider it to be a discovery that is as important as the formal aspects of quantum theory. Examples of the latter include Leon Rosenfeld, who claimed that "[C]omplementarity is not a philosophical superstructure invented by Bohr to be placed as a decoration on top of the quantal formalism, it is the bedrock of the quantal description.",[16] and John Wheeler, who opined that "Bohr's principle of complementarity is the most revolutionary scientific concept of this century and the heart of his fifty-year search for the full significance of the quantum idea."[17]

5.4 Experiments

The quintessential example of wave–particle complementarity in the laboratory is the double slit. The crux of the complementary behavior is the question: "What information exists – embedded in the constituents of the universe – that can reveal the history of the signal particles as they pass through the double slit?" If information exists (even if it is not measured by a conscious observer) that reveals "which slit" each particle traversed, then each particle will exhibit no wave interference with the other slit. This is the particle-like behavior. But if *no information* exists about which slit – so that no conscious observer, no matter how well equipped, will ever be able to determine which slit each particle traverses – then the signal particles will interfere with themselves as if they traveled through both slits at the same time, as a wave. This is the wave-like behavior. These behaviors are complementary, according to the Englert–Greenberger duality relation, because when one behavior is observed the other is absent. Both behaviors *can* be observed at the same time, but each only as lesser manifestations of their full behavior (as determined by the duality relation). This superposition of complementary behaviors exists whenever there is partial "which slit" information. While there is some contention to the duality relation, and thus complementarity itself, the contrary position is not accepted by mainstream physics.[18]:35–40

Various neutron interferometry experiments demonstrate the subtlety of the notions of duality and complementarity. By passing through the interferometer, the neutron appears to act as a wave. Yet upon passage, the neutron is subject to gravitation. As the neutron interferometer is rotated through Earth's gravitational field a phase change between the two arms of the interferometer can be observed, accompanied by a change in the constructive and destructive interference of the neutron waves on exit from the interferometer. Some interpretations claim that understanding the interference effect requires one to concede that a single neutron takes both paths through the interferometer at the same time; a single neutron would "be in two places at once", as it were. Since the two paths through a neutron interferometer can be as far as 5 cm to 15 cm apart, the effect is hardly microscopic. This is similar to traditional double-slit and mirror interferometer experiments where the slits (or mirrors) can be arbitrarily far apart. So, in interference and diffraction experiments, neutrons behave the same way as photons (or electrons) of corresponding wavelength.[19][20]:211–213

5.5 History

Niels Bohr apparently conceived of the principle of complementarity during a skiing vacation in Norway in February and March 1927, during which he received a letter from Werner Heisenberg regarding the latter's newly discovered (and not yet published) uncertainty principle. Upon returning from his vacation, by which time Heisenberg had already submitted his paper on the uncertainty principle for publication, he convinced Heisenberg that the uncertainty principle was a manifestation of the deeper concept of complementarity.[11] Heisenberg duly appended a note to this effect to his paper on the uncertainty principle, before its publication, stating:

> Bohr has brought to my attention [that] the uncertainty in our observation does not arise exclusively from the occurrence of discontinuities, but is tied directly to the demand that we ascribe equal validity to the quite different experiments which show up in the [particulate] theory on one hand, and in the wave theory on the other hand.

Bohr publicly introduced the principle of complementarity in a lecture he delivered on 16 September 1927 at the International Physics Congress held in Como, Italy, attended by most of the leading physicists of the era, with the notable exceptions of Einstein, Schrödinger, and Dirac. However, these three were in attendance one month later when Bohr again presented the principle at the Fifth Solvay Congress in Brussels, Belgium. The lecture was published in the proceedings of both of these conferences, and was republished the following year in *Naturwissenschaften* (in German) and in *Nature* (in English).[21]

An article written by Bohr in 1949 titled "Discussions with Einstein on Epistemological Problems in Atomic Physics"[10] is considered by many to be a definitive description of the notion of complementarity.[22]

5.6 See also

- Afshar experiment

- Bohr–Einstein debates

- Copenhagen interpretation

- Englert–Greenberger duality relation

- Ehrenfest's theorem

- Interpretation of quantum mechanics

- Quantum entanglement

- Quantum indeterminacy

- Transactional interpretation

- Wheeler–Feynman absorber theory

5.7 References

[1] Hall, George M. (1997). *The Ingenious Mind of Nature: Deciphering the Patterns of Man, Society, and the*. Springer. p. 409. ISBN 978-0-306-45571-1.

[2] Whitaker, Andrew (2006). *Einstein, Bohr and the Quantum Dilemma: From Quantum Theory to Quantum Dillema*. Cambridge. p. 414. ISBN 9780521671026.

[3] Selleri, Franco (2012). *Wave-Particle Duality*. Springer. p. 55. ISBN 978-1461364689.

[4] Walker, Evan Harris (2000). *The Physics of Consciousness*. Cambridge, Massachusetts: Perseus. p. 271. ISBN 0-7382-0436-6. ...the founders of quantum mechanics -- Heisenberg, Schrodinger and Bohr...

[5] Cramer, J. G. et al. "An Inquiry into the Possibility of Nonlocal Quantum Communication". Foundation of Phys. Submitted. Revised Feb., 2015

[6] Ozawa, M. "Universally valid reformulation of the Heisenberg uncertainty principle on noise and disturbance in measurement". Phys. Rev. A 67, 042105 (2003).

[7] Hall, M. J. W. "Prior information: How to circumvent the standard jointmeasurement uncertainty relation". Phys. Rev. A 69, 052113 (2004).

[8] Erhart, J. et al. "Experimental demonstration of a universally valid error-disturbance uncertainty relation in spin measurements". Nature Phys. 8, 185–189 (2012).

[9] Shadbolt,P et al. "Testing foundations of quantum mechanics with photons". Nat.Phys. v10. DOI:10.1038/NPHYS2931

[10] Niels Bohr (1949). "Discussions with Einstein on Epistemological Problems in Atomic Physics". In P. Schilpp. *Albert Einstein: Philosopher-Scientist*. Open Court.

[11] Jim Baggott (2011). *The Quantum Story: A History in*. Oxford University Press. p. 97.

[12] Boscá Díaz-Pintado, María C. (29–31 March 2007). "Updating the wave-particle duality". *15th UK and European Meeting on the Foundations of Physics*. Leeds, UK. Retrieved 2008-06-21.

[13] F. A. M. Frescura, B. J. Hiley: *Algebras, quantum theory and pre-space*, published in Revista Brasileira de Fisica, Volume Especial, Julho 1984, Os 70 anos de Mario Schonberg, pp. 49–86, p. 2

[14] Jørgen Kalckar, Niels Bohr, Léon Rosenfeld, Erik Rüdinger, Finn Aaserud (1996). *Foundations of Quantum Physics II (1933-1958)*. Elsevier. p. 210. ISBN 978-0-444-89892-0. Retrieved 2011-10-24.

[15] Bandyopadhyay, Supriyo (2000). "Welcher Weg Experiments and the Orthodox Bohr's Complementarity Principle". *Physics Letters A* 276 (5–6): 233–239. arXiv:quant-ph/0003073. Bibcode:2000PhLA..276..233B. doi:10.1016/S0375-9601(00)00670-8.

[16] Niels Bohr; fwd. Léon Rosenfeld; ed. Kalckar; et al. (1996). "Complementarity: Bedrock of the Quantal Description". *Foundations of Quantum Physics II (1933–1958)*. Niels Bohr Collected Works 7. Elsevier. pp. 284–285. ISBN 978-0-444-89892-0.

[17] John Wheeler, *Physics Today*, January 1963, p. 30.

[18] Haroche, Serge; Raimond, Jean-Michel (2006). *Exploring the Quantum: Atoms, Cavities, and Photons* (1st ed.). Oxford University Press. ISBN 978-0198509141.

[19] Colella, R.; Overhauser, A. W.; Werner, S. A. (1975). "Observation of gravitationally induced quantum interference" (PDF). *Phys. Rev. Lett.* 34 (23): 1472–1474. Bibcode:1975PhRvL..34.1472C. doi:10.1103/physrevlett.34.1472.

[20] Helmut Rauch; Samuel A. Werner (2000). *Neutron Interferometry: Lessons in Experimental Quantum Mechanics*. Oxford University Press. ISBN 978-0-19-850027-8.

[21] Bohr N (1928). "The Quantum Postulate and the Recent Development of Atomic Theory". *Nature* 121: 580–590. Bibcode doi:10.1038/121580a0. Available in the collection of Bohr's early writings, *Atomic Theory and the Description of Nature* (1934).

[22] Saunders S (2005). "Complementarity and Scientific Rationality". *Foundations of Physics* 35 (3): 417–447. arXiv:quant-ph/0412195. Bibcode:2005FoPh...35..417S. doi:10.1007/s10701-004-1982-x.

5.8 Further reading

- Berthold-Georg Englert, Marlan O. Scully & Herbert Walther, *Quantum Optical Tests of Complementarity*, Nature, Vol 351, pp 111–116 (9 May 1991) and (same authors) *The Duality in Matter and Light* Scientific American, pg 56–61, (December 1994). Demonstrates that complementarity is enforced, and quantum interference effects destroyed, by decoherence (irreversible object-apparatus correlations), and not, as was previously popularly believed, by Heisenberg's uncertainty principle itself.

- Niels Bohr, *Causality and Complementarity: Supplementary papers edited by Jan Faye and Henry J. Folse. The Philosophical Writings of Niels Bohr, Volume IV*. Ox Bow Press. 1998.

5.9 External links

- Discussions with Einstein on Epistemological Problems in Atomic Physics

- Einstein's Reply to Criticisms

Chapter 6

Second quantization

Second quantization is a formalism used to describe and analyze quantum many-body systems. It is also known as canonical quantization in quantum field theory, in which the fields (typically as the wave functions of matters) are thought of as field operators, in a similar manner to how the physical quantities (position, momentum etc.) are thought of as operators in first quantization. The key ideas of this method were introduced in 1927 by Dirac,[1] and were developed, most notably, by Fock and Jordan later.[2][3]

In this approach, the quantum many-body states are represented in the Fock state basis, which are constructed by filling up each single-particle state with a certain number of identical particles. The second quantization formalism introduces the creation and annihilation operators to construct and handle the Fock states, providing useful tools to the study of the quantum many-body theory.

6.1 Quantum many-body states

The starting point of the second quantization formalism is the notion of indistinguishability of particles in quantum mechanics. Unlike in classical mechanics, where each particle is labeled by a distinct position vector $\mathbf{r}_i$ and different configurations of the set of $\mathbf{r}_i$'s correspond to different many-body states, *in quantum mechanics, the particles are identical, such that exchanging two particles, i.e. $\mathbf{r}_i \leftrightarrow \mathbf{r}_j$, does not lead to a different many-body quantum state*. This implies that the quantum many-body wave function must be invariant (up to a phase factor) under the exchange of two particles. According to the statistics of the particles, the many-body wave function can either be symmetric or antisymmetric under the particle exchange:

$$\Psi_B(\cdots, \mathbf{r}_i, \cdots, \mathbf{r}_j, \cdots) = +\Psi_B(\cdots, \mathbf{r}_j, \cdots, \mathbf{r}_i, \cdots) \text{ if the particles are bosons,}$$

$$\Psi_F(\cdots, \mathbf{r}_i, \cdots, \mathbf{r}_j, \cdots) = -\Psi_F(\cdots, \mathbf{r}_j, \cdots, \mathbf{r}_i, \cdots) \text{ if the particles are fermions.}$$

This exchange symmetry property imposes a constraint on the many-body wave function. Each time a particle is added or removed from the many-body system, the wave function must be properly symmetrized or anti-symmetrized to satisfy the symmetry constraint. In the first quantization formalism, this constraint is guaranteed by representing the wave function as linear combination of permanents (for bosons) or determinants (for fermions) of single-particle states. In the second quantization formalism, the issue of symmetrization is automatically taken care of by the creation and annihilation operators, such that its notation can be much simpler.

6.1.1 First-quantized many-body wave function

Consider a complete set of single-particle wave functions $\psi_\alpha(\mathbf{r})$ labeled by α (which may be a combined index of a bunch of quantum numbers). The following wave function

$$\Psi[\mathbf{r}_i] = \prod_{i=1}^{N} \psi_{\alpha_i}(\mathbf{r}_i) \equiv \psi_{\alpha_1} \otimes \psi_{\alpha_2} \otimes \cdots \otimes \psi_{\alpha_N}$$

represents an N-particle state with the ith particle occupying the single-particle state $|\alpha_i\rangle$. In the shorthanded notation, the position argument of the wave function may be omitted, and it is assumed that the ith single-particle wave function describes the state of the ith particle. The wave function Ψ has not been symmetrized or anti-symmetrized, thus in general not qualified as a many-body wave function for identical particles. However, it can be brought to the symmetrized (anti-symmetrized) form by the symmetrization (anti-symmetrization) operators, denoted $\mathcal{S}$ ($\mathcal{A}$).

For bosons, the many-body wave function must be symmetrized,

$$\Psi_B[\mathbf{r}_i] = \mathcal{N}\mathcal{S}\Psi[\mathbf{r}_i] = \mathcal{N} \sum_{\pi \in S_N} \prod_{i=1}^{N} \psi_{\alpha_{\pi(i)}}(\mathbf{r}_i) = \mathcal{N} \sum_{\pi \in S_N} \psi_{\alpha_{\pi(1)}} \otimes \psi_{\alpha_{\pi(2)}} \otimes \cdots \otimes \psi_{\alpha_{\pi(N)}};$$

while for fermions, the many-body wave function must be anti-symmetrized,

$$\Psi_F[\mathbf{r}_i] = \mathcal{N}\mathcal{A}\Psi[\mathbf{r}_i] = \mathcal{N} \sum_{\pi \in S_N} (-1)^{\pi} \prod_{i=1}^{N} \psi_{\alpha_{\pi(i)}}(\mathbf{r}_i) = \mathcal{N} \sum_{\pi \in S_N} (-1)^{\pi} \psi_{\alpha_{\pi(1)}} \otimes \psi_{\alpha_{\pi(2)}} \otimes \cdots \otimes \psi_{\alpha_{\pi(N)}}.$$

Here π is an element in the N-body permutation group (or symmetric group) S_N, which performs a permutation among the state labels α_i, and $(-1)^{\pi}$ denotes the corresponding permutation sign. $\mathcal{N}$ is the normalization operator that normalizes the wave function. (It is the operator that applies a suitable numerical normalization factor to the symmetrized tensors of degree n; see the next section for its value.)

If one arranges the single-particle wave functions in a matrix U, such that the row-i column-j matrix element is $U_{ij} = \psi_{\alpha_j}(\mathbf{r}_i) \equiv \langle \mathbf{r}_i | \alpha_j \rangle$, then the boson many-body wave function can be simply written as a permanent $\Psi_B = \mathcal{N} \operatorname{perm} U$, and the fermion many-body wave function as a determinant $\Psi_F = \mathcal{N} \det U$ (also known as the Slater determinant).

6.1.2   Second-quantized Fock states

First quantized wave functions involve complicated symmetrization procedures to describe physically realizable many-body states because the language of first quantization is redundant for indistinguishable particles. **In the first quantization language**, the many-body state is described by answering a series of questions like *"which particle is on which state"*. However these are not physical questions, because the particles are identical, and it is impossible to tell which particle is which in the first place. The seemingly different states $\psi_1 \otimes \psi_2$ and $\psi_2 \otimes \psi_1$ are actually redundant names of the same quantum many-body state. So the symmetrization (or anti-symmetrization) must be introduced to eliminate this redundancy in the first quantization description.

In the second quantization language, instead of asking "each particle on which state", one asks *"how many particles are there on each state"*. Because this description does not refer to the labeling of particles, it contains no redundant information, and hence leads to a precise and simpler description of the quantum many-body state. In this approach, the many-body state is represented in the occupation number basis, and the basis state is labeled by the set of occupation numbers, denoted

$$|[n_\alpha]\rangle \equiv |n_1, n_2, \cdots, n_\alpha, \cdots\rangle,$$

meaning that there are n_α particles in the single-particle state $|\alpha\rangle$ (or as ψ_α). The occupation numbers sum up to the total number of particles, i.e. $\sum_\alpha n_\alpha = N$. For fermions, the occupation number n_α can only be 0 or 1, due to the Pauli exclusion principle; while for bosons it can be any non negative integer

$$n_\alpha = \begin{cases} 0, 1 & \text{fermions,} \\ 0, 1, 2, 3, \ldots & \text{bosons.} \end{cases}$$

The occupation number states $|[n_\alpha]\rangle$ are also known as the Fock states. All the Fock states form a complete set of basis of the many-body Hilbert space, or the Fock space. Any generic quantum many-body state can be expressed as a linear combination of Fock states.

Note that besides providing a more efficient language, Fock space allows for a variable number of particles. As a Hilbert space, it is isomorphic to the sum of the n-particle bosonic or fermionic tensor spaces described in the previous section, including a one-dimensional zero-particle space $\mathbb{C}$.

The Fock state with all occupation numbers equal to zero is called the vacuum state, denoted $|0\rangle \equiv |\cdots, 0_\alpha, \cdots\rangle$. The Fock state with only one non-zero occupation number is a single-mode Fock state, denoted $|n_\alpha\rangle \equiv |\cdots, 0, n_\alpha, 0, \cdots\rangle$. In terms of the first quantized wave function, the vacuum state is the unit of tensor product, and can be denoted as $|0\rangle = 1$. The single-particle state is reduced to its wave function $|1_\alpha\rangle = \psi_\alpha$. Other single-mode many-body (boson) state are just the tensor product of the wave function of that mode, such as $|2_\alpha\rangle = \psi_\alpha \otimes \psi_\alpha$ and $|n_\alpha\rangle = \psi_\alpha^{\otimes n}$. For multi-mode Fock states (meaning more than one single-particle state $|\alpha\rangle$ is involved), the corresponding first-quantized wave function will require proper symmetrization according to the particle statistics, e.g. $|1_1, 1_2\rangle = (\psi_1\psi_2 + \psi_2\psi_1)/\sqrt{2}$ for a boson state, and $|1_1, 1_2\rangle = (\psi_1\psi_2 - \psi_2\psi_1)/\sqrt{2}$ for a fermion state (the symbol $\otimes$ between ψ_1 and ψ_2 is omitted for simplicity). In general, the normalization is found to be $\sqrt{\frac{\prod_\alpha n_\alpha!}{N!}}$, where N is the total number of particles. For fermion, this expression reduces to $\frac{1}{\sqrt{N!}}$ as n_α can only be either zero or one. So the first-quantized wave function corresponding to the Fock state reads

$$|[n_\alpha]\rangle_B = \left(\frac{\prod_\alpha n_\alpha!}{N!}\right)^{1/2} \mathcal{S}\bigotimes_\alpha \psi_\alpha^{\otimes n_\alpha}$$

for bosons and

$$|[n_\alpha]\rangle_F = \frac{1}{\sqrt{N!}} \mathcal{A}\bigotimes_\alpha \psi_\alpha^{\otimes n_\alpha}$$

for fermions. Note that for fermions, $n_\alpha = 0, 1$ only, so the tensor product above is effectively just a product over all occupied single-particle states.

6.2 Creation and annihilation operators

The creation and annihilation operators are introduced to add or remove a particle from the many-body system. These operators lie at the core of the second quantization formalism, bridging the gap between the first- and the second-quantized states. Applying the creation (annihilation) operator to a first-quantized many-body wave function will insert (delete) a single-particle state from the wave function in a symmetrized way depending on the particle statistics. On the other hand, all the second-quantized Fock states can be constructed by applying the creation operators to the vacuum state repeatedly.

The creation and annihilation operators (for bosons) are originally constructed in the context of the quantum harmonic oscillator as the raising and lowering operators, which are then generalized to the field operators in the quantum field theory.[4] They are fundamental to the quantum many-body theory, in the sense that every many-body operator (including the Hamiltonian of the many-body system and all the physical observables) can be expressed in terms of them.

6.2.1 Insertion and deletion operation

The creation and annihilation of a particle is implemented by the insertion and deletion of the single-particle state from the first quantized wave function in an either symmetric or anti-symmetric manner. Let ψ_α be a single-particle state, let

1 be the tensor identity (it is the generator of the zero-particle space $\mathbb{C}$ and satisfies $\psi_\alpha \equiv 1 \otimes \psi_\alpha \equiv \psi_\alpha \otimes 1$ in the tensor algebra over the fundamental Hilbert space), and let $\Psi = \psi_{\alpha_1} \otimes \psi_{\alpha_2} \otimes \cdots$ be a generic tensor product state. The insertion $\otimes_\pm$ and the deletion $\oslash_\pm$ operators are linear operators defined by the following recursive equations

$$\psi_\alpha \otimes_\pm 1 = \psi_\alpha, \quad \psi_\alpha \otimes_\pm (\psi_\beta \otimes \Psi) = \psi_\alpha \otimes \psi_\beta \otimes \Psi \pm \psi_\beta \otimes (\psi_\alpha \otimes_\pm \Psi);$$

$$\psi_\alpha \oslash_\pm 1 = 0, \quad \psi_\alpha \oslash_\pm (\psi_\beta \otimes \Psi) = \delta_{\alpha\beta}\Psi \pm \psi_\beta \otimes (\psi_\alpha \oslash_\pm \Psi).$$

Here $\delta_{\alpha\beta}$ is the Kronecker delta symbol, which gives 1 if $\alpha = \beta$, and 0 otherwise. The subscript $\pm$ of the insertion or deletion operators indicates whether symmetrization (for bosons) or anti-symmetrization (for fermions) is implemented.

6.2.2   Boson creation and annihilation operators

The boson creation (annihilation) operator is usually denoted as $b_\alpha^\dagger$ (b_α). The creation operator $b_\alpha^\dagger$ adds a boson to the single-particle state $|\alpha\rangle$, and the annihilation operator b_α removes a boson from the single-particle state $|\alpha\rangle$. The creation and annihilation operators are Hermitian conjugate to each other, but neither of them are Hermitian operators ($b_\alpha \neq b_\alpha^\dagger$).

Definition

The boson creation (annihilation) operator is a linear operator, whose action on a N-particle first-quantized wave function Ψ is defined as

$$b_\alpha^\dagger \Psi = \frac{1}{\sqrt{N+1}} \psi_\alpha \otimes_+ \Psi,$$

$$b_\alpha \Psi = \frac{1}{\sqrt{N}} \psi_\alpha \oslash_+ \Psi,$$

where $\psi_\alpha \otimes_+$ inserts the single-particle state ψ_α in $N+1$ possible insertion positions symmetrically, and $\psi_\alpha \oslash_+$ deletes the single-particle state ψ_α from N possible deletion positions symmetrically.

Examples (click *show* to view)

Hereinafter the tensor symbol $\otimes$ between single-particle states is omitted for simplicity. Take the state $|1_1, 1_2\rangle = (\psi_1\psi_2 + \psi_2\psi_1)/\sqrt{2}$, create one more boson on the state ψ_1,

$$
\begin{aligned}
b_1^\dagger |1_1, 1_2\rangle &= \tfrac{1}{\sqrt{2}}(b_1^\dagger \psi_1\psi_2 + b_1^\dagger \psi_2\psi_1) \\
&= \tfrac{1}{\sqrt{2}}\left(\tfrac{1}{\sqrt{3}}\psi_1 \otimes_+ \psi_1\psi_2 + \tfrac{1}{\sqrt{3}}\psi_1 \otimes_+ \psi_2\psi_1\right) \\
&= \tfrac{1}{\sqrt{2}}\left(\tfrac{1}{\sqrt{3}}(\psi_1\psi_1\psi_2 + \psi_1\psi_1\psi_2 + \psi_1\psi_2\psi_1) + \tfrac{1}{\sqrt{3}}(\psi_1\psi_2\psi_1 + \psi_2\psi_1\psi_1 + \psi_2\psi_1\psi_1)\right) \\
&= \tfrac{\sqrt{2}}{\sqrt{3}}(\psi_1\psi_1\psi_2 + \psi_1\psi_2\psi_1 + \psi_2\psi_1\psi_1) \\
&= \sqrt{2}|2_1, 1_2\rangle.
\end{aligned}
$$

Then annihilate one boson from the state ψ_1,

$$
\begin{aligned}
b_1 |2_1, 1_2\rangle &= \tfrac{1}{\sqrt{3}}(b_1 \psi_1\psi_1\psi_2 + b_1 \psi_1\psi_2\psi_1 + b_1 \psi_2\psi_1\psi_1) \\
&= \tfrac{1}{\sqrt{3}}\left(\tfrac{1}{\sqrt{3}}\psi_1 \oslash_+ \psi_1\psi_1\psi_2 + \tfrac{1}{\sqrt{3}}\psi_1 \oslash_+ \psi_1\psi_2\psi_1 + \tfrac{1}{\sqrt{3}}\psi_1 \oslash_+ \psi_2\psi_1\psi_1\right) \\
&= \tfrac{1}{\sqrt{3}}\left(\tfrac{1}{\sqrt{3}}(\psi_1\psi_2 + \psi_1\psi_2 + 0) + \tfrac{1}{\sqrt{3}}(\psi_2\psi_1 + 0 + \psi_1\psi_2) + \tfrac{1}{\sqrt{3}}(0 + \psi_2\psi_1 + \psi_2\psi_1)\right) \\
&= \psi_1\psi_2 + \psi_2\psi_1 \\
&= \sqrt{2}|1_1, 1_2\rangle.
\end{aligned}
$$

Action on Fock states

Starting from the single-mode vacuum state $|0_\alpha\rangle = 1$, applying the creation operator $b_\alpha^\dagger$ repeatedly, one finds

$$b_\alpha^\dagger|0_\alpha\rangle = \psi_\alpha \otimes_+ 1 = \psi_\alpha = |1_\alpha\rangle,$$

$$b_\alpha^\dagger|n_\alpha\rangle = \frac{1}{\sqrt{n_\alpha + 1}}\psi_\alpha \otimes_+ \psi_\alpha^{\otimes n_\alpha} = \sqrt{n_\alpha + 1}\psi_\alpha^{\otimes(n_\alpha+1)} = \sqrt{n_\alpha + 1}|n_\alpha + 1\rangle.$$

The creation operator raises the boson occupation number by 1. Therefore all the occupation number states can be constructed by the boson creation operator from the vacuum state

$$|n_\alpha\rangle = \frac{1}{\sqrt{n_\alpha!}}(b_\alpha^\dagger)^{n_\alpha}|0_\alpha\rangle.$$

On the other hand, the annihilation operator b_α lowers the boson occupation number by 1

$$b_\alpha|n_\alpha\rangle = \frac{1}{\sqrt{n_\alpha}}\psi_\alpha \oslash_+ \psi_\alpha^{\otimes n_\alpha} = \sqrt{n_\alpha}\psi_\alpha^{\otimes(n_\alpha-1)} = \sqrt{n_\alpha}|n_\alpha - 1\rangle.$$

It will also quench the vacuum state $b_\alpha|0_\alpha\rangle = 0$ as there has been no boson left in the vacuum state to be annihilated. Using the above formulae, it can be shown that

$$b_\alpha^\dagger b_\alpha|n_\alpha\rangle = n_\alpha|n_\alpha\rangle,$$

meaning that $\hat{n}_\alpha = b_\alpha^\dagger b_\alpha$ defines the boson number operator.

The above result can be generalized to any Fock state of bosons.

$$b_\alpha^\dagger|\cdots, n_\beta, n_\alpha, n_\gamma, \cdots\rangle = \sqrt{n_\alpha + 1}|\cdots, n_\beta, n_\alpha + 1, n_\gamma, \cdots\rangle.$$

$$b_\alpha|\cdots, n_\beta, n_\alpha, n_\gamma, \cdots\rangle = \sqrt{n_\alpha}|\cdots, n_\beta, n_\alpha - 1, n_\gamma, \cdots\rangle.$$

These two equations can be considered as the defining properties of boson creation and annihilation operators in the second-quantization formalism. The complicated symmetrization of the underlying first-quantized wave function is automatically taken care of by the creation and annihilation operators (when acting on the first-quantized wave function), so that the complexity is not revealed on the second-quantized level, and the second-quantization formulae are simple and neat.

Operator identities

The following operator identities follows from the action of the boson creation and annihilation operators on the Fock state,

$$[b_\alpha^\dagger, b_\beta^\dagger] = [b_\alpha, b_\beta] = 0, \quad [b_\alpha, b_\beta^\dagger] = \delta_{\alpha\beta}.$$

These commutation relations can be considered as the algebraic definition of the boson creation and annihilation operators. The fact that the boson many-body wave function is symmetric under particle exchange is also manifested by the commutation of the boson operators.

The raising and lowering operators of the quantum harmonic oscillator also satisfies the same set of commutation relations, implying that the bosons can be interpreted as the energy quanta (phonons) of an oscillator. This is indeed the idea of quantum field theory, which considers each mode of the matter field as an oscillator subject to quantum fluctuations, and the bosons are treated as the excitations (or energy quanta) of the field.

6.2.3   Fermion creation and annihilation operators

The fermion creation (annihilation) operator is usually denoted as $c_\alpha^\dagger$ (c_α). The creation operator $c_\alpha^\dagger$ adds a fermion to the single-particle state $|\alpha\rangle$, and the annihilation operator c_α removes a fermion from the single-particle state $|\alpha\rangle$. The creation and annihilation operators are Hermitian conjugate to each other, but neither of them are Hermitian operators ($c_\alpha \neq c_\alpha^\dagger$). The Hermitian combination of the fermion creation and annihilation operators

$$\chi_{\alpha,\mathrm{Re}} = (c_\alpha + c_\alpha^\dagger)/2, \quad \chi_{\alpha,\mathrm{Im}} = (c_\alpha - c_\alpha^\dagger)/(2i),$$

are called Majorana fermion operators.

Definition

The fermion creation (annihilation) operator is a linear operator, whose action on a N-particle first-quantized wave function Ψ is defined as

$$c_\alpha^\dagger \Psi = \frac{1}{\sqrt{N+1}} \psi_\alpha \otimes_- \Psi,$$

$$c_\alpha \Psi = \frac{1}{\sqrt{N}} \psi_\alpha \oslash_- \Psi,$$

where $\psi_\alpha \otimes_-$ inserts the single-particle state ψ_α in $N + 1$ possible insertion positions anti-symmetrically, and $\psi_\alpha \oslash_-$ deletes the single-particle state ψ_α from N possible deletion positions anti-symmetrically.

Examples (click *show* to view)

Hereinafter the tensor symbol $\otimes$ between single-particle states is omitted for simplicity. Take the state $|1_1, 1_2\rangle = (\psi_1\psi_2 - \psi_2\psi_1)/\sqrt{2}$, attempt to create one more fermion on the occupied ψ_1 state will quench the whole many-body wave function,

$$
\begin{aligned}
c_1^\dagger |1_1, 1_2\rangle &= \tfrac{1}{\sqrt{2}}(c_1^\dagger \psi_1 \psi_2 - c_1^\dagger \psi_2 \psi_1)\\
&= \tfrac{1}{\sqrt{2}}\left(\tfrac{1}{\sqrt{3}}\psi_1 \otimes_- \psi_1\psi_2 - \tfrac{1}{\sqrt{3}}\psi_1 \otimes_- \psi_2\psi_1\right)\\
&= \tfrac{1}{\sqrt{2}}\left(\tfrac{1}{\sqrt{3}}(\psi_1\psi_1\psi_2 - \psi_1\psi_1\psi_2 + \psi_1\psi_2\psi_1) - \tfrac{1}{\sqrt{3}}(\psi_1\psi_2\psi_1 - \psi_2\psi_1\psi_1 + \psi_2\psi_1\psi_1)\right)\\
&= 0.
\end{aligned}
$$

Annihilate a fermion on the ψ_2 state, take the state $|1_1, 1_2\rangle = (\psi_1\psi_2 - \psi_2\psi_1)/\sqrt{2}$,

$$
\begin{aligned}
c_2 |1_1, 1_2\rangle &= \tfrac{1}{\sqrt{2}}(c_2\psi_1\psi_2 - c_2\psi_2\psi_1)\\
&= \tfrac{1}{\sqrt{2}}\left(\tfrac{1}{\sqrt{2}}\psi_2 \oslash_- \psi_1\psi_2 - \tfrac{1}{\sqrt{2}}\psi_2 \oslash_- \psi_2\psi_1\right)\\
&= \tfrac{1}{\sqrt{2}}\left(\tfrac{1}{\sqrt{2}}(0 - \psi_1) - \tfrac{1}{\sqrt{2}}(\psi_1 - 0)\right)\\
&= -\psi_1\\
&= -|1_1, 0_2\rangle.
\end{aligned}
$$

The minus sign (known as the fermion sign) appears due to the anti-symmetric property of the fermion wave function.

Action on Fock states

Starting from the single-mode vacuum state $|0_\alpha\rangle = 1$, applying the fermion creation operator $c_\alpha^\dagger$,

$$c_\alpha^\dagger |0_\alpha\rangle = \psi_\alpha \otimes_- 1 = \psi_\alpha = |1_\alpha\rangle,$$

$$c_\alpha^\dagger |1_\alpha\rangle = \frac{1}{\sqrt{2}}\psi_\alpha \otimes_- \psi_\alpha = 0.$$

If the single-particle state $|\alpha\rangle$ is empty, the creation operator will fill the state with a fermion. However if the state is already occupied by a fermion, further application of the creation operator will quench the state, demonstrating the Pauli exclusion principle that two identical fermions can not occupy the same state simultaneously. Nevertheless the fermion can be removed from the occupied state by the fermion annihilation operator c_α ,

$$c_\alpha |1_\alpha\rangle = \psi_\alpha \oslash_- \psi_\alpha = 1 = |0_\alpha\rangle,$$

$$c_\alpha |0_\alpha\rangle = 0.$$

The vacuum state is quenched by the action of the annihilation operator.

Similar to the boson case, the fermion Fock state can be constructed from the vacuum state using the fermion creation operator

$$|n_\alpha\rangle = (c_\alpha^\dagger)^{n_\alpha} |0_\alpha\rangle.$$

It is easy to check (by enumeration) that

$$c_\alpha^\dagger c_\alpha |n_\alpha\rangle = n_\alpha |n_\alpha\rangle,$$

meaning that $\hat{n}_\alpha = c_\alpha^\dagger c_\alpha$ defines the fermion number operator.

The above result can be generalized to any Fock state of fermions.

$$c_\alpha^\dagger |\cdots, n_\beta, n_\alpha, n_\gamma, \cdots\rangle = (-1)^{\sum_{\beta<\alpha} n_\beta}(1 - n_\alpha)|\cdots, n_\beta, 1 - n_\alpha, n_\gamma, \cdots\rangle.$$

$$c_\alpha |\cdots, n_\beta, n_\alpha, n_\gamma, \cdots\rangle = (-1)^{\sum_{\beta<\alpha} n_\beta} n_\alpha |\cdots, n_\beta, 1 - n_\alpha, n_\gamma, \cdots\rangle.$$

Recall that the occupation number n_α can only take 0 or 1 for fermions. These two equations can be considered as the defining properties of fermion creation and annihilation operators in the second quantization formalism. Note that the fermion sign structure $(-1)^{\sum_{\beta<\alpha} n_\beta}$, also known as the Jordan-Wigner string, requires there to exist a predefined ordering of the single-particle states (the spin structure) and involves a counting of the fermion occupation numbers of all the preceding states; therefore the fermion creation and annihilation operators are considered non-local in some sense. This observation leads to the idea that fermions are emergent particles in the long-range entangled local qubit system.[5]

Operator identities

The following operator identities follow from the action of the fermion creation and annihilation operators on the Fock state,

$$\{c_\alpha^\dagger, c_\beta^\dagger\} = \{c_\alpha, c_\beta\} = 0, \quad \{c_\alpha, c_\beta^\dagger\} = \delta_{\alpha\beta}.$$

These anti-commutation relations can be considered as the algebraic definition of the fermion creation and annihilation operators. The fact that the fermion many-body wave function is anti-symmetric under particle exchange is also manifested by the anti-commutation of the fermion operators.

6.3   Quantum field operators

Defining $a_\nu^\dagger$ as a general annihilation(creation) operator for a single-particle state ν that could be either fermionic $(c_\nu^\dagger)$ or bosonic $(b_\nu^\dagger)$, the real space representation of the operators defines the quantum field operators $\Psi(\mathbf{r})$ and $\Psi^\dagger(\mathbf{r})$ by

$$\Psi(\mathbf{r}) = \sum_\nu \psi_\nu(\mathbf{r}) a_\nu$$

$$\Psi^\dagger(\mathbf{r}) = \sum_\nu \psi_\nu^*(\mathbf{r}) a_\nu^\dagger$$

These are second quantization operators, with coefficients $\psi_\nu(\mathbf{r})$ and $\psi_\nu^*(\mathbf{r})$ that are ordinary first-quantization wavefunctions. Thus, for example, any expectation values will be ordinary first-quantization wavefunctions. Loosely speaking, $\Psi^\dagger(\mathbf{r})$ is the sum of all possible ways to add a particle to the system at position $\mathbf{r}$ through any of the basis states $\psi_\nu(\mathbf{r})$.

Since $\Psi(\mathbf{r})$ and $\Psi^\dagger(\mathbf{r})$ are second quantization operators defined in every point in space they are called quantum field operators. They obey the following fundamental commutator and anti-commutator relations,

$$\left[\Psi(\mathbf{r}_1), \Psi^\dagger(\mathbf{r}_2)\right] = \delta(\mathbf{r}_1 - \mathbf{r}_2)$$

$$\{\Psi(\mathbf{r}_1), \Psi^\dagger(\mathbf{r}_2)\} = \delta(\mathbf{r}_1 - \mathbf{r}_2)$$

In homogeneous systems it is often desirable to transform between real space and the momentum representations, hence, the quantum fields operators in Fourier basis yields:

$$\Psi(\mathbf{r}) = \frac{1}{\sqrt{V}} \sum_{\mathbf{k}} e^{i\mathbf{k}\cdot\mathbf{r}} a_{\mathbf{k}}$$

$$\Psi^\dagger(\mathbf{r}) = \frac{1}{\sqrt{V}} \sum_{\mathbf{k}} e^{-i\mathbf{k}\cdot\mathbf{r}} a^\dagger_{\mathbf{k}}$$

6.4   Comment on nomenclature

The term "second quantization" is a misnomer that has persisted for historical reasons. One is not quantizing "again", as the term "second" might suggest; the field that is being quantized is not a Schrödinger wave function that was produced as the result of quantizing a particle, but is a classical field (such as the electromagnetic field or Dirac spinor field) that was not previously quantized. One is merely shifting from a semiclassical treatment of the system to a fully quantum-mechanical one.

6.5   See also

- Fock state

- Fock space

- Canonical quantization

6.6 References

[1] Dirac, P. A. M. (1927). "The Quantum Theory of the Emission and Absorption of Radiation". *Proceedings of the Royal Society A: Mathematical, Physical and Engineering Sciences* 114 (767): 243. Bibcode:1927RSPSA.114..243D. doi:10.1098/rspa.1927.0039.

[2] V. Fock, *Z. Phys.* 75 (1932), 622-647

[3] M.C. Reed, B. Simon, "Methods of Modern Mathematical Physics, Volume II", Academic Press 1975. Page 328.

[4] Mahan, GD (1981). *Many Particle Physics.* New York: Springer. ISBN 0306463385.

[5] Levin, M.; Wen, X. G. (2003). "Fermions, strings, and gauge fields in lattice spin models". *Physical Review B* 67 (24). doi:10.1103/PhysRevB.67.245316.

6.7 Further reading

- Second quantization Carlo Maria Becchi, Scholarpedia, 5(6):7902. doi:10.4249/scholarpedia.7902

6.8 External links

- Many-Electron States in E. Pavarini, E. Koch, and U. Schollwöck: Emergent Phenomena in Correlated Matter, Jülich 2013, ISBN 978-3-89336-884-6

Chapter 7

Planck's law

Not to be confused with Planck relation.

See also: Black body radiation and Thermal radiation

Planck's law describes the electromagnetic radiation emitted by a black body in thermal equilibrium at a definite

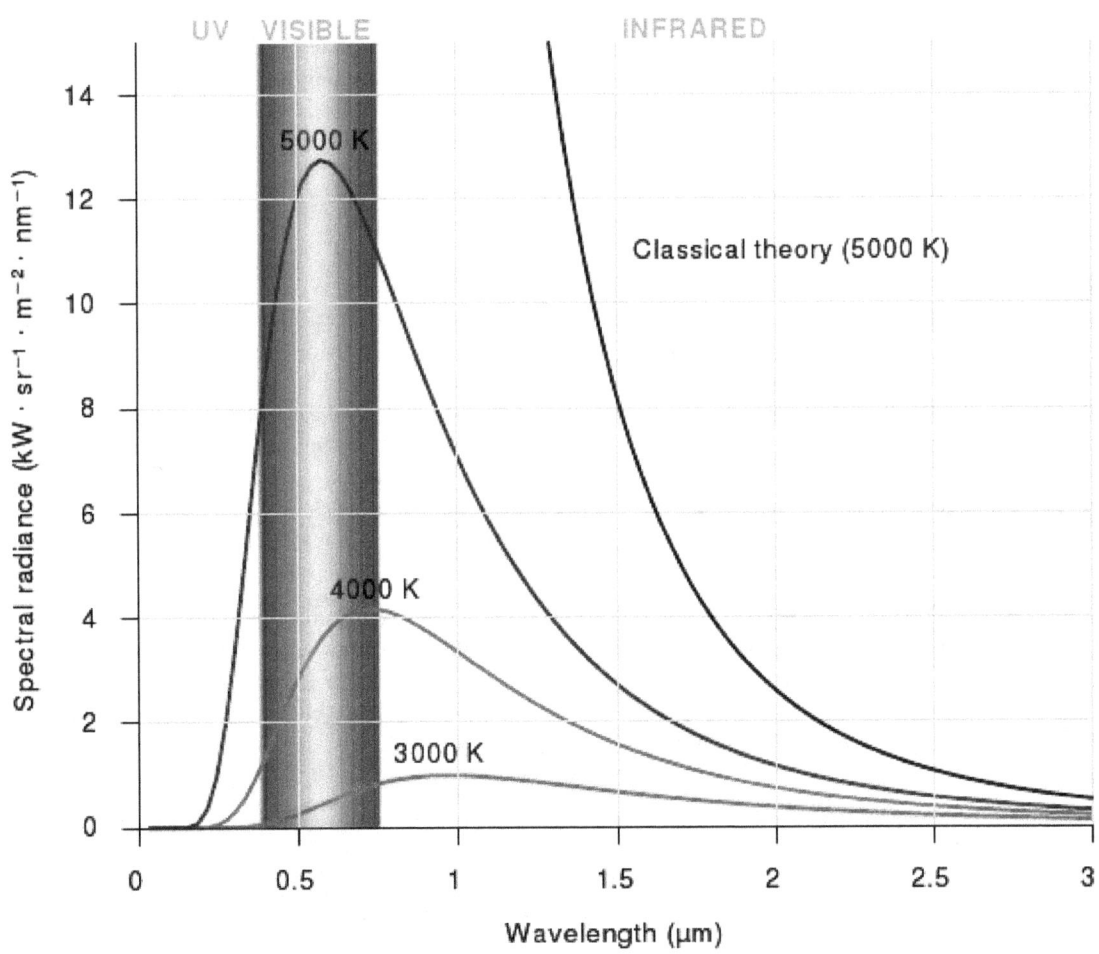

Planck's law (colored curves) accurately described black body radiation and resolved the ultraviolet catastrophe (black curve).

temperature. The law is named after Max Planck, who originally proposed it in 1900. It is a pioneering result of modern physics and quantum theory.

The spectral radiance of a body, $B\nu$, describes the amount of energy it gives off as radiation of different frequencies. It is measured in terms of the power emitted per unit area of the body, per unit solid angle that the radiation is measured over, per unit frequency. Planck showed that the spectral radiance of a body at absolute temperature T is given by

$$B_\nu(\nu, T) = \frac{2h\nu^3}{c^2} \frac{1}{e^{\frac{h\nu}{k_B T}} - 1}$$

where k_B the Boltzmann constant, h the Planck constant, and c the speed of light in the medium, whether material or vacuum.[1][2][3] The spectral radiance can also be measured per unit wavelength instead of per unit frequency. In this case, it is given by

$$B_\lambda(\lambda, T) = \frac{2hc^2}{\lambda^5} \frac{1}{e^{\frac{hc}{\lambda k_B T}} - 1}$$

The law may also be expressed in other terms, such as the number of photons emitted at a certain wavelength, or the energy density in a volume of radiation. The SI units of $B\nu$ are $W \cdot sr^{-1} \cdot m^{-2} \cdot Hz^{-1}$, while those of $B\lambda$ are $W \cdot sr^{-1} \cdot m^{-3}$.

In the limit of low frequencies (i.e. long wavelengths), Planck's law tends to the Rayleigh–Jeans law, while in the limit of high frequencies (i.e. small wavelengths) it tends to the Wien approximation.

Max Planck developed the law in 1900, originally with only empirically determined constants, and later showed that, expressed as an energy distribution, it is the unique stable distribution for radiation in thermodynamic equilibrium.[4] As an energy distribution, it is one of a family of thermal equilibrium distributions which include the Bose–Einstein distribution, the Fermi–Dirac distribution and the Maxwell–Boltzmann distribution.

7.1 Introduction

Every physical body spontaneously and continuously emits electromagnetic radiation. Near thermodynamic equilibrium, the emitted radiation is nearly described by Planck's law. Because of its dependence on temperature, Planck radiation is said to be thermal radiation. The higher the temperature of a body the more radiation it emits at every wavelength. Planck radiation has a maximum intensity at a specific wavelength that depends on the temperature. For example, at room temperature (~300 K), a body emits thermal radiation that is mostly infrared and invisible. At higher temperatures the amount of infrared radiation increases and can be felt as heat, and the body glows visibly red. At even higher temperatures, a body is dazzlingly bright yellow or blue-white and emits significant amounts of short wavelength radiation, including ultraviolet and even x-rays. The surface of the sun (~6000 K) emits large amounts of both infrared and ultraviolet radiation; its emission is peaked in the visible spectrum.

Planck radiation is the greatest amount of radiation that any body at thermal equilibrium can emit from its surface, whatever its chemical composition or surface structure.[5] The passage of radiation across an interface between media can be characterized by the emissivity of the interface (the ratio of the actual radiance to the theoretical Planck radiance), usually denoted by the symbol ε. It is in general dependent on chemical composition and physical structure, on temperature, on the wavelength, on the angle of passage, and on the polarization.[6] The emissivity of a natural interface is always between $\varepsilon = 0$ and 1.

A body that interfaces with another medium which both has $\varepsilon = 1$ and absorbs all the radiation incident upon it, is said to be a black body. The surface of a black body can be modelled by a small hole in the wall of a large enclosure which is maintained at a uniform temperature with opaque walls that, at every wavelength, are not perfectly reflective. At equilibrium, the radiation inside this enclosure follows Planck's law, and so will the radiation coming out of the small hole.

Just as the Maxwell–Boltzmann distribution is the unique maximum entropy energy distribution for a gas of material particles at thermal equilibrium, so is Planck's distribution for a gas of photons.[7][8] By contrast to a material gas where the masses and number of particles play a role, the spectral radiance, pressure and energy density of a photon gas at thermal equilibrium are entirely determined by the temperature.

If the photon gas is not initially Planckian, the second law of thermodynamics guarantees that interactions (between photons and other particles or even, at sufficiently high temperatures, between the photons themselves) will cause the photon energy distribution to change and approach the Planck distribution. In such an approach to thermodynamic equilibrium, photons are created or annihilated in the right numbers and with the right energies to fill the cavity with a Planck distribution until they reach the equilibrium temperature. It is as if the gas is a mixture of sub-gases, one for every band of wavelengths, and each sub-gas eventually attains the common temperature.

The quantity $B\nu(\nu, T)$ is the spectral radiance as a function of temperature and frequency. It has units of $W \cdot m^{-2} \cdot sr^{-1} \cdot Hz^{-1}$ in the SI system. An infinitesimal amount of power $B\nu(\nu, T) \cos \theta \, dA \, d\Omega \, d\nu$ is radiated in the direction described by the angle θ from the surface normal from infinitesimal surface area dA into infinitesimal solid angle $d\Omega$ in an infinitesimal frequency band of width $d\nu$ centered on frequency ν. The total power radiated into any solid angle is the integral of $B\nu(\nu, T)$ over those three quantities, and is given by the Stefan–Boltzmann law. The spectral radiance of Planckian radiation from a black body has the same value for every direction and angle of polarization, and so the black body is said to be a Lambertian radiator.

7.2   Different forms

Planck's law can be encountered in several forms depending on the conventions and preferences of different scientific fields. The various forms of the law for spectral radiance are summarized in the table below. Forms on the left are most often encountered in experimental fields, while those on the right are most often encountered in theoretical fields.

These distributions represent the spectral radiance of blackbodies—the power emitted from the emitting surface, per unit projected area of emitting surface, per unit solid angle, per spectral unit (frequency, wavelength, wavenumber or their angular equivalents). Since the radiance is isotropic (i.e. independent of direction), the power emitted at an angle to the normal is proportional to the projected area, and therefore to the cosine of that angle as per Lambert's cosine law, and is unpolarized.

7.2.1   Correspondence between spectral variable forms

Different spectral variables require different corresponding forms of expression of the law. In general, one may not convert between the various forms of Planck's law simply by substituting one variable for another, because this would not take into account that the different forms have different units. Wavelength and frequency units are reciprocal.

Corresponding forms of expression are related because they express one and the same physical fact: for a particular physical spectral increment, a corresponding particular physical energy increment is radiated.

This is so whether it is expressed in terms of an increment of frequency, $d\nu$, or, correspondingly, of wavelength, $d\lambda$. Introduction of a minus sign can indicate that an increment of frequency corresponds with decrement of wavelength. For the above corresponding forms of expression of the spectral radiance, one may use an obvious expansion of notation, temporarily for the present calculation only. Then, for a particular spectral increment, the particular physical energy increment may be written

$$B_\lambda(\lambda, T) \, d\lambda = -B_\nu(\nu(\lambda), T) \, d\nu \text{ , which leads to } B_\lambda(\lambda, T) = -\frac{d\nu}{d\lambda} B_\nu(\nu(\lambda), T).$$

Also, $\nu(\lambda) = c/\lambda$, so that $d\nu/d\lambda = -c/\lambda^2$. Substitution gives the correspondence between the frequency and wavelength forms, with their different dimensions and units.[11][12] Consequently

$$\frac{B_\lambda(T)}{B_\nu(T)} = \frac{c}{\lambda^2} = \frac{\nu^2}{c}.$$

Evidently, the location of the peak of the spectral distribution for Planck's law depends on the choice of spectral variable. Nevertheless, in a manner of speaking, this formula means that the shape of the spectral distribution is independent of temperature, according to Wien's displacement law, as detailed below in the sub-section **Percentiles** of the section **Properties**.

7.2.2 Spectral energy density form

Planck's law can also be written in terms of the spectral energy density (u) by multiplying B by $4\pi/c$:[13]

$$u_i(T) = \frac{4\pi}{c} B_i(T).$$

These distributions have units of energy per volume per spectral unit.

7.2.3 First and second radiation constants

In the above variants of Planck's law, the *Wavelength* and *Wavenumber* variants use the terms $2hc^2$ and hc/kB which comprise physical constants only. Consequently, these terms can be considered as physical constants themselves,[14] and are therefore referred to as the **first radiation constant** c_1L and the **second radiation constant** c_2 with

$$c1L = 2hc^2$$

and

$$c_2 = hc/kB$$

Using the radiation constants, the *Wavelength* variant of Planck's law can be simplified to

$$L(\lambda, T) = \frac{c_{1L}}{\lambda^5} \frac{1}{\exp\left(\frac{c_2}{\lambda T}\right) - 1}$$

and the *Wavenumber* variant can be simplified correspondingly.

L is used here instead of B because it is the SI symbol for *spectral radiance*. The L in $c1L$ refers to that. This reference is necessary because Planck's law can be reformulated to give spectral radiant exitance $M(\lambda, T)$ rather than *spectral radiance* $L(\lambda, T)$, in which case c_1 replaces c_1L, with

$$c_1 = 2\pi hc^2$$

so that Planck's law for *spectral radiant exitance* can be written as

$$M(\lambda, T) = \frac{c_1}{\lambda^5} \frac{1}{\exp\left(\frac{c_2}{\lambda T}\right) - 1}$$

7.3 Derivation

See also: Gas in a box and Photon gas

Consider a cube of side L with conducting walls filled with electromagnetic radiation in thermal equilibrium at temperature T. If there is a small hole in one of the walls, the radiation emitted from the hole will be characteristic of a perfect black body. We will first calculate the spectral energy density within the cavity and then determine the spectral radiance of the emitted radiation.

At the walls of the cube, the parallel component of the electric field and the orthogonal component of the magnetic field must vanish. Analogous to the wave function of a particle in a box, one finds that the fields are superpositions of periodic functions. The three wavelengths λ_1, λ_2, and λ_3, in the three directions orthogonal to the walls can be:

$$\lambda_i = \frac{2L}{n_i},$$

where the n_i are positive integers. For each set of integers n_i there are two linear independent solutions (modes). According to quantum theory, the energy levels of a mode are given by:

$$E_{n_1,n_2,n_3}(r) = \left(r + \frac{1}{2}\right) \frac{hc}{2L} \sqrt{n_1^2 + n_2^2 + n_3^2}. \qquad (1)$$

The quantum number r can be interpreted as the number of photons in the mode. The two modes for each set of n_i correspond to the two polarization states of the photon which has a spin of 1. Note that for $r = 0$ the energy of the mode is not zero. This vacuum energy of the electromagnetic field is responsible for the Casimir effect. In the following we will calculate the internal energy of the box at absolute temperature T.

According to statistical mechanics, the probability distribution over the energy levels of a particular mode is given by:

$$P_r = \frac{\exp(-\beta E(r))}{Z(\beta)}.$$

Here

$$\beta \stackrel{\text{def}}{=} 1/(k_\mathrm{B}T).$$

The denominator $Z(\beta)$, is the partition function of a single mode and makes Pr properly normalized:

$$Z(\beta) = \sum_{r=0}^{\infty} e^{-\beta E(r)} = \frac{e^{-\beta\varepsilon/2}}{1 - e^{-\beta\varepsilon}}.$$

Here we have implicitly defined

$$\varepsilon \stackrel{\text{def}}{=} \frac{hc}{2L}\sqrt{n_1^2 + n_2^2 + n_3^2},$$

which is the energy of a single photon. As explained here, the average energy in a mode can be expressed in terms of the partition function:

$$\langle E \rangle = -\frac{d\log(Z)}{d\beta} = \frac{\varepsilon}{2} + \frac{\varepsilon}{e^{\beta\varepsilon} - 1}.$$

This formula, apart from the first vacuum energy term, is a special case of the general formula for particles obeying Bose–Einstein statistics. Since there is no restriction on the total number of photons, the chemical potential is zero.

If we measure the energy relative to the ground state, the total energy in the box follows by summing $\langle E \rangle - \frac{\varepsilon}{2}$ over all allowed single photon states. This can be done exactly in the thermodynamic limit as L approaches infinity. In this limit, ε becomes continuous and we can then integrate $\langle E \rangle - \frac{\varepsilon}{2}$ over this parameter. To calculate the energy in the box in this way, we need to evaluate how many photon states there are in a given energy range. If we write the total number of single photon states with energies between ε and $\varepsilon + d\varepsilon$ as $g(\varepsilon)d\varepsilon$, where $g(\varepsilon)$ is the density of states (which we'll evaluate in a moment), then we can write:

$$U = \int_0^\infty \frac{\varepsilon}{e^{\beta\varepsilon} - 1} g(\varepsilon) \, d\varepsilon. \quad (2)$$

To calculate the density of states we rewrite equation (1) as follows:

$$\varepsilon \overset{\text{def}}{=} \frac{hc}{2L} n,$$

where n is the norm of the vector $\mathbf{n} = (n_1, n_2, n_3)$:

$$n = \sqrt{n_1^2 + n_2^2 + n_3^2}.$$

For every vector $\mathbf{n}$ with integer components larger than or equal to zero, there are two photon states. This means that the number of photon states in a certain region of n-space is twice the volume of that region. An energy range of $d\varepsilon$ corresponds to shell of thickness $dn = (2L/hc)d\varepsilon$ in n-space. Because the components of $\mathbf{n}$ have to be positive, this shell spans an octant of a sphere. The number of photon states $g(\varepsilon)d\varepsilon$, in an energy range $d\varepsilon$, is thus given by:

$$g(\varepsilon) \, d\varepsilon = 2 \frac{1}{8} 4\pi n^2 \, dn = \frac{8\pi L^3}{h^3 c^3} \varepsilon^2 \, d\varepsilon.$$

Inserting this in Eq. (2) gives:

$$U = L^3 \frac{8\pi}{h^3 c^3} \int_0^\infty \frac{\varepsilon^3}{e^{\beta\varepsilon} - 1} \, d\varepsilon. \quad (3)$$

From this equation one can derive the spectral energy density as a function of frequency $u_\nu(T)$ and as a function of wavelength $u\lambda(T)$:

$$\frac{U}{L^3} = \int_0^\infty u_\nu(T) \, d\nu,$$

where:

$$u_\nu(T) = \frac{8\pi h\nu^3}{c^3} \frac{1}{e^{h\nu/k_B T} - 1}.$$

And:

$$\frac{U}{L^3} = \int_0^\infty u_\lambda(T) \, d\lambda,$$

where

$$u_\lambda(T) = \frac{8\pi hc}{\lambda^5} \frac{1}{e^{hc/\lambda k_b T} - 1}.$$

This is also a spectral energy density function with units of energy per unit wavelength per unit volume. Integrals of this type for Bose and Fermi gases can be expressed in terms of polylogarithms. In this case, however, it is possible to calculate the integral in closed form using only elementary functions. Substituting

$$\varepsilon = k_B T x,$$

in Eq. (3), makes the integration variable dimensionless giving:

$$u(T) = \frac{8\pi (k_B T)^4}{(hc)^3} J,$$

where J is a Bose–Einstein integral given by:

$$J = \int_0^\infty \frac{x^3}{e^x - 1}\, dx = \frac{\pi^4}{15}.$$

The total electromagnetic energy inside the box is thus given by:

$$\frac{U}{V} = \frac{8\pi^5 (k_B T)^4}{15(hc)^3},$$

where $V = L^3$ is the volume of the box.

The combination hc/kB has the value 14 387.770 μm·K.

This is **not** the Stefan–Boltzmann law (which provides the total energy *radiated* by a black body per unit surface area per unit time), but it can be written more compactly using the Stefan–Boltzmann constant σ, giving

$$\frac{U}{V} = \frac{4\sigma T^4}{c}.$$

The constant $4\sigma/c$ is sometimes called the radiation constant.

Since the radiation is the same in all directions, and propagates at the speed of light (c), the spectral radiance of radiation exiting the small hole is

$$B_\nu(T) = \frac{u_\nu(T)\, c}{4\pi},$$

which yields

$$B_\nu(T) = \frac{2h\nu^3}{c^2} \frac{1}{e^{h\nu/k_b T} - 1}.$$

It can be converted to an expression for $B\lambda(T)$ in wavelength units by substituting ν by c/λ and evaluating

$$B_\lambda(T) = B_\nu(T) \left| \frac{d\nu}{d\lambda} \right|.$$

Note that dimensional analysis shows that the unit of steradians, shown in the denominator of left hand side of the equation above, is generated in and carried through the derivation but does not appear in any of the dimensions for any element on the left-hand-side of the equation.

This derivation is based on Brehm & Mullin 1989.

7.4 Physics

7.4.1 Outline

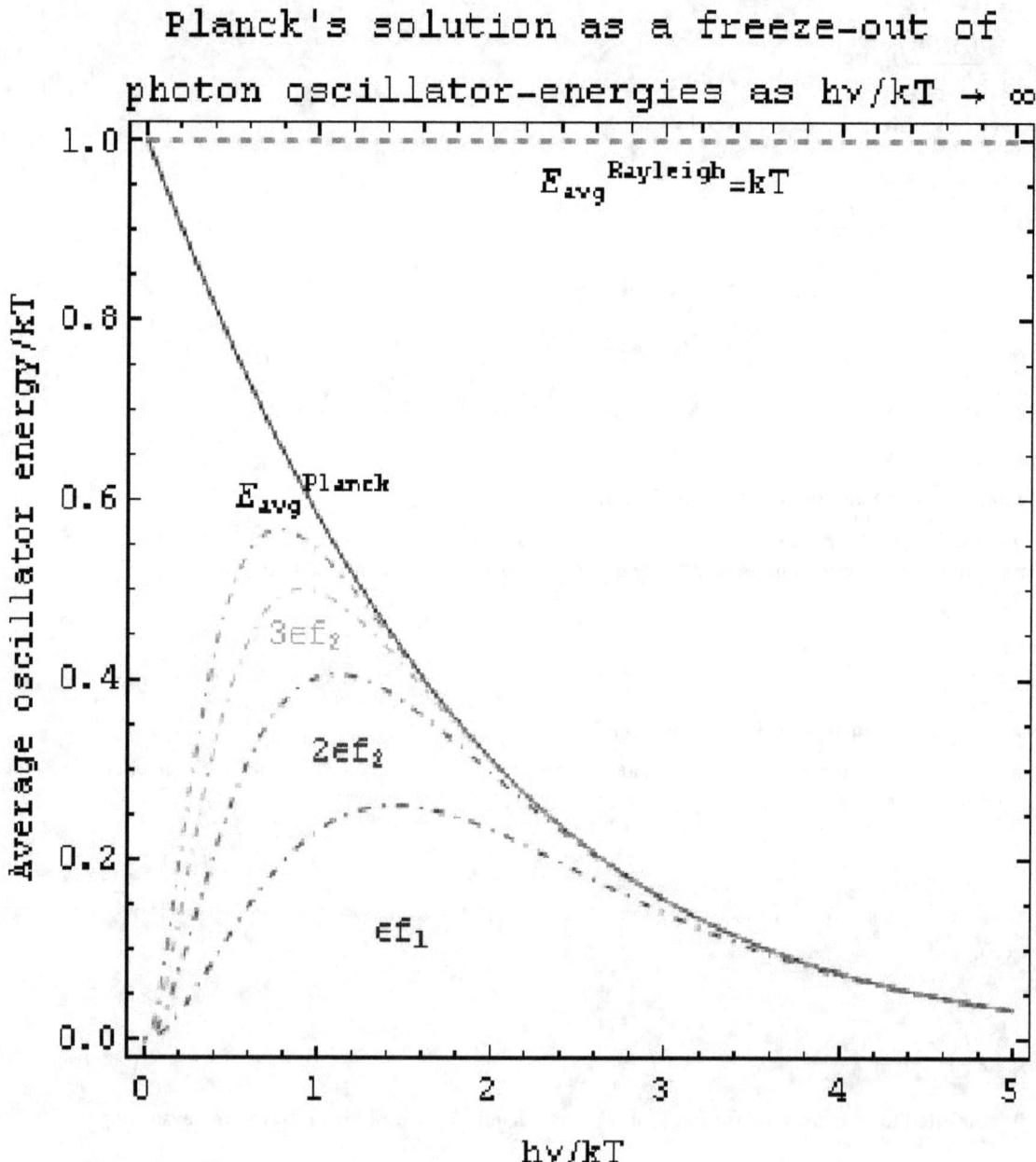

Freeze-out of high-energy oscillators.

Planck's law describes the unique and characteristic spectral distribution for electromagnetic radiation in thermodynamic equilibrium, when there is no net flow of matter or energy.[4] Its physics is most easily understood by considering the radiation in a cavity with rigid opaque walls. Motion of the walls can affect the radiation. If the walls are not opaque, then the thermodynamic equilibrium is not isolated. It is of interest to explain how the thermodynamic equilibrium is attained. There are two main cases: (a) when the approach to thermodynamic equilibrium is in the presence of matter, when the walls of the cavity are imperfectly reflective for every wavelength or when the walls are perfectly reflective while the cavity contains a small black body (this was the main case considered by Planck); or (b) when the approach to equilibrium is in the absence of matter, when the walls are perfectly reflective for all wavelengths and the cavity contains no matter. For matter not enclosed in such a cavity, thermal radiation can be approximately explained by appropriate use of Planck's law.

Classical physics led, via the Equipartition theorem, to the Ultraviolet catastrophe, a prediction that the total blackbody radiation intensity was infinite. If supplemented by the classically unjustifiable assumption that for some reason the radiation is finite, classical thermodynamics provides an account of some aspects of the Planck distribution, such as the Stefan–Boltzmann law, and the Wien displacement law. For the case of the presence of matter, quantum mechanics provides a good account, as found below in the section headed Einstein coefficients. This was the case considered by Einstein, and is nowadays used for quantum optics.[15][16] For the case of the absence of matter, quantum field theory is necessary, because non-relativistic quantum mechanics with fixed particle numbers does not provide a sufficient account.

7.4.2   Photons

Quantum theoretical explanation of Planck's law views the radiation as a gas of massless, uncharged, bosonic particles, namely photons, in thermodynamic equilibrium. Photons are viewed as the carriers of the electromagnetic interaction between electrically charged elementary particles. Photon numbers are not conserved. Photons are created or annihilated in the right numbers and with the right energies to fill the cavity with the Planck distribution. For a photon gas in thermodynamic equilibrium, the internal energy density is entirely determined by the temperature; moreover, the pressure is entirely determined by the internal energy density. This is unlike the case of thermodynamic equilibrium for material gases, for which the internal energy is determined not only by the temperature, but also, independently, by the respective numbers of the different molecules, and independently again, by the specific characteristics of the different molecules. For different material gases at given temperature, the pressure and internal energy density can vary independently, because different molecules can carry independently different excitation energies.

Planck's law arises as a limit of the Bose–Einstein distribution, the energy distribution describing non-interactive bosons in thermodynamic equilibrium. In the case of massless bosons such as photons and gluons, the chemical potential is zero and the Bose-Einstein distribution reduces to the Planck distribution. There is another fundamental equilibrium energy distribution: the Fermi–Dirac distribution, which describes fermions, such as electrons, in thermal equilibrium. The two distributions differ because multiple bosons can occupy the same quantum state, while multiple fermions cannot. At low densities, the number of available quantum states per particle is large, and this difference becomes irrelevant. In the low density limit, the Bose-Einstein and the Fermi-Dirac distribution each reduce to the Maxwell–Boltzmann distribution.

7.4.3   Kirchhoff's law of thermal radiation

Main article: Kirchhoff's law of thermal radiation

Kirchhoff's law of thermal radiation is a succinct and brief account of a complicated physical situation. The following is an introductory sketch of that situation, and is very far from being a rigorous physical argument. The purpose here is only to summarize the main physical factors in the situation, and the main conclusions.

Spectral dependence of thermal radiation

There is a difference between conductive heat transfer and radiative heat transfer. Radiative heat transfer can be filtered to pass only a definite band of radiative frequencies.

It is generally known that the hotter a body becomes the more heat it radiates at every frequency.

In a cavity in an opaque body with rigid walls that are not perfectly reflective at any frequency, in thermodynamic equilibrium, there is only one temperature, and it must be shared in common by the radiation of every frequency.

One may imagine two such cavities, each in its own isolated radiative and thermodynamic equilibrium. One may imagine an optical device that allows radiative heat transfer between the two cavities, filtered to pass only a definite band of radiative frequencies. If the values of the spectral radiances of the radiations in the cavities differ in that frequency band, heat may be expected to pass from the hotter to the colder. One might propose to use such a filtered transfer of heat in such a band to drive a heat engine. If the two bodies are at the same temperature, the second law of thermodynamics does not allow the heat engine to work. It may be inferred that for a temperature common to the two bodies, the values of the spectral radiances in the pass-band must also be common. This must hold for every frequency band.[17][18][19] This became clear to Balfour Stewart and later to Kirchhoff. Balfour Stewart found experimentally that of all surfaces, one of lamp-black emitted the greatest amount of thermal radiation for every quality of radiation, judged by various filters.

Thinking theoretically, Kirchhoff went a little further, and pointed out that this implied that the spectral radiance, as a function of radiative frequency, of any such cavity in thermodynamic equilibrium must be a unique universal function of temperature. He postulated an ideal black body that interfaced with its surrounds in just such a way as to absorb all the radiation that falls on it. By the Helmholtz reciprocity principle, radiation from the interior of such a body would pass unimpeded, directly to its surrounds without reflection at the interface. In thermodynamic equilibrium, the thermal radiation emitted from such a body would have that unique universal spectral radiance as a function of temperature. This insight is the root of Kirchhoff's law of thermal radiation.

Relation between absorptivity and emissivity

One may imagine a small homogeneous spherical material body labeled X at a temperature TX, lying in a radiation field within a large cavity with walls of material labeled Y at a temperature TY. The body X emits its own thermal radiation. At a particular frequency v, the radiation emitted from a particular cross-section through the centre of X in one sense in a direction normal to that cross-section may be denoted Iv, X (TX), characteristically for the material of X. At that frequency v, the radiative power from the walls into that cross-section in the opposite sense in that direction may be denoted Iv, Y (TY), for the wall temperature TY. For the material of X, defining the absorptivity av, X,Y (TX, TY) as the fraction of that incident radiation absorbed by X, that incident energy is absorbed at a rate av, X,Y (TX, TY) Iv, Y (TY).

The rate $q(v, TX, TY)$ of accumulation of energy in one sense into the cross-section of the body can then be expressed

$$q(\nu, T_X, T_Y) = \alpha_{\nu, X, Y}(T_X, T_Y) I_{\nu, Y}(T_Y) - I_{\nu, X}(T_X).$$

Kirchhoff's seminal insight, mentioned just above, was that, at thermodynamic equilibrium at temperature T, there exists a unique universal radiative distribution, nowadays denoted $Bv(T)$, that is independent of the chemical characteristics of the materials X and Y, that leads to a very valuable understanding of the radiative exchange equilibrium of any body at all, as follows.

When there is thermodynamic equilibrium at temperature T, the cavity radiation from the walls has that unique universal value, so that Iv, Y $(TY) = Bv(T)$. Further, one may define the emissivity $εv$, X (TX) of the material of the body X just so that at thermodynamic equilibrium at temperature $TX = T$, one has Iv, X $(TX) = Iv$, X $(T) = εv$, X (T) $Bv(T)$.

When thermal equilibrium prevails at temperature $T = TX = TY$, the rate of accumulation of energy vanishes so that $q(v, TX, TY) = 0$. It follows that:

in thermodynamic equilibrium, when $T = T_X = T_Y$, it is true that $0 = \alpha_{\nu, X, Y}(T, T) B_\nu(T) - \epsilon_{\nu, X}(T) B_\nu(T).$

Kirchhoff pointed out that it follows that

in thermodynamic equilibrium, when $T = T_X = T_Y$, it is true that $\alpha_{\nu, X, Y}(T, T) = \epsilon_{\nu, X}(T).$

Introducing the special notation av, X (T) for the absorptivity of material X at thermodynamic equilibrium at temperature T (justified by a discovery of Einstein, as indicated below), one further has the equality

$\alpha_{\nu,X}(T) = \epsilon_{\nu,X}(T)$ at thermodynamic equilibrium.

The equality of absorptivity and emissivity here demonstrated is specific for thermodynamic equilibrium at temperature T and is in general not to be expected to hold when conditions of thermodynamic equilibrium do not hold. The emissivity and absorptivity are each separately properties of the molecules of the material but they depend differently upon the distributions of states of molecular excitation on the occasion, because of a phenomenon known as "stimulated emission", that was discovered by Einstein. On occasions when the material is in thermodynamic equilibrium or in a state known as local thermodynamic equilibrium, the emissivity and absorptivity become equal. Very strong incident radiation or other factors can disrupt thermodynamic equilibrium or local thermodynamic equilibrium. Local thermodynamic equilibrium in a gas means that molecular collisions far outweigh light emission and absorption in determining the distributions of states of molecular excitation.

Kirchhoff pointed out that he did not know the precise character of $B\nu(T)$, but he thought it important that it should be found out. Four decades after Kirchhoff's insight of the general principles of its existence and character, Planck's contribution was to determine the precise mathematical expression of that equilibrium distribution $B\nu(T)$.

7.4.4 Black body

Main article: Black body

In physics, one considers an ideal black body, here labeled B, defined as one that completely absorbs all of the electromagnetic radiation falling upon it at every frequency ν (hence the term "black"). According to Kirchhoff's law of thermal radiation, this entails that, for every frequency ν, at thermodynamic equilibrium at temperature T, one has $\alpha\nu, B(T) = \epsilon\nu, B(T) = 1$, so that the thermal radiation from a black body is always equal to the full amount specified by Planck's law. No physical body can emit thermal radiation that exceeds that of a black body, since if it were in equilibrium with a radiation field, it would be emitting more energy than was incident upon it.

Though perfectly black materials do not exist, in practice a black surface can be accurately approximated.[4] As to its material interior, a body of condensed matter, liquid, solid, or plasma, with a definite interface with its surroundings, is completely black to radiation if it is completely opaque. That means that it absorbs all of the radiation that penetrates the interface of the body with its surroundings, and enters the body. This is not too difficult to achieve in practice. On the other hand, a perfectly black interface is not found in nature. A perfectly black interface reflects no radiation, but transmits all that falls on it, from either side. The best practical way to make an effectively black interface is to simulate an 'interface' by a small hole in the wall of a large cavity in a completely opaque rigid body of material that does not reflect perfectly at any frequency, with its walls at a controlled temperature. Beyond these requirements, the component material of the walls is unrestricted. Radiation entering the hole has almost no possibility of escaping the cavity without being absorbed by multiple impacts with its walls.[20]

7.4.5 Lambert's cosine law

Main article: Lambert's cosine law

As explained by Planck,[21] a radiating body has an interior consisting of matter, and an interface with its contiguous neighbouring material medium, which is usually the medium from within which the radiation from the surface of the body is observed. The interface is not composed of physical matter but is a theoretical conception, a mathematical two-dimensional surface, a joint property of the two contiguous media, strictly speaking belonging to neither separately. Such an interface can neither absorb nor emit, because it is not composed of physical matter; but it is the site of reflection and transmission of radiation, because it is a surface of discontinuity of optical properties. The reflection and transmission of radiation at the interface obey the Stokes–Helmholtz reciprocity principle.

At any point in the interior of a black body located inside a cavity in thermodynamic equilibrium at temperature T the radiation is homogeneous, isotropic and unpolarized. A black body absorbs all and reflects none of the electromagnetic radiation incident upon it. According to the Helmholtz reciprocity principle, radiation from the interior of a black body is not reflected at its surface, but is fully transmitted to its exterior. Because of the isotropy of the radiation in the body's interior, the spectral radiance of radiation transmitted from its interior to its exterior through its surface is independent of direction.[22]

This is expressed by saying that radiation from the surface of a black body in thermodynamic equilibrium obeys Lambert's cosine law.[23][24] This means that the spectral flux $d\Phi(dA, \theta, d\Omega, d\nu)$ from a given infinitesimal element of area dA of the actual emitting surface of the black body, detected from a given direction that makes an angle θ with the normal to the actual emitting surface at dA, into an element of solid angle of detection $d\Omega$ centred on the direction indicated by θ, in an element of frequency bandwidth $d\nu$, can be represented as[25]

$$\frac{d\Phi(dA, \theta, d\Omega, d\nu)}{d\Omega} = L^0(dA, d\nu)\, dA\, d\nu\, \cos\theta$$

where $L^0(dA, d\nu)$ denotes the flux, per unit area per unit frequency per unit solid angle, that area dA would show if it were measured in its normal direction $\theta = 0$.

The factor $\cos\theta$ is present because the area to which the spectral radiance refers directly is the projection, of the actual emitting surface area, onto a plane perpendicular to the direction indicated by θ. This is the reason for the name *cosine law*.

Taking into account the independence of direction of the spectral radiance of radiation from the surface of a black body in thermodynamic equilibrium, one has $L^0(dA, d\nu) = B\nu(T)$ and so

$$\frac{d\Phi(dA, \theta, d\Omega, d\nu)}{d\Omega} = B_\nu(T)\, dA\, d\nu\, \cos\theta.$$

Thus Lambert's cosine law expresses the independence of direction of the spectral radiance $B\nu(T)$ of the surface of a black body in thermodynamic equilibrium.

7.4.6 Stefan–Boltzmann law

Main article: Stefan–Boltzmann law

The total power emitted per unit area at the surface of a black body (P) may be found by integrating the black body spectral flux found from Lambert's law over all frequencies, and over the solid angles corresponding to a hemisphere (h) above the surface.

$$P = \int_0^\infty d\nu \int_h d\Omega\, B_\nu \cos(\theta)$$

The infinitesimal solid angle can be expressed in spherical polar coordinates:

$$d\Omega = \sin(\theta)\, d\theta\, d\phi.$$

So that:

$$P = \int_0^\infty d\nu \int_0^{\pi/2} d\theta \int_0^{2\pi} d\phi\, B_\nu(T) \cos(\theta) \sin(\theta) = \sigma T^4$$

where

$$\sigma = \frac{2k_B^4 \pi^5}{15c^2 h^3} \approx 5.670400 \times 10^{-8} \, \mathrm{J\,s^{-1}m^{-2}K^{-4}}$$

is known as the Stefan–Boltzmann constant.[26]

7.4.7 Radiative transfer

Main article: Radiative transfer

The equation of radiative transfer describes the way in which radiation is affected as it travels through a material medium. For the special case in which the material medium is in thermodynamic equilibrium in the neighborhood of a point in the medium, Planck's law is of special importance.

For simplicity, we can consider the linear steady state, without scattering. The equation of radiative transfer states that for a beam of light going through a small distance ds, energy is conserved: The change in the (spectral) radiance of that beam (I_ν) is equal to the amount removed by the material medium plus the amount gained from the material medium. If the radiation field is in equilibrium with the material medium, these two contributions will be equal. The material medium will have a certain emission coefficient and absorption coefficient.

The absorption coefficient α is the fractional change in the intensity of the light beam as it travels the distance ds, and has units of 1/length. It is composed of two parts, the decrease due to absorption and the increase due to stimulated emission. Stimulated emission is emission by the material body which is caused by and is proportional to the incoming radiation. It is included in the absorption term because, like absorption, it is proportional to the intensity of the incoming radiation. Since the amount of absorption will generally vary linearly as the density ρ of the material, we may define a "mass absorption coefficient" $\kappa_\nu = \alpha/\rho$ which is a property of the material itself. The change in intensity of a light beam due to absorption as it traverses a small distance ds will then be $dI_\nu = -\kappa_\nu \, \rho \, I_\nu \, ds$ [2]

The "mass emission coefficient" j_ν is equal to the radiance per unit volume of a small volume element divided by its mass (since, as for the mass absorption coefficient, the emission is proportional to the emitting mass) and has units of power/solid angle/frequency/density. Like the mass absorption coefficient, it too is a property of the material itself. The change in a light beam as it traverses a small distance ds will then be $dI_\nu = j_\nu \, \rho \, ds$ [27]

The equation of radiative transfer will then be the sum of these two contributions:[28]

$$\frac{dI_\nu}{ds} = j_\nu \rho - \kappa_\nu \rho I_\nu.$$

If the radiation field is in equilibrium with the material medium, then the radiation will be homogeneous (independent of position) so that $dI_\nu = 0$ and:

$$\kappa_\nu B_\nu = j_\nu$$

which is another statement of Kirchhoff's law, relating two material properties of the medium, and which yields the radiative transfer equation at a point around which the medium is in thermodynamic equilibrium:

$$\frac{dI_\nu}{ds} = \kappa_\nu \rho (B_\nu - I_\nu).$$

7.4.8 Einstein coefficients

Main article: Atomic spectral line

The principle of detailed balance states that, at thermodynamic equilibrium, each elementary process is equilibrated by its reverse process.

In 1916, Albert Einstein applied this principle on an atomic level to the case of an atom radiating and absorbing radiation due to transitions between two particular energy levels,[29] giving a deeper insight into the equation of radiative transfer and Kirchhoff's law for this type of radiation. If level 1 is the lower energy level with energy E_1, and level 2 is the upper energy level with energy E_2, then the frequency ν of the radiation radiated or absorbed will be determined by Bohr's frequency condition: $E_2 - E_1 = h\nu$.[30][31]

If n_1 and n_2 are the number densities of the atom in states 1 and 2 respectively, then the rate of change of these densities in time will be due to three processes:

where $I_\nu(T)$ is the spectral radiance of the radiation field. The three parameters A_{21}, B_{21} and B_{12}, known as the Einstein coefficients, are associated with the photon frequency (ν) produced by the transition between two energy levels (states). As a result, each line in a spectra has it own set of associated coefficients. When the atoms and the radiation field are in equilibrium, the radiance will be given by Planck's law and, by the principle of detailed balance, the sum of these rates must be zero:

$$0 = A_{21}n_2 + B_{21}n_2 B_\nu(T) - B_{12}n_1 B_\nu(T)$$

Since the atoms are also in equilibrium, the populations of the two levels are related by the Boltzmann factor:

$$\frac{n_2}{n_1} = \frac{g_2}{g_1} e^{-h\nu/k_{\rm B}T}$$

where g_1 and g_2 are the multiplicities of the respective energy levels. Combining the above two equations with the requirement that they be valid at any temperature yields two relationships between the Einstein coefficients:

$$\frac{A_{21}}{B_{21}} = \frac{2h\nu^3}{c^2}$$

$$\frac{B_{21}}{B_{12}} = \frac{g_1}{g_2}$$

so that knowledge of one coefficient will yield the other two. For the case of isotropic absorption and emission, the emission coefficient (j_ν) and absorption coefficient (κ_ν) defined in the radiative transfer section above, can be expressed in terms of the Einstein coefficients. The relationships between the Einstein coefficients will yield the expression of Kirchhoff's law expressed in the *Radiative transfer* section above, namely that

$$j_\nu = \kappa_\nu B_\nu.$$

These coefficients apply to both atoms and molecules.

7.5   Properties

7.5.1   Peaks

The distributions B_ν, B_ω, $B_{\tilde{\nu}}$ and B_k peak at a photon energy of[32]

$$E = \left[3 + W\left(\frac{-3}{e^3}\right)\right] k_B T \approx 2.821\, k_B T,$$

where W is the Lambert W function and e is Euler's number.

The distributions B_λ and B_y however, peak at a different energy[32]

$$E = \left[5 + W\left(\frac{-5}{e^5}\right)\right] k_B T \approx 4.965\, k_B T,$$

The reason for this is that, as mentioned above, one cannot go from (for example) B_ν to B_λ simply by substituting ν by λ. In addition, one must also multiply the result of the substitution by $\left|\frac{d\nu}{d\lambda}\right| = c/\lambda^2$. This $1/\lambda^2$ factor shifts the peak of the distribution to higher energies.

Dividing hc by this energy expression gives the wavelength of the peak. For this one can use $hc/k_B = 387.770\,14 \mu m \cdot K$. The spectral radiance at these peaks is given by:

$$B_{\nu,\max}(T) = \frac{2k_B^3 T^3 (3 + W(-3\exp(-3)))^3}{h^2 c^2} \frac{1}{e^{3+W(-3\exp(-3))} - 1} \approx \left(1.896 \times 10^{-19} \frac{W}{m^2 \cdot Hz \cdot K^3}\right) \times T^3$$

$$B_{\lambda,\max}(T) = \frac{2k_B^5 T^5 (5 + W(-5\exp(-5)))^5}{h^4 c^3} \frac{1}{e^{5+W(-5\exp(-5))} - 1} \approx \left(4.096 \times 10^{-6} \frac{W}{m^3 \cdot K^5}\right) \times T^5$$

7.5.2 Approximations

In the limit of low frequencies (i.e. long wavelengths), Planck's law becomes the Rayleigh–Jeans law[33][34][35]

$$B_\nu(T) \approx \frac{2\nu^2}{c^2} k_B T \text{ or } \qquad B_\lambda(T) \approx \frac{2c}{\lambda^4} k_B T.$$

The radiance increases as the square of the frequency, illustrating the ultraviolet catastrophe. In the limit of high frequencies (i.e. small wavelengths) Planck's law tends to the Wien approximation:[35][36][37]

$$B_\nu(T) \approx \frac{2h\nu^3}{c^2} e^{-\frac{h\nu}{k_B T}} \text{ or } B_\lambda(T) \approx \frac{2hc^2}{\lambda^5} e^{-\frac{hc}{\lambda k_B T}}.$$

Both approximations were known to Planck before he developed his law. He was led by these two approximations to develop a law which incorporated both limits, which ultimately became Planck's law.

7.5.3 Percentiles

Wien's displacement law in its stronger form states that the shape of Planck's law is independent of temperature. It is therefore possible to list the percentile points of the total radiation as well as the peaks for wavelength and frequency, in a form which gives the wavelength λ when divided by temperature T.[38] The second row of the following table lists the corresponding values of λT, that is, those values of x for which the wavelength λ is x/T micrometers at the radiance percentile point given by the corresponding entry in the first row.

That is, 0.01% of the radiation is at a wavelength below $910/T$ μm, 20% below $2676/T$ μm, etc. The wavelength and frequency peaks are in bold and occur at 25.0% and 64.6% respectively. The 41.8% point is the wavelength-frequency-neutral peak. These are the points at which the respective Planck-law functions $1/\lambda^5$, ν^3, and ν^2/λ^2 divided by

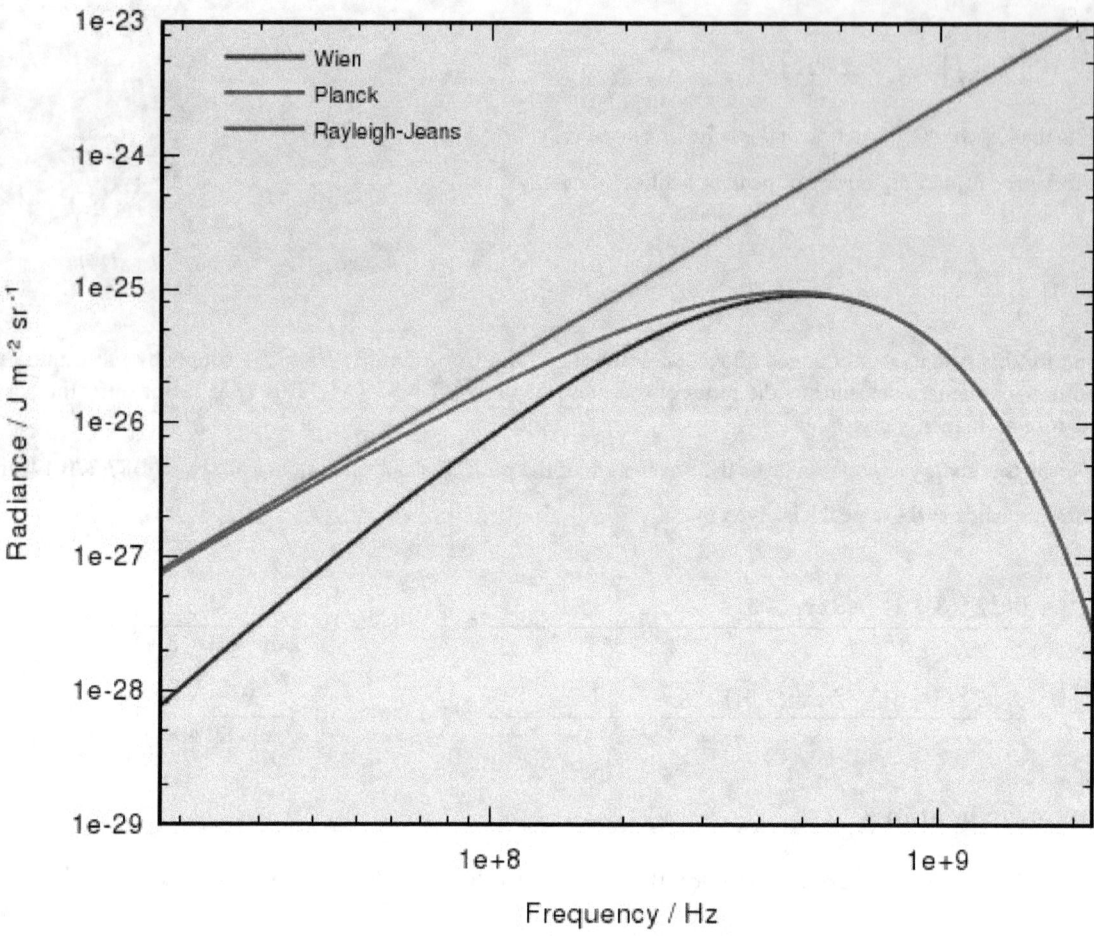

Log-log plots of radiance vs. frequency for Planck's law (green), compared with the Rayleigh–Jeans law (red) and the Wien approximation (blue) for a black body at 8 mK temperature.

$\exp(h\nu/kBT) - 1$ attain their maxima. Also note the much smaller gap in ratio of wavelengths between 0.1% and 0.01% (1110 is 22% more than 910) than between 99.9% and 99.99% (113374 is 120% more than 51613), reflecting the exponential decay of energy at short wavelengths (left end) and polynomial decay at long.

Which peak to use depends on the application. The conventional choice is the wavelength peak at 25.0% given by Wien's displacement law in its weak form. For some purposes the median or 50% point dividing the total radiation into two halves may be more suitable. The latter is closer to the frequency peak than to the wavelength peak because the radiance drops exponentially at short wavelengths and only polynomially at long. The neutral peak occurs at a shorter wavelength than the median for the same reason.

For the Sun, T is 5778 K, allowing the percentile points of the Sun's radiation, in nanometers, to be tabulated as follows when modeled as a black body radiator, to which the Sun is a fair approximation. For comparison a planet modeled as a black body radiating at a nominal 288 K (15 °C) as a representative value of the Earth's highly variable temperature has wavelengths more than twenty times that of the Sun, tabulated in the third row in micrometers (thousands of nanometers).

That is, only 1% of the Sun's radiation is at wavelengths shorter than 251 nm, and only 1% at longer than 3961 nm. Expressed in micrometers this puts 98% of the Sun's radiation in the range from 0.251 to 3.961 μm. The corresponding 98% of energy radiated from a 288 K planet is from 5.03 to 79.5 μm, well above the range of solar radiation (or below if expressed in terms of frequencies $\nu = c/\lambda$ instead of wavelengths λ).

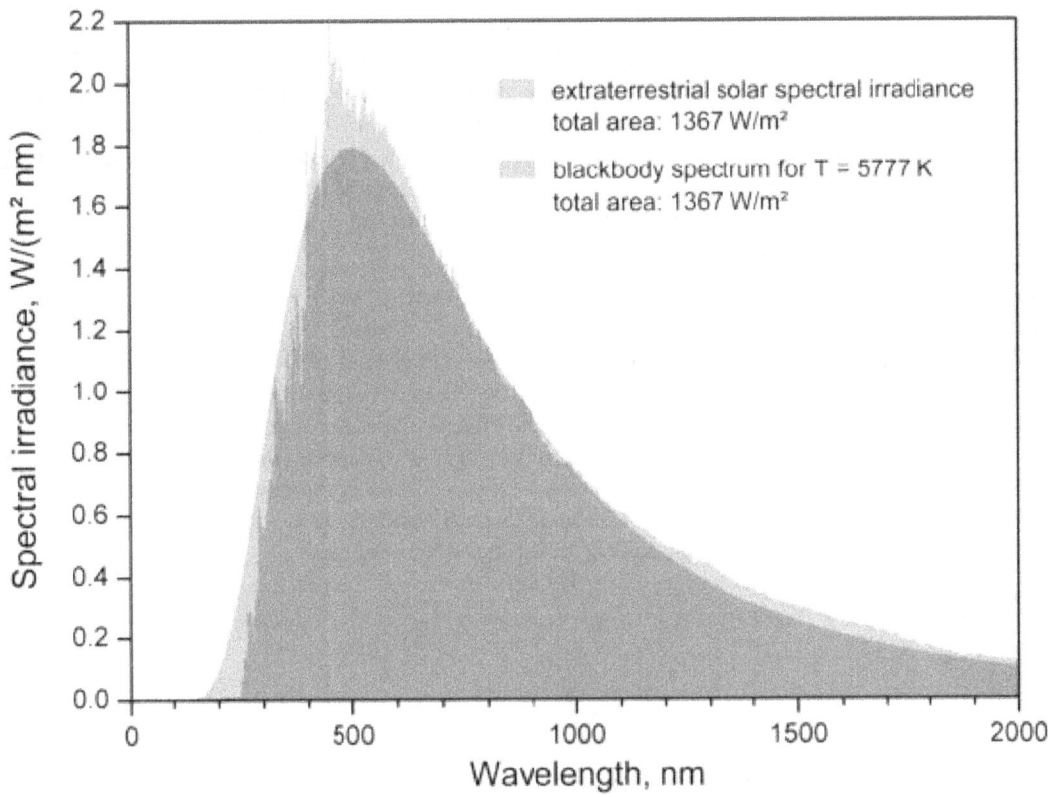

The Sun is an excellent approximation of a black body. Its effective temperature is ~5777 K.

A consequence of this more-than-order-of-magnitude difference in wavelength between solar and planetary radiation is that filters designed to pass one and block the other are easy to construct. For example windows fabricated of ordinary glass or transparent plastic pass at least 80% of the incoming 5778 K solar radiation, which is below 1.2 µm in wavelength, while blocking over 99% of the outgoing 288 K thermal radiation from 5 µm upwards, wavelengths at which most kinds of glass and plastic of construction-grade thickness are effectively opaque.

The Sun's radiation is that arriving at the top of the atmosphere (TOA). As can be read from the table, radiation below 400 nm, or ultraviolet, is about 12%, while that above 700 nm, or infrared, starts at about the 49% point and so accounts for 51% of the total. Hence only 37% of the TOA insolation is visible to the human eye. The atmosphere shifts these percentages substantially in favor of visible light as it absorbs most of the ultraviolet and significant amounts of infrared.

7.6   History

7.6.1   Forerunners

Balfour Stewart

In 1858, Balfour Stewart described his experiments on the thermal radiative emissive and absorptive powers of polished plates of various substances, compared with the powers of lamp-black surfaces, at the same temperature.[5] Stewart chose lamp-black surfaces as his reference because of various previous experimental findings, especially those of Pierre Prevost and of John Leslie. He wrote "Lamp-black, which absorbs all the rays that fall upon it, and therefore possesses the greatest possible absorbing power, will possess also the greatest possible radiating power."

Stewart measured radiated power with a thermo-pile and sensitive galvanometer read with a microscope. He was concerned with selective thermal radiation, which he investigated with plates of substances that radiated and absorbed selectively for different qualities of radiation rather than maximally for all qualities of radiation. He discussed the experiments in terms of rays which could be reflected and refracted, and which obeyed the Helmholtz reciprocity principle (though he did not use an eponym for it). He did not in this paper mention that the qualities of the rays might be described by their wavelengths, nor did he use spectrally resolving apparatus such as prisms or diffraction gratings. His work was quantitative within these constraints. He made his measurements in a room temperature environment, and quickly so as to catch his bodies in a condition near the thermal equilibrium in which they had been prepared by heating to equilibrium with boiling water. His measurements confirmed that substances that emit and absorb selectively respect the principle of selective equality of emission and absorption at thermal equilibrium.

Stewart offered a theoretical proof that this should be the case separately for every selected quality of thermal radiation, but his mathematics was not rigorously valid. According to historian D.M Siegel: "He was not a practitioner of the more sophisticated techniques of nineteenth-century mathematical physics; he did not even make use of the functional notation in dealing with spectral distributions."[39] He made no mention of thermodynamics in this paper, though he did refer to conservation of *vis viva*. He proposed that his measurements implied that radiation was both absorbed and emitted by particles of matter throughout depths of the media in which it propagated. He applied the Helmholtz reciprocity principle to account for the material interface processes as distinct from the processes in the interior material. He concluded that his experiments showed that, in the interior of an enclosure in thermal equilibrium, the radiant heat, reflected and emitted combined, leaving any part of the surface, regardless of its substance, was the same as would have left that same portion of the surface if it had been composed of lamp-black. He did not mention the possibility of ideally perfectly reflective walls; in particular he noted that highly polished real physical metals absorb very slightly.

Gustav Kirchhoff

In 1859, not knowing of Stewart's work, Gustav Robert Kirchhoff reported the coincidence of the wavelengths of spectrally resolved lines of absorption and of emission of visible light. Importantly for thermal physics, he also observed that bright lines or dark lines were apparent depending on the temperature difference between emitter and absorber.[40]

Kirchhoff then went on to consider bodies that emit and absorb heat radiation, in an opaque enclosure or cavity, in equilibrium at temperature T.

Here is used a notation different from Kirchhoff's. Here, the emitting power $E(T, i)$ denotes a dimensioned quantity, the total radiation emitted by a body labeled by index i at temperature T. The total absorption ratio $a(T, i)$ of that body is dimensionless, the ratio of absorbed to incident radiation in the cavity at temperature T. (In contrast with Balfour Stewart's, Kirchhoff's definition of his absorption ratio did not refer in particular to a lamp-black surface as the source of the incident radiation.) Thus the ratio $E(T, i) / a(T, i)$ of emitting power to absorption ratio is a dimensioned quantity, with the dimensions of emitting power, because $a(T, i)$ is dimensionless. Also here the wavelength-specific emitting power of the body at temperature T is denoted by $E(\lambda, T, i)$ and the wavelength-specific absorption ratio by $a(\lambda, T, i)$. Again, the ratio $E(\lambda, T, i) / a(\lambda, T, i)$ of emitting power to absorption ratio is a dimensioned quantity, with the dimensions of emitting power.

In a second report made in 1859, Kirchhoff announced a new general principle or law for which he offered a theoretical and mathematical proof, though he did not offer quantitative measurements of radiation powers.[41] His theoretical proof was and still is considered by some writers to be invalid.[39][42] His principle, however, has endured: it was that for heat rays of the same wavelength, in equilibrium at a given temperature, the wavelength-specific ratio of emitting power to absorption ratio has one and the same common value for all bodies that emit and absorb at that wavelength. In symbols, the law stated that the wavelength-specific ratio $E(\lambda, T, i) / a(\lambda, T, i)$ has one and the same value for all bodies, that is for all values of index i. In this report there was no mention of black bodies.

In 1860, still not knowing of Stewart's measurements for selected qualities of radiation, Kirchhoff pointed out that it was long established experimentally that for total heat radiation, of unselected quality, emitted and absorbed by a body in equilibrium, the dimensioned total radiation ratio $E(T, i) / a(T, i)$, has one and the same value common to all bodies, that is, for every value of the material index i.[43] Again without measurements of radiative powers or other new experimental data, Kirchhoff then offered a fresh theoretical proof of his new principle of the universality of the value of the wavelength-specific ratio $E(\lambda, T, i) / a(\lambda, T, i)$ at thermal equilibrium. His fresh theoretical proof was and still is considered by some writers to be invalid.[39][42]

But more importantly, it relied on a new theoretical postulate of "**perfectly black bodies**", which is the reason why one speaks of Kirchhoff's law. Such black bodies showed complete absorption in their infinitely thin most superficial surface. They correspond to Balfour Stewart's reference bodies, with internal radiation, coated with lamp-black. They were not the more realistic perfectly black bodies later considered by Planck. Planck's black bodies radiated and absorbed only by the material in their interiors; their interfaces with contiguous media were only mathematical surfaces, capable neither of absorption nor emission, but only of reflecting and transmitting with refraction.[44]

Kirchhoff's proof considered an arbitrary non-ideal body labeled i as well as various perfect black bodies labeled BB . It required that the bodies be kept in a cavity in thermal equilibrium at temperature T . His proof intended to show that the ratio $E(\lambda, T, i) / a(\lambda, T, i)$ was independent of the nature i of the non-ideal body, however partly transparent or partly reflective it was.

His proof first argued that for wavelength λ and at temperature T, at thermal equilibrium, all perfectly black bodies of the same size and shape have the one and the same common value of emissive power $E(\lambda, T, \text{BB})$, with the dimensions of power. His proof noted that the dimensionless wavelength-specific absorption ratio $a(\lambda, T, \text{BB})$ of a perfectly black body is by definition exactly 1. Then for a perfectly black body, the wavelength-specific ratio of emissive power to absorption ratio $E(\lambda, T, \text{BB}) / a(\lambda, T, \text{BB})$ is again just $E(\lambda, T, \text{BB})$, with the dimensions of power. Kirchhoff considered, successively, thermal equilibrium with the arbitrary non-ideal body, and with a perfectly black body of the same size and shape, in place in his cavity in equilibrium at temperature T . He argued that the flows of heat radiation must be the same in each case. Thus he argued that at thermal equilibrium the ratio $E(\lambda, T, i) / a(\lambda, T, i)$ was equal to $E(\lambda, T, \text{BB})$, which may now be denoted $B\lambda\ (\lambda, T)$, a continuous function, dependent only on λ at fixed temperature T, and an increasing function of T at fixed wavelength λ, at low temperatures vanishing for visible but not for longer wavelengths, with positive values for visible wavelengths at higher temperatures, which does not depend on the nature i of the arbitrary non-ideal body. (Geometrical factors, taken into detailed account by Kirchhoff, have been ignored in the foregoing.)

Thus **Kirchhoff's law of thermal radiation** can be stated: *For any material at all, radiating and absorbing in thermodynamic equilibrium at any given temperature T, for every wavelength λ, the ratio of emissive power to absorptive ratio has one universal value, which is characteristic of a perfect black body, and is an emissive power which we here represent by* $B\lambda\ (\lambda, T)$ *.* (For our notation $B\lambda\ (\lambda, T)$, Kirchhoff's original notation was simply e.)[2][43][45][46][47][48]

Kirchhoff announced that the determination of the function $B\lambda\ (\lambda, T)$ was a problem of the highest importance, though he recognized that there would be experimental difficulties to be overcome. He supposed that like other functions that do not depend on the properties of individual bodies, it would be a simple function. That function $B\lambda\ (\lambda, T)$ has occasionally been called 'Kirchhoff's (emission, universal) function',[49][50][51][52] though its precise mathematical form would not be known for another forty years, till it was discovered by Planck in 1900. The theoretical proof for Kirchhoff's universality principle was worked on and debated by various physicists over the same time, and later.[42] Kirchhoff stated later in 1860 that his theoretical proof was better than Balfour Stewart's, and in some respects it was so.[39] Kirchhoff's 1860 paper did not mention the second law of thermodynamics, and of course did not mention the concept of entropy which had not at that time been established. In a more considered account in a book in 1862, Kirchhoff mentioned the connection of his law with "Carnot's principle", which is a form of the second law.[53]

According to Helge Kragh, "Quantum theory owes its origin to the study of thermal radiation, in particular to the "black-body" radiation that Robert Kirchhoff had first defined in 1859–1860."[54]

7.6.2   Empirical and theoretical ingredients for the scientific induction of Planck's law

In 1860, Kirchhoff predicted experimental difficulties for the empirical determination of the function that described the dependence of the black-body spectrum as a function only of temperature and wavelength. And so it turned out. It took some forty years of development of improved methods of measurement of electromagnetic radiation to get a reliable result.[55]

In 1865, John Tyndall described radiation from electrically heated filaments and from carbon arcs as visible and invisible.[56] Tyndall spectrally decomposed the radiation by use of a rock salt prism, which passed heat as well as visible rays, and measured the radiation intensity by means of a thermopile.[57][58]

In 1880, André-Prosper-Paul Crova published a diagram of the three-dimensional appearance of the graph of the strength of thermal radiation as a function of wavelength and temperature.[59] He determined the spectral variable by use of prisms. He analyzed the surface through what he called "isothermal" curves, sections for a single temperature, with a spectral variable on the abscissa and a power variable on the ordinate. He put smooth curves through his experimental data points. They had one peak at a spectral value characteristic for the temperature, and fell either side of it towards the horizontal axis.[60][61] Such spectral sections are widely shown even today.

In a series of papers from 1881 to 1886, Langley reported measurements of the spectrum of heat radiation, using diffraction gratings and prisms, and the most sensitive detectors that he could make. He reported that there was a peak intensity that increased with temperature, that the shape of the spectrum was not symmetrical about the peak, that there was a strong fall-off of intensity when the wavelength was shorter than an approximate cut-off value for each temperature, that the approximate cut-off wavelength decreased with increasing temperature, and that the wavelength of the peak intensity decreased with temperature, so that the intensity increased strongly with temperature for short wavelengths that were longer than the approximate cut-off for the temperature.[62]

Having read Langley, in 1888, Russian physicist V.A. Michelson published a consideration of the idea that the unknown Kirchhoff radiation function could be explained physically and stated mathematically in terms of "complete irregularity of the vibrations of ... atoms".[63][64] At this time, Planck was not studying radiation closely, and believed in neither atoms nor statistical physics.[65] Michelson produced a formula for the spectrum for temperature:

$$I_\lambda = B_1 \theta^{\frac{3}{2}} \exp\left(-\frac{c}{\lambda^2 \theta}\right) \lambda^{-6},$$

where $I\lambda$ denotes specific radiative intensity at wavelength λ and temperature θ, and where B_1 and c are empirical constants.

In 1898, Otto Lummer and Ferdinand Kurlbaum published an account of their cavity radiation source.[66] Their design has been used largely unchanged for radiation measurements to the present day. It was a platinum box, divided by diaphragms, with its interior blackened with iron oxide. It was an important ingredient for the progressively improved measurements that led to the discovery of Planck's law.[67] A version described in 1901 had its interior blackened with a mixture of chromium, nickel, and cobalt oxides.[68]

The importance of the Lummer and Kurlbaum cavity radiation source was that it was an experimentally accessible source of black-body radiation, as distinct from radiation from a simply exposed incandescent solid body, which previously was the nearest available experimental approximation to black-body radiation over a suitable range of temperatures. The simply exposed incandescent solid bodies, that had previously been used, emitted radiation with departures from the black-body spectrum that made it impossible to find the true black-body spectrum from experiments.[69][70]

7.6.3 Planck's views just before the empirical facts led him to find his eventual law

Theoretical and empirical progress enabled Lummer and Pringsheim to write in 1899 that available experimental evidence was approximately consistent with the specific intensity law $C\lambda^{-5}e^{(-c/\lambda T)}$ where C and c denote empirically measurable constants, and where λ and T denote wavelength and temperature respectively.[71][72] For theoretical reasons, Planck at that time accepted this formulation, which has an effective cut-off of short wavelengths.[73][74][75]

7.6.4 Finding the empirical law

Max Planck originally produced his law on 19 October 1900[76][77] as an improvement upon the Wien approximation, published in 1896 by Wilhelm Wien, which fit the experimental data at short wavelengths (high frequencies) but deviated from it at long wavelengths (low frequencies).[36] In June 1900, based on heuristic theoretical considerations, Rayleigh had suggested a formula[78] that he proposed might be checked experimentally. The suggestion was that the Stewart–Kirchhoff universal function might be of the form $c_1 T \lambda^{-4} \exp(-c_2/(\lambda T))$. This was not the celebrated Rayleigh–Jeans formula $8\pi kBT \lambda^{-4}$, which did not emerge until 1905,[33] though it did reduce to the latter for long wavelengths, which are the relevant ones here. According to Klein,[79] one may speculate that it is likely that Planck had seen this suggestion though he did not mention it in his papers of 1900 and 1901. Planck would have been aware of various other proposed formulas which had been offered.[55][80] On 7 October 1900, Rubens told Planck that in the complementary domain (long wavelength, low frequency), and only there, Rayleigh's 1900 formula fitted the observed data well.[80]

For long wavelengths, Rayleigh's 1900 heuristic formula approximately meant that energy was proportional to temperature, $U\lambda = \text{const. } T$.[79][80][81] It is known that $dS/dU\lambda = 1/T$ and this leads to $dS/dU\lambda = \text{const.}/U\lambda$ and thence to $d^2S/dU\lambda^2 = -\text{const. }/U\lambda^2$ for long wavelengths. But for short wavelengths, the Wien formula leads to $1/T = -\text{const. } \ln U\lambda + \text{const.}$ and thence to $d^2S/dU\lambda^2 = -\text{const. }/U\lambda$ for short wavelengths. Planck perhaps patched together these two heuristic formulas, for long and for short wavelengths,[80][82] to produce a formula

$$d^2S/dU_\lambda^2 = \frac{\alpha}{U_\lambda(\beta + U_\lambda)}.\ [76]$$

This led Planck to the formula

$$B_\lambda(T) = \frac{C\lambda^{-5}}{e^{\frac{c}{\lambda T}} - 1},$$

where Planck used the symbols C and c to denote empirical fitting constants.

Planck sent this result to Rubens, who compared it with his and Kurlbaum's observational data and found that it fitted for all wavelengths remarkably well. On 19 October 1900, Rubens and Kurlbaum briefly reported the fit to the data,[83] and Planck added a short presentation to give a theoretical sketch to account for his formula.[76] Within a week, Rubens and Kurlbaum gave a fuller report of their measurements confirming Planck's law. Their technique for spectral resolution of the longer wavelength radiation was called the residual ray method. The rays were repeatedly reflected from polished crystal surfaces, and the rays that made it all the way through the process were 'residual', and were of wavelengths preferentially reflected by crystals of suitably specific materials.[84][85][86]

7.6.5 Trying to find a physical explanation of the law

See also: Planck–Einstein relation

Once Planck had discovered the empirically fitting function, he constructed a physical derivation of this law. His thinking revolved around entropy rather than being directly about temperature. Planck considered a cavity with perfectly reflective walls; the cavity contained finitely many hypothetical well separated and recognizable but identically constituted, of definite magnitude, resonant oscillatory bodies, several such oscillators at each of finitely many characteristic frequencies. The hypothetical oscillators were for Planck purely imaginary theoretical investigative probes, and he said of them that such oscillators do not need to "really exist somewhere in nature, provided their existence and their properties are consistent with the laws of thermodynamics and electrodynamics.".[87] Planck did not attribute any definite physical significance to his hypothesis of resonant oscillators, but rather proposed it as a mathematical device that enabled him to derive a single expression for the black body spectrum that matched the empirical data at all wavelengths.[88] He tentatively mentioned the possible connection of such oscillators with atoms. In a sense, the oscillators corresponded to Planck's speck of carbon; the size of the speck could be small regardless of the size of the cavity, provided the speck effectively transduced energy between radiative wavelength modes.[80]

Partly following a heuristic method of calculation pioneered by Boltzmann for gas molecules, Planck considered the possible ways of distributing electromagnetic energy over the different modes of his hypothetical charged material oscillators. This acceptance of the probabilistic approach, following Boltzmann, for Planck was a radical change from his former position, which till then had deliberately opposed such thinking proposed by Boltzmann.[89] Heuristically, Boltzmann had distributed the energy in arbitrary merely mathematical quanta ϵ, which he had proceeded to make tend to zero in magnitude, because the finite magnitude ϵ had served only to allow definite counting for the sake of mathematical calculation of probabilities, and had no physical significance. Referring to a new universal constant of nature, h,[90] Planck supposed that, in the several oscillators of each of the finitely many characteristic frequencies, the total energy was distributed to each in an integer multiple of a definite physical unit of energy, ϵ, not arbitrary as in Boltzmann's method, but now for Planck, in a new departure, characteristic of the respective characteristic frequency.[77][91][92][93] His new universal constant of nature, h, is now known as Planck's constant.

Planck explained further[77] that the respective definite unit, ϵ, of energy should be proportional to the respective characteristic oscillation frequency ν of the hypothetical oscillator, and in 1901 he expressed this with the constant of proportionality h:[94][95]

$$\epsilon = h\nu.$$

Planck did not propose that light propagating in free space is quantized.[96][97][98] The idea of quantization of the free electromagnetic field was developed later, and eventually incorporated into what we now know as quantum field theory.[99]

In 1906 Planck acknowledged that his imaginary resonators, having linear dynamics, did not provide a physical explanation for energy transduction between frequencies.[100][101] Present-day physics explains the transduction between frequencies in the presence of atoms by their quantum excitability, following Einstein. Planck believed that in a cavity with perfectly reflecting walls and with no matter present, the electromagnetic field cannot exchange energy between frequency components.[102] This is because of the linearity of Maxwell's equations.[103] Present-day quantum field theory predicts that, in the absence of matter, the electromagnetic field obeys nonlinear equations and in that sense does self-interact.[104][105] Such interaction in the absence of matter has not yet been directly measured because it would require very high intensities and very sensitive and low-noise detectors, which are still in the process of being constructed.[104][106] Planck believed that a field with no interactions neither obeys nor violates the classical principle of equipartition of energy,[107][108] and instead remains exactly as it was when introduced, rather than evolving into a black body field.[109] Thus, the linearity of his mechanical assumptions precluded Planck from having a mechanical explanation of the maximization of the entropy of the thermodynamic equilibrium thermal radiation field. This is why he had to resort to Boltzmann's probabilistic arguments.[110][111]

Planck's law may be regarded as fulfilling the prediction of Gustav Kirchhoff that his law of thermal radiation was of the highest importance. In his mature presentation of his own law, Planck offered a thorough and detailed theoretical proof for Kirchhoff's law,[112] theoretical proof of which until then had been sometimes debated, partly because it was said to rely on unphysical theoretical objects, such as Kirchhoff's perfectly absorbing infinitely thin black surface.[113]

7.6.6   Subsequent events

It was not till five years after Planck made his heuristic assumption of abstract elements of energy or of action that Albert Einstein conceived of really existing quanta of light in 1905[114] as a revolutionary explanation of black-body radiation, of photoluminescence, of the photoelectric effect, and of the ionization of gases by ultraviolet light. In 1905, "Einstein believed that Planck's theory could not be made to agree with the idea of light quanta, a mistake he corrected in 1906."[115] Contrary to Planck's beliefs of the time, Einstein proposed a model and formula whereby light was emitted, absorbed, and propagated in free space in energy quanta localized in points of space.[114] As an introduction to his reasoning, Einstein recapitulated Planck's model of hypothetical resonant material electric oscillators as sources and sinks of radiation, but then he offered a new argument, disconnected from that model, but partly based on a thermodynamic argument of Wien, in which Planck's formula $\epsilon = h\nu$ played no role.[116] Einstein gave the energy content of such quanta in the form $R\beta\nu/N$. Thus Einstein was contradicting the undulatory theory of light held by Planck. In 1910, criticizing a manuscript sent to him by Planck, knowing that Planck was a steady supporter of Einstein's theory of special relativity, Einstein wrote to Planck: "To me it seems absurd to have energy continuously distributed in space without assuming an aether."[117]

According to Thomas Kuhn, it was not till 1908 that Planck more or less accepted part of Einstein's arguments for physical as distinct from abstract mathematical discreteness in thermal radiation physics. Still in 1908, considering Einstein's proposal of quantal propagation, Planck opined that such a revolutionary step was perhaps unnecessary.[118] Until then, Planck had been consistent in thinking that discreteness of action quanta was to be found neither in his resonant oscillators nor in the propagation of thermal radiation. Kuhn wrote that, in Planck's earlier papers and in his 1906 monograph,[119] there is no "mention of discontinuity, [nor] of talk of a restriction on oscillator energy, [nor of] any formula like $U = nh\nu$."[120] Kuhn pointed out that his study of Planck's papers of 1900 and 1901, and of his monograph of 1906,[119] had led him to "heretical" conclusions, contrary to the widespread assumptions of others who saw Planck's writing only from the perspective of later, anachronistic, viewpoints.[121][122] Kuhn's conclusions, finding a period till 1908, when Planck consistently held his 'first theory', have been accepted by other historians.[123][124]

In the second edition of his monograph, in 1912, Planck sustained his dissent from Einstein's proposal of light quanta. He proposed in some detail that absorption of light by his virtual material resonators might be continuous, occurring at a constant rate in equilibrium, as distinct from quantal absorption. Only emission was quantal.[103][125] This has at times been called Planck's "second theory".[126]

It was not till 1919 that Planck in the third edition of his monograph more or less accepted his 'third theory', that both emission and absorption of light were quantal.[127]

The colourful term "ultraviolet catastrophe" was given by Paul Ehrenfest in 1911 to the paradoxical result that the total energy in the cavity tends to infinity when the equipartition theorem of classical statistical mechanics is (mistakenly) applied to black body radiation.[128][129] But this had not been part of Planck's thinking, because he had not tried to apply the doctrine of equipartition: when he made his discovery in 1900, he had not noticed any sort of "catastrophe".[73][74][75][79][130] It was first noted by Lord Rayleigh in 1900,[78][131][132] and then in 1901[133] by Sir James Jeans; and later, in 1905, by Einstein when he wanted to support the idea that light propagates as discrete packets, later called 'photons', and by Rayleigh[34] and by Jeans.[33][134][135][136]

In 1913, Bohr gave another formula with a further different physical meaning to the quantity $h\nu$.[29][30][31][137][138][139] In contrast to Planck's and Einstein's formulas, Bohr's formula referred explicitly and categorically to energy levels of atoms. Bohr's formula was $W_{\tau_2} - W_{\tau_1} = h\nu$ where W_{τ_2} and W_{τ_1} denote the energy levels of quantum states of an atom, with quantum numbers τ_2 and τ_1 . The symbol ν denotes the frequency of a quantum of radiation that can be emitted or absorbed as the atom passes between those two quantum states. In contrast to Planck's model, the frequency ν has no immediate relation to frequencies that might describe those quantum states themselves.

Later, in 1924, Satyendra Nath Bose developed the theory of the statistical mechanics of photons, which allowed a theoretical derivation of Planck's law. The actual word 'photon' was invented still later, by G.N. Lewis in 1926,[140] who mistakenly believed that photons were conserved, contrary to Bose–Einstein statistics; nevertheless the word 'photon' was adopted to express the Einstein postulate of the packet nature of light propagation. In an electromagnetic field isolated in a vacuum in a vessel with perfectly reflective walls, such as was considered by Planck, indeed the photons would be conserved according to Einstein's 1905 model, but Lewis was referring to a field of photons considered as a system closed with respect to ponderable matter but open to exchange of electromagnetic energy with a surrounding system of ponderable matter, and he mistakenly imagined that still the photons were conserved, being stored inside atoms.

Ultimately, Planck's law of black-body radiation contributed to Einstein's concept of quanta of light carrying linear momentum,[29][114] which became the fundamental basis for the development of quantum mechanics.

The above-mentioned linearity of Planck's mechanical assumptions, not allowing for energetic interactions between frequency components, was superseded in 1925 by Heisenberg's original quantum mechanics. In his paper submitted on 29 July 1925, Heisenberg's theory accounted for Bohr's above-mentioned formula of 1913. It admitted non-linear oscillators as models of atomic quantum states, allowing energetic interaction between their own multiple internal discrete Fourier frequency components, on the occasions of emission or absorption of quanta of radiation. The frequency of a quantum of radiation was that of a definite coupling between internal atomic meta-stable oscillatory quantum states.[141][142] At that time, Heisenberg knew nothing of matrix algebra, but Max Born read the manuscript of Heisenberg's paper and recognized the matrix character of Heisenberg's theory. Then Born and Jordan published an explicitly matrix theory of quantum mechanics, based on, but in form distinctly different from, Heisenberg's original quantum mechanics; it is the Born and Jordan matrix theory that is today called matrix mechanics.[143][144][145] Heisenberg's explanation of the Planck oscillators, as non-linear effects apparent as Fourier modes of transient processes of emission or absorption of radiation, showed why Planck's oscillators, viewed as enduring physical objects such as might be envisaged by classical physics, did not give an adequate explanation of the phenomena.

Nowadays, as a statement of the energy of a light quantum, often one finds the formula $E = \hbar\omega$, where $\hbar = h/2\pi$, and $\omega = 2\pi\nu$ denotes angular frequency,[146][147][148][149][150] and less often the equivalent formula $E = h\nu$.[149][150][151][152][153] This statement about a really existing and propagating light quantum, based on Einstein's, has a physical meaning different from that of Planck's above statement $\epsilon = h\nu$ about the abstract energy units to be distributed amongst his hypothetical resonant material oscillators.

An article by Helge Kragh published in *Physics World* gives an account of this history.[93]

7.7 See also

- Emissivity

- Radiance

- Sakuma–Hattori equation
- Wien's displacement law

7.8   References

[1]   Planck 1914, pp. 6, 168

[2]   Chandrasekhar 1960, p. 8

[3]   Rybicki & Lightman 1979, p. 22

[4]   Planck 1914, p. 42

[5]   Stewart 1858

[6]   Hapke 1993, pp. 362–373

[7]   Planck 1914

[8]   Loudon 2000, pp. 3–45

[9]   Caniou 1999, p. 117

[10] Kramm & Mölders 2009

[11] Sharkov 2003, p. 210

[12] Goody & Yung 1989, p. 16.

[13] Fischer 2011

[14] Mohr, Peter J.; Taylor, Barry N.; Newell, David B. (2012). "CODATA Recommended Values of the Fundamental Physical Constants: 2010" (PDF). p. 1591.

[15] Loudon 2000

[16] Mandel & Wolf 1995

[17] Wilson 1957, p. 182

[18] Adkins 1983, pp. 147–148

[19] Landsberg 1978, p. 208

[20] Siegel & Howell 2002, p. 25

[21] Planck 1914, pp. 9–11

[22] Planck 1914, p. 35

[23] Landsberg 1961, pp. 273–274

[24] Born & Wolf 1999, pp. 194–199

[25] Born & Wolf 1999, p. 195

[26] Rybicki & Lightman 1979, p. 19

[27] Chandrasekhar 1960, p. 7

[28] Chandrasekhar 1960, p. 9

[29] Einstein 1916

[30] Bohr 1913

[31] Jammer 1989, pp. 113, 115

[32] Kittel & Kroemer 1980, p. 98

[33] Jeans 1905a, p. 98

[34] Rayleigh 1905

[35] Rybicki & Lightman 1979, p. 23

[36] Wien 1896, p. 667

[37] Planck 1906, p. 158

[38] Lowen & Blanch 1940

[39] Siegel 1976

[40] Kirchhoff 1860a

[41] Kirchhoff 1860b

[42] Schirrmacher 2001

[43] Kirchhoff 1860c

[44] Planck 1914, p. 11

[45] Milne 1930, p. 80

[46] Rybicki & Lightman 1979, pp. 16–17

[47] Mihalas & Weibel-Mihalas 1984, p. 328

[48] Goody & Yung 1989, pp. 27–28

[49] Paschen, F. (1896), personal letter cited by Hermann 1971, p. 6

[50] Hermann 1971, p. 7

[51] Kuhn 1978, pp. 8, 29

[52] Mehra and Rechenberg 1982, pp. 26, 28, 31, 39

[53] Kirchhoff 1862/1882, p. 573

[54] Kragh 1999, p. 58

[55] Kangro 1976

[56] Tyndall 1865a

[57] Tyndall 1865b

[58] Kangro 1976, pp. 8–10

[59] Crova 1880

[60] Crova 1880, p. 577, Plate I

[61] Kangro 1976, pp. 10–15

[62] Kangro 1976, pp. 15–26

[63] Michelson 1888

[64] Kangro 1976, pp. 30–36

[65] Kangro 1976, pp. 122–123

[66] Lummer & Kurlbaum 1898

[67] Kangro 1976, p. 159

[68] Lummer & Kurlbaum 1901

[69] Kangro 1976, pp. 75–76

[70] Paschen 1895, pp. 297–301

[71] Lummer & Pringsheim 1899, p. 225

[72] Kangro 1976, p. 174

[73] Planck 1900d

[74] Rayleigh 1900, p. 539

[75] Kangro 1976, pp. 181–183

[76] Planck 1900a

[77] Planck 1900b

[78] Rayleigh 1900

[79] Klein 1962

[80] Dougal 1976

[81] Planck 1943, p. 156

[82] Hettner 1922

[83] Rubens & Kurlbaum 1900a

[84] Rubens & Kurlbaum 1900b

[85] Kangro 1976, p. 165

[86] Mehra & Rechenberg 1982, p. 41

[87] Planck 1914, p. 135

[88] Kuhn 1978, pp. 117–118

[89] Hermann 1971, p. 16

[90] Planck 1900c

[91] Kangro 1976, p. 214

[92] Kuhn 1978, p. 106

[93] Kragh 2000

[94] Planck 1901

[95] Planck 1915, p. 89

[96] Ehrenfest & Kamerlingh Onnes 1914, p. 873

[97] ter Haar 1967, p. 14

[98] Stehle 1994, p. 128

[99] Scully & Zubairy 1997, p. 21.

[100] Planck 1906, p. 220

[101] Kuhn 1978, p. 162

[102] Planck 1914, pp. 44–45, 113–114

[103] Stehle 1994, p. 150

[104] Jauch & Rohrlich 1980, Chapter 13

[105] Karplus & Neuman 1951

[106] Tommasini et al. 2008

[107] Jeffreys 1973, p. 223

[108] Planck 1906, p. 178

[109] Planck 1914, p. 26

[110] Boltzmann 1878

[111] Kuhn 1978, pp. 38–39

[112] Planck 1914, pp. 1–45

[113] Cotton 1899

[114] Einstein 1905

[115] Kragh 1999, p. 67

[116] Stehle 1994, pp. 132–137

[117] Einstein 1993, p. 143, letter of 1910.

[118] Planck 1915, p. 95

[119] Planck 1906

[120] Kuhn 1984, p. 236

[121] Kuhn 1978, pp. 196–202

[122] Kuhn 1984

[123] Darrigol 1992, p. 76

[124] Kragh 1999, pp. 63–66

[125] Planck 1914, p. 161

[126] Kuhn 1978, pp. 235–253

[127] Kuhn 1978, pp. 253–254

[128] Ehrenfest 1911

[129] Kuhn 1978, p. 152

[130] Kuhn 1978, pp. 151–152

[131] Kangro 1976, p. 190

[132] Kuhn 1978, pp. 144–145

[133] See footnote on p. 398 in Jeans 1901.

[134] Jeans 1905b

[135] Jeans 1905c

[136] Jeans 1905d

[137] Sommerfeld 1923, p. 43

[138] Heisenberg 1925, p. 108

[139] Brillouin 1970, p. 31

[140] Lewis 1926

[141] Heisenberg 1925

[142] Razavy 2011, pp. 39–41

[143] Born & Jordan 1925

[144] Stehle 1994, p. 286

[145] Razavy 2011, pp. 42–43

[146] Messiah 1958, p. 14

[147] Pauli 1973, p. 1

[148] Feynman, Leighton & Sands 1963, p. 38-1

[149] Schwinger 2001, p. 203

[150] Bohren & Clothiaux 2006, p. 2

[151] Schiff 1949, p. 2

[152] Mihalas & Weibel-Mihalas 1984, p. 143

[153] Rybicki & Lightman 1979, p. 20

7.8.1   Bibliography

- Adkins, C. J. (1983). *Equilibrium Thermodynamics* (3rd ed.). Cambridge University Press. ISBN 0-521-25445-0.

- Bohr, N. (1913). "On the constitution of atoms and molecules" (PDF). *Philosophical Magazine* **26**: 1–25. doi:.

- Bohren, C. F.; Clothiaux, E. E. (2006). *Fundamentals of Atmospheric Radiation*. Wiley-VCH. ISBN 3-527-40503-8.

- Boltzmann, L. (1878). "Über die Beziehung zwischen dem zweiten Hauptsatze der mechanischen Wärmetheorie und der Wahrscheinlichkeitsrechnung, respective den Sätzen über das Wärmegleichgewicht". *Sitzungsberichte Mathematisch-Naturwissenschaftlichen Classe der kaiserlichen Akademie der Wissenschaften in Wien* **76** (2): 373–435.

- Born, M.; Wolf, E. (1999). *Principles of Optics* (7th ed.). Cambridge University Press. ISBN 0-521-64222-1.

- Born, M.; Jordan, P. (1925). "Zur Quantenmechanik". *Zeitschrif t für Physik* **34**: 858–888. Bibcode:1925ZPhy...34doi:10.1007/BF01328531. Translated in part as "On quantum mechanics" in van der Waerden, B.L. (1967). *Sources of Quantum Mechanics*. North-Holland Publishing. pp. 277–306.

- Brehm, J. J.; Mullin, W. J. (1989). *Introduction to the Structure of Matter*. Wiley. ISBN 0-471-60531-X.

- Brillouin, L. (1970). *Relativity Reexamined*. Academic Press. ISBN 978-0-12-134945-5.

- Caniou, J. (1999). *Passive Infrared Detection: Theory and Applications*. Springer. ISBN 978-0-7923-8532-5.

- Chandrasekhar, S. (1960) [1950]. *Radiative Transfer* (Revised reprint ed.). Dover Publications. ISBN 978-0-486-60590-6.

- Cotton, A. (1899). "The present status of Kirchhoff's law". *The Astrophysical Journal* **9**: 237–268. Bibcode: doi:10.1086/140585.

- Crova, A. P. P. (1880). "Étude des radiations émises par les corps incandescents. Mesure optique des hautes températures". *Annales de chimie et de physique*. Série 5 **19**: 472–550.

- Dougal, R. C. (September 1976). "The presentation of the Planck radiation formula (tutorial)". *Physics Education* **11** (6): 438–443. Bibcode:1976PhyEd..11..438D. doi:10.1088/0031-9120/11/6/008.

- Ehrenfest, P. (1911). "Welche Züge der Lichtquantenhypothese spielen in der Theorie der Wärmestrahlung eine wesentliche Rolle?". *Annalen der Physik* **36**: 91–118. Bibcode:1911AnP...341...91E. doi:10.1002/andp. 19113411

- Ehrenfest, P.; Kamerlingh Onnes, H. (1914). "Simplified deduction of the formula from the theory of combinations which Planck uses as the basis of his radiation theory". *Proceedings of the Royal Dutch Academy of Sciences in Amsterdam* **17** (2): 870–873. Bibcode:1914KNAB...17..870E.

- Einstein, A. (1905). "Über einen die Erzeugung und Verwandlung des Lichtes betreffenden heuristischen Gesichtspunkt". *Annalen der Physik* **17** (6): 132–148. Bibcode:1905AnP...322..132E. doi:10.1002/andp.19053220607. Translated in Arons, A. B.; Peppard, M. B. (1965). "Einstein's proposal of the photon concept: A translation of the *Annalen der Physik* paper of 1905" (PDF). *American Journal of Physics* **33** (5): 367. Bibcode:1965AmJPh..33..367A. doi:10.1119/1.1971542.

- Einstein, A. (1916). "Zur Quantentheorie der Strahlung". *Mitteilungen der Physikalischen Gesellschaft Zürich* **18**: 47–62. and a nearly identical version Einstein, A. (1917). "Zur Quantentheorie der Strahlung". *Physikalische Zeitschrift* **18**: 121–128. Bibcode:1917PhyZ...18..121E. Translated in ter Haar, D. (1967). *The Old Quantum Theory*. Pergamon Press. pp. 167–183. LCCN 66029628. See also .

- Einstein, A. (1993). *The Collected Papers of Albert Einstein* 3. English translation by Beck, A. Princeton University Press. ISBN 0-691-10250-3.

- Feynman, R. P.; Leighton, R. B.; Sands, M. (1963). *The Feynman Lectures on Physics, Volume 1*. Addison-Wesley. ISBN 0-201-02010-6.

- Fischer, T. (1 November 2011). "Topics: Derivation of Planck's Law". ThermalHUB. Retrieved 2015-06-19.

- Goody, R. M.; Yung, Y. L. (1989). *Atmospheric Radiation: Theoretical Basis* (2nd ed.). Oxford University Press. ISBN 978-0-19-510291-8.

- Guggenheim, E.A. (1967). *Thermodynamics. An Advanced Treatment for Chemists and Physicists* (fifth revised ed.). North-Holland Publishing Company.

- Haken, H. (1981). *Light* (Reprint ed.). Amsterdam: North-Holland Publishing. ISBN 0-444-86020-7.

- Hapke, B. (1993). *Theory of Reflectance and Emittance Spectroscopy*. Cambridge University Press, Cambridge UK. ISBN 0-521-30789-9.

- Heisenberg, W. (1925). "Über quantentheoretische Umdeutung kinematischer und mechanischer Beziehungen". *Zeitschrift für Physik* **33**: 879–893. Bibcode:1925ZPhy...33..879H. doi:10.1007/BF01328377. Translated as "Quantum-theoretical Re-interpretation of kinematic and mechanical relations" in van der Waerden, B.L. (1967). *Sources of Quantum Mechanics*. North-Holland Publishing. pp. 261–276.

- Heisenberg, W. (1930). *The Physical Principles of the Quantum Theory*. Eckart, C.; Hoyt, F. C. (transl.). University of Chicago Press.

- Hermann, A. (1971). *The Genesis of Quantum Theory*. Nash, C.W. (transl.). MIT Press. ISBN 0-262-08047-8. a translation of *Frühgeschichte der Quantentheorie (1899–1913)*, Physik Verlag, Mosbach/Baden.

- Hettner, G. (1922). "Die Bedeutung von Rubens Arbeiten für die Plancksche Strahlungsformel". *Naturwissenschaften* **10**: 1033–1038. Bibcode:1922NW.....10.1033H. doi:10.1007/BF01565205.

- Jammer, M. (1989). *The Conceptual Development of Quantum Mechanics* (second ed.). Tomash Publishers/American Institute of Physics. ISBN 0-88318-617-9.

- Jauch, J. M.; Rohrlich, F. (1980) [1955]. *The Theory of Photons and Electrons. The Relativistic Quantum Field Theory of Charged Particles with Spin One-half* (second printing of second ed.). Springer. ISBN 0-387-07295-0.

- Jeans, J. H. (1901). "The Distribution of Molecular Energy". *Philosophical Transactions of the Royal Society A* **196** (274–286): 397. Bibcode:1901RSPTA.196..397J. doi:10.1098/rsta.1901.0008. JSTOR 90811.

- Jeans, J. H. (1905a). "XI. On the partition of energy between matter and æther". *Philosophical Magazine* **10** (55): 91. doi:10.1080/14786440509463348.

- Jeans, J. H. (1905b). "On the Application of Statistical Mechanics to the General Dynamics of Matter and Ether". *Proceedings of the Royal Society A* **76** (510): 296. Bibcode:1905RSPSA..76..296J. doi:10.1098/rspa.1905.0029. JSTOR 92714.

- Jeans, J. H. (1905c). "A Comparison between Two Theories of Radiation". *Nature* **72** (1865): 293. Bibcode: doi:10.1038/072293d0.

- Jeans, J. H. (1905d). "On the Laws of Radiation". *Proceedings of the Royal Society A* **76** (513): 545 doi:10.1098/rspa.1905.0060. JSTOR 92704.

- Jeffreys, H. (1973). *Scientific Inference* (3rd ed.). Cambridge University Press. ISBN 978-0-521-08446-8.

- Kangro, H. (1976). *Early History of Planck's Radiation Law*. Taylor & Francis. ISBN 0-85066-063-7.

- Karplus, R.; Neuman, M. (1951). "The Scattering of Light by Light". *Physical Review* **83** (4): 776–784 doi:10.1103/PhysRev.83.776.

- Kirchhoff, G. R.; [27 October 1859] (1860a). "Über die Fraunhofer'schen Linien". *Monatsberichte der Königlich Preussischen Akademie der Wissenschaften zu Berlin*: 662–665.

- Kirchhoff, G. R.; [11 December 1859] (1860b). "Über den Zusammenhang zwischen Emission und Absorption von Licht und Wärme". *Monatsberichte der Königlich Preussischen Akademie der Wissenschaften zu Berlin*: 783–787.

- Kirchhoff, G. R. (1860c). "Über das Verhältniss zwischen dem Emissionsvermögen und dem Absorptionsvermögen der Körper für Wärme and Licht". *Annalen der Physik und Chemie* **109**: 275–301. Bibcode:1860AnP...185..275K. doi:10.1002/andp.18601850205. Translated by Guthrie, F. as Kirchhoff, G. R. (1860). "On the relation between the radiating and absorbing powers of different bodies for light and heat". *Philosophical Magazine*. Series 4 **20**: 1–21.

- Kirchhoff, G. R. (1882) [1862], "Über das Verhältniss zwischen dem Emissionsvermögen und dem Absorptionsver-mögen der Körper für Wärme und Licht", *Gessamelte Abhandlungen*, Johann Ambrosius Barth, pp. 571–598

- Kittel, C.; Kroemer, H. (1980). *Thermal Physics* (2nd ed.). W. H. Freeman. ISBN 0-7167-1088-9.

- Klein, M.J. (1962). "Max Planck and the beginnings of the quantum theory". *Archive for History of Exact Sciences* **1** (5): 459–479. doi:10.1007/BF00327765.

- Kragh, H. (1999). *Quantum Generations. A History of Physics in the Twentieth Century*. Princeton University Press. ISBN 0-691-01206-7.

- Kragh, H. (December 2000). "Max Planck: The reluctant revolutionary". *Physics World*.

- Kramm, Gerhard; Mölders, N. (2009). "Planck's Blackbody Radiation Law: Presentation in Different Domains and Determination of the Related Dimensional Constant". *Journal of the Calcutta Mathematical Society* **5** (1–2): 27–61. arXiv:0901.1863. Bibcode:2009arXiv0901.1863K.

- Kuhn, T. S. (1978). *Black–Body Theory and the Quantum Discontinuity*. Oxford University Press. ISBN 0-19-502383-8.

- Landsberg, P.T. (1961). *Thermodynamics with Quantum Statistical Illustrations*. Interscience Publishers.

- Landsberg, P.T. (1978). *Thermodynamics and Statistical Mechanics*. Oxford University Press. ISBN 0-19-851142-6.

- Lewis, G. N. (1926). "The Conservation of Photons". *Nature* **118** (2981): 874. Bibcode:1926Natur.118..874L. doi:10.1038/118874a0.

- Loudon, R. (2000). *The Quantum Theory of Light* (3rd ed.). Oxford University Press. ISBN 0-19-850177-3.

- Lowen, A. N.; Blanch, G. (1940). "Tables of Planck's radiation and photon functions". *Journal of the Optical Society of America* **30** (2): 70. doi:10.1364/JOSA.30.000070.

- Lummer, O.; Kurlbaum, F. (1898). "Der electrisch geglühte "absolut schwarze" Körper und seine Temperaturmessung". *Verhandlungen der Deutschen Physikalischen Gesellschaft* **17**: 106–111.

- Lummer, O.; Pringsheim, E. (1899). "1. Die Vertheilung der Energie in Spectrum des schwarzen Körpers und des blanken Platins; 2. Temperaturbestimmung fester glühender Körper". *Verhandlungen der Deutschen Physikalischen Gesellschaft* **1**: 215–235.

- Lummer, O.; Kurlbaum, F. (1901). "Der elektrisch geglühte "schwarze" Körper". *Annalen der Physik* **310** (8): 829–836. Bibcode:1901AnP...310..829L. doi:10.1002/andp.19013100809.

- Mandel, L.; Wolf, E. (1995). *Optical Coherence and Quantum Optics*. Cambridge University Press. ISBN 0-521-41711-2.

- Mehra, J.; Rechenberg, H. (1982). *The Historical Development of Quantum Theory* **1**. Springer-Verlag. ISBN 0-387-90642-8.

- Messiah, A. (1958). *Quantum Mechanics*. Temmer, G. G. (transl.). Wiley.

- Michelson, V. A. (1888). "Theoretical essay on the distribution of energy in the spectra of solids". *Philosophical Magazine*. Series 5 **25**: 425–435. doi:10.1080/14786448808628207.

- Mihalas, D.; Weibel-Mihalas, B. (1984). *Foundations of Radiation Hydrodynamics*. Oxford University Press. ISBN 0-19-503437-6.

- Milne, E.A. (1930). "Thermodynamics of the Stars". *Handbuch der Astrophysik* **3** (1): 63–255.

- Paltridge, G. W.; Platt, C. M. R. (1976). *Radiative Processes in Meteorology and Climatology*. Elsevier. ISBN 0-444-41444-4.

- Paschen, F. (1895). "Über Gesetzmäßigkeiten in den Spectren fester Körper und über ein neue Bestimmung der Sonnentemperatur". *Nachrichten von der Königlichen Gesellschaft der Wissenschaften zu Göttingen (Mathematisch-Physikalische Klasse)*: 294–304.

- Pauli, W. (1973). Enz, C. P., ed. *Wave Mechanics*. Margulies, S.; Lewis, H. R. (transl.). MIT Press. ISBN 0-262-16050-1.

- Planck, M. (1900a). "Über eine Verbesserung der Wienschen Spektralgleichung". *Verhandlungen der Deutschen Physikalischen Gesellschaft* **2**: 202–204. Translated in ter Haar, D. (1967). "On an Improvement of Wien's Equation for the Spectrum". *The Old Quantum Theory* (PDF). Pergamon Press. pp. 79–81. LCCN 66029628.

- Planck, M. (1900b). "Zur Theorie des Gesetzes der Energieverteilung im Normalspektrum". *Verhandlungen der Deutschen Physikalischen Gesellschaft* **2**: 237. Translated in ter Haar, D. (1967). "The Old Quantum Theory" (PDF). Pergamon Press: 82. LCCN 66029628. |chapter= ignored (help)

- Planck, M. (1900c). "Entropie und Temperatur strahlender Wärme". *Annalen der Physik* **306** (4): 719–737. Bibcode:1900AnP...306..719P. doi:10.1002/andp.19003060410.

- Planck, M. (1900d). "Über irreversible Strahlungsvorgänge". *Annalen der Physik* **306** (1): 69–122. Bibcode: doi:10.1002/andp.19003060105.

- Planck, M. (1901). "Über das Gesetz der Energieverteilung im Normalspektrum". *Annalen der Physik* 4: 553. Bibcode:1901AnP...309..553P. doi:10.1002/andp.19013090310. Translated in Ando, K. "On the Law of Distribution of Energy in the Normal Spectrum" (PDF). Retrieved 2011-10-13.

- Planck, M. (1906). *Vorlesungen über die Theorie der Wärmestrahlung*. Johann Ambrosius Barth. LCCN 07004527.

- Planck, M. (1914). *The Theory of Heat Radiation*. Masius, M. (transl.) (2nd ed.). P. Blakiston's Son & Co. OL 7154661M.

- Planck, M. (1915). *Eight Lectures on Theoretical Physics*. Wills, A. P. (transl.). Dover Publications. ISBN 0-486-69730-4.

- Planck, M. (1943). "Zur Geschichte der Auffindung des physikalischen Wirkungsquantums". *Naturwissenschaften* 31 (14–15): 153–159. Bibcode:1943NW.....31..153P. doi:10.1007/BF01475738.

- Rayleigh, Lord (1900). "LIII. Remarks upon the law of complete radiation". *Philosophical Magazine*. Series 5 49 (301): 539. doi:10.1080/14786440009463878.

- Rayleigh, Lord (1905). "The Dynamical Theory of Gases and of Radiation". *Nature* 72 (1855): 54–55. doi:10.1038/072054c0.

- Razavy, M. (2011). *Heisenberg's Quantum Mechanics*. World Scientific. ISBN 978-981-4304-10-8.

- Rubens, H.; Kurlbaum, F. (1900a). "Über die Emission langer Wellen durch den schwarzen Körper". *Verhandlungen der Deutschen Physikalischen Gesellschaft* 2: 181.

- Rubens, H.; Kurlbaum, F. (1900b). "Über di e Emission langwelliger Wärmestrahlen durc h den schwarzen Kör-per be i verschiedenen Temperaturen". *Sitzungsberichte der Königlich Preussischen Akademie der Wissenschaften zuBerlin*: 929–941. Translated in Rubens, H.; Kurlbaum, F. (1901). "On the heat-radiation of long wave-length emit-ted b y blac k bodies at diff erent temperatures". *The Astrophysical Journal* 14: 335–348. Bibcode:1901ApJ....14..335R.doi:10.1086/140874.

- Rybicki, G. B.; Lightman, A. P. (1979). *Radiative Processes in Astrophysics*. John Wiley & Sons. ISBN 0-471-82759-2.

- Sharkov, E. A. (2003). "Black-body radiation". *Passive Microwave Remote Sensing of the Earth* (PDF). Springer. ISBN 978-3-540-43946-2.

- Schiff, L. I. (1949). *Quantum Mechanics*. McGraw-Hill.

- Schirrmacher, A. (2001). *Experimenting theory: the proofs of Kirchhoff's radiation law before and after Planck*. Münchner Zentrum für Wissenschafts und Technikgeschichte.

- Schwinger, J. (2001). Englert, B.-G., ed. *Quantum Mechanics: Symbolism of Atomic Measurements*. Springer. ISBN 3-540-41408-8.

- Scully, M. O.; Zubairy, M.S. (1997). *Quantum Optics*. Cambridge University Press. ISBN 0-521-43458-0.

- Siegel, D.M. (1976). "Balfour Stewart and Gustav Robert Kirchhoff: two independent approaches to "Kirchhoff's radiation law"". *Isis* 67: 565–600. doi:10.1086/351669.

- Siegel, R.; Howell, J. R. (2002). *Thermal Radiation Heat Transfer, Volume 1* (4th ed.). Taylor & Francis. ISBN 978-1-56032-839-1.

- Sommerfeld, A. (1923). *Atomic Structure and Spectral Lines*. Brose, H. L. (transl.) (from 3rd German ed.). Methuen.

- Stehle, P. (1994). *Order, Chaos, Order. The Transition from Classical to Quantum Physics*. Oxford University Press. ISBN 0-19-507513-7.

- Stewart, B. (1858). "An account of some experiments on radiant heat". *Transactions of the Royal Society of Edinburgh* 22: 1–20.

- ter Haar, D. (1967). *The Old Quantum Theory*. Pergamon Press. LCCN 66-029628.

- Thornton, S. T.; Rex, A. F. (2002). *Modern Physics*. Thomson Learning. ISBN 0-03-006049-4.

- Tisza, L. (1966). *Generalized Thermodynamics*. MIT Press.

- Tommasini, D.; Ferrando, F.; Michinel, H.; Seco, M. (2008). "Detecting photon-photon scattering in vacuum at exawatt lasers". *Physical Review A* 77: 042101. arXiv:quant-ph/0703076. Bibcode:2008PhRvA..77a2101M. doi:10.1103/PhysRevA.77.012101.

- Tyndall, J. (1865a). "Über leuchtende und dunkle Strahlung". *Annalen der Physik und Chemie* **200**: 36–53. Bibcode:1865AnP...200...36T. doi:10.1002/andp.18652000103.

- Tyndall, J. (1865b). *Heat considered as a Mode of Motion* (PDF). D. Appleton & Company.

- Wien, W. (1896). "Über die Energievertheilung im Emissionsspectrum eines schwarzen Körpers". *Annalen der Physik und Chemie* **294**: 662–669. Bibcode:1896AnP...294..662W. doi:10.1002/andp.18962940803.

- Wilson, A. H. (1957). *Thermodynamics and Statistical Mechanics*. Cambridge University Press.

7.9   External links

- Summary of Radiation

- Radiation of a Blackbody – interactive simulation to play with Planck's law

- Scienceworld entry on Planck's Law

Chapter 8

Photoelectric effect

The **photoelectric effect** is the observation that many metals emit electrons when light shines upon them. Electrons emitted in this manner can be called *photoelectrons*. The phenomenon is commonly studied in electronic physics, as well as in fields of chemistry, such as quantum chemistry or electrochemistry.

According to classical electromagnetic theory, this effect can be attributed to the transfer of energy from the light to an electron in the metal. From this perspective, an alteration in either the intensity or wavelength of light would induce changes in the rate of emission of electrons from the metal. Furthermore, according to this theory, a sufficiently dim light would be expected to show a time lag between the initial shining of its light and the subsequent emission of an electron. However, the experimental results did not correlate with either of the two predictions made by classical theory.

Instead, electrons are only dislodged by the impingement of photons when those photons reach or exceed a threshold frequency. Below that threshold, no electrons are emitted from the metal regardless of the light intensity or the length of time of exposure to the light. To make sense of the fact that light can eject electrons even if its intensity is low, Albert Einstein proposed that a beam of light is not a wave propagating through space, but rather a collection of discrete wave packets (photons), each with energy hf. This shed light on Max Planck's previous discovery of the Planck relation ($E = hf$) linking energy (E) and frequency (f) as arising from quantization of energy. The factor h is known as the Planck constant.[1][2]

In 1887, Heinrich Hertz[2][3] discovered that electrodes illuminated with ultraviolet light create electric sparks more easily. In 1905 Albert Einstein published a paper that explained experimental data from the photoelectric effect as the result of light energy being carried in discrete quantized packets. This discovery led to the quantum revolution. In 1914, Robert Millikan's experiment confirmed Einstein's law on photoelectric effect. Einstein was awarded the Nobel Prize in 1921 for "his discovery of the law of the photoelectric effect",[4] and Millikan was awarded the Nobel Prize in 1923 for "his work on the elementary charge of electricity and on the photoelectric effect".[5]

The photoelectric effect requires photons with energies from a few electronvolts to over 1 MeV in elements with a high atomic number. Study of the photoelectric effect led to important steps in understanding the quantum nature of light and electrons and influenced the formation of the concept of wave–particle duality.[1] Other phenomena where light affects the movement of electric charges include the photoconductive effect (also known as photoconductivity or photoresistivity), the photovoltaic effect, and the photoelectrochemical effect.

8.1   Emission mechanism

The photons of a light beam have a characteristic energy proportional to the frequency of the light. In the photoemission process, if an electron within some material absorbs the energy of one photon and acquires more energy than the work function (the electron binding energy) of the material, it is ejected. If the photon energy is too low, the electron is unable to escape the material. Since an increase in the intensity of low-frequency light will only increase the number of low-energy photons sent over a given interval of time, this change in intensity will not create any single photon with enough energy to dislodge an electron. Thus, the energy of the emitted electrons does not depend on the intensity of the incoming light, but only on the energy (equivalently frequency) of the individual photons. It is an interaction between the incident photon and the outermost electrons.

Electrons can absorb energy from photons when irradiated, but they usually follow an "all or nothing" principle. All of the energy from one photon must be absorbed and used to liberate one electron from atomic binding, or else the energy is re-emitted. If the photon energy is absorbed, some of the energy liberates the electron from the atom, and the rest contributes to the electron's kinetic energy as a free particle.[6][7][8]

8.1.1   Experimental observations of photoelectric emission

The theory of the photoelectric effect must explain the experimental observations of the emission of electrons from an illuminated metal surface.

For a given metal, there exists a certain minimum frequency of incident radiation below which no photoelectrons are emitted. This frequency is called the *threshold frequency*. Increasing the frequency of the incident beam, keeping the number of incident photons fixed (this would result in a proportionate increase in energy) increases the maximum kinetic energy of the photoelectrons emitted. Thus the stopping voltage increases. The number of electrons also changes because the probability that each photon results in an emitted electron is a function of photon energy. If the intensity of the incident radiation of a given frequency is increased, there is no effect on the kinetic energy of each photoelectron.

Above the threshold frequency, the maximum kinetic energy of the emitted photoelectron depends on the frequency of the incident light, but is independent of the intensity of the incident light so long as the latter is not too high.[9]

For a given metal and frequency of incident radiation, the rate at which photoelectrons are ejected is directly proportional to the intensity of the incident light. An increase in the intensity of the incident beam (keeping the frequency fixed) increases the magnitude of the photoelectric current, although the stopping voltage remains the same.

The time lag between the incidence of radiation and the emission of a photoelectron is very small, less than 10^{-9} second.

The direction of distribution of emitted electrons peaks in the direction of polarization (the direction of the electric field) of the incident light, if it is linearly polarized.[10]

8.1.2   Mathematical description

The maximum kinetic energy K_{max} of an ejected electron is given by

$$K_{\mathrm{max}} = h f - \varphi,$$

where h is the Planck constant and f is the frequency of the incident photon. The term φ is the work function (sometimes denoted W, or ϕ [11]), which gives the minimum energy required to remove a delocalised electron from the surface of the metal. The work function satisfies

$$\varphi = h f_0,$$

where f_0 is the threshold frequency for the metal. The maximum kinetic energy of an ejected electron is then

$$K_{\mathrm{max}} = h \left(f - f_0 \right).$$

Kinetic energy is positive, so we must have $f > f_0$ for the photoelectric effect to occur.[12]

8.1.3   Stopping potential

The relation between current and applied voltage illustrates the nature of the photoelectric effect. For discussion, a light source illuminates a plate P, and another plate electrode Q collects any emitted electrons. We vary the potential between P and Q and measure the current flowing in the external circuit between the two plates.

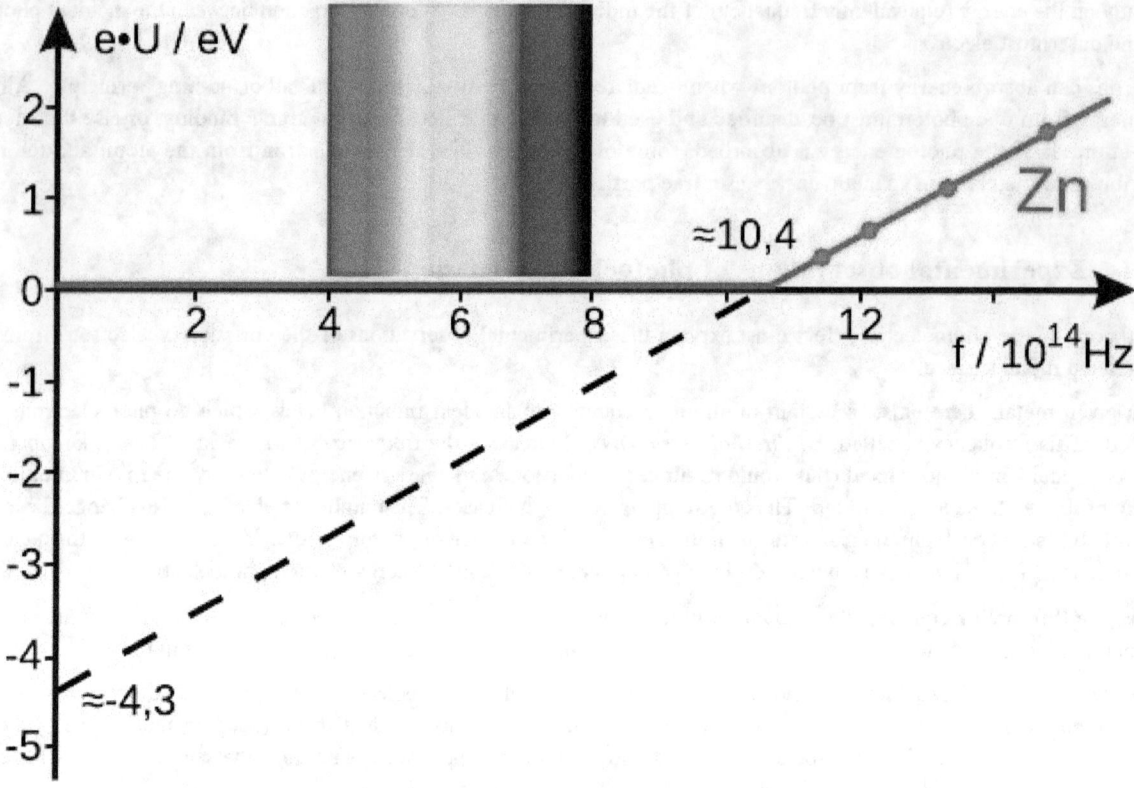

Diagram of the maximum kinetic energy as a function of the frequency of light on zinc

$$E = \frac{hc}{\lambda}$$

Work function = E,

Cut-off wavelength = λ

Work function and cut off frequency

If the frequency and the intensity of the incident radiation are fixed, the photoelectric current increases gradually with an increase in the positive potential on the collector electrode until all the photoelectrons emitted are collected. The photoelectric current attains a saturation value and does not increase further for any increase in the positive potential. The saturation current increases with the increase of the light intensity. It also increases with greater frequencies due to a greater probability of electron emission when collisions happen with higher energy photons.

If we apply a negative potential to the collector plate Q with respect to the plate P and gradually increase it, the photoelectric current decreases, becoming zero at a certain negative potential. The negative potential on the collector at which the photoelectric current becomes zero is called the *stopping potential* or *cut off* potential[13]

i. For a given frequency of incident radiation, the stopping potential is independent of its intensity.

ii. For a given frequency of incident radiation, the stopping potential is determined by the maximum kinetic energy K_{max} of the photoelectrons that are emitted. If qe is the charge on the electron and V_0 is the stopping potential, then the work done by the retarding potential in stopping the electron is $q_e V_0$, so we have

$$q_e V_0 = K_{max}.$$

Recalling

$$K_{max} = h\left(f - f_0\right),$$

we see that the stopping voltage varies linearly with frequency of light, but depends on the type of material. For any particular material, there is a threshold frequency that must be exceeded, independent of light intensity, to observe any electron emission.

8.1.4 Three-step model

In the X-ray regime, the photoelectric effect in crystalline material is often decomposed into three steps:[14]:50-51

1. Inner photoelectric effect (see photodiode below). The hole left behind can give rise to Auger effect, which is visible even when the electron does not leave the material. In molecular solids phonons are excited in this step and may be visible as lines in the final electron energy. The inner photoeffect has to be dipole allowed. The transition rules for atoms translate via the tight-binding model onto the crystal. They are similar in geometry to plasma oscillations in that they have to be transversal.

2. Ballistic transport of half of the electrons to the surface. Some electrons are scattered.

3. Electrons escape from the material at the surface.

In the three-step model, an electron can take multiple paths through these three steps. All paths can interfere in the sense of the path integral formulation. For surface states and molecules the three-step model does still make some sense as even most atoms have multiple electrons which can scatter the one electron leaving.

8.2 History

When a surface is exposed to electromagnetic radiation above a certain threshold frequency (typically visible light for alkali metals, near ultraviolet for other metals, and extreme ultraviolet for non-metals), the radiation is absorbed and electrons are emitted. Light, and especially ultra-violet light, discharges negatively electrified bodies with the production of rays of the same nature as cathode rays.[15] Under certain circumstances it can directly ionize gases.[15] The first of these phenomena was discovered by Hertz and Hallwachs in 1887.[15] The second was announced first by Philipp Lenard in 1900.[15]

The ultra-violet light to produce these effects may be obtained from an arc lamp, or by burning magnesium, or by sparking with an induction coil between zinc or cadmium terminals, the light from which is very rich in ultra-violet rays. Sunlight is not rich in ultra-violet rays, as these have been absorbed by the atmosphere, and it does not produce nearly so large an effect as the arc-light. Many substances besides metals discharge negative electricity under the action of ultraviolet light: lists of these substances will be found in papers by G. C. Schmidt[16] and O. Knoblauch.[17]

8.2.1   19th century

In 1839, Alexandre Edmond Becquerel discovered the photovoltaic effect while studying the effect of light on electrolytic cells.[18] Though not equivalent to the photoelectric effect, his work on photovoltaics was instrumental in showing a strong relationship between light and electronic properties of materials. In 1873, Willoughby Smith discovered photoconductivity in selenium while testing the metal for its high resistance properties in conjunction with his work involving submarine telegraph cables.[19]

Johann Elster (1854–1920) and Hans Geitel (1855–1923), students in Heidelberg, developed the first practical photo-electric cells that could be used to measure the intensity of light.[20][21]:458 Elster and Geitel had investigated with great success the effects produced by light on electrified bodies.[22]

In 1887, Heinrich Hertz observed the photoelectric effect and the production and reception of electromagnetic waves.[15] He published these observations in the journal Annalen der Physik. His receiver consisted of a coil with a spark gap, where a spark would be seen upon detection of electromagnetic waves. He placed the apparatus in a darkened box to see the spark better. However, he noticed that the maximum spark length was reduced when in the box. A glass panel placed between the source of electromagnetic waves and the receiver absorbed ultraviolet radiation that assisted the electrons in jumping across the gap. When removed, the spark length would increase. He observed no decrease in spark length when he replaced glass with quartz, as quartz does not absorb UV radiation. Hertz concluded his months of investigation and reported the results obtained. He did not further pursue investigation of this effect.

The discovery by Hertz[23] in 1887 that the incidence of ultra-violet light on a spark gap facilitated the passage of the spark, led immediately to a series of investigations by Hallwachs,[24] Hoor,[25] Righi[26] and Stoletow.[27][28][29][30][31][32][33] on the effect of light, and especially of ultra-violet light, on charged bodies. It was proved by these investigations that a newly cleaned surface of zinc, if charged with negative electricity, rapidly loses this charge however small it may be when ultra-violet light falls upon the surface; while if the surface is uncharged to begin with, it acquires a positive charge when exposed to the light, the negative electrification going out into the gas by which the metal is surrounded; this positive electrification can be much increased by directing a strong airblast against the surface. If however the zinc surface is positively electrified it suffers no loss of charge when exposed to the light: this result has been questioned, but a very careful examination of the phenomenon by Elster and Geitel[34] has shown that the loss observed under certain circumstances is due to the discharge by the light reflected from the zinc surface of negative electrification on neighbouring conductors induced by the positive charge, the negative electricity under the influence of the electric field moving up to the positively electrified surface.[35]

With regard to the *Hertz effect*, the researches from the start showed a great complexity of the phenomenon of photoelectric fatigue — that is, the progressive diminution of the effect observed upon fresh metallic surfaces. According to an important research by Wilhelm Hallwachs, ozone played an important part in the phenomenon.[36] However, other elements enter such as oxidation, the humidity, the mode of polish of the surface, etc. It was at the time not even sure that the fatigue is absent in a vacuum.

In the period from February 1888 and until 1891, a detailed analysis of photoeffect was performed by Aleksandr Stoletov with results published in 6 works; four of them in *Comptes Rendus*, one review in *Physikalische Revue* (translated from Russian), and the last work in *Journal de Physique*. First, in these works Stoletov invented a new experimental setup which was more suitable for a quantitative analysis of photoeffect. Using this setup, he discovered the direct proportionality between the intensity of light and the induced photo electric current (the first law of photoeffect or Stoletov's law). One of his other findings resulted from measurements of the dependence of the intensity of the electric photo current on the gas pressure, where he found the existence of an optimal gas pressure P_m corresponding to a maximum photocurrent; this property was used for a creation of solar cells.

In 1899, J. J. Thomson investigated ultraviolet light in Crookes tubes.[37] Thomson deduced that the ejected particles were the same as those previously found in the cathode ray, later called electrons, which he called "corpuscles". In the research, Thomson enclosed a metal plate (a cathode) in a vacuum tube, and exposed it to high frequency radiation.[38] It was thought that the oscillating electromagnetic fields caused the atoms' field to resonate and, after reaching a certain amplitude, caused a subatomic "corpuscle" to be emitted, and current to be detected. The amount of this current varied with the intensity and colour of the radiation. Larger radiation intensity or frequency would produce more current.

Heinrich Rudolf Hertz

8.2.2   20th century

The discovery of the ionization of gases by ultra-violet light was made by Philipp Lenard in 1900. As the effect was produced across several centimeters of air and made very great positive and small negative ions, it was natural to interpret the phenomenon, as did J. J. Thomson, as a *Hertz effect* upon the solid or liquid particles present in the gas.[15]

In 1902, Lenard observed that the energy of individual emitted electrons increased with the frequency (which is related to the color) of the light.[6]

German physicist Philipp Lenard

This appeared to be at odds with Maxwell's wave theory of light, which predicted that the electron energy would be proportional to the intensity of the radiation.

Lenard observed the variation in electron energy with light frequency using a powerful electric arc lamp which enabled him to investigate large changes in intensity, and that had sufficient power to enable him to investigate the variation of potential with light frequency. His experiment directly measured potentials, not electron kinetic energy: he found the electron energy by relating it to the maximum stopping potential (voltage) in a phototube. He found that the calculated maximum electron kinetic energy is determined by the frequency of the light. For example, an increase in frequency results in an increase in the maximum kinetic energy calculated for an electron upon liberation – ultraviolet radiation would require a higher applied stopping potential to stop current in a phototube than blue light. However Lenard's results were qualitative rather than quantitative because of the difficulty in performing the experiments: the experiments needed to be done on freshly cut metal so that the pure metal was observed, but it oxidised in a matter of minutes even in the partial vacuums he used. The current emitted by the surface was determined by the light's intensity, or brightness: doubling the intensity of the light doubled the number of electrons emitted from the surface.

The researches of Langevin and those of Eugene Bloch[39] have shown that the greater part of the Lenard effect is certainly due to this 'Hertz effect'. The Lenard effect upon the gas itself nevertheless does exist. Refound by J. J. Thomson[40] and then more decisively by Frederic Palmer, Jr.,[41][42] it was studied and showed very different characteristics than those at first attributed to it by Lenard.[15]

In 1905, Albert Einstein solved this apparent paradox by describing light as composed of discrete quanta, now called photons, rather than continuous waves. Based upon Max Planck's theory of black-body radiation, Einstein theorized that the energy in each quantum of light was equal to the frequency multiplied by a constant, later called Planck's constant. A photon above a threshold frequency has the required energy to eject a single electron, creating the observed effect. This discovery led to the quantum revolution in physics and earned Einstein the Nobel Prize in Physics in 1921.[43] By wave-particle duality the effect can be analyzed purely in terms of waves though not as conveniently.[44]

Albert Einstein's mathematical description of how the photoelectric effect was caused by absorption of quanta of light was in one of his 1905 papers, named *"On a Heuristic Viewpoint Concerning the Production and Transformation of Light"*. This paper proposed the simple description of "light quanta", or photons, and showed how they explained such phenomena as the photoelectric effect. His simple explanation in terms of absorption of discrete quanta of light explained the features of the phenomenon and the characteristic frequency.

The idea of light quanta began with Max Planck's published law of black-body radiation (*"On the Law of Distribution of Energy in the Normal Spectrum"*[45]) by assuming that Hertzian oscillators could only exist at energies E proportional to the frequency f of the oscillator by $E = hf$, where h is Planck's constant. By assuming that light actually consisted of discrete energy packets, Einstein wrote an equation for the photoelectric effect that agreed with experimental results. It explained why the energy of photoelectrons was dependent only on the *frequency* of the incident light and not on its *intensity*: a low-intensity, high-frequency source could supply a few high energy photons, whereas a high-intensity, low-frequency source would supply no photons of sufficient individual energy to dislodge any electrons. This was an enormous theoretical leap, but the concept was strongly resisted at first because it contradicted the wave theory of light that followed naturally from James Clerk Maxwell's equations for electromagnetic behavior, and more generally, the assumption of infinite divisibility of energy in physical systems. Even after experiments showed that Einstein's equations for the photoelectric effect were accurate, resistance to the idea of photons continued, since it appeared to contradict Maxwell's equations, which were well-understood and verified.

Einstein's work predicted that the energy of individual ejected electrons increases linearly with the frequency of the light. Perhaps surprisingly, the precise relationship had not at that time been tested. By 1905 it was known that the energy of photoelectrons increases with increasing *frequency* of incident light and is independent of the *intensity* of the light. However, the manner of the increase was not experimentally determined until 1914 when Robert Andrews Millikan showed that Einstein's prediction was correct.[7]

The photoelectric effect helped to propel the then-emerging concept of wave–particle duality in the nature of light. Light simultaneously possesses the characteristics of both waves and particles, each being manifested according to the circumstances. The effect was impossible to understand in terms of the classical wave description of light,[46][47][48] as the energy of the emitted electrons did not depend on the intensity of the incident radiation. Classical theory predicted that the electrons would 'gather up' energy over a period of time, and then be emitted.[47][49]

Einstein, in 1905, when he wrote the Annus Mirabilis *papers*

8.3 Uses and effects

8.3.1   Photomultipliers

Main article: Photomultiplier

These are extremely light-sensitive vacuum tubes with a photocathode coated onto part (an end or side) of the inside of the envelope. The photocathode contains combinations of materials such as caesium, rubidium and antimony specially selected to provide a low work function, so when illuminated even by very low levels of light, the photocathode readily releases electrons. By means of a series of electrodes (dynodes) at ever-higher potentials, these electrons are accelerated and substantially increased in number through secondary emission to provide a readily detectable output current. Photomultipliers are still commonly used wherever low levels of light must be detected.[50]

8.3.2   Image sensors

Video camera tubes in the early days of television used the photoelectric effect, for example, Philo Farnsworth's "Image dissector" used a screen charged by the photoelectric effect to transform an optical image into a scanned electronic signal.[51]

8.3.3   Gold-leaf electroscope

Gold-leaf electroscopes are designed to detect static electricity. Charge placed on the metal cap spreads to the stem and the gold leaf of the electroscope. Because they then have the same charge, the stem and leaf repel each other. This will cause the leaf to bend away from the stem.

The electroscope is an important tool in illustrating the photoelectric effect. For example, if the electroscope is negatively charged throughout, there is an excess of electrons and the leaf is separated from the stem. If high-frequency light shines on the cap, the electroscope discharges and the leaf will fall limp. This is because the frequency of the light shining on the cap is above the cap's threshold frequency. The photons in the light have enough energy to liberate electrons from the cap, reducing its negative charge. This will discharge a negatively charged electroscope and further charge a positive electroscope. However, if the electromagnetic radiation hitting the metal cap does not have a high enough frequency (its frequency is below the threshold value for the cap), then the leaf will never discharge, no matter how long one shines the low-frequency light at the cap.[52]:389–390

8.3.4   Photoelectron spectroscopy

Since the energy of the photoelectrons emitted is exactly the energy of the incident photon minus the material's work function or binding energy, the work function of a sample can be determined by bombarding it with a monochromatic X-ray source or UV source, and measuring the kinetic energy distribution of the electrons emitted.[14]:14–20

Photoelectron spectroscopy is usually done in a high-vacuum environment, since the electrons would be scattered by gas molecules if they were present. However, some companies are now selling products that allow photoemission in air. The light source can be a laser, a discharge tube, or a synchrotron radiation source.[53]

The concentric hemispherical analyser (CHA) is a typical electron energy analyzer, and uses an electric field to change the directions of incident electrons, depending on their kinetic energies. For every element and core (atomic orbital) there will be a different binding energy. The many electrons created from each of these combinations will show up as spikes in the analyzer output, and these can be used to determine the elemental composition of the sample.

8.3.5   Spacecraft

The photoelectric effect will cause spacecraft exposed to sunlight to develop a positive charge. This can be a major problem, as other parts of the spacecraft in shadow develop a negative charge from nearby plasma, and the imbalance can discharge through delicate electrical components. The static charge created by the photoelectric effect is self-limiting, though, because a more highly charged object gives up its electrons less easily.[54] [55]

8.3.6 Moon dust

Light from the sun hitting lunar dust causes it to become charged through the photoelectric effect. The charged dust then repels itself and lifts off the surface of the Moon by electrostatic levitation.[56][57] This manifests itself almost like an "atmosphere of dust", visible as a thin haze and blurring of distant features, and visible as a dim glow after the sun has set. This was first photographed by the Surveyor program probes in the 1960s. It is thought that the smallest particles are repelled up to kilometers high, and that the particles move in "fountains" as they charge and discharge.

8.3.7 Night vision devices

Photons hitting a thin film of alkali metal or semiconductor material such as gallium arsenide in an image intensifier tube cause the ejection of photoelectrons due to the photoelectric effect. These are accelerated by an electrostatic field where they strike a phosphor coated screen, converting the electrons back into photons. Intensification of the signal is achieved either through acceleration of the electrons or by increasing the number of electrons through secondary emissions, such as with a micro-channel plate. Sometimes a combination of both methods is used. Additional kinetic energy is required to move an electron out of the conduction band and into the vacuum level. This is known as the electron affinity of the photocathode and is another barrier to photoemission other than the forbidden band, explained by the band gap model. Some materials such as Gallium Arsenide have an effective electron affinity that is below the level of the conduction band. In these materials, electrons that move to the conduction band are all of sufficient energy to be emitted from the material and as such, the film that absorbs photons can be quite thick. These materials are known as negative electron affinity materials.

8.4   Cross section

The photoelectric effect is one interaction mechanism between photons and atoms. It is one of 12 theoretically possible interactions.[58]

At the high photon energies comparable to the electron rest energy of 511 keV, Compton scattering, another process, may take place. Above twice this (1.022 MeV) pair production may take place.[59] Compton scattering and pair production are examples of two other competing mechanisms.

Indeed, even if the photoelectric effect is the favoured reaction for a particular single-photon bound-electron interaction, the result is also subject to statistical processes and is not guaranteed, albeit the photon has certainly disappeared and a bound electron has been excited (usually K or L shell electrons at gamma ray energies). The probability of the photoelectric effect occurring is measured by the cross section of interaction, σ. This has been found to be a function of the atomic number of the target atom and photon energy. A crude approximation, for photon energies above the highest atomic binding energy, is given by:[60]

$$\sigma = \text{constant} \cdot \frac{Z^n}{E^3}$$

Here Z is atomic number and n is a number which varies between 4 and 5. (At lower photon energies a characteristic structure with edges appears, K edge, L edges, M edges, etc.) The obvious interpretation follows that the photoelectric effect rapidly decreases in significance, in the gamma ray region of the spectrum, with increasing photon energy, and that photoelectric effect increases steeply with atomic number. The corollary is that high-Z materials make good gamma-ray shields, which is the principal reason that lead ($Z = 82$) is a preferred and ubiquitous gamma radiation shield.[61]

8.5   See also

Electronics:

- Photocurrent

- Photomultiplier

- Solar cell

- Solar energy

Physics:

- Anomalous photovoltaic effect

- Dember effect

- Photo-Dember

- Photomagnetic effect

- Corona discharge

- Photoelectron spectroscopy

- Planck's law of black body radiation

- Quantum mechanics

Chemistry:

- Photochemistry

- Electrochemistry

- Quantum Chemistry

- Silicon

- Boron

- Phosphorus

- Semi-conductor

- Metalloid

- Non-metal

- Ionization

- Crystallized atoms

Lists:

- Timeline of mechanics and physics

8.6 References

[1] Serway, R. A. (1990). *Physics for Scientists & Engineers* (3rd ed.). Saunders. p. 1150. ISBN 0-03-030258-7.

[2] Sears, F. W.; Zemansky, M. W.; Young, H. D. (1983). *University Physics* (6th ed.). Addison-Wesley. pp. 843–844. ISBN 0-201-07195-9.

[3] Hertz, H. (1887). "Ueber den Einfluss des ultravioletten Lichtes auf die electrische Entladung". *Annalen der Physik* 267 (8): S. 983–1000. Bibcode:1887AnP...267..983H. doi:10.1002/andp.18872670827.

[4] "The Nobel Prize in Physics 1921". Nobel Foundation. Retrieved 2013-03-16.

[5] "The Nobel Prize in Physics 1923". Nobel Foundation. Retrieved 2015-03-29.

[6] Lenard, P. (1902). "Ueber die lichtelektrische Wirkung". *Annalen der Physik* 313 (5): 149–198. Bibcode:1902AnP...313..149L. doi:10.1002/andp.19023130510.

[7] Millikan, R. (1914). "A Direct Determination of "h."". *Physical Review* 4 (1): 73–75. Bibcode:1914PhRv....4R..73M. doi:10.1103/PhysRev.4.73.2.

[8] Millikan, R. (1916). "A Direc t Photoelectri c Determination of Planck's "h"" (PDF). *Physical Review* 7 (3): 355–388. doi:10.1103/PhysRev.7.355.

[9] Zhang, Q. (1996). "Intensity dependence of the photoelectric effect induced by a circularly polarized laser beam". *Physics Letters A* 216 (1–5): 125. Bibcode:1996PhLA..216..125Z. doi:10.1016/0375-9601(96)00259-9.

[10] Bubb, F. (1924). "Direction of Ejection of Photo-Electrons b y Polarized X-rays". *Physical Review* 23 (2): 137–143. Bib doi:10.1103/PhysRev.23.137.

[11] Mee, C.; Crundell, M.; Arnold, B.; Brown, W. (2011). *International A/AS Level Physics*. Hodder Education. p. 241. ISBN 978-0-340-94564-3.

[12] Fromhold, A. T. (1991). *Quantum Mechanics for Applied Physics and Engineering*. Courier Dover Publications. pp. 5–6. ISBN 978-0-486-66741-6.

[13] Gautreau, R.; Savin, W. (1999). *Schaum's Outline of Modern Physics* (2nd ed.). McGraw-Hill. pp. 60–61. ISBN 0-07-024830-3.

[14] Hüfner, S. (2003). *Photoelectron Spectroscopy: Principles and Applications*. Springer. ISBN 3-540-41802-4.

[15] Report of the Board of Regents By Smithsonian Institution. Board of Regents, United States National Museum, Smithsonian Institution. p. 239.

[16] Schmidt, G. C. (1898) Wied. Ann. Uiv. p. 708.

[17] Knoblauch, O. (1899). *Zeitschrift Für Physikalische Chemie* xxix. p. 527.

[18] Vesselinka Petrova-Koch; Rudolf Hezel; Adolf Goetzberger (2009). *High-Efficient Low-Cost Photovoltaics: Recent Developments*. Springer. pp. 1–. ISBN 978-3-540-79358-8.

[19] Smith, W. (1873). "Eff ec t of Light on Selenium during the passage of an Electri c Current". *Nature* 7 (173): 303. Bibcode doi:10.1038/007303e0.

[20] Asimov, A. (1964) *Asimov's Biographical Encyclopedia of Science and Technology*, Doubleday, ISBN ISBN 0-385-04693-6.

[21] Robert Bud; Deborah Jean Warner (1998). *Instruments of Science: An Historical Encyclopedia*. Science Museum, London, and National Museum of American History, Smithsonian Institution. ISBN 978-0-8153-1561-2.

[22] Elster and Geitel arrange the metals in the following order with respect to their power of discharging negative electricity: rubidium, potassium, alloy of potassium and sodium, sodium, lithium, magnesium, thallium and zinc. For copper, platinum, lead, iron, cadmium, carbon, and mercury the effects with ordinary light are too small to be measurable. The order of the metals for this effect is the same as in Volta's series for contact-electricity, the most electropositive metals giving the largest photo-electric effect.

[23] Hertz, Wied. Ann. xxxi. p. 983, 1887.

[24] Hallwachs, Wied. Ann. xxxiii. p. 301, 1888.

[25] Hoor, Repertorium des Physik. xxv. p. 91, 1889.

[26] Bighi, C. R. cvi. p. 1349; cvii. p. 559, 1888

[27] Stoletow. C. R. cvi. pp. 1149, 1593; cvii. p. 91; cviii. p. 1241; Physikalische Revue, Bd. i., 1892.

[28] Stoletow, A. (1888). "Sur une sorte de courants electriques provoques par les rayons ultraviolets". *Comptes Rendus* **CVI**: 1149. (Reprinted in Stoletow, M.A. (1888). "On a kind of electric current produced by ultra-violet rays". *Philosophical Magazine Series 5* **26** (160): 317. doi:10.1080/14786448808628270.; abstract in Beibl. Ann. d. Phys. 12, 605, 1888).

[29] Stoletow, A. (1888). "Sur les courants actino-electriqies au travers deTair". *Comptes Rendus* **CVI**: 1593. (Abstract in Beibl. Ann. d. Phys. 12, 723, 1888).

[30] Stoletow, A. (1888). "Suite des recherches actino-electriques". *Comptes Rendus* **CVII**: 91. (Abstract in Beibl. Ann. d. Phys. 12, 723, 1888).

[31] Stoletow, A. (1889). "Sur les phénomènes actino-électriques". *Comptes Rendus.* CVIII: 1241.

[32] Stoletow, A. (1889). *Journal of the Russian Physico-chemical Society* (in Russian) **21**: 159. Missing or empty |title= (help)

[33] Stoletow, A. (1890). "Sur les courants actino-électriques dans l'air raréfié". *Journal de Physique* **9**: 468. doi:10.1051/jphystap:018900090046800.

[34] Elster and Geitel, Wied. Ann. xxxviii. pp. 40, 497, 1889; xli. p. 161, 1890; xlii. p. 564, 1891; xliii. p. 225, 1892; lii. p. 433, 1894 ; lv. p. 684, 1895.

[35] Thomson, J. J. (2005). *Conduction of Electricity Through Gases.* Watchmaker Publishing. ISBN 978-1-929148-49-3. Retrieved 9 July 2011.

[36] Hallwachs, W. (1907). "Über di e lichtelektrische Ermüdung". *Annalen der Physik* **328** (8): 459–516. Bibcode:1907AnP...328.. doi:10.1002/andp.19073280807.

[37] *The International Year Book.* (1900). New York: Dodd, Mead & Company. p. 659.

[38] Buchwald, Jed; Warwick, Andrew, eds. (2004). *Histories of the Electron: The Birth of Microphysics* (PDF) (illustrated, reprint ed.). MIT Press. pp. 21–23. ISBN 9780262524247.

[39] Bloch, E. (1908). "L'ionisation de l'air par la lumière ultra-violette". *Le Radium* **5** (8): 240. doi:10.1051/radium:019080050802

[40] Thomson, J. J. (1907). "On the Ionisation of Gases by Ultra-Violet Light and on the evidence as to the Structure of Light afforded by its Electrical Effects". *Proc. Cambr. Phil. Soc.* **14**: 417.

[41] Palmer, Frederic (1908). "Ionisation of Air by Ultra-violet Light". *Nature* **77** (2008): 582–582. Bibcode:1908Natur..77..582P. doi:10.1038/077582b0.

[42] Palmer, Frederic (1911). "Volume Ionization Produced by Light of Extremely Short Wave-Length". *Physical Review (Series I)* **32**: 1–22. Bibcode:1911PhRvI..32....1P. doi:10.1103/PhysRevSeriesI.32.1.

[43] "The Nobel Prize in Physics 1921". Nobel Foundation. Retrieved 2008-10-09.

[44] Lamb, Willis E.; Scully, Marlan O. (1968). "Photoelectric effect without photons, discussing classical field falling on quantized atomic electron" (PDF).

[45] Planck, Max (1901). "Ueber das Gesetz der Energieverteilung im Normalspectrum (On the Law of Distribution of Energy in the Normal Spectrum)". *Annalen der Physik* **4** (3): 553. Bibcode:1901AnP...309..553P. doi:10.1002/andp.19013090310.

[46] Resnick, Robert (1972) *Basic Concepts in Relativity and Early Quantum Theory,* Wiley, p. 137, ISBN 0471717029.

[47] Knight, Randall D. (2004) *Physics for Scientists and Engineers With Modern Physics: A Strategic Approach,* Pearson-Addison-Wesley, p. 1224, ISBN 0805386858.

[48] Penrose, Roger (2005) *The Road to Reality: A Complete Guide to the Laws of the Universe,* Knopf, p. 502, ISBN 0-679-45443-8

[49] Resnick, Robert (1972) *Basic Concepts in Relativity and Early Quantum Theory,* Wiley, p. 138, ISBN 0471717029.

[50] Timothy, J. Gethyn (2010) in Huber, Martin C.E. (ed.) *Observing Photons in Space*, ISSI Scientific Report 009, ESA Communications, pp. 365–408, ISBN 978-92-9221-938-3

[51] Burns, R. W. (1998) *Television: An International History of the Formative Years*, IET, p. 358, ISBN 0-85296-914-7.

[52] Tsokos, K. A. (2010). *Cambridge Physics for the IB Diploma* (revised ed.). Cambridge University Press. ISBN 9780521138215.

[53] Weaver, J. H.; Margaritondo, G. (1979). "Solid-State Photoelectron Spectroscopy with Synchrotron Radiation". *Science* **206** (4415): 151–156. Bibcode:1979Sci...206..151W. doi:10.1126/science.206.4415.151. PMID 17801770.

[54] Lai, Shu T. (2011). *Fundamentals of Spacecraft Charging: Spacecraft Interactions with Space Plasmas* (illustrated ed.). Princeton University Press. pp. 1–6. ISBN 9780691129471.

[55] "Spacecraft charging". *Arizona State University*.

[56] Bell, Trudy E., "Moon fountains", NASA.gov, 2005-03-30.

[57] Dust gets a charge in a vacuum. spacedaily.com, July 14, 2000.

[58] Evans, R. D. (1955). *The Atomic Nucleus*. Malabar, Fla.: Krieger. p. 673. ISBN 0-89874-414-8.

[59] Evans, R. D. (1955). *The Atomic Nucleus*. Malabar, Fla.: Krieger. p. 712. ISBN 0-89874-414-8.

[60] Davisson, C. M. (1965). *Interaction of gamma-radiation with matter*. pp. 37–78. Bibcode:1965abgs.conf...37D.

[61] Knoll, Glenn F. (1999). *Radiation Detection and Measurement*. New York: Wiley. p. 49. ISBN 0-471-49545-X.

8.7 External links

- AstronomyCast "http://www.astronomycast.com/2014/02/ep-335-photoelectric-effect/". AstronomyCast.

- Nave, R., "*Wave-Particle Duality*". HyperPhysics.

- "*Photoelectric effect*". Physics 2000. University of Colorado, Boulder, Colorado.

- ACEPT W3 Group, "*The Photoelectric Effect*". Department of Physics and Astronomy, Arizona State University, Tempe, AZ.

- Haberkern, Thomas, and N Deepak "*Grains of Mystique: Quantum Physics for the Layman*". Einstein Demystifies Photoelectric Effect, Chapter 3.

- Department of Physics, "*The Photoelectric effect*". Physics 320 Laboratory, Davidson College, Davidson.

- Fowler, Michael, "*The Photoelectric Effect*". Physics 252, University of Virginia.

- Go to "*Concerning an Heuristic Point of View Toward the Emission and Transformation of Light*" to read an English translation of Einstein's 1905 paper. (Retrieved: 2014 Apr 11)

- http://www.chemistryexplained.com/Ru-Sp/Solar-Cells.html

- Photo-electric transducers: http://sensorse.com/page4en.html

Applets

- "*Photoelectric Effect*". The Physics Education Technology (PhET) project. (Java)

- Fendt, Walter, "*The Photoelectric Effect*". (Java)

- "*Applet: Photo Effect*". Open Source Distributed Learning Content Management and Assessment System. (Java)

R. Millikan (picture around 1923), who first experimentally showed Einstein's prediction on photoelectric effect was correct.

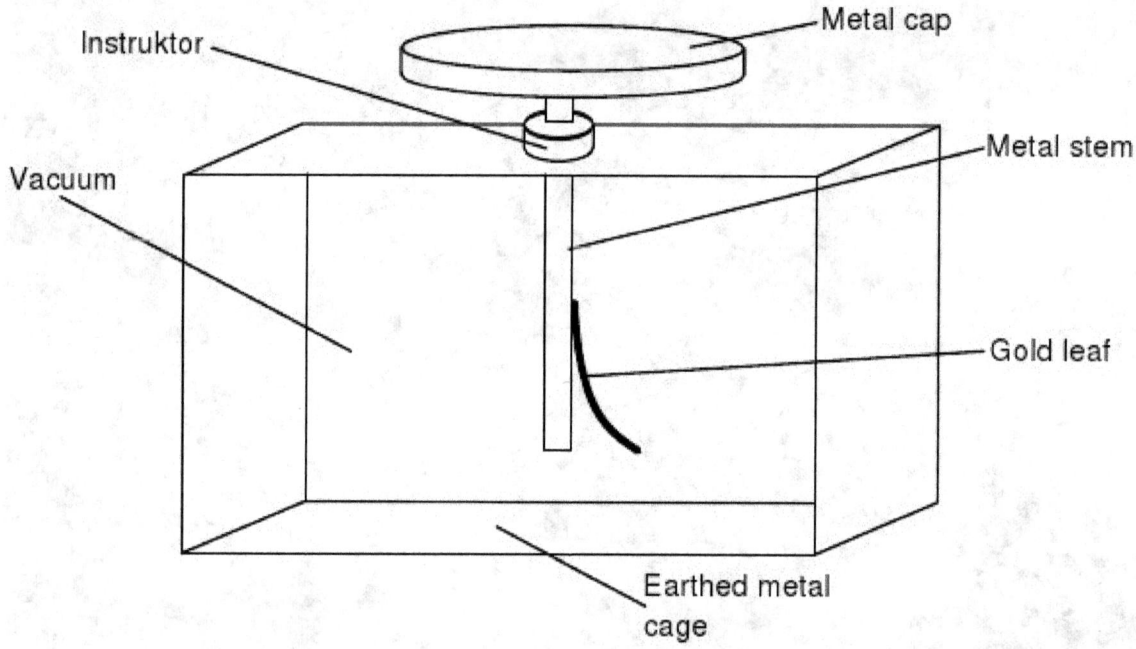

The gold leaf electroscope

Chapter 9

Matter wave

The de Broglie relations redirect here.
This article is about wave-like phenomena exhibited by particles of matter. For the ordinary type of wave propagating through material media, see Mechanical wave.

All matter can exhibit wave-like behaviour. For example a beam of electrons can be diffracted just like a beam of light or a water wave. **Matter waves** are a central part of the theory of quantum mechanics, being an example of wave–particle duality. The concept that matter behaves like a wave is also referred to as the **de Broglie hypothesis** (/də'brɔɪ/) due to having been proposed by Louis de Broglie in 1924.[1] Matter waves are often referred to as **de Broglie waves**.

The **de Broglie wavelength** is the wavelength, λ, associated with a massive particle and is related to its momentum, p, through the Planck constant, h:

$$\lambda = \frac{h}{p}.$$

Wave-like behaviour of matter was first experimentally demonstrated in the Davisson–Germer experiment using electrons, and it has also been confirmed for other elementary particles, neutral atoms and even molecules. The wave-like behaviour of matter is crucial to the modern theory of atomic structure and particle physics.

9.1   Historical context

At the end of the 19th century, light was thought to consist of waves of electromagnetic fields which propagated according to Maxwell's equations, while matter was thought to consist of localized particles (See history of wave and particle viewpoints). In 1900, this division was exposed to doubt, when, investigating the theory of black body thermal radiation, Max Planck proposed that light is emitted in discrete quanta of energy. It was thoroughly challenged in 1905. Extending Planck's investigation in several ways, including its connection with the photoelectric effect, Albert Einstein proposed that light is also propagated and absorbed in quanta. Light quanta are now called photons. These quanta would have an energy given by the Planck–Einstein relation:

$$E = h\nu$$

and a momentum

$$p = \frac{E}{c} = \frac{h}{\lambda}$$

where ν (lowercase Greek letter nu) and λ (lowercase Greek letter lambda) denote the frequency and wavelength of the light, c the speed of light, and h Planck's constant.[2] In the modern convention, frequency is symbolized by f as is done in the rest of this article. Einstein's postulate was confirmed experimentally by Robert Millikan and Arthur Compton over the next two decades.

9.2 The de Broglie hypothesis

De Broglie, in his 1924 PhD thesis, proposed that just as light has both wave-like and particle-like properties, electrons also have wave-like properties. By rearranging the momentum equation stated in the above section, we find a relationship between the wavelength, λ associated with an electron and its momentum, p, through the Planck constant, h:[3]

$$\lambda = \frac{h}{p}.$$

The relationship is now known to hold for all types of matter: all matter exhibits properties of both particles and waves.

In 1926, Erwin Schrödinger published an equation describing how a matter wave should evolve—the matter wave analogue of Maxwell's equations—and used it to derive the energy spectrum of hydrogen.

9.3 Experimental confirmation

Matter waves were first experimentally confirmed to occur in the Davisson-Germer experiment for electrons, and the de Broglie hypothesis has been confirmed for other elementary particles. Furthermore, neutral atoms and even molecules have been shown to be wave-like.

9.3.1 Electrons

Further information: Davisson–Germer experiment and Electron diffraction

In 1927 at Bell Labs, Clinton Davisson and Lester Germer fired slow-moving electrons at a crystalline nickel target. The angular dependence of the diffracted electron intensity was measured, and was determined to have the same diffraction pattern as those predicted by Bragg for x-rays. Before the acceptance of the de Broglie hypothesis, diffraction was a property that was thought to be only exhibited by waves. Therefore, the presence of any diffraction effects by matter demonstrated the wave-like nature of matter. When the de Broglie wavelength was inserted into the Bragg condition, the observed diffraction pattern was predicted, thereby experimentally confirming the de Broglie hypothesis for electrons.[5]

This was a pivotal result in the development of quantum mechanics. Just as the photoelectric effect demonstrated the particle nature of light, the Davisson–Germer experiment showed the wave-nature of matter, and completed the theory of wave-particle duality. For physicists this idea was important because it meant that not only could any particle exhibit wave characteristics, but that one could use wave equations to describe phenomena in matter if one used the de Broglie wavelength.

9.3.2 Neutral atoms

Further information: Atom optics

Experiments with Fresnel diffraction[6] and an atomic mirror for specular reflection[7][8] of neutral atoms confirm the application of the de Broglie hypothesis to atoms, i.e. the existence of atomic waves which undergo diffraction, interference and allow quantum reflection by the tails of the attractive potential.[9] Advances in laser cooling have allowed cooling of neutral atoms down to nanokelvin temperatures. At these temperatures, the thermal de Broglie wavelengths come into the micrometre range. Using Bragg diffraction of atoms and a Ramsey interferometry technique, the de Broglie wavelength of cold sodium atoms was explicitly measured and found to be consistent with the temperature measured by a different method.[10]

This effect has been used to demonstrate atomic holography, and it may allow the construction of an atom probe imaging system with nanometer resolution.[11][12] The description of these phenomena is based on the wave properties of neutral atoms, confirming the de Broglie hypothesis.

The effect has also been used to explain the spatial version of the quantum Zeno effect, in which an otherwise unstable object may be stabilised by rapidly-repeated observations.[8]

9.3.3   Molecules

Recent experiments even confirm the relations for molecules and even macromolecules that otherwise might be supposed too large to undergo quantum mechanical effects. In 1999, a research team in Vienna demonstrated diffraction for molecules as large as fullerenes.[13] The researchers calculated a De Broglie wavelength of the most probable C_{60} velocity as 2.5 pm. More recent experiments prove the quantum nature of molecules with a mass up to 6910 amu.[14]

9.4   de Broglie relations

The de Broglie equations relate the wavelength λ to the momentum p, and frequency f to the total energy E of a particle:[15]

where h is Planck's constant. The equations can also be written as

where $\hbar$ is the reduced Planck's constant, **k** is the wave vector, and ω is the angular frequency.

In each pair, the second equation is also referred to as the Planck-Einstein relation, since it was also proposed by Planck and Einstein.

9.4.1   Special relativity

Using two formulas from special relativity, one for the relativistic momentum and one for the energy

$$E = mc^2 = \gamma m_0 c^2$$
$$\vec{p} = m\vec{v} = \gamma m_0 \vec{v}$$

allows the equations to be written as

$$\lambda = \frac{h}{\gamma m_0 v} = \frac{h}{m_0 v} \sqrt{1 - \frac{v^2}{c^2}}$$
$$f = \frac{\gamma m_0 c^2}{h} = \frac{m_0 c^2}{h} \bigg/ \sqrt{1 - \frac{v^2}{c^2}}$$

where m_0 denotes the particle's rest mass, v its velocity, γ the Lorentz factor, and c the speed of light in a vacuum.[16][17][18] See below for details of the derivation of the de Broglie relations. Group velocity (equal to the particle's speed) should not be confused with phase velocity (equal to the product of the particle's frequency and its wavelength). In the case of a non-dispersive medium, they happen to be equal, but otherwise they are not.

Group velocity

Albert Einstein first explained the wave–particle duality of light in 1905. Louis de Broglie hypothesized that any particle should also exhibit such a duality. The velocity of a particle, he concluded, should always equal the group velocity of the corresponding wave. The magnitude of the group velocity is equal to the particle's speed.

Both in relativistic and non-relativistic quantum physics, we can identify the group velocity of a particle's wave function with the particle velocity. Quantum mechanics has very accurately demonstrated this hypothesis, and the relation has been shown explicitly for particles as large as molecules.

De Broglie deduced that if the duality equations already known for light were the same for any particle, then his hypothesis would hold. This means that

$$v_g = \frac{\partial \omega}{\partial k} = \frac{\partial (E/\hbar)}{\partial (p/\hbar)} = \frac{\partial E}{\partial p}$$

where E is the total energy of the particle, p is its momentum, $\hbar$ is the reduced Planck constant. For a free non-relativistic particle it follows that

$$v_g = \frac{\partial E}{\partial p} = \frac{\partial}{\partial p} \left(\frac{1}{2} \frac{p^2}{m} \right)$$
$$= \frac{p}{m}$$
$$= v$$

where m is the mass of the particle and v its velocity.

Also in special relativity we find that

$$v_g = \frac{\partial E}{\partial p} = \frac{\partial}{\partial p} \left(\sqrt{p^2 c^2 + m_0^2 c^4} \right)$$
$$= \frac{pc^2}{\sqrt{p^2 c^2 + m_0^2 c^4}}$$
$$= \frac{pc^2}{E}$$

where m_0 is the rest mass of the particle and c is the speed of light in a vacuum. But (see below), using that the phase velocity is $v_p = E/p = c^2/v$, therefore

$$v_g = \frac{pc^2}{E}$$
$$= \frac{c^2}{v_p}$$
$$= v$$

where v is the velocity of the particle regardless of wave behavior.

Phase velocity

In quantum mechanics, particles also behave as waves with complex phases. The phase velocity is equal to the product of the frequency multiplied by the wavelength.

By the de Broglie hypothesis, we see that

$$v_{\mathrm{p}} = \frac{\omega}{k} = \frac{E/\hbar}{p/\hbar} = \frac{E}{p}.$$

Using relativistic relations for energy and momentum, we have

$$v_{\mathrm{p}} = \frac{E}{p} = \frac{\gamma m_0 c^2}{\gamma m_0 v} = \frac{c^2}{v} = \frac{c}{\beta}$$

where E is the total energy of the particle (i.e. rest energy plus kinetic energy in kinematic sense), p the momentum, γ the Lorentz factor, c the speed of light, and β the speed as a fraction of c. The variable v can either be taken to be the speed of the particle or the group velocity of the corresponding matter wave. Since the particle speed $v < c$ for any particle that has mass (according to special relativity), the phase velocity of matter waves always exceeds c, i.e.

$$v_{\mathrm{p}} > c,$$

and as we can see, it approaches c when the particle speed is in the relativistic range. The superluminal phase velocity does not violate special relativity, because phase propagation carries no energy. See the article on *Dispersion (optics)* for details.

9.4.2   Four-vectors

Main article: Four-vector

Using 4-Vectors, the De Broglie relations form a single equation:

which is frame-independent.

Likewise, the relation between group/particle velocity and phase velocity is given in frame-independent form by:

where

4-Momentum $\mathbf{P} = \left(\frac{E}{c}, \vec{\mathbf{p}} \right)$

4-WaveVector $\mathbf{K} = \left(\frac{\omega}{c}, \vec{\mathbf{k}} \right) = \left(\frac{\omega}{c}, \frac{\omega}{v_p}\hat{\mathbf{n}} \right)$

4-Velocity $\mathbf{U} = \gamma(c, \vec{\mathbf{u}}) = \gamma(c, v_g \hat{\mathbf{n}})$

9.5   Interpretations

The physical reality underlying de Broglie waves is a subject of ongoing debate. Some theories treat either the particle or the wave aspect as its fundamental nature, seeking to explain the other as an emergent property. Some, such as the hidden variable theory, treat the wave and the particle as distinct entities. Yet others propose some intermediate entity that is neither quite wave nor quite particle but only appears as such when we measure one or the other property. The Copenhagen interpretation states that the nature of the underlying reality is unknowable and beyond the bounds of scientific enquiry.

Schrödinger's quantum mechanical waves are conceptually different from ordinary physical waves such as water or sound. Ordinary physical waves are characterized by undulating real-number 'displacements' of dimensioned physical variables at each point of ordinary physical space at each instant of time. Schrödinger's "waves" are characterized by the undulating value of a dimensionless complex number at each point of an abstract multi-dimensional space, for example of configuration space.

At the Fifth Solvay Conference in 1927, Max Born and Werner Heisenberg reported as follows

At the same conference, Erwin Schrödinger reported likewise.

In 1955, Heisenberg reiterated this.

It is mentioned above that the "displaced quantity" of the Schrödinger wave has values that are dimensionless complex numbers. One may ask what is the physical meaning of those numbers. According to Heisenberg, rather than being of some ordinary physical quantity such as for example Maxwell's electric field intensity, or for example mass density, the Schrödinger-wave packet's "displaced quantity" is probability amplitude. He wrote that instead of using the term 'wave packet', it is preferable to speak of a probability packet. [22] The probability amplitude supports calculation of probability of location or momentum of discrete particles. Heisenberg recites Duane's account of particle diffraction by probabilistic quantal translation momentum transfer, which allows, for example in Young's two-slit experiment, each diffracted particle probabilistically to pass discretely through a particular slit.[23] Thus one does not need necessarily think of the matter wave, as it were, as 'composed of smeared matter'.

These ideas may be expressed in ordinary language as follows. In the account of ordinary physical waves, a 'point' refers to a position in ordinary physical space at an instant of time, at which there is specified a 'displacement' of some physical quantity. But in the account of quantum mechanics, a 'point' refers to a configuration of the system at an instant of time, every particle of the system being in a sense present in every 'point' of configuration space, each particle at such a 'point' being located possibly at a different position in ordinary physical space. There is no explicit definite indication that, at an instant, this particle is 'here' and that particle is 'there' in some separate 'location' in configuration space. This conceptual difference entails that, in contrast to de Broglie's pre-quantum mechanical wave description, the quantum mechanical probability packet description does not directly and explicitly express the Aristotelian idea, referred to by Newton, that causal efficacy propagates through ordinary space by contact, nor the Einsteinian idea that such propagation is no faster than light. In contrast, these ideas are so expressed in the classical wave account, through the Green's function, though it is inadequate for the observed quantal phenomena. The physical reasoning for this was first recognized by Einstein.[24][25]

9.6 De Broglie's phase wave and periodic phenomenon

De Broglie's thesis started from the hypothesis, "that to each portion of energy with a proper mass m_0 one may associate a periodic phenomenon of the frequency ν_0 , such that one finds: $h\nu_0 = m_0 c^2$. The frequency ν_0 is to be measured, of course, in the rest frame of the energy packet. This hypothesis is the basis of our theory."[26][27][28][29][30][31]

De Broglie followed his initial hypothesis of a periodic phenomenon, with frequency ν_0 , associated with the energy packet. He used the special theory of relativity to find, in the frame of the observer of the electron energy packet that is moving with velocity v , that its frequency was apparently reduced to

$$f = \nu_0 \sqrt{1 - \frac{v^2}{c^2}} \,.$$

Then

$$\lambda f = E/p = v_{\mathrm{p}} \,.$$

using the same notation as above. The quantity v_{p} is the velocity of what de Broglie called the "phase wave". Its wavelength is λ and frequency f . De Broglie reasoned that his hypothetical intrinsic particle periodic phenomenon is in phase with that phase wave. This was his basic matter wave conception. He noted, as above, that $v_p > c$, and the phase wave does not transfer energy.[28][32]

While the concept of waves being associated with matter is correct, de Broglie did not leap directly to the final understanding of quantum mechanics with no missteps. There are conceptual problems with the approach that de Broglie took in his thesis that he was not able to resolve, despite trying a number of different fundamental hypotheses in different papers published while working on, and shortly after publishing, his thesis.[29][33] These difficulties were resolved by Erwin Schrödinger, who developed the wave mechanics approach, starting from a somewhat different basic hypothesis.

9.7   See also

- Bohr model

- Faraday wave

- Kapitsa–Dirac effect

- Matter wave clock

- Schrödinger equation

- Theoretical and experimental justification for the Schrödinger equation

- Thermal de Broglie wavelength

- De Broglie–Bohm theory

9.8   References

[1] Feynman, R.; *QED the Strange Theory of Light and matter*, Penguin 1990 Edition, page 84.

[2] Einstein, A. (1917). Zur Quantentheorie der Strahlung. *Physicalische Zeitschrift* **18**: 121–128. Translated in ter Haar, D. (1967). *The Old Quantum Theory*. Pergamon Press. pp. 167–183. LCCN 66029628.

[3] J. P. McEvoy & Oscar Zarate (2004). *Introducing Quantum Theory*. Totem Books. pp. 110–114. ISBN 1-84046-577-8.

[4] Louis de Broglie "The Reinterpretation of Wave Mechanics" Foundations of Physics, Vol. 1 No. 1 (1970)

[5] Mauro Dardo, *Nobel Laureates and Twentieth-Century Physics*, Cambridge University Press 2004, pp. 156–157

[6] R.B.Doak; R.E.Grisenti; S.Rehbein; G.Schmahl; J.P.Toennies; Ch. Wöll (1999). "Towards Realization of an Atomic de Broglie Microscope: Helium Atom Focusing Using Fresnel Zone Plates". *Physical Review Letters* **83** (21): 4229–4232. Bibcode:1999PhRvL..83.4229D. doi:10.1103/PhysRevLett.83.4229.

[7] F. Shimizu (2000). "Specular Reflection of Very Slow Metastable Neon Atoms from a Solid Surface". *Physical Review Letters* **86** (6): 987–990. Bibcode:2001PhRvL..86..987S. doi:10.1103/PhysRevLett.86.987. PMID 11177991.

[8] D. Kouznetsov; H. Oberst (2005). "Reflection of Waves from a Ridged Surface and the Zeno Effect". *Optical Review* **12** (5): 1605–1623. Bibcode:2005OptRv..12..363K. doi:10.1007/s10043-005-0363-9.

[9] H.Friedrich; G.Jacoby; C.G.Meister (2002). "quantum reflection by Casimir–van der Waals potential tails". *Physical Review A* **65** (3): 032902. Bibcode:2002PhRvA..65c2902F. doi:10.1103/PhysRevA.65.032902.

[10] Pierre Cladé; Changhyun Ryu; Anand Ramanathan; Kristian Helmerson; William D. Phillips (2008). "Observation of a 2D Bose Gas: From thermal to quasi-condensate to superfluid". arXiv:0805.3519.

[11] Shimizu; J.Fujita (2002). "Re flection-Type Hologram for Atoms". *Physical Review Letters* **88** (12): 123201. Bibcode: doi:10.1103/PhysRevLett.88.123201. PMID 11909457.

[12] D. Kouznetsov; H. Oberst; K. Shimizu; A. Neumann; Y. Kuznetsova; J.-F. Bisson; K. Ueda; S. R. J. Brueck (2006). "Ridged atomic mirrors and atomic nanoscope". *Journal of Physics B* **39** (7): 1605–1623. Bibcode:2006JPhB...39.1605K. doi:10.1088/0953-4075/39/7/005.

[13] Arndt, M.; O. Nairz; J. Voss-Andreae; C. Keller; G. van der Zouw; A. Zeilinger (14 October 1999). "Wave-particle duality of C60". *Nature* **401** (6754): 680–682. Bibcode:1999Natur.401..680A. doi:10.1038/44348. PMID 18494170.

[14] Gerlich, S.; S. Eibenberger; M. Tomandl; S. Nimmrichter; K. Hornberger; P. J. Fagan; J. Tüxen; M. Mayor & M. Arndt (5 April 2011). "Quantum interference of large organic molecules". *Nature Communications* **2** (263): 263–. Bibcode:2011NatCo...2E.263G. doi:10.1038/ncomms1263. PMC 3104521. PMID 21468015.

[15] Resnick, R.; Eisberg, R. (1985). *Quantum Physics of Atoms, Molecules, Solids, Nuclei and Particles* (2nd ed.). New York: John Wiley & Sons. ISBN 0-471-87373-X.

[16] Holden, Alan (1971). *Stationary states*. New York: Oxford University Press. ISBN 0-19-501497-9.

[17] Williams, W.S.C. (2002). *Introducing Special Relativity*, Taylor & Francis, London, ISBN 0-415-27761-2, p. 192.

[18] de Broglie, L. (1970). The reinterpretation of wave mechanics, *Foundations of Physics* 1(1): 5–15, p. 9.

[19] Born, M., Heisenberg, W. (1928). Quantum mechanics, pp. 143–181 of *Électrons et Photons: Rapports et Discussions du Cinquième Conseil de Physique, tenu à Bruxelles du 24 au 29 Octobre 1927, sous les Auspices de l'Institut International de Physique Solvay*, Gauthier-Villars, Paris, p. 166; this translation at p. 425 of Bacciagaluppi, G., Valentini, A. (2009), *Quantum Theory at the Crossroads: Reconsidering the 1927 Solvay Conference*, Cambridge University Press, Cambridge UK, ISBN 978-0-521-81421-8.

[20] Schrödinger, E. (1928). Wave mechanics, pp. 185–206 of *Électrons et Photons: Rapports et Discussions du Cinquième Conseil de Physique, tenu à Bruxelles du 24 au 29 Octobre 1927, sous les Auspices de l'Institut International de Physique Solvay*, Gauthier-Villars, Paris, pp. 185–186; this translation at p. 447 of Bacciagaluppi, G., Valentini, A. (2009), *Quantum Theory at the Crossroads: Reconsidering the 1927 Solvay Conference*, Cambridge University Press, Cambridge UK, ISBN 978-0-521-81421-8.

[21] Heisenberg, W. (1955). The development of the interpretation of the quantum theory, pp. 12–29, in *Niels Bohr and the Development of Physics: Essays dedicated to Niels Bohr on the occasion of his seventieth birthday*, edited by W. Pauli, with the assistance of L. Rosenfeld and V. Weisskopf, Pergamon Press, London, p. 13.

[22] Heisenberg, W. (1927). Über den anschaulichen Inhalt der quantentheoretischen Kinematik und Mechanik, *Z. Phys.* **43**: 172–198, translated by eds. Wheeler, J.A., Zurek, W.H. (1983), at pp. 62–84 of *Quantum Theory and Measurement*, Princeton University Press, Princeton NJ, p. 73. Also translated as 'The actual content of quantum theoretical kinematics and mechanics' here

[23] Heisenberg, W. (1930). *The Physical Principles of the Quantum Theory*, translated by C. Eckart, F. C. Hoyt, University of Chicago Press, Chicago IL, pp. 77–78.

[24] Fine, A. (1986). *The Shaky Game: Einstein Realism and the Quantum Theory*, University of Chicago, Chicago, ISBN 0-226-24946-8

[25] Howard, D. (1990). "Nicht sein kann was nicht sein darf", or the prehistory of the EPR, 1909–1935; Einstein's early worries about the quantum mechanics of composite systems, pp. 61–112 in *Sixty-two Years of Uncertainty: Historical Philosophical and Physical Inquiries into the Foundations of Quantum Mechanics*, edited by A.I. Miller, Plenum Press, New York, ISBN 978-1-4684-8773-2.

[26] de Broglie, L. (1923). Waves and quanta, *Nature* **112**: 540.

[27] de Broglie, L. (1924). Thesis, p. 8 of Kracklauer's translation.

[28] Medicus, H.A. (1974). Fifty years of matter waves, *Physics Today* 27(2): 38–45.

[29] MacKinnon, E. (1976). De Broglie's thesis: a critical retrospective. *Am. J. Phys.* **44**: 1047–1055.

[30] Espinosa, J.M. (1982). Physical properties of de Broglie's phase waves, *Am. J. Phys.* **50**: 357–362.

[31] Brown, H.R., Martins, R.deA. (1984). De Broglie's relativistic phase waves and wave groups, *Am. J. Phys.* **52**: 1130–1140.

[32] Bacciagaluppi, G., Valentini, A. (2009). *Quantum Theory at the Crossroads: Reconsidering the 1927 Solvay Conference*, Cambridge University Press, Cambridge UK, ISBN 978-0-521-81421-8, pp. 30–88.

[33] Martins, Roberto de Andrade (2010). "Louis de Broglie's Struggle with the Wave-Particle Dualism, 1923-1925". Quantum History Project, Fritz Haber Institute of the Max Planck Society and the Max Planck Institute for the History of Science. Retrieved 2015-01-03.

9.9   Further reading

- L. de Broglie, *Recherches sur la théorie des quanta* (Researches on the quantum theory), Thesis (Paris), 1924; L. de Broglie, *Ann. Phys.* (Paris) 3, 22 (1925). English translation by A.F. Kracklauer. And here.

- Broglie, Louis de, *The wave nature of the electron* Nobel Lecture, 12, 1929

- Tipler, Paul A. and Ralph A. Llewellyn (2003). *Modern Physics*. 4th ed. New York; W. H. Freeman and Co. ISBN 0-7167-4345-0. pp. 203–4, 222–3, 236.

- Zumdahl, Steven S. (2005). *Chemical Principles* (5th ed.). Boston: Houghton Mifflin. ISBN 0-618-37206-7.

- An extensive review article "Optics and interferometry with atoms and molecules" appeared in July 2009: http://www.atomwave.org/rmparticle/RMPLAO.pdf.

- "Scientific Papers Presented to Max Born on his retirement from the Tait Chair of Natural Philosophy in the University of Edinburgh", 1953 (Oliver and Boyd)

9.10   External links

- Bowley, Roger. "de Broglie Waves". *Sixty Symbols*. Brady Haran for the University of Nottingham.

$$\Psi = Ae^{i(px - \omega t)}$$

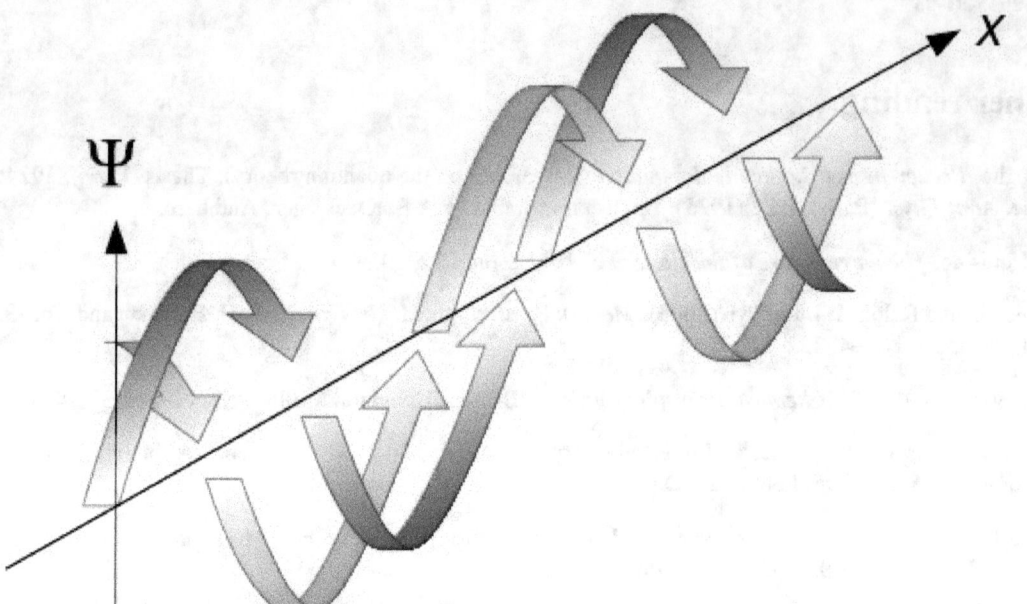

*Propagation of **de Broglie** waves in 1d – real part of the complex amplitude is blue, imaginary part is green. The probability (shown as the colour opacity) of finding the particle at a given point x is spread out like a waveform, there is no definite position of the particle. As the amplitude increases above zero the curvature decreases, so the amplitude decreases again, and vice versa – the result is an alternating amplitude: a wave. Top: plane wave. Bottom: wave packet.*

$$\Psi = \sum_n A_n e^{i(p_n x - \omega_n t)}$$

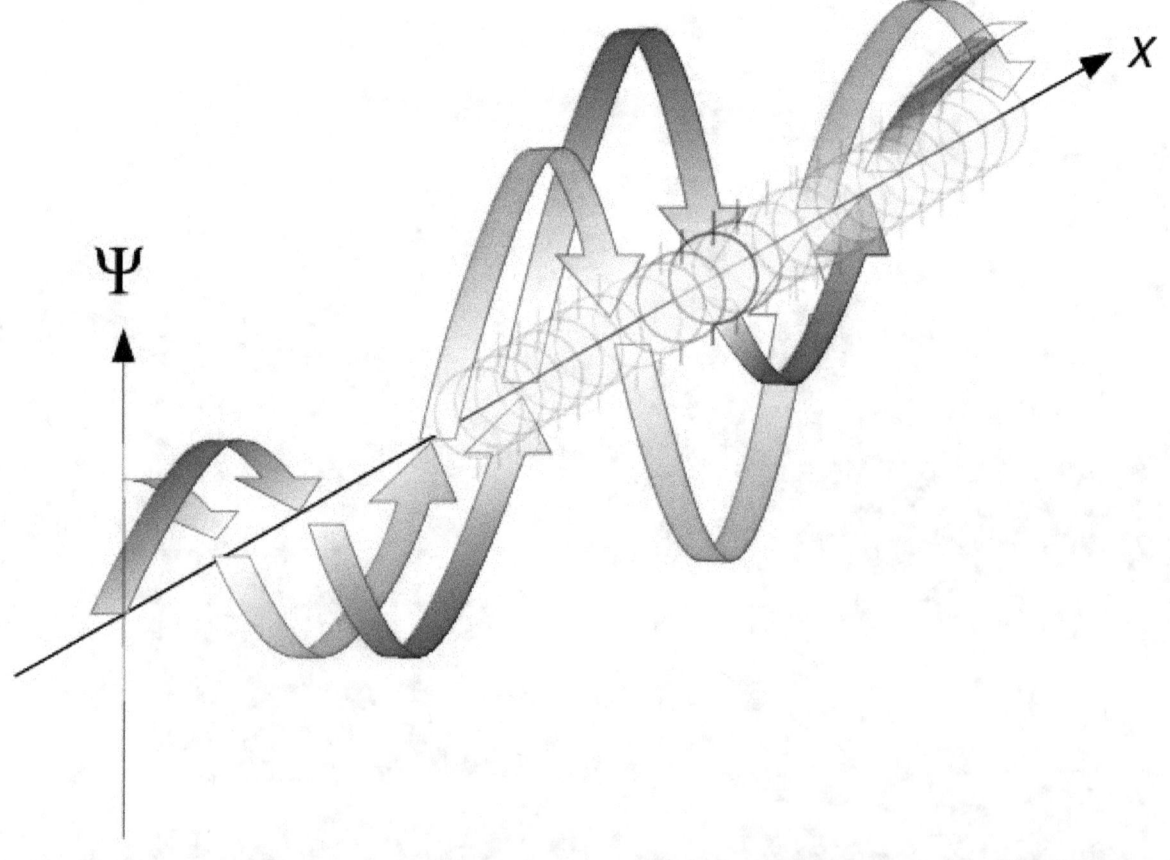

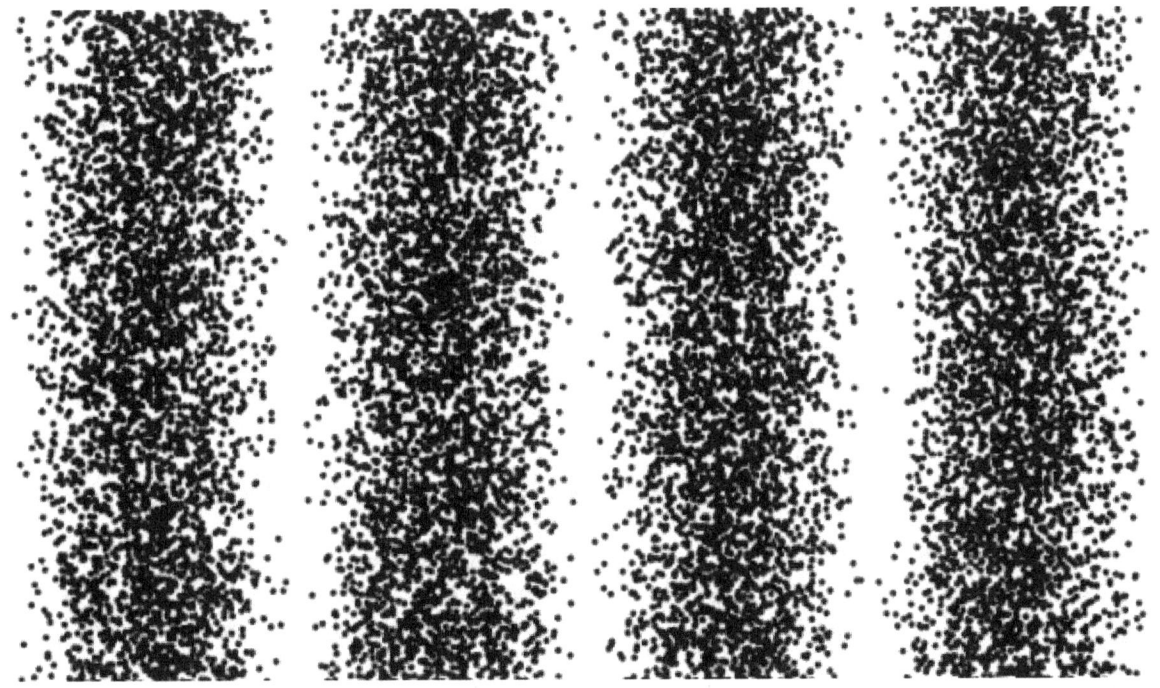

Demonstration of a matter wave in diffraction of electrons

Chapter 10

Uncertainty principle

For other uses, see Uncertainty principle (disambiguation).

In quantum mechanics, the **uncertainty principle**, also known as **Heisenberg's uncertainty principle**, is any of a variety of mathematical inequalities asserting a fundamental limit to the precision with which certain pairs of physical properties of a particle, known as complementary variables, such as position x and momentum p, can be known simultaneously.

Introduced first in 1927, by the German physicist Werner Heisenberg, it states that the more precisely the position of some particle is determined, the less precisely its momentum can be known, and vice versa.[1] The formal inequality relating the standard deviation of position σx and the standard deviation of momentum σp was derived by Earle Hesse Kennard[2] later that year and by Hermann Weyl[3] in 1928:

($\hbar$ is the reduced Planck constant, $h\,/\,2\pi$).

Historically, the uncertainty principle has been confused[4][5] with a somewhat similar effect in physics, called the observer effect, which notes that measurements of certain systems cannot be made without affecting the systems. Heisenberg offered such an observer effect at the quantum level (see below) as a physical "explanation" of quantum uncertainty.[6] It has since become clear, however, that the uncertainty principle is inherent in the properties of all wave-like systems,[7] and that it arises in quantum mechanics simply due to the matter wave nature of all quantum objects. Thus, *the uncertainty principle actually states a fundamental property of quantum systems, and is not a statement about the observational success of current technology.*[8] It must be emphasized that *measurement* does not mean only a process in which a physicist-observer takes part, but rather any interaction between classical and quantum objects regardless of any observer.[9]

Since the uncertainty principle is such a basic result in quantum mechanics, typical experiments in quantum mechanics routinely observe aspects of it. Certain experiments, however, may deliberately test a particular form of the uncertainty principle as part of their main research program. These include, for example, tests of number–phase uncertainty relations in superconducting[10] or quantum optics[11] systems. Applications dependent on the uncertainty principle for their operation include extremely low noise technology such as that required in gravitational-wave interferometers.[12]

10.1   Introduction

Main article: Introduction to quantum mechanics

As a fundamental constraint, higher level descriptions of the universe must supervene on quantum mechanical descriptions which includes Heisenberg's uncertainty relationship. However, humans do not form an intuitive understanding of this uncertainty principle in everyday life. This is because the constraint is not readily apparent on the macroscopic scales of everyday experience. So it may be helpful to demonstrate how it is integral to more easily understood physical situations. Two alternative conceptualizations of quantum physics can be examined with the goal of demonstrating the key role the

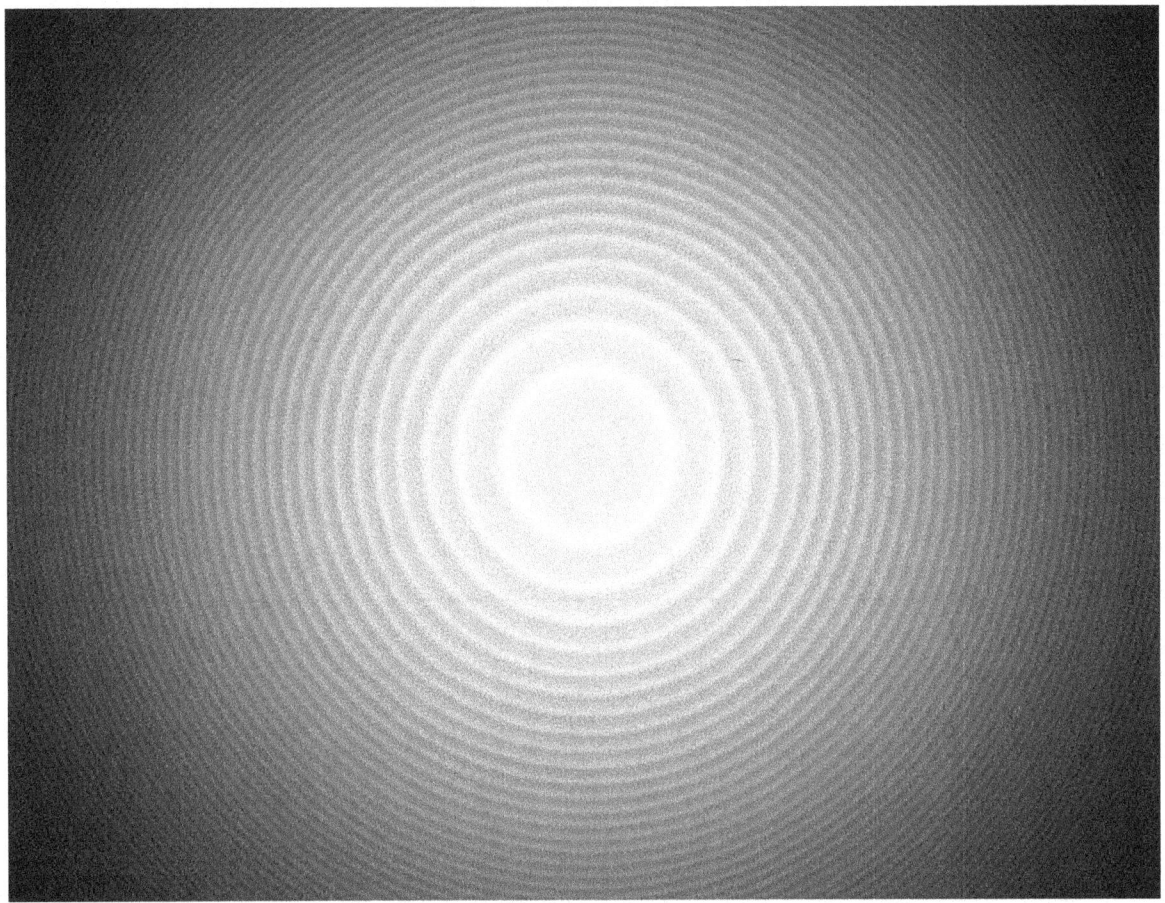

The evolution o f an initiall y very localized gaussian wave function o f a free particle in two -dimensional space, with colour and intensity indicating phase and amplitude. The spreading o f the wave function in all directions shows that the initial momentum has a spread o f values, unmodi fied in time; while the spread in position increases in time; as a result, the uncertainty Δ x Δp increases in time.

uncertainty principle plays. A wave mechanics picture of the uncertainty principle provides for a more visually intuitive demonstration, and the somewhat more abstract matrix mechanics picture provides for a demonstration of the uncertainty principle that is more easily generalized to cover a multitude of physical contexts.

Mathematically, in wave mechanics, the uncertainty relation between position and momentum arises because the expressions of the wavefunction in the two corresponding orthonormal bases in Hilbert space are Fourier transforms of one another (i.e., position and momentum are conjugate variables). A nonzero function and its Fourier transform cannot both be sharply localized. A similar tradeoff between the variances of Fourier conjugates arises in all systems underlain by Fourier analysis, for example in sound waves: A pure tone is a sharp spike at a single frequency, while its Fourier transform gives the shape of the sound wave in the time domain, which is a completely delocalized sine wave. In quantum mechanics, the two key points are that the position of the particle takes the form of a matter wave, and momentum is its Fourier conjugate, assured by the de Broglie relation $p = \hbar k$, where k is the wavenumber.

In matrix mechanics, the mathematical formulation of quantum mechanics, any pair of non-commuting self-adjoint operators representing observables are subject to similar uncertainty limits. An eigenstate of an observable represents the state of the wavefunction for a certain measurement value (the eigenvalue). For example, if a measurement of an observable A is performed, then the system is in a particular eigenstate Ψ of that observable. However, the particular eigenstate of the observable A need not be an eigenstate of another observable B: If so, then it does not have a unique associated measurement for it, as the system is not in an eigenstate of that observable.[13]

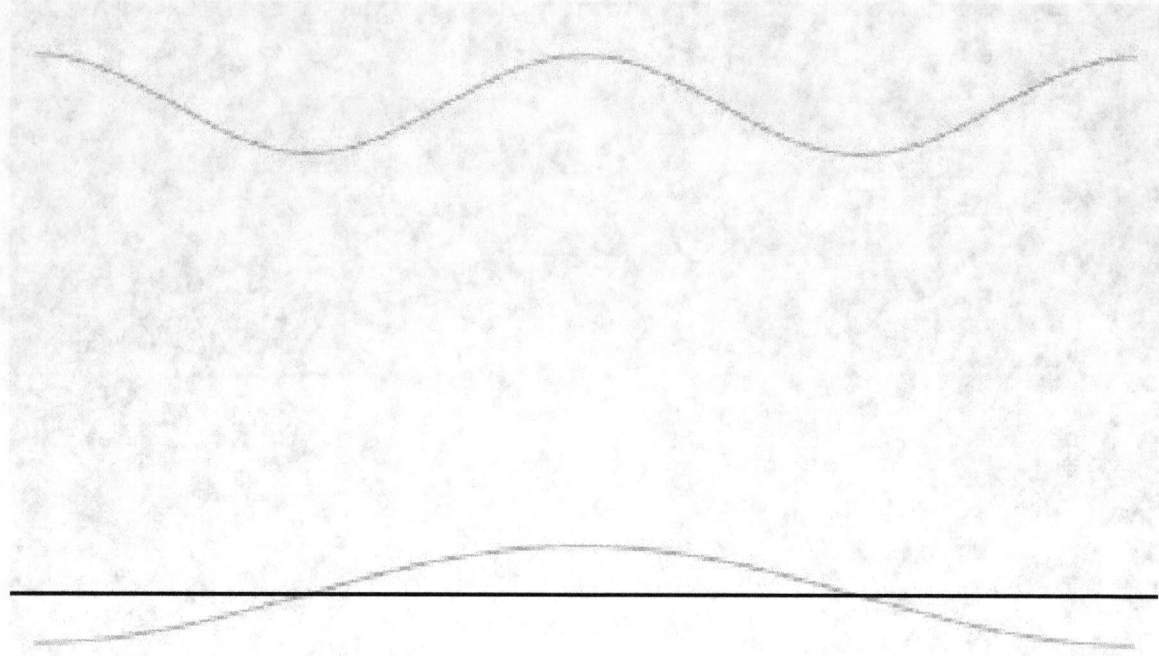

The superposition of several plane waves to form a wave packet. This wave packet becomes increasingly localized with the addition of many waves. The Fourier transform is a mathematical operation that separates a wave packet into its individual plane waves. Note that the waves shown here are real for illustrative purposes only, whereas in quantum mechanics the wave function is generally complex.

10.1.1 Wave mechanics interpretation

(Ref [9])

$$\Psi = Ae^{i(px - \omega t)}$$

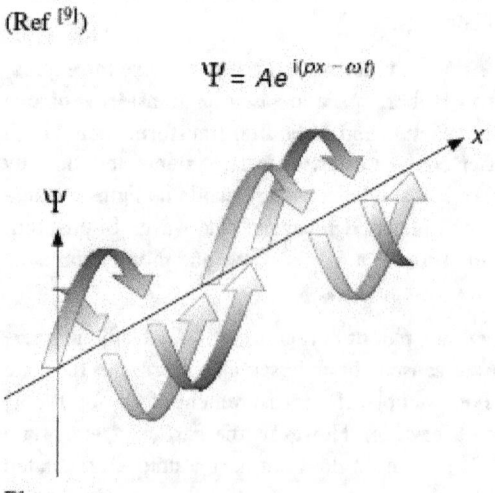

Plane wave

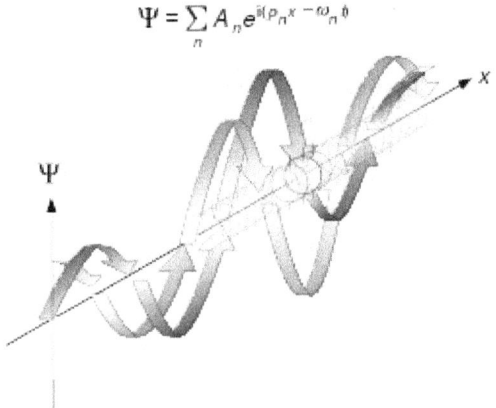

$$\Psi = \sum_n A_n e^{i(p_n x - \omega_n t)}$$

Wave packet

Propagation of de Broglie waves in 1d—real part of the complex amplitude is blue, imaginary part is green. The probability (shown as the colour opacity) of finding the particle at a given point x is spread out like a waveform, there is no definite position of the particle. As the amplitude increases above zero the curvature reverses sign, so the amplitude begins to decrease again, and vice versa—the result is an alternating amplitude: a wave.

Main article: Wave packet

Main article: Schrödinger equation

According to the de Broglie hypothesis, every object in the universe is a wave, a situation which gives rise to this phenomenon. The position of the particle is described by a wave function $\Psi(x, t)$. The time-independent wave function of a single-moded plane wave of wavenumber k_0 or momentum p_0 is

$$\psi(x) \propto e^{ik_0 x} = e^{ip_0 x/\hbar} .$$

The Born rule states that this should be interpreted as a probability density function in the sense that the probability of finding the particle between a and b is

$$P[a \le X \le b] = \int_a^b |\psi(x)|^2 \, \mathrm{d}x .$$

In the case of the single-moded plane wave, $|\psi(x)|^2$ is a uniform distribution. In other words, the particle position is extremely uncertain in the sense that it could be essentially anywhere along the wave packet. Consider a wave function that is a sum of many waves, however, we may write this as

$$\psi(x) \propto \sum_n A_n e^{ip_n x/\hbar} ,$$

where An represents the relative contribution of the mode pn to the overall total. The figures to the right show how with the addition of many plane waves, the wave packet can become more localized. We may take this a step further to the continuum limit, where the wave function is an integral over all possible modes

$$\psi(x) = \frac{1}{\sqrt{2\pi\hbar}} \int_{-\infty}^{\infty} \phi(p) \cdot e^{ipx/\hbar} \, dp ,$$

with $\phi(p)$ representing the amplitude of these modes and is called the wave function in momentum space. In mathematical terms, we say that $\phi(p)$ is the *Fourier transform* of $\psi(x)$ and that x and p are conjugate variables. Adding together all of these plane waves comes at a cost, namely the momentum has become less precise, having become a mixture of waves of many different momenta.

One way to quantify the precision of the position and momentum is the standard deviation σ. Since $|\psi(x)|^2$ is a probability density function for position, we calculate its standard deviation.

The precision of the position is improved, i.e. reduced σ_x, by using many plane waves, thereby weakening the precision of the momentum, i.e. increased σ_p. Another way of stating this is that σ_x and σ_p have an inverse relationship or are at least bounded from below. This is the uncertainty principle, the exact limit of which is the Kennard bound. Click the *show* button below to see a semi-formal derivation of the Kennard inequality using wave mechanics.

10.1.2 Matrix mechanics interpretation

(Ref [9])

Main article: Matrix mechanics

In matrix mechanics, observables such as position and momentum are represented by self-adjoint operators. When considering pairs of observables, an important quantity is the *commutator*. For a pair of operators $\hat{A}$ and $\hat{B}$, one defines their commutator as

$$[\hat{A}, \hat{B}] = \hat{A}\hat{B} - \hat{B}\hat{A}.$$

In the case of position and momentum, the commutator is the canonical commutation relation

$$[\hat{x}, \hat{p}] = i\hbar.$$

The physical meaning of the non-commutativity can be understood by considering the effect of the commutator on position and momentum eigenstates. Let $|\psi\rangle$ be a right eigenstate of position with a constant eigenvalue x_0. By definition, this means that $\hat{x}|\psi\rangle = x_0|\psi\rangle$. Applying the commutator to $|\psi\rangle$ yields

$$[\hat{x}, \hat{p}]|\psi\rangle = (\hat{x}\hat{p} - \hat{p}\hat{x})|\psi\rangle = (\hat{x} - x_0\hat{I}) \cdot \hat{p}|\psi\rangle = i\hbar|\psi\rangle,$$

where $\hat{I}$ is the identity operator.

Suppose, for the sake of proof by contradiction, that $|\psi\rangle$ is also a right eigenstate of momentum, with constant eigenvalue p_0. If this were true, then one could write

$$(\hat{x} - x_0\hat{I}) \cdot \hat{p}|\psi\rangle = (\hat{x} - x_0\hat{I}) \cdot p_0|\psi\rangle = (x_0\hat{I} - x_0\hat{I}) \cdot p_0|\psi\rangle = 0.$$

On the other hand, the above canonical commutation relation requires that

$$[\hat{x}, \hat{p}]|\psi\rangle = i\hbar|\psi\rangle \neq 0.$$

This implies that no quantum state can simultaneously be both a position and a momentum eigenstate.

When a state is measured, it is projected onto an eigenstate in the basis of the relevant observable. For example, if a particle's position is measured, then the state amounts to a position eigenstate. This means that the state is *not* a momentum eigenstate, however, but rather it can be represented as a sum of multiple momentum basis eigenstates. In other words, the momentum must be less precise. This precision may be quantified by the standard deviations,

$$\sigma_x = \sqrt{\langle \hat{x}^2 \rangle - \langle \hat{x} \rangle^2}$$

$$\sigma_p = \sqrt{\langle \hat{p}^2 \rangle - \langle \hat{p} \rangle^2}.$$

As in the wave mechanics interpretation above, one sees a tradeoff between the respective precisions of the two, quantified by the uncertainty principle.

10.2 Robertson–Schrödinger uncertainty relations

The most common general form of the uncertainty principle is the *Robertson uncertainty relation*.[14]

For an arbitrary Hermitian operator $\hat{\mathcal{O}}$ we can associate a standard deviation

$$\sigma_{\mathcal{O}} = \sqrt{\langle \hat{\mathcal{O}}^2 \rangle - \langle \hat{\mathcal{O}} \rangle^2}$$

where the brackets $\langle \mathcal{O} \rangle$ indicate an expectation value. For a pair of operators $\hat{A}$ and $\hat{B}$, we may define their *commutator* as

$$[\hat{A}, \hat{B}] = \hat{A}\hat{B} - \hat{B}\hat{A}$$

In this notation, the Robertson uncertainty relation is given by

$$\sigma_A \sigma_B \geq \left| \frac{1}{2i} \langle [\hat{A}, \hat{B}] \rangle \right| = \frac{1}{2} \left| \langle [\hat{A}, \hat{B}] \rangle \right|$$

The Robertson uncertainty relation immediately follows from a slightly stronger inequality, the *Schrödinger uncertainty relation*,[15]

where we have introduced the *anticommutator*,

$$\{\hat{A}, \hat{B}\} = \hat{A}\hat{B} + \hat{B}\hat{A}$$

Since the Robertson and Schrödinger relations are for general operators, the relations can be applied to any two observables to obtain specific uncertainty relations. A few of the most common relations found in the literature are given below.

- For position and linear momentum, the canonical commutation relation $[\hat{x}, \hat{p}] = i\hbar$ implies the Kennard inequality from above:

$$\sigma_x \sigma_p \geq \frac{\hbar}{2}$$

- For two orthogonal components of the total angular momentum operator of an object:

  $\sigma_{J_i} \sigma_{J_j} \geq \frac{\hbar}{2} |\langle J_k \rangle|$,

  where i, j, k are distinct and Ji denotes angular momentum along the xi axis. This relation implies that unless all three components vanish together, only a single component of a system's angular momentum can be defined with arbitrary precision, normally the component parallel to an external (magnetic or electric) field. Moreover, for $[J_x, J_y] = i\hbar\epsilon_{xyz}J_z$, a choice $\hat{A} = J_x$, $\hat{B} = J_y$, in angular momentum multiplets, $\psi = |j, m \rangle$, bounds the Casimir invariant (angular momentum squared, $\langle J_x^2 + J_y^2 + J_z^2 \rangle$) from below and thus yields useful constraints such as $j(j+1) \geq m(m+1)$, and hence $j \geq m$, among others.

- In non-relativistic mechanics, time is privileged as an independent variable. Nevertheless, in 1945, L. I. Mandelshtam and I. E. Tamm derived a non-relativistic *time–energy uncertainty relation*, as follows.[23] For a quantum system in a non-stationary state ψ and an observable B represented by a self-adjoint operator $\hat{B}$, the following formula holds:

$$\sigma_E \frac{\sigma_B}{\left|\frac{d\langle\hat{B}\rangle}{dt}\right|} \geq \frac{\hbar}{2}$$

where σE is the standard deviation of the energy operator (Hamiltonian) in the state ψ, σB stands for the standard deviation of B. Although the second factor in the left-hand side has dimension of time, it is different from the time parameter that enters the Schrödinger equation. It is a *lifetime* of the state ψ with respect to the observable B: In other words, this is the *time interval* (Δt) after which the expectation value $\langle\hat{B}\rangle$ changes appreciably.

An informal, heuristic meaning of the principle is the following: A state that only exists for a short time cannot have a definite energy. To have a definite energy, the frequency of the state must be defined accurately, and this requires the state to hang around for many cycles, the reciprocal of the required accuracy. For example, in spectroscopy, excited states have a finite lifetime. By the time–energy uncertainty principle, they do not have a definite energy, and, each time they decay, the energy they release is slightly different. The average energy of the outgoing photon has a peak at the theoretical energy of the state, but the distribution has a finite width called the *natural linewidth*. Fast-decaying states have a broad linewidth, while slow decaying states have a narrow linewidth.[24]

The same linewidth effect also makes it difficult to specify the rest mass of unstable, fast-decaying particles in particle physics. The faster the particle decays (the shorter its lifetime), the less certain is its mass (the larger the particle's width).

- For the number of electrons in a superconductor and the phase of its Ginzburg–Landau order parameter[25][26]

$$\Delta N \Delta \phi \geq 1$$

10.3 Examples

(Refs [9][16])

10.3.1 Quantum harmonic oscillator stationary states

Main articles: Quantum harmonic oscillator and Stationary state

Consider a one-dimensional quantum harmonic oscillator (QHO). It is possible to express the position and momentum operators in terms of the creation and annihilation operators:

$$\hat{x} = \sqrt{\frac{\hbar}{2m\omega}}(a + a^\dagger)$$

$$\hat{p} = i\sqrt{\frac{m\omega\hbar}{2}}(a^\dagger - a)$$

Using the standard rules for creation and annihilation operators on the eigenstates of the QHO,

$$a^\dagger |n\rangle = \sqrt{n+1}|n+1\rangle$$

$$a|n\rangle = \sqrt{n}|n-1\rangle$$

the variances may be computed directly,

$$\sigma_x^2 = \frac{\hbar}{m\omega}\left(n+\frac{1}{2}\right)$$

$$\sigma_p^2 = \hbar m\omega \left(n+\frac{1}{2}\right).$$

The product of these standard deviations is then

$$\sigma_x \sigma_p = \hbar \left(n+\frac{1}{2}\right) \geq \frac{\hbar}{2}$$

In particular, the above Kennard bound[2] is saturated for the ground state $n=0$, for which the probability density is just the normal distribution.

10.3.2   Quantum harmonic oscillator with Gaussian initial condition

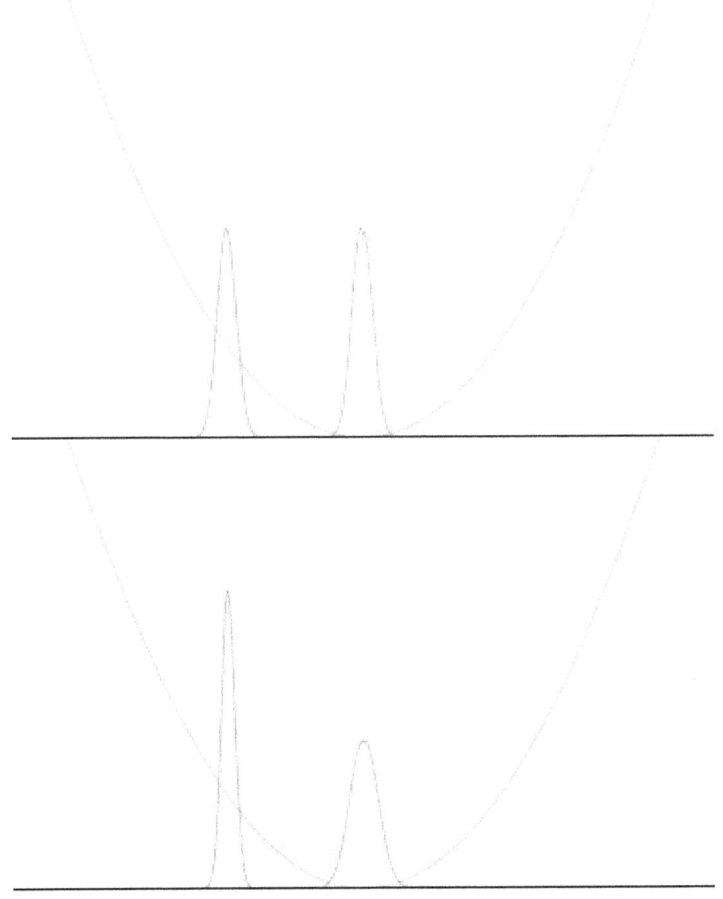

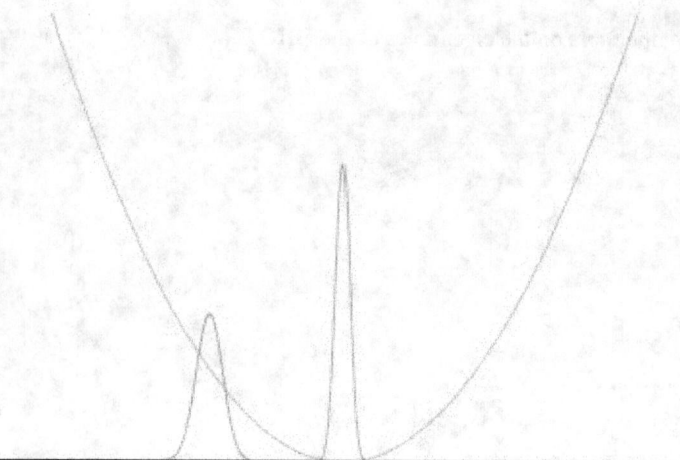

Position (blue) and momentum (red) probability densities for an initially Gaussian distribution. From top to bottom, the animations show the cases $\Omega=\omega$, $\Omega=2\omega$, and $\Omega=\omega/2$. Note the tradeoff between the widths of the distributions.

In a quantum harmonic oscillator of characteristic angular frequency ω, place a state that is offset from the bottom of the potential by some displacement x_0 as

$$\psi(x) = \left(\frac{m\Omega}{\pi\hbar}\right)^{1/4} \exp\left(-\frac{m\Omega(x-x_0)^2}{2\hbar}\right)$$

where Ω describes the width of the initial state but need not be the same as ω. Through integration over the propagator, we can solve for the full time-dependent solution. After many cancelations, the probability densities reduce to

$$|\Psi(x,t)|^2 \sim \mathcal{N}\left(x_0\cos(\omega t), \frac{\hbar}{2m\Omega}\left(\cos^2(\omega t) + \frac{\Omega^2}{\omega^2}\sin^2(\omega t)\right)\right)$$

$$|\Phi(p,t)|^2 \sim \mathcal{N}\left(-mx_0\omega\sin(\omega t), \frac{\hbar m\Omega}{2}\left(\cos^2(\omega t) + \frac{\omega^2}{\Omega^2}\sin^2(\omega t)\right)\right)$$

where we have used the notation $\mathcal{N}(\mu, \sigma^2)$ to denote a normal distribution of mean μ and variance σ^2. Copying the variances above and applying trigonometric identities, we can write the product of the standard deviations as

$$\sigma_x\sigma_p = \frac{\hbar}{2}\sqrt{\left(\cos^2(\omega t) + \frac{\Omega^2}{\omega^2}\sin^2(\omega t)\right)\left(\cos^2(\omega t) + \frac{\omega^2}{\Omega^2}\sin^2(\omega t)\right)}$$

$$= \frac{\hbar}{4}\sqrt{3 + \frac{1}{2}\left(\frac{\Omega^2}{\omega^2} + \frac{\omega^2}{\Omega^2}\right) - \left(\frac{1}{2}\left(\frac{\Omega^2}{\omega^2} + \frac{\omega^2}{\Omega^2}\right) - 1\right)\cos(4\omega t)}$$

From the relations

$$\frac{\Omega^2}{\omega^2} + \frac{\omega^2}{\Omega^2} \geq 2, \quad |\cos(4\omega t)| \leq 1$$

we can conclude

$$\sigma_x\sigma_p \geq \frac{\hbar}{4}\sqrt{3 + \frac{1}{2}\left(\frac{\Omega^2}{\omega^2} + \frac{\omega^2}{\Omega^2}\right) - \left(\frac{1}{2}\left(\frac{\Omega^2}{\omega^2} + \frac{\omega^2}{\Omega^2}\right) - 1\right)} = \frac{\hbar}{2}$$

10.3.3   Coherent states

Main article: Coherent state

A coherent state is a right eigenstate of the annihilation operator,

$$\hat{a}|\alpha\rangle = \alpha|\alpha\rangle$$

which may be represented in terms of Fock states as

$$|\alpha\rangle = e^{-\frac{|\alpha|^2}{2}} \sum_{n=0}^{\infty} \frac{\alpha^n}{\sqrt{n!}} |n\rangle$$

In the picture where the coherent state is a massive particle in a QHO, the position and momentum operators may be expressed in terms of the annihilation operators in the same formulas above and used to calculate the variances,

$$\sigma_x^2 = \frac{\hbar}{2m\omega}$$

$$\sigma_p^2 = \frac{\hbar m\omega}{2}$$

Therefore, every coherent state saturates the Kennard bound

$$\sigma_x \sigma_p = \sqrt{\frac{\hbar}{2m\omega}} \sqrt{\frac{\hbar m\omega}{2}} = \frac{\hbar}{2}$$

with position and momentum each contributing an amount $\sqrt{\hbar/2}$ in a "balanced" way. Moreover, every squeezed coherent state also saturates the Kennard bound although the individual contributions of position and momentum need not be balanced in general.

10.3.4   Particle in a box

Main article: Particle in a box

Consider a particle in a one-dimensional box of length L . The eigenfunctions in position and momentum space are

$$\psi_n(x,t) = \begin{cases} A\sin(k_n x)e^{-i\omega_n t}, & 0 < x < L, \\ 0, & \text{otherwise,} \end{cases}$$

and

$$\phi_n(p,t) = \sqrt{\frac{\pi L}{\hbar}} \frac{n\left(1-(-1)^n e^{-ikL}\right) e^{-i\omega_n t}}{\pi^2 n^2 - k^2 L^2}$$

where $\omega_n = \frac{\pi^2 \hbar n^2}{8L^2 m}$ and we have used the de Broglie relation $p = \hbar k$. The variances of x and p can be calculated explicitly:

$$\sigma_x^2 = \frac{L^2}{12}\left(1 - \frac{6}{n^2\pi^2}\right)$$

$$\sigma_p^2 = \left(\frac{\hbar n\pi}{L}\right)^2$$

The product of the standard deviations is therefore

$$\sigma_x\sigma_p = \frac{\hbar}{2}\sqrt{\frac{n^2\pi^2}{3} - 2}.$$

For all $n = 1, 2, 3\dots$, the quantity $\sqrt{\frac{n^2\pi^2}{3} - 2}$ is greater than 1, so the uncertainty principle is never violated. For numerical concreteness, the smallest value occurs when $n = 1$, in which case

$$\sigma_x\sigma_p = \frac{\hbar}{2}\sqrt{\frac{\pi^2}{3} - 2} \approx 0.568\hbar > \frac{\hbar}{2}$$

10.3.5 Constant momentum

Main article: Wave packet

Assume a particle initially has a momentum space wave function described by a normal distribution around some constant

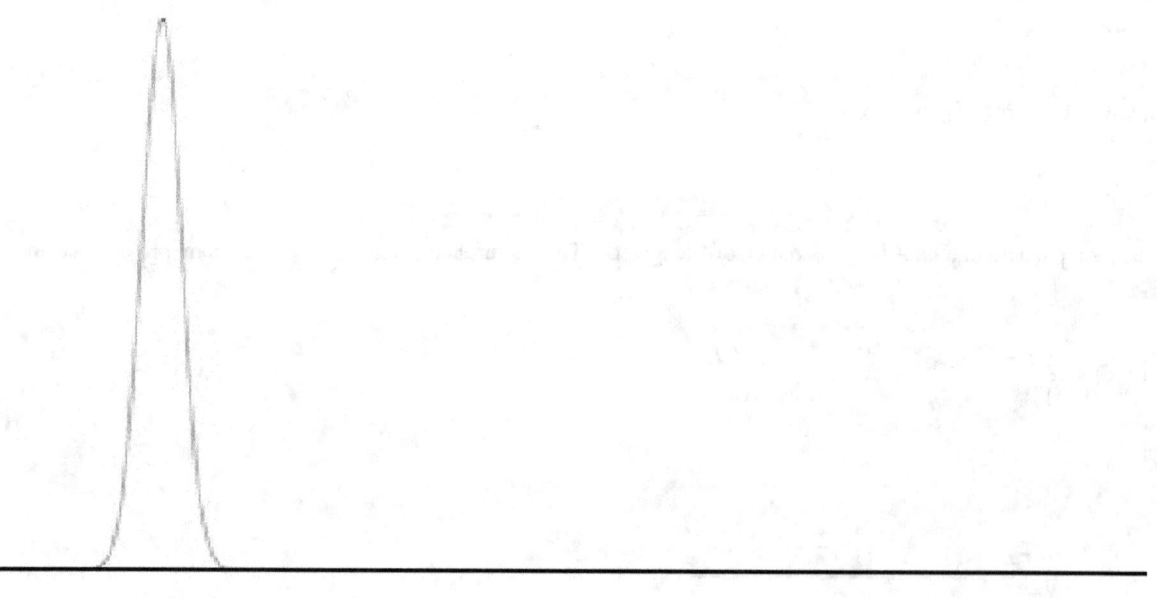

Position space probability density of an initially Gaussian state moving at minimally uncertain, constant momentum in free space

momentum p_0 according to

$$\phi(p) = \left(\frac{x_0}{\hbar\sqrt{\pi}}\right)^{1/2} \cdot \exp\left(\frac{-x_0^2(p-p_0)^2}{2\hbar^2}\right),$$

where we have introduced a reference scale $x_0 = \sqrt{\hbar/m\omega_0}$, with $\omega_0 > 0$ describing the width of the distribution—cf. nondimensionalization. If the state is allowed to evolve in free space, then the time-dependent momentum and position space wave functions are

$$\Phi(p,t) = \left(\frac{x_0}{\hbar\sqrt{\pi}}\right)^{1/2} \cdot \exp\left(\frac{-x_0^2(p-p_0)^2}{2\hbar^2} - \frac{ip^2t}{2m\hbar}\right),$$

$$\Psi(x,t) = \left(\frac{1}{x_0\sqrt{\pi}}\right)^{1/2} \cdot \frac{e^{-x_0^2p_0^2/2\hbar^2}}{\sqrt{1+i\omega_0 t}} \cdot \exp\left(-\frac{(x-ix_0^2p_0/\hbar)^2}{2x_0^2(1+i\omega_0 t)}\right).$$

Since $\langle p(t)\rangle = p_0$ and $\sigma_p(t) = \hbar/x_0\sqrt{2}$, this can be interpreted as a particle moving along with constant momentum at arbitrarily high precision. On the other hand, the standard deviation of the position is

$$\sigma_x = \frac{x_0}{\sqrt{2}}\sqrt{1+\omega_0^2 t^2}$$

such that the uncertainty product can only increase with time as

$$\sigma_x(t)\sigma_p(t) = \frac{\hbar}{2}\sqrt{1+\omega_0^2 t^2}$$

10.4 Additional uncertainty relations

10.4.1 Mixed states

The Robertson–Schrödinger uncertainty relation may be generalized in a straightforward way to describe mixed states.[27]

$$\sigma_A^2\sigma_B^2 \geq \left(\frac{1}{2}\mathrm{tr}(\rho\{A,B\}) - \mathrm{tr}(\rho A)\mathrm{tr}(\rho B)\right)^2 + \left(\frac{1}{2i}\mathrm{tr}(\rho[A,B])\right)^2$$

10.4.2 Phase space

In the phase space formulation of quantum mechanics, the Robertson–Schrödinger relation follows from a positivity condition on a real star-square function. Given a Wigner function $W(x,p)$ with star product $\star$ and a function f, the following is generally true:[28]

$$\langle f^* \star f\rangle = \int (f^* \star f)\, W(x,p)\, dx\, dp \geq 0.$$

Choosing $f = a + bx + cp$, we arrive at

$$\langle f^* \star f\rangle = \begin{bmatrix} a^* & b^* & c^* \end{bmatrix} \begin{bmatrix} 1 & \langle x\rangle & \langle p\rangle \\ \langle x\rangle & \langle x\star x\rangle & \langle x\star p\rangle \\ \langle p\rangle & \langle p\star x\rangle & \langle p\star p\rangle \end{bmatrix} \begin{bmatrix} a \\ b \\ c \end{bmatrix} \geq 0.$$

Since this positivity condition is true for *all* a, b, and c, it follows that all the eigenvalues of the matrix are positive. The positive eigenvalues then imply a corresponding positivity condition on the determinant:

$$\det \begin{bmatrix} 1 & \langle x \rangle & \langle p \rangle \\ \langle x \rangle & \langle x \star x \rangle & \langle x \star p \rangle \\ \langle p \rangle & \langle p \star x \rangle & \langle p \star p \rangle \end{bmatrix} = \det \begin{bmatrix} 1 & \langle x \rangle & \langle p \rangle \\ \langle x \rangle & \langle x^2 \rangle & \langle xp + \frac{i\hbar}{2} \rangle \\ \langle p \rangle & \langle xp - \frac{i\hbar}{2} \rangle & \langle p^2 \rangle \end{bmatrix} \geq 0,$$

or, explicitly, after algebraic manipulation,

$$\sigma_x^2 \sigma_p^2 = \left(\langle x^2 \rangle - \langle x \rangle^2 \right) \left(\langle p^2 \rangle - \langle p \rangle^2 \right) \geq \left(\langle xp \rangle - \langle x \rangle \langle p \rangle \right)^2 + \frac{\hbar^2}{4}.$$

10.4.3 Systematic error

The inequalities above focus on the statistical imprecision of observables as quantified by the standard deviation. Heisenberg's original version, however, was interested in *systematic error*, incurred by a disturbance of a quantum system by the measuring apparatus, i.e., an observer effect. If we let $\epsilon_{\mathcal{O}}$ represent the error (i.e., accuracy) of a measurement of an observable $\mathcal{O}$ and $\eta_{\mathcal{O}}$ represent its disturbance by the measurement process, then the following inequality holds:[5]

In fact, Heisenberg's uncertainty principle as originally described in the 1927 formulation mentions only the first term. Applying the notation above to Heisenberg's position–momentum relation, Heisenberg's argument could be rewritten as

Such a formulation is both mathematically incorrect and experimentally refuted.[29] It is also possible to derive a similar uncertainty relation combining both the statistical and systematic error components.[30] Nevertheless, with sufficient care, Heisenberg's intuitive observation may be formulated and proven in a mathematically consistent manner.[31]

10.4.4 Entropic uncertainty principle

Main article: Entropic uncertainty

For many distributions, the standard deviation is not a particularly natural way of quantifying the structure. For example, uncertainty relations in which one of the observables is an angle has little physical meaning for fluctuations larger than one period.[21][32][33][34] Other examples include highly bimodal distributions, or unimodal distributions with divergent variance.

A solution that overcomes these issues is an uncertainty based on entropic uncertainty instead of the product of variances. While formulating the many-worlds interpretation of quantum mechanics in 1957, Hugh Everett III conjectured a stronger extension of the uncertainty principle based on entropic certainty.[35] This conjecture, also studied by Hirschman[36] and proven in 1975 by Beckner[37] and by Iwo Bialynicki-Birula and Jerzy Mycielski[38] is

where we have used the Shannon entropy (*not* the quantum von Neumann entropy)

$$H_x = - \int |\psi(x)|^2 \ln(|\psi(x)|^2 \cdot \ell) \, dx = - \left\langle \ln(|\psi(x)|^2 \cdot \ell) \right\rangle$$

$$H_p = -\int |\phi(p)|^2 \ln(|\phi(p)|^2 \cdot \hbar/\ell)\, dp = -\left\langle \ln(|\phi(p)|^2 \cdot \hbar/\ell) \right\rangle$$

for some arbitrary fixed length scale ℓ .

From the inverse logarithmic Sobolev inequalities[39]

$$H_x \leq \frac{1}{2}\ln(2e\pi\sigma_x^2/\ell^2) \, ,$$

$$H_p \leq \frac{1}{2}\ln(2e\pi\sigma_p^2\ell^2/\hbar^2) \, ,$$

(equivalently, from the fact that normal distributions maximize the entropy of all such with a given variance), it readily follows that this entropic uncertainty principle is *stronger than the one based on standard deviations*, because

$$\sigma_x\sigma_p \geq \frac{\hbar}{2}\cdot \exp\left(H_x + H_p - \ln(e\pi)\right) \geq \frac{\hbar}{2} \, .$$

A few remarks on these inequalities. First, the choice of base e is a matter of popular convention in physics. The logarithm can alternatively be in any base, provided that it be consistent on both sides of the inequality. Second, the numerical value on the right hand side assumes the unitary convention of the Fourier transform, used throughout physics and elsewhere in this article. Third, the normal distribution saturates the inequality, and it is the only distribution with this property, because it is the maximum entropy probability distribution among those with fixed variance (cf. here for proof).

10.5 Harmonic analysis

Main article: Fourier transform § Uncertainty principle

In the context of harmonic analysis, a branch of mathematics, the uncertainty principle implies that one cannot at the same time localize the value of a function and its Fourier transform. To wit, the following inequality holds,

$$\left(\int_{-\infty}^{\infty} x^2|f(x)|^2\, dx\right)\left(\int_{-\infty}^{\infty} \xi^2|\hat{f}(\xi)|^2\, d\xi\right) \geq \frac{\|f\|_2^4}{16\pi^2}.$$

Further mathematical uncertainty inequalities, including the above entropic uncertainty, hold between a function f and its Fourier transform $\hat{f}$.[40][41][42]

10.5.1 Signal processing

In the context of signal processing, and in particular time–frequency analysis, uncertainty principles are referred to as the **Gabor limit**, after Dennis Gabor, or sometimes the *Heisenberg–Gabor limit*. The basic result, which follows from "Benedicks's theorem", below, is that a function cannot be both time limited and band limited (a function and its Fourier transform cannot both have bounded domain)—see bandlimited versus timelimited.

Stated alternatively, "One cannot simultaneously sharply localize a signal (function f) in both the time domain and frequency domain ($\hat{f}$, its Fourier transform)".

When applied to filters, the result implies that one cannot achieve high temporal resolution and frequency resolution at the same time; a concrete example are the resolution issues of the short-time Fourier transform—if one uses a wide window, one achieves good frequency resolution at the cost of temporal resolution, while a narrow window has the opposite trade-off.

Alternate theorems give more precise quantitative results, and, in time–frequency analysis, rather than interpreting the (1-dimensional) time and frequency domains separately, one instead interprets the limit as a lower limit on the support of a function in the (2-dimensional) time–frequency plane. In practice, the Gabor limit limits the *simultaneous* time–frequency resolution one can achieve without interference; it is possible to achieve higher resolution, but at the cost of different components of the signal interfering with each other.

10.5.2 Benedicks's theorem

Amrein-Berthier[43] and Benedicks's theorem[44] intuitively says that the set of points where f is non-zero and the set of points where f is nonzero cannot both be small.

Specifically, it is impossible for a function f in $L^2(\mathbf{R})$ and its Fourier transform f to both be supported on sets of finite Lebesgue measure. A more quantitative version is[45][46]

$$\|f\|_{L^2(\mathbf{R}^d)} \le Ce^{C|S||\Sigma|}\left(\|f\|_{L^2(S^c)} + \|\hat{f}\|_{L^2(\Sigma^c)}\right).$$

One expects that the factor $Ce^{C|S||\Sigma|}$ may be replaced by $Ce^{C(|S||\Sigma|)^{1/d}}$, which is only known if either S or Σ is convex.

10.5.3 Hardy's uncertainty principle

The mathematician G. H. Hardy formulated the following uncertainty principle:[47] it is not possible for f and f to both be "very rapidly decreasing." Specifically, if f in $L^2(\mathbf{R})$ is such that

$$|f(x)| \le C(1+|x|)^N e^{-a\pi x^2}$$

and

$$|\hat{f}(\xi)| \le C(1+|\xi|)^N e^{-b\pi\xi^2} \ (C > 0, N \text{ an integer}),$$

then, if $ab > 1$, $f = 0$, while if $ab = 1$, then there is a polynomial P of degree $\le N$ such that

$$f(x) = P(x)e^{-a\pi x^2}.$$

This was later improved as follows: if $f \in L^2(\mathbf{R}^d)$ is such that

$$\int_{\mathbf{R}^d} \int_{\mathbf{R}^d} |f(x)||\hat{f}(\xi)| \frac{e^{\pi|\langle x,\xi \rangle|}}{(1+|x|+|\xi|)^N} \, dx \, d\xi < +\infty,$$

then

$$f(x) = P(x)e^{-\pi\langle Ax,x \rangle},$$

where P is a polynomial of degree $(N - d)/2$ and A is a real $d \times d$ positive definite matrix.

This result was stated in Beurling's complete works without proof and proved in Hörmander[48] (the case $d = 1, N = 0$) and Bonami, Demange, and Jaming[49] for the general case. Note that Hörmander–Beurling's version implies the case

$ab > 1$ in Hardy's Theorem while the version by Bonami–Demange–Jaming covers the full strength of Hardy's Theorem. A different proof of Beurling's theorem based on Liouville's theorem appeared in ref.[50]

A full description of the case $ab < 1$ as well as the following extension to Schwarz class distributions appears in ref.[51]

Theorem. If a tempered distribution $f \in \mathcal{S}'(\mathbb{R}^d)$ is such that

$$e^{\pi|x|^2} f \in \mathcal{S}'(\mathbb{R}^d)$$

and

$$e^{\pi|\xi|^2} \hat{f} \in \mathcal{S}'(\mathbb{R}^d) \,,$$

then

$$f(x) = P(x) e^{-\pi\langle Ax, x\rangle} \,,$$

for some convenient polynomial P and real positive definite matrix A of type $d \times d$.

10.6 History

Werner Heisenberg formulated the Uncertainty Principle at Niels Bohr's institute in Copenhagen, while working on the mathematical foundations of quantum mechanics.[52]

In 1925, following pioneering work with Hendrik Kramers, Heisenberg developed matrix mechanics, which replaced the ad hoc old quantum theory with modern quantum mechanics. The central premise was that the classical concept of motion does not fit at the quantum level, as electrons in an atom do not travel on sharply defined orbits. Rather, their motion is smeared out in a strange way: the Fourier transform of its time dependence only involves those frequencies that could be observed in the quantum jumps of their radiation.

Heisenberg's paper did not admit any unobservable quantities like the exact position of the electron in an orbit at any time; he only allowed the theorist to talk about the Fourier components of the motion. Since the Fourier components were not defined at the classical frequencies, they could not be used to construct an exact trajectory, so that the formalism could not answer certain overly precise questions about where the electron was or how fast it was going.

In March 1926, working in Bohr's institute, Heisenberg realized that the non-commutativity implies the uncertainty principle. This implication provided a clear physical interpretation for the non-commutativity, and it laid the foundation for what became known as the Copenhagen interpretation of quantum mechanics. Heisenberg showed that the commutation relation implies an uncertainty, or in Bohr's language a complementarity.[53] Any two variables that do not commute cannot be measured simultaneously—the more precisely one is known, the less precisely the other can be known. Heisenberg wrote:

> It can be expressed in its simplest form as follows: One can never know with perfect accuracy both of those two important factors which determine the movement of one of the smallest particles—its position and its velocity. It is impossible to determine accurately *both* the position and the direction and speed of a particle *at the same instant*.[54]

In his celebrated 1927 paper, "Über den anschaulichen Inhalt der quantentheoretischen Kinematik und Mechanik" ("On the Perceptual Content of Quantum Theoretical Kinematics and Mechanics"), Heisenberg established this expression as the minimum amount of unavoidable momentum disturbance caused by any position measurement,[1] but he did not give a precise definition for the uncertainties Δx and Δp. Instead, he gave some plausible estimates in each case separately. In his Chicago lecture[55] he refined his principle:

Werner Heisenberg and Niels Bohr

Kennard[2] in 1927 first proved the modern inequality:

where $\hbar = h/2\pi$, and σx, σp are the standard deviations of position and momentum. Heisenberg only proved relation (2) for the special case of Gaussian states.[55]

10.6.1 Terminology and translation

Throughout the main body of his original 1927 paper, written in German, Heisenberg used the word, "Ungenauigkeit" ("indeterminacy"),[1] to describe the basic theoretical principle. Only in the endnote did he switch to the word, "Unsicherheit" ("uncertainty"). When the English-language version of Heisenberg's textbook, *The Physical Principles of the Quantum Theory*, was published in 1930, however, the translation "uncertainty" was used, and it became the more commonly used term in the English language thereafter.[56]

10.6.2 Heisenberg's microscope

Main article: Heisenberg's microscope

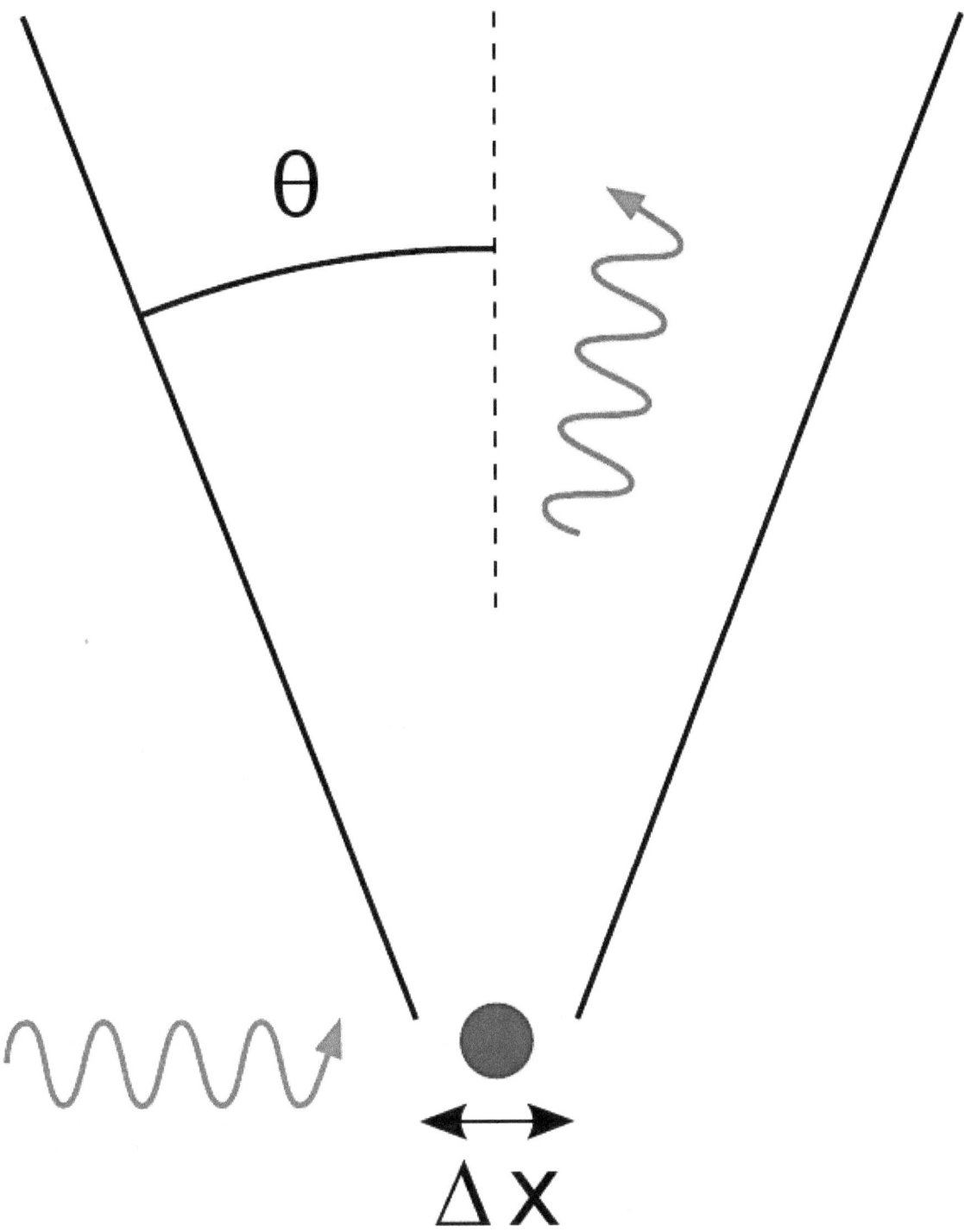

Heisenberg's gamma-ray microscope for locating an electron (shown in blue). The incoming gamma ray (shown in green) is scattered by the electron up into the microscope's aperture angle θ. The scattered gamma-ray is shown in red. Classical optics shows that the electron position can be resolved only up to an uncertainty Δx that depends on θ and the wavelength λ of the incoming light.

The principle is quite counter-intuitive, so the early students of quantum theory had to be reassured that naive measurements to violate it were bound always to be unworkable. One way in which Heisenberg originally illustrated the intrinsic impossibility of violating the uncertainty principle is by using an imaginary microscope as a measuring device.[55]

He imagines an experimenter trying to measure the position and momentum of an electron by shooting a photon at it.

> Problem 1 – If the photon has a short wavelength, and therefore, a large momentum, the position can be measured accurately. But the photon scatters in a random direction, transferring a large and uncertain amount of momentum to the electron. If the photon has a long wavelength and low momentum, the collision does not disturb the electron's momentum very much, but the scattering will reveal its position only vaguely.

> Problem 2 – If a large aperture is used for the microscope, the electron's location can be well resolved (see Rayleigh criterion); but by the principle of conservation of momentum, the transverse momentum of the incoming photon affects the electrons beamline momentum and hence, the new momentum of the electron resolves poorly. If a small aperture is used, the accuracy of both resolutions is the other way around.

The combination of these trade-offs imply that no matter what photon wavelength and aperture size are used, the product of the uncertainty in measured position and measured momentum is greater than or equal to a lower limit, which is (up to a small numerical factor) equal to Planck's constant.[57] Heisenberg did not care to formulate the uncertainty principle as an exact limit (which is elaborated below), and preferred to use it instead, as a heuristic quantitative statement, correct up to small numerical factors, which makes the radically new noncommutativity of quantum mechanics inevitable.

10.7 Critical reactions

Main article: Bohr–Einstein debates

The Copenhagen interpretation of quantum mechanics and Heisenberg's Uncertainty Principle were, in fact, seen as twin targets by detractors who believed in an underlying determinism and realism. According to the Copenhagen interpretation of quantum mechanics, there is no fundamental reality that the quantum state describes, just a prescription for calculating experimental results. There is no way to say what the state of a system fundamentally is, only what the result of observations might be.

Albert Einstein believed that randomness is a reflection of our ignorance of some fundamental property of reality, while Niels Bohr believed that the probability distributions are fundamental and irreducible, and depend on which measurements we choose to perform. Einstein and Bohr debated the uncertainty principle for many years. Some experiments within the first decade of the twenty-first century have cast doubt on how extensively the uncertainty principle applies.[58]

10.7.1 Einstein's slit

The first of Einstein's thought experiments challenging the uncertainty principle went as follows:

> Consider a particle passing through a slit of width d. The slit introduces an uncertainty in momentum of approximately h/d because the particle passes through the wall. But let us determine the momentum of the particle by measuring the recoil of the wall. In doing so, we find the momentum of the particle to arbitrary accuracy by conservation of momentum.

Bohr's response was that the wall is quantum mechanical as well, and that to measure the recoil to accuracy Δp, the momentum of the wall must be known to this accuracy before the particle passes through. This introduces an uncertainty in the position of the wall and therefore the position of the slit equal to $h/\Delta p$, and if the wall's momentum is known precisely enough to measure the recoil, the slit's position is uncertain enough to disallow a position measurement.

A similar analysis with particles diffracting through multiple slits is given by Richard Feynman.[59]

In another thought experiment Lawrence Marq Goldberg theorized that one could, for example, determine the position of a particle and then travel back in time to a point before the first reading to measure the velocity, then time travel back to a point before the second (earlier) reading was taken to deliver the resulting measurements before the particle was disturbed so that the measurements did not need to be taken. This, of course, would result in a temporal paradox. But it does support his contention that "the problems inherent to the uncertainly principle lay in the measuring not in the "uncertainty" of physics."

10.7.2 Einstein's box

Bohr was present when Einstein proposed the thought experiment which has become known as Einstein's box. Einstein argued that "Heisenberg's uncertainty equation implied that the uncertainty in time was related to the uncertainty in energy, the product of the two being related to Planck's constant."[60] Consider, he said, an ideal box, lined with mirrors so that it can contain light indefinitely. The box could be weighed before a clockwork mechanism opened an ideal shutter at a chosen instant to allow one single photon to escape. "We now know, explained Einstein, precisely the time at which the photon left the box."[61] "Now, weigh the box again. The change of mass tells the energy of the emitted light. In this manner, said Einstein, one could measure the energy emitted and the time it was released with any desired precision, in contradiction to the uncertainty principle."[60]

Bohr spent a sleepless night considering this argument, and eventually realized that it was flawed. He pointed out that if the box were to be weighed, say by a spring and a pointer on a scale, "since the box must move vertically with a change in its weight, there will be uncertainty in its vertical velocity and therefore an uncertainty in its height above the table. ... Furthermore, the uncertainty about the elevation above the earth's surface will result in an uncertainty in the rate of the clock,"[62] because of Einstein's own theory of gravity's effect on time. "Through this chain of uncertainties, Bohr showed that Einstein's light box experiment could not simultaneously measure exactly both the energy of the photon and the time of its escape."[63]

10.7.3 EPR paradox for entangled particles

Bohr was compelled to modify his understanding of the uncertainty principle after another thought experiment by Einstein. In 1935, Einstein, Podolsky and Rosen (see EPR paradox) published an analysis of widely separated entangled particles. Measuring one particle, Einstein realized, would alter the probability distribution of the other, yet here the other particle could not possibly be disturbed. This example led Bohr to revise his understanding of the principle, concluding that the uncertainty was not caused by a direct interaction.[64]

But Einstein came to much more far-reaching conclusions from the same thought experiment. He believed the "natural basic assumption" that a complete description of reality, would have to predict the results of experiments from "locally changing deterministic quantities", and therefore, would have to include more information than the maximum possible allowed by the uncertainty principle.

In 1964, John Bell showed that this assumption can be falsified, since it would imply a certain inequality between the probabilities of different experiments. Experimental results confirm the predictions of quantum mechanics, ruling out Einstein's basic assumption that led him to the suggestion of his *hidden variables*. Ironically this fact is one of the best pieces of evidence supporting Karl Popper's philosophy of invalidation of a theory by falsification-experiments. That is to say, here Einstein's "basic assumption" became falsified by experiments based on Bell's inequalities. For the objections of Karl Popper to the Heisenberg inequality itself, see below.

While it is possible to assume that quantum mechanical predictions are due to nonlocal, hidden variables, and in fact David Bohm invented such a formulation, this resolution is not satisfactory to the vast majority of physicists. The question of whether a random outcome is predetermined by a nonlocal theory can be philosophical, and it can be potentially intractable. If the hidden variables are not constrained, they could just be a list of random digits that are used to produce the measurement outcomes. To make it sensible, the assumption of nonlocal hidden variables is sometimes augmented by a second assumption—that the size of the observable universe puts a limit on the computations that these variables can do. A nonlocal theory of this sort predicts that a quantum computer would encounter fundamental obstacles when attempting to factor numbers of approximately 10,000 digits or more; a potentially achievable task in quantum mechanics.[65]

10.7.4 Popper's criticism

Main article: Popper's experiment

Karl Popper approached the problem of indeterminacy as a logician and metaphysical realist.[66] He disagreed with the application of the uncertainty relations to individual particles rather than to ensembles of identically prepared particles, referring to them as "statistical scatter relations".[66][67] In this statistical interpretation, a *particular* measurement may be made to arbitrary precision without invalidating the quantum theory. This directly contrasts with the Copenhagen interpretation of quantum mechanics, which is non-deterministic but lacks local hidden variables.

In 1934, Popper published *Zur Kritik der Ungenauigkeitsrelationen* (*Critique o f the Uncertainty Relations*) and in the same year *Logik der Forschung* (translated and updated by the author as *The Logic of Scientific Discovery* in 1959), outlining his arguments for the statistical interpretation. In 1982, he further developed his theory in *Quantum theory and the schism in Physics*, writing:

> [Heisenberg's] formulae are, beyond all doubt, derivable *statistical formulae* of the quantum theory. But they have been *habitually misinterpreted* by those quantum theorists who said that these formulae can be interpreted as determining some upper limit to the *precision of our measurements*.[original emphasis][69]

Popper proposed an experiment to falsify the uncertainty relations, although he later withdrew his initial version after discussions with Weizsäcker, Heisenberg, and Einstein; this experiment may have influenced the formulation of the EPR experiment.[66][70]

10.7.5 Many-worlds uncertainty

Main article: Many-worlds interpretation

The many-worlds interpretation originally outlined by Hugh Everett III in 1957 is partly meant to reconcile the differences between the Einstein and Bohr's views by replacing Bohr's wave function collapse with an ensemble of deterministic and independent universes whose *distribution* is governed by wave functions and the Schrödinger equation. Thus, uncertainty in the many-worlds interpretation follows from each observer within any universe having no knowledge of what goes on in the other universes.

10.7.6 Free will

Some scientists including Arthur Compton[71] and Martin Heisenberg[72] have suggested that the uncertainty principle, or at least the general probabilistic nature of quantum mechanics, could be evidence for the two-stage model of free will. The standard view, however, is that apart from the basic role of quantum mechanics as a foundation for chemistry, nontrivial biological mechanisms requiring quantum mechanics are unlikely, due to the rapid decoherence time of quantum systems at room temperature.[73]

10.8 See also

- Canonical commutation relation

- Correspondence principle

- Correspondence rules

- Gromov's non-squeezing theorem

- Discrete Fourier transform#Uncertainty principle

- Heisenbug

- Introduction to quantum mechanics

- Operationalization

- Observer effect (information technology)

- Observer effect (physics)

- Quantum indeterminacy

- Quantum non-equilibrium

- Quantum tunnelling

- *The Part and The Whole* (book)

- Weak measurement

10.9   Notes

[1] Heisenberg, W. (1927), "Über den anschaulichen Inhalt der quantentheoretischen Kinematik und Mechanik", *Zeitschrift für Physik* (in German) 43 (3–4): 172–198, Bibcode:1927ZPhy...43..172H, doi:10.1007/BF01397280.. Annotated pre-publication proof sheet of Über den anschaulichen Inhalt der quantentheoretischen Kinematik und Mechanik, March 23, 1927.

[2] Kennard, E. H. (1927), "Zur Quantenmechanik einfacher Bewegungstypen", *Zeitschrift für Physik* (in German) 44 (4–5): 326, Bibcode:1927ZPhy...44..326K, doi:10.1007/BF01391200.

[3] Weyl, H. (1928), *Gruppentheorie und Quantenmechanik*, Leipzig: Hirzel

[4] Furuta, Aya (2012), "One Thing Is Certain: Heisenberg's Uncertainty Principle Is Not Dead", *Scientific American*

[5] Ozawa, Masanao (2003), "Universally valid reformulation of the Heisenberg uncertainty principle on noise and disturbance in measurement", *Physical Review A* 67 (4): 42105, arXiv:quant-ph/0207121, Bibcode:2003PhRvA..67d2105O, doi:10.1103

[6] Werner Heisenberg, *The Physical Principles of the Quantum Theory*, p. 20

[7] Rozema, L. A.; Darabi, A.; Mahler, D. H.; Hayat, A.; Soudagar, Y.; Steinberg, A. M. (2012). "Violation of Heisenberg's Measurement-Disturbance Relationship b y Weak Measurements". *Physical Review Letters* 109 (10). doi:10.1103/Phys

[8] Indian Institute of Technology Madras, Professor V. Balakrishnan, Lecture 1 – Introduction to Quantum Physics; Heisenberg's uncertainty principle, National Programme of Technology Enhanced Learning on YouTube

[9] L.D. Landau, E.M. Lifshitz (1977). *Quantum Mechanics: Non-Relativistic Theory*. Vol. 3 (3rd ed.). Pergamon Press. ISBN 978-0-08-020940-1. Online copy.

[10] Elion, W. J.; M. Matters, U. Geigenmüller & J. E. Mooij; Geigenmüller, U.; Mooij, J. E. (1994), "Direct demonstration of Heisenberg's uncertainty principle in a superconductor", *Nature* 371 (6498): 594–595, Bibcode:1994Natur.371..594E, doi:10.1038/371594a0

[11] Smithey, D. T.; M. Beck, J. Cooper, M. G. Raymer; Cooper, J.; Raymer, M. G. (1993), "Measurement of number–phase uncertainty relations of optical fields", *Phys. Rev. A* 48 (4): 3159–3167, Bibcode:1993PhRvA..48.3159S, doi:10.1103/PhysRevA.48 .3159,PMID 9909968

[12] Caves, Carlton (1981), "Quantum-mechanical noise in an interferometer", *Phys. Rev. D* 23 (8): 1693–1708, Bibcode:1981PhRvD.. 23.1693C,   doi:10.1103/PhysRevD.23.1693

[13] Claude Cohen-Tannoudji, Bernard Diu, Franck Laloë (1996), *Quantum mechanics*, Wiley-Interscience: Wiley, pp. 231–233, ISBN 978-0-471-56952-7

[14] Robertson, H. P. (1929), "The Uncertainty Principle", *Phys. Rev.* 34: 163–64, Bibcode:1929PhRv...34..163R, doi:10.1103

[15] Schrödinger, E. (1930), "Zum Heisenbergschen Unschärfeprinzip", *Sitzungsberichte der Preussischen Akademie der Wissenschaften, Physikalisch-mathematische Klasse* 14: 296–303

[16] Griffiths, David (2005), *Quantum Mechanics*, New Jersey: Pearson

[17] Riley, K. F.; M. P. Hobson and S. J. Bence (2006), *Mathematical Methods for Physics and Engineering*, Cambridge, p. 246

[18] Davidson, E. R. (1965), "On Derivations of the Uncertainty Principle", *J. Chem. Phys.* 42 (4): 1461, Bibcode:1965JChPh.42.1 doi:10.1063/1.1696139

[19] Hall, B. C. (2013), *Quantum Theory for Mathematicians*, Springer, p. 245

[20] Jackiw, Roman (1968), "Minimum Uncertainty Product, Number-Phase Uncertainty Product, and Coherent States", *J. Math. Phys.* 9 (3): 339, Bibcode:1968JMP.....9..339J, doi:10.1063/1.1664585

[21] Carruthers, P.; Nieto, M. M. (1968), "Phase and Angle Variables in Quantum Mechanics", *Rev. Mod. Phys.* 40 (2): 411, Bibcode:1968RvMP...40..411C, doi:10.1103/RevModPhys.40.411

[22] Hall, B. C. (2013), *Quantum Theory for Mathematicians*, Springer

[23] L. I. Mandelshtam, I. E. Tamm, *The uncertainty relation between energy and time in nonrelativistic quantum mechanics*, 1945

[24] The broad linewidth of fast decaying states makes it difficult to accurately measure the energy of the state, and researchers have even used detuned microwave cavities to slow down the decay rate, to get sharper peaks. Gabrielse, Gerald; H. Dehmelt (1985), "Observation of Inhibited Spontaneous Emission", *Physical Review Letters* 55 (1): 67–70, Bibcode:1985PhRvL..55...67G, doi:10.1103/PhysRevLett.55.67, PMID 10031682

[25] Likharev, K.K.; A.B. Zorin (1985), "Theory of Bloch-Wave Oscillations in Small Josephson Junctions", *J. Low Temp. Phys.* 59 (3/4): 347–382, Bibcode:1985JLTP...59..347L, doi:10.1007/BF00683782

[26] Anderson, P.W. (1964), "Special Effects in Superconductivity", in Caianiello, E.R., *Lectures on the Many-Body Problem, Vol. 2*, New York: Academic Press

[27] Steiger, Nathan. "Quantum Uncertainty and Conservation Law Restrictions on Gate Fidelity". Brigham Young University. Retrieved 19 June 2011.

[28] Curtright, T.; Zachos, C. (2001). "Negative Probability and Uncertainty Relations". *Modern Physics Letters A* 16 (37): 2381–2385. doi:10.1142/S021773230100576X.

[29] Erhart, Jacqueline; Stephan Sponar, Georg Sulyok, Gerald Badurek, Masanao Ozawa, Yuji Hasegawa (2012), "Experimental demonstration of a universally valid error-disturbance uncertainty relation in spin measurements", *Nature Physics* 8 (3): 185–189, arXiv:1201.1833, Bibcode:2012NatPh...8..185E, doi:10.1038/nphys2194 Cite uses deprecated parameter lcoauthors= (help)

[30] Fujikawa, Kazuo (2012), "Universall y vali d Heisenber g uncertainty relation", *Phys. Rev. A* 85 (6), arXiv:1205.1360, doi:10.1103/PhysRevA.85.062117

[31] Busch, P.; Lahti, P.; Werner, R. F. (2013). "Proof of Heisenberg's Error-Disturbance Relation". *Physical Review Letters* 111 (16). doi:10.1103/PhysRevLett.111.160405.; Busch, P.; Lahti, P.; Werner, R. F. (2014). "Heisenberg uncertainty for qubit measurements". *Physical Review A* 89. doi:10.1103/PhysRevA.89.012129.

[32] Judge, D. (1964), "On the uncertainty relation for angle variables", *Il Nuovo Cimento* 31 (2): 332–340, doi:10.1007/BF02733639

[33] Bouten, M.; N. Maene, P. Van Leuven (1965), "On an uncertainty relation for angle variables", *Il Nuovo Cimento* 37 (3): 1119–1125, doi:10.1007/BF02773197 Cite uses deprecated parameter lcoauthors= (help)

[34] Louisell, W. H. (1963), "Amplitude and phase uncertainty relations", *Physics Letters* 7 (1): 60–61, Bibcode:1963PhL......7...60L, doi:10.1016/0031-9163(63)90442-6

[35] DeWitt, B. S.; Graham, N. (1973), *The Many-Worlds Interpretation of Quantum Mechanics*, Princeton: Princeton University Press, pp. 52–53, ISBN 0-691-08126-3

[36] Hirschman, I. I., Jr. (1957), "A note on entropy", *American Journal of Mathematics* 79 (1): 152–156, doi:10.2307/2372390, JSTOR 2372390.

[37] Beckner, W. (1975), "Inequalities in Fourier analysis", *Annals of Mathematics* **102** (6): 159–182, doi:10.2307/1970980, JSTOR 1970980.

[38] Bialynicki-Birula, I.; Mycielski, J. (1975), "Uncertainty Relations for Information Entropy in Wave Mechanics", *Communications in Mathematical Physics* **44** (2): 129, Bibcode:1975CMaPh..44..129B, doi:10.1007/BF01608825

[39] Chafaï, D. (2003), *Gaussian maximum of entropy and reversed log-Sobolev inequality*, arXiv:math/0102227, doi:10.1007/978-3-540-36107-7_5, ISBN 978-3-540-00072-3

[40] Havin, V.; Jöricke, B. (1994), *The Uncertainty Principle in Harmonic Analysis*, Springer-Verlag

[41] Folland, Gerald; Sitaram, Alladi (May 1997), "The Uncertainty Principle: A Mathematical Survey", *Journal of Fourier Analysis and Applications* **3** (3): 207–238, doi:10.1007/BF02649110, MR 98f:42006

[42] Sitaram, A (2001), "Uncertainty principle, mathematical", in Hazewinkel, Michiel, *Encyclopedia of Mathematics*, Springer, ISBN 978-1-55608-010-4

[43] Amrein, W.O.; Berthier, A.M. (1977), "On support properties of L^p-functions and their Fourier transforms", *Journal of Functional Analysis* **24** (3): 258–267, doi:10.1016/0022-1236(77)90056-8.

[44] Benedicks, M. (1985), "On Fourier transforms of functions supported on sets of finite Lebesgue measure", *J. Math. Anal. Appl.* **106** (1): 180–183, doi:10.1016/0022-247X(85)90140-4

[45] Nazarov, F. (1994), "Local estimates for exponential polynomials and their applications to inequalities of the uncertainty principle type.", *St. Petersburg Math. J.* **5**: 663–717

[46] Jaming, Ph. (2007), "Nazarov's uncertainty principles in higher dimension", *J. Approx. Theory* **149** (1): 30–41, doi:10.1016

[47] Hardy, G.H. (1933), "A theorem concerning Fourier transforms", *Journal of the London Mathematical Society* **8** (3): 227–231, doi:10.1112/jlms/s1-8.3.227

[48] Hörmander, L. (1991), "A uniqueness theorem of Beurling for Fourier transform pairs", *Ark. Mat.* **29**: 231–240, Bibcode: doi:10.1007/BF02384339

[49] Bonami, A.; Demange, B.; Jaming, Ph. (2003), "Hermite functions and uncertainty principles for the Fourier and the windowed Fourier transforms", *Rev. Mat. Iberoamericana* **19**: 23–55., arXiv:math/0102111, Bibcode:2001math.......2111B, doi:10.

[50] Hedenmalm, H. (2012), "Heisenberg's uncertainty principle in the sense of Beurling", *J. Anal. Math.* **118** (2): 691–702, doi:10.1007/s11854-012-0048-9

[51] Demange, Bruno (2009), *Uncertainty Principles Associated to Non-degenerate Quadratic Forms*, Société Mathématique de France, ISBN 978-2-85629-297-6

[52] American Physical Society online exhibit on the Uncertainty Principle

[53] Bohr, Niels; Noll, Waldemar (1958), "Atomic Physics and Human Knowledge", *American Journal of Physics* (New York: Wiley) **26** (8): 38, Bibcode:1958AmJPh..26..596B, doi:10.1119/1.1934707

[54] Heisenberg, W., *Die Physik der Atomkerne*, Taylor & Francis, 1952, p. 30.

[55] Heisenberg, W. (1930), *Physikalische Prinzipien der Quantentheorie* (in German), Leipzig: Hirzel English translation *The Physical Principles of Quantum Theory*. Chicago: University of Chicago Press, 1930.

[56] Cassidy, David; Saperstein, Alvin M. (2009), "Beyond Uncertainty: Heisenberg, Quantum Physics, and the Bomb", *Physics Today* (New York: Bellevue Literary Press) **63**: 185, Bibcode:2010PhT....63a..49C, doi:10.1063/1.3293416

[57] Tipler, Paul A.; Llewellyn, Ralph A. (1999), "5–5", *Modern Physics* (3rd ed.), W. H. Freeman and Co., ISBN 1-57259-164-1

[58] R&D Magazine & University of Toronto, September 10, 2012 Scientists cast doubt on the uncertainty principle retrieved Sept 10, 2012

[59] Feynman lectures on Physics, vol 3, 2–2

[60] Gamow, G., *The great physicists from Galileo to Einstein*, Courier Dover, 1988, p.260.

[61] Kumar, M., *Quantum: Einstein, Bohr and the Great Debate About the Nature of Reality*, Icon, 2009, p. 282.

[62] Gamow, G., *The great physicists from Galileo to Einstein*, Courier Dover, 1988, p. 260–261.

[63] Kumar, M., *Quantum: Einstein, Bohr and the Great Debate About the Nature of Reality*, Icon, 2009, p. 287.

[64] Isaacson, Walter (2007), *Einstein: His Life and Universe*, New York: Simon & Schuster, p. 452, ISBN 978-0-7432-6473-0

[65] Gerardus 't Hooft has at times advocated this point of view.

[66] Popper, Karl (1959), *The Logic of Scientific Discovery*, Hutchinson & Co.

[67] Jarvie, Ian Charles; Milford, Karl; Miller, David W (2006), *Karl Popper: a centenary assessment, 3*, Ashgate Publishing, ISBN 978-0-7546-5712-5

[68] Popper, Karl; Carl Friedrich von Weizsäcker (1934), "Zur Kritik der Ungenauigkeitsrelationen (Critique of the Uncertainty Relations)", *Naturwissenschaften* 22 (48): 807–808, Bibcode:1934NW.....22..807P, doi:10.1007/BF01496543.

[69] Popper, K. *Quantum theory and the schism in Physics*, Unwin Hyman Ltd, 1982, pp. 53–54.

[70] Mehra, Jagdish; Rechenberg, Helmut (2001), *The Historical Development of Quantum Theory*, Springer, ISBN 978-0-387-95086-0

[71] Compton, A. H. (1931). "The Uncertainty Principle and Free Will". *Science* 74 (1911): 172. doi:10.1126/science.74.1911.172. PMID 17808216.

[72] Heisenberg, M. (2009). "Is free will an illusion?". *Nature* 459 (7244): 164. doi:10.1038/459164a.

[73] Davies, P. C. W. (2004). "Does quantum mechanic s pla y a non-trivial role in life?". *Biosystems* 78 (1–3): 69–79. doi:10. PMID 15555759.

10.10 External links

- Hazewinkel, Michiel, ed. (2001), "Uncertainty principle", *Encyclopedia of Mathematics*, Springer, ISBN 978-1-55608-010-4

- Matter as a Wave – a chapter from an online textbook

- Quantum mechanics: Myths and facts

- Stanford Encyclopedia of Philosophy entry

- Fourier Transforms and Uncertainty at MathPages

- aip.org: Quantum mechanics 1925–1927 – The uncertainty principle

- Eric Weisstein's World of Physics – Uncertainty principle

- John Baez on the time–energy uncertainty relation

- The certainty principle

- Common Interpretation of Heisenberg's Uncertainty Principle Is Proved False

Chapter 11

Quantum mechanics

For a more accessible and less technical introduction to this topic, see Introduction to quantum mechanics.

Quantum mechanics (QM; also known as **quantum physics** or **quantum theory**) including quantum field theory, is a fundamental branch of physics concerned with processes involving, for example, atoms and photons. In such processes, said to be quantized, the action has been observed to be only in integer multiples of the Planck constant, a physical quantity that is exceedingly, indeed perhaps ultimately, small. This is utterly inexplicable in classical physics.

Quantum mechanics gradually arose from Max Planck's solution in 1900 to the black-body radiation problem (reported 1859) and Albert Einstein's 1905 paper which offered a quantum-based theory to explain the photoelectric effect (reported 1887). Early quantum theory was profoundly reconceived in the mid-1920s.

The reconceived theory is formulated in various specially developed mathematical formalisms. In one of them, a mathematical function, the wave function, provides information about the probability amplitude of position, momentum, and other physical properties of a particle.

Important applications of quantum mechanical theory include superconducting magnets, light-emitting diodes and the laser, the transistor and semiconductors such as the microprocessor, medical and research imaging such as magnetic resonance imaging and electron microscopy, and explanations for many biological and physical phenomena.

11.1 History

Main article: History of quantum mechanics

Scientific inquiry into the wave nature of light began in the 17th and 18th centuries, when scientists such as Robert Hooke, Christiaan Huygens and Leonhard Euler proposed a wave theory of light based on experimental observations.[1] In 1803, Thomas Young, an English polymath, performed the famous double-slit experiment that he later described in a paper entitled *On the nature of light and colours*. This experiment played a major role in the general acceptance of the wave theory of light.

In 1838, Michael Faraday discovered cathode rays. These studies were followed by the 1859 statement of the black-body radiation problem by Gustav Kirchhoff, the 1877 suggestion by Ludwig Boltzmann that the energy states of a physical system can be discrete, and the 1900 quantum hypothesis of Max Planck.[2] Planck's hypothesis that energy is radiated and absorbed in discrete "quanta" (or energy elements) precisely matched the observed patterns of black-body radiation.

In 1896, Wilhelm Wien empirically determined a distribution law of black-body radiation,[3] known as Wien's law in his honor. Ludwig Boltzmann independently arrived at this result by considerations of Maxwell's equations. However, it was valid only at high frequencies and underestimated the radiance at low frequencies. Later, Planck corrected this model using Boltzmann's statistical interpretation of thermodynamics and proposed what is now called Planck's law, which led to the development of quantum mechanics.

Following Max Planck's solution in 1900 to the black-body radiation problem (reported 1859), Albert Einstein offered a

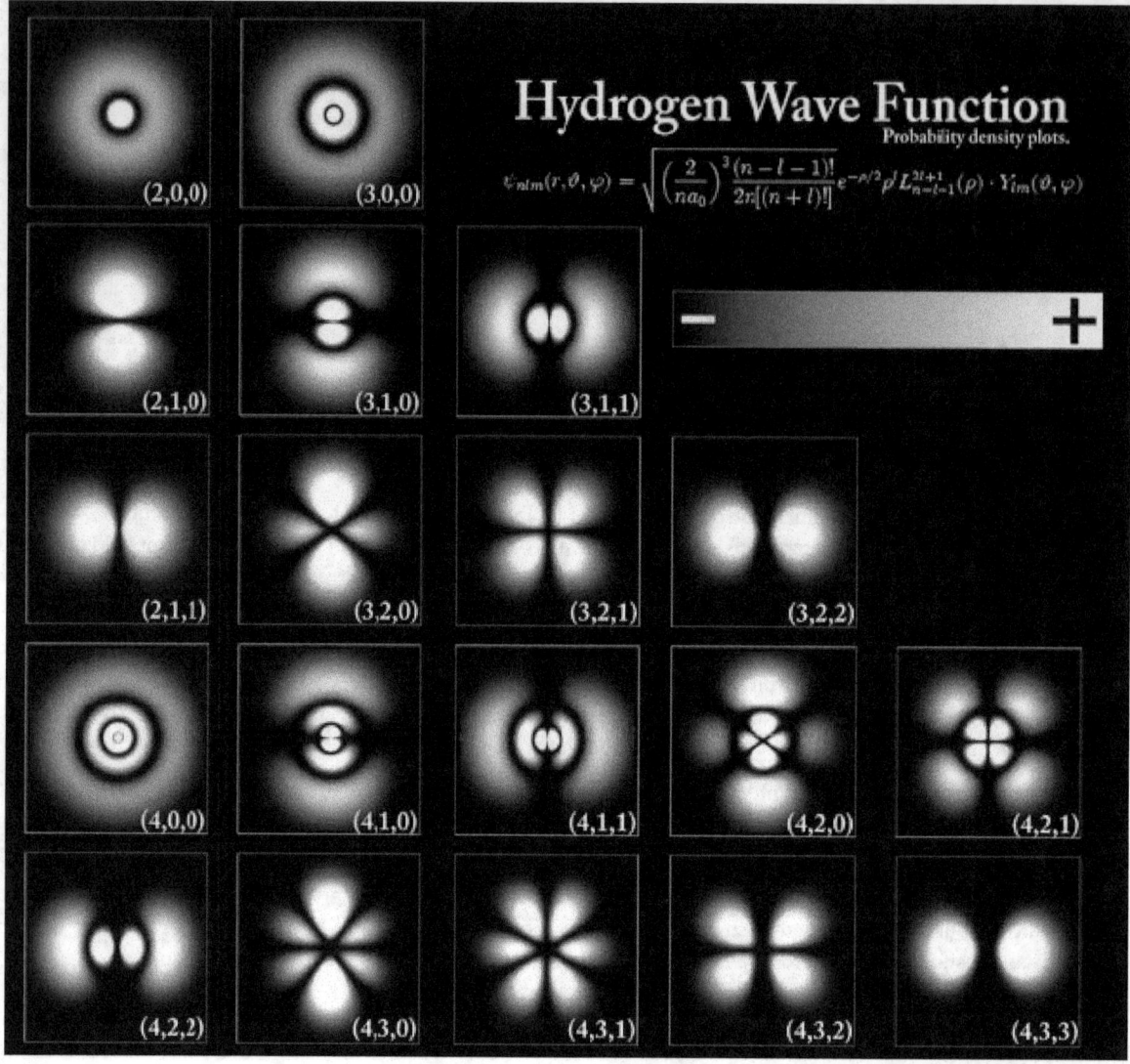

$$\psi_{nlm}(r,\vartheta,\varphi) = \sqrt{\left(\frac{2}{na_0}\right)^3 \frac{(n-l-1)!}{2n[(n+l)!]}} e^{-\rho/2} \rho^l L_{n-l-1}^{2l+1}(\rho) \cdot Y_{lm}(\vartheta,\varphi)$$

Solution to Schrödinger's equation for the hydrogen atom at different energy levels. The brighter areas represent a higher probability of finding an electron

quantum-based theory to explain the photoelectric effect (1905, reported 1887). Around 1900-1910, the atomic theory and the corpuscular theory of light[4] first came to be widely accepted as scientific fact; these latter theories can be viewed as quantum theories of matter and electromagnetic radiation, respectively.

Among the first to study quantum phenomena in nature were Arthur Compton, C. V. Raman, and Pieter Zeeman, each of whom has a quantum effect named after him. Robert Andrews Millikan studied the photoelectric effect experimentally, and Albert Einstein developed a theory for it. At the same time, Niels Bohr developed his theory of the atomic structure, which was later confirmed by the experiments of Henry Moseley. In 1913, Peter Debye extended Niels Bohr's theory of atomic structure, introducing elliptical orbits, a concept also introduced by Arnold Sommerfeld.[5] This phase is known as old quantum theory.

According to Planck, each energy element (E) is proportional to its frequency (ν):

$$E = h\nu$$

where h is Planck's constant.

Planck cautiously insisted that this was simply an aspect of the *processes* of absorption and emission of radiation and had nothing to do with the *physical reality* of the radiation itself.[6] In fact, he considered his quantum hypothesis a mathematical trick to get the right answer rather than a sizable discovery.[7] However, in 1905 Albert Einstein interpreted Planck's quantum hypothesis realistically and used it to explain the photoelectric effect, in which shining light on certain materials can eject electrons from the material. He won the 1921 Nobel Prize in Physics for this work.

Einstein further developed this idea to show that an electromagnetic wave such as light could also be described as a particle (later called the photon), with a discrete quantum of energy that was dependent on its frequency.[8]

The foundations of quantum mechanics were established during the first half of the 20th century by Max Planck, Niels Bohr, Werner Heisenberg, Louis de Broglie, Arthur Compton, Albert Einstein, Erwin Schrödinger, Max Born, John von Neumann, Paul Dirac, Enrico Fermi, Wolfgang Pauli, Max von Laue, Freeman Dyson, David Hilbert, Wilhelm Wien, Satyendra Nath Bose, Arnold Sommerfeld, and others. The Copenhagen interpretation of Niels Bohr became widely accepted.

In the mid-1920s, developments in quantum mechanics led to its becoming the standard formulation for atomic physics. In the summer of 1925, Bohr and Heisenberg published results that closed the old quantum theory. Out of deference to their particle-like behavior in certain processes and measurements, light quanta came to be called photons (1926). From Einstein's simple postulation was born a flurry of debating, theorizing, and testing. Thus, the entire field of quantum physics emerged, leading to its wider acceptance at the Fifth Solvay Conference in 1927.

It was found that subatomic particles and electromagnetic waves are neither simply particle nor wave but have certain properties of each. This originated the concept of wave–particle duality.

By 1930, quantum mechanics had been further unified and formalized by the work of David Hilbert, Paul Dirac and John von Neumann[9] with greater emphasis on measurement, the statistical nature of our knowledge of reality, and philosophical speculation about the 'observer'. It has since permeated many disciplines including quantum chemistry, quantum electronics, quantum optics, and quantum information science. Its speculative modern developments include string theory and quantum gravity theories. It also provides a useful framework for many features of the modern periodic table of elements, and describes the behaviors of atoms during chemical bonding and the flow of electrons in computer semiconductors, and therefore plays a crucial role in many modern technologies.

While quantum mechanics was constructed to describe the world of the very small, it is also needed to explain some macroscopic phenomena such as superconductors,[10] and superfluids.[11]

The word *quantum* derives from the Latin, meaning "how great" or "how much".[12] In quantum mechanics, it refers to a discrete unit assigned to certain physical quantities such as the energy of an atom at rest (see Figure 1). The discovery that particles are discrete packets of energy with wave-like properties led to the branch of physics dealing with atomic and subatomic systems which is today called quantum mechanics. It underlies the mathematical framework of many fields of physics and chemistry, including condensed matter physics, solid-state physics, atomic physics, molecular physics, computational physics, computational chemistry, quantum chemistry, particle physics, nuclear chemistry, and nuclear physics.[13] Some fundamental aspects of the theory are still actively studied.[14]

Quantum mechanics is essential to understanding the behavior of systems at atomic length scales and smaller. If the physical nature of an atom was solely described by classical mechanics, electrons would not *orbit* the nucleus, since orbiting electrons emit radiation (due to circular motion) and would eventually collide with the nucleus due to this loss of energy. This framework was unable to explain the stability of atoms. Instead, electrons remain in an uncertain, non-deterministic, *smeared*, probabilistic wave–particle orbital about the nucleus, defying the traditional assumptions of classical mechanics and electromagnetism.[15]

Quantum mechanics was initially developed to provide a better explanation and description of the atom, especially the differences in the spectra of light emitted by different isotopes of the same chemical element, as well as subatomic particles. In short, the quantum-mechanical atomic model has succeeded spectacularly in the realm where classical mechanics and electromagnetism falter.

Broadly speaking, quantum mechanics incorporates four classes of phenomena for which classical physics cannot account:

- quantization of certain physical properties

- quantum entanglement

- principle of uncertainty

- wave–particle duality

11.2   Mathematical formulations

Main article: Mathematical formulation of quantum mechanics
See also: Quantum logic

In the mathematically rigorous formulation of quantum mechanics developed by Paul Dirac,[16] David Hilbert,[17] John von Neumann,[18] and Hermann Weyl,[19] the possible states of a quantum mechanical system are symbolized[20] as unit vectors (called *state vectors*). Formally, these reside in a complex separable Hilbert space—variously called the *state space* or the *associated Hilbert space* of the system—that is well defined up to a complex number of norm 1 (the phase factor). In other words, the possible states are points in the projective space of a Hilbert space, usually called the complex projective space. The exact nature of this Hilbert space is dependent on the system—for example, the state space for position and momentum states is the space of square-integrable functions, while the state space for the spin of a single proton is just the product of two complex planes. Each observable is represented by a maximally Hermitian (precisely: by a self-adjoint) linear operator acting on the state space. Each eigenstate of an observable corresponds to an eigenvector of the operator, and the associated eigenvalue corresponds to the value of the observable in that eigenstate. If the operator's spectrum is discrete, the observable can attain only those discrete eigenvalues.

In the formalism of quantum mechanics, the state of a system at a given time is described by a complex wave function, also referred to as state vector in a complex vector space.[21] This abstract mathematical object allows for the calculation of probabilities of outcomes of concrete experiments. For example, it allows one to compute the probability of finding an electron in a particular region around the nucleus at a particular time. Contrary to classical mechanics, one can never make simultaneous predictions of conjugate variables, such as position and momentum, with accuracy. For instance, electrons may be considered (to a certain probability) to be located somewhere within a given region of space, but with their exact positions unknown. Contours of constant probability, often referred to as "clouds", may be drawn around the nucleus of an atom to conceptualize where the electron might be located with the most probability. Heisenberg's uncertainty principle quantifies the inability to precisely locate the particle given its conjugate momentum.[22]

According to one interpretation, as the result of a measurement the wave function containing the probability information for a system collapses from a given initial state to a particular eigenstate. The possible results of a measurement are the eigenvalues of the operator representing the observable—which explains the choice of *Hermitian* operators, for which all the eigenvalues are real. The probability distribution of an observable in a given state can be found by computing the spectral decomposition of the corresponding operator. Heisenberg's uncertainty principle is represented by the statement that the operators corresponding to certain observables do not commute.

The probabilistic nature of quantum mechanics thus stems from the act of measurement. This is one of the most difficult aspects of quantum systems to understand. It was the central topic in the famous Bohr–Einstein debates, in which the two scientists attempted to clarify these fundamental principles by way of thought experiments. In the decades after the formulation of quantum mechanics, the question of what constitutes a "measurement" has been extensively studied. Newer interpretations of quantum mechanics have been formulated that do away with the concept of "wave function collapse" (see, for example, the relative state interpretation). The basic idea is that when a quantum system interacts with a measuring apparatus, their respective wave functions become entangled, so that the original quantum system ceases to exist as an independent entity. For details, see the article on measurement in quantum mechanics.[23]

Generally, quantum mechanics does not assign definite values. Instead, it makes a prediction using a probability distribution; that is, it describes the probability of obtaining the possible outcomes from measuring an observable. Often these results are skewed by many causes, such as dense probability clouds. Probability clouds are approximate (but better than the Bohr model) whereby electron location is given by a probability function, the wave function eigenvalue, such that the probability is the squared modulus of the complex amplitude, or quantum state nuclear attraction.[24][25] Naturally, these probabilities will depend on the quantum state at the "instant" of the measurement. Hence, uncertainty is involved in the value. There are, however, certain states that are associated with a definite value of a particular observable. These are known as eigenstates of the observable ("eigen" can be translated from German as meaning "inherent" or "characteristic").[26]

In the everyday world, it is natural and intuitive to think of everything (every observable) as being in an eigenstate. Everything appears to have a definite position, a definite momentum, a definite energy, and a definite time of occurrence. However, quantum mechanics does not pinpoint the exact values of a particle's position and momentum (since they are conjugate pairs) or its energy and time (since they too are conjugate pairs); rather, it provides only a range of probabilities in which that particle might be given its momentum and momentum probability. Therefore, it is helpful to use different words to describe states having *uncertain* values and states having *definite* values (eigenstates). Usually, a system will not be in an eigenstate of the observable (particle) we are interested in. However, if one measures the observable, the wave function will instantaneously be an eigenstate (or "generalized" eigenstate) of that observable. This process is known as wave function collapse, a controversial and much-debated process[27] that involves expanding the system under study to include the measurement device. If one knows the corresponding wave function at the instant before the measurement, one will be able to compute the probability of the wave function collapsing into each of the possible eigenstates. For example, the free particle in the previous example will usually have a wave function that is a wave packet centered around some mean position x_0 (neither an eigenstate of position nor of momentum). When one measures the position of the particle, it is impossible to predict with certainty the result.[23] It is probable, but not certain, that it will be near x_0, where the amplitude of the wave function is large. After the measurement is performed, having obtained some result x, the wave function collapses into a position eigenstate centered at x.[28]

The time evolution of a quantum state is described by the Schrödinger equation, in which the Hamiltonian (the operator corresponding to the total energy of the system) generates the time evolution. The time evolution of wave functions is deterministic in the sense that - given a wave function at an *initial* time - it makes a definite prediction of what the wave function will be at any *later* time.[29]

During a measurement, on the other hand, the change of the initial wave function into another, later wave function is not deterministic, it is unpredictable (i.e., random). A time-evolution simulation can be seen here.[30][31]

Wave functions change as time progresses. The Schrödinger equation describes how wave functions change in time, playing a role similar to Newton's second law in classical mechanics. The Schrödinger equation, applied to the aforementioned example of the free particle, predicts that the center of a wave packet will move through space at a constant velocity (like a classical particle with no forces acting on it). However, the wave packet will also spread out as time progresses, which means that the position becomes more uncertain with time. This also has the effect of turning a position eigenstate (which can be thought of as an infinitely sharp wave packet) into a broadened wave packet that no longer represents a (definite, certain) position eigenstate.[32]

Some wave functions produce probability distributions that are constant, or independent of time—such as when in a stationary state of constant energy, time vanishes in the absolute square of the wave function. Many systems that are treated dynamically in classical mechanics are described by such "static" wave functions. For example, a single electron in an unexcited atom is pictured classically as a particle moving in a circular trajectory around the atomic nucleus, whereas in quantum mechanics it is described by a static, spherically symmetric wave function surrounding the nucleus (Fig. 1) (note, however, that only the lowest angular momentum states, labeled s, are spherically symmetric).[33]

The Schrödinger equation acts on the *entire* probability amplitude, not merely its absolute value. Whereas the absolute value of the probability amplitude encodes information about probabilities, its phase encodes information about the interference between quantum states. This gives rise to the "wave-like" behavior of quantum states. As it turns out, analytic solutions of the Schrödinger equation are available for only a very small number of relatively simple model Hamiltonians, of which the quantum harmonic oscillator, the particle in a box, the dihydrogen cation, and the hydrogen atom are the most important representatives. Even the helium atom—which contains just one more electron than does the hydrogen atom—has defied all attempts at a fully analytic treatment.

There exist several techniques for generating approximate solutions, however. In the important method known as perturbationtheory, one uses the analyti c result for a simple quantum mechanical mode l to generate a result for a more complicatedmode l that is related to the simpler mode l b y (for one example) the addition of a weak potential energy. Another method isthe "semi-classical equation of motion" approach, whic h applies to systems for whic h quantum mechanic s produces onlyweak (small) deviations from classical behavior. These deviations can then be computed based on the classical motion.This approac h is particularl y important in the fi eld of quantum chaos.

11.3 Mathematically equivalent formulations of quantum mechanics

There are numerous mathematically equivalent formulations of quantum mechanics. One of the oldest and most commonly used formulations is the "transformation theory" proposed by Paul Dirac, which unifies and generalizes the two earliest formulations of quantum mechanics - matrix mechanics (invented by Werner Heisenberg) and wave mechanics (invented by Erwin Schrödinger).[34]

Especially since Werner Heisenberg was awarded the Nobel Prize in Physics in 1932 for the creation of quantum mechanics, the role of Max Born in the development of QM was overlooked until the 1954 Nobel award. The role is noted in a 2005 biography of Born, which recounts his role in the matrix formulation of quantum mechanics, and the use of probability amplitudes. Heisenberg himself acknowledges having learned matrices from Born, as published in a 1940 *festschrift* honoring Max Planck.[35] In the matrix formulation, the instantaneous state of a quantum system encodes the probabilities of its measurable properties, or "observables". Examples of observables include energy, position, momentum, and angular momentum. Observables can be either continuous (e.g., the position of a particle) or discrete (e.g., the energy of an electron bound to a hydrogen atom).[36] An alternative formulation of quantum mechanics is Feynman's path integral formulation, in which a quantum-mechanical amplitude is considered as a sum over all possible classical and non-classical paths between the initial and final states. This is the quantum-mechanical counterpart of the action principle in classical mechanics.

11.4 Interactions with other scientific theories

The rules of quantum mechanics are fundamental. They assert that the state space of a system is a Hilbert space and that observables of that system are Hermitian operators acting on that space—although they do not tell us which Hilbert space or which operators. These can be chosen appropriately in order to obtain a quantitative description of a quantum system. An important guide for making these choices is the correspondence principle, which states that the predictions of quantum mechanics reduce to those of classical mechanics when a system moves to higher energies or, equivalently, larger quantum numbers, i.e. whereas a single particle exhibits a degree of randomness, in systems incorporating millions of particles averaging takes over and, at the high energy limit, the statistical probability of random behaviour approaches zero. In other words, classical mechanics is simply a quantum mechanics of large systems. This "high energy" limit is known as the *classical* or *correspondence limit*. One can even start from an established classical model of a particular system, then attempt to guess the underlying quantum model that would give rise to the classical model in the correspondence limit.

When quantum mechanics was originally formulated, it was applied to models whose correspondence limit was non-relativistic classical mechanics. For instance, the well-known model of the quantum harmonic oscillator uses an explicitly non-relativistic expression for the kinetic energy of the oscillator, and is thus a quantum version of the classical harmonic oscillator.

Early attempts to merge quantum mechanics with special relativity involved the replacement of the Schrödinger equation with a covariant equation such as the Klein–Gordon equation or the Dirac equation. While these theories were successful in explaining many experimental results, they had certain unsatisfactory qualities stemming from their neglect of the relativistic creation and annihilation of particles. A fully relativistic quantum theory required the development of quantum field theory, which applies quantization to a field (rather than a fixed set of particles). The first complete quantum field theory, quantum electrodynamics, provides a fully quantum description of the electromagnetic interaction. The full apparatus of quantum field theory is often unnecessary for describing electrodynamic systems. A simpler approach, one that has been employed since the inception of quantum mechanics, is to treat charged particles as quantum mechanical objects being acted on by a classical electromagnetic field. For example, the elementary quantum model of the hydrogen atom describes the electric field of the hydrogen atom using a classical $-e^2/(4\pi \epsilon_0 r)$ Coulomb potential. This "semi-classical" approach fails if quantum fluctuations in the electromagnetic field play an important role, such as in the emission of photons by charged particles.

Quantum field theories for the strong nuclear force and the weak nuclear force have also been developed. The quantum field theory of the strong nuclear force is called quantum chromodynamics, and describes the interactions of subnuclear particles such as quarks and gluons. The weak nuclear force and the electromagnetic force were unified, in their quantized forms, into a single quantum field theory (known as electroweak theory), by the physicists Abdus Salam, Sheldon Glashow and Steven Weinberg. These three men shared the Nobel Prize in Physics in 1979 for this work.[37]

It has proven difficult to construct quantum models of gravity, the remaining fundamental force. Semi-classical approximations are workable, and have led to predictions such as Hawking radiation. However, the formulation of a complete theory of quantum gravity is hindered by apparent incompatibilities between general relativity (the most accurate theory of gravity currently known) and some of the fundamental assumptions of quantum theory. The resolution of these incompatibilities is an area of active research, and theories such as string theory are among the possible candidates for a future theory of quantum gravity.

Classical mechanics has also been extended into the complex domain, with complex classical mechanics exhibiting behaviors similar to quantum mechanics.[38]

11.4.1   Quantum mechanics and classical physics

Predictions of quantum mechanics have been verified experimentally to an extremely high degree of accuracy.[39] According to the correspondence principle between classical and quantum mechanics, all objects obey the laws of quantum mechanics, and classical mechanics is just an approximation for large systems of objects (or a statistical quantum mechanics of a large collection of particles).[40] The laws of classical mechanics thus follow from the laws of quantum mechanics as a statistical average at the limit of large systems or large quantum numbers.[41] However, chaotic systems do not have good quantum numbers, and quantum chaos studies the relationship between classical and quantum descriptions in these systems.

Quantum coherence is an essential difference between classical and quantum theories as illustrated by the Einstein–Podolsky–Rosen (EPR) paradox — an attack on a certain philosophical interpretation of quantum mechanics by an appeal to local realism.[42] Quantum interference involves adding together *probability amplitudes*, whereas classical "waves" infer that there is an adding together of *intensities*. For microscopic bodies, the extension of the system is much smaller than the coherence length, which gives rise to long-range entanglement and other nonlocal phenomena characteristic of quantum systems.[43] Quantum coherence is not typically evident at macroscopic scales, though an exception to this rule may occur at extremely low temperatures (i.e. approaching absolute zero) at which quantum behavior may manifest itself macroscopically.[44] This is in accordance with the following observations:

- Many macroscopic properties of a classical system are a direct consequence of the quantum behavior of its parts. For example, the stability of bulk matter (consisting of atoms and molecules which would quickly collapse under electric forces alone), the rigidity of solids, and the mechanical, thermal, chemical, optical and magnetic properties of matter are all results of the interaction of electric charges under the rules of quantum mechanics.[45]

- While the seemingly "exotic" behavior of matter posited by quantum mechanics and relativity theory become more apparent when dealing with particles of extremely small size or velocities approaching the speed of light, the laws of classical, often considered "Newtonian", physics remain accurate in predicting the behavior of the vast majority of "large" objects (on the order of the size of large molecules or bigger) at velocities much smaller than the velocity of light.[46]

11.4.2   Copenhagen interpretation of quantum versus classical kinematics

A big difference between classical and quantum mechanics is that they use very different kinematic descriptions.[47]

In Niels Bohr's mature view, quantum mechanical phenomena are required to be experiments, with complete descriptions of all the devices for the system, preparative, intermediary, and finally measuring. The descriptions are in macroscopic terms, expressed in ordinary language, supplemented with the concepts of classical mechanics.[48][49][50][51] The initial condition and the final condition of the system are respectively described by values in a configuration space, for example a position space, or some equivalent space such as a momentum space. Quantum mechanics does not admit a completely precise description, in terms of both position and momentum, of an initial condition or "state" (in the classical sense of the word) that would support a precisely deterministic and causal prediction of a final condition.[52][53] In this sense, advocated by Bohr in his mature writings, a quantum phenomenon is a process, a passage from initial to final condition, not an instantaneous "state" in the classical sense of that word.[54][55] Thus there are two kinds of processes in quantum mechanics:

stationary and transitional. For a stationary process, the initial and final condition are the same. For a transition, they are different. Obviously by definition, if only the initial condition is given, the process is not determined.[52] Given its initial condition, prediction of its final condition is possible, causally but only probabilistically, because the Schrödinger equation is deterministic for wave function evolution, but the wave function describes the system only probabilistically.[56][57]

For many experiments, it is possible to think of the initial and final conditions of the system as being a particle. In some cases it appears that there are potentially several spatially distinct pathways or trajectories by which a particle might pass from initial to final condition. It is an important feature of the quantum kinematic description that it does not permit a unique definite statement of which of those pathways is actually followed. Only the initial and final conditions are definite, and, as stated in the foregoing paragraph, they are defined only as precisely as allowed by the configuration space description or its equivalent. In every case for which a quantum kinematic description is needed, there is always a compelling reason for this restriction of kinematic precision. An example of such a reason is that for a particle to be experimentally found in a definite position, it must be held motionless; for it to be experimentally found to have a definite momentum, it must have free motion; these two are logically incompatible.[58][59]

Classical kinematics does not primarily demand experimental description of its phenomena. It allows completely precise description of an instantaneous state by a value in phase space, the Cartesian product of configuration and momentum spaces. This description simply assumes or imagines a state as a physically existing entity without concern about its experimental measurability. Such a description of an initial condition, together with Newton's laws of motion, allows a precise deterministic and causal prediction of a final condition, with a definite trajectory of passage. Hamiltonian dynamics can be used for this. Classical kinematics also allows the description of a process analogous to the initial and final condition description used by quantum mechanics. Lagrangian mechanics applies to this.[60] For processes that need account to be taken of actions of a small number of Planck constants, classical kinematics is not adequate; quantum mechanics is needed.

11.4.3 Relativity and quantum mechanics

Main article: Relativistic quantum mechanics

Even with the defining postulates of both Einstein's theory of general relativity and quantum theory being indisputably supported by rigorous and repeated empirical evidence, and while they do not directly contradict each other theoretically (at least with regard to their primary claims), they have proven extremely difficult to incorporate into one consistent, cohesive model.[61]

Einstein himself is well known for rejecting some of the claims of quantum mechanics. While clearly contributing to the field, he did not accept many of the more "philosophical consequences and interpretations" of quantum mechanics, such as the lack of deterministic causality. He is famously quoted as saying, in response to this aspect, "My God does not play with dice". He also had difficulty with the assertion that a single subatomic particle can occupy numerous areas of space at one time. However, he was also the first to notice some of the apparently exotic consequences of entanglement, and used them to formulate the Einstein–Podolsky–Rosen paradox in the hope of showing that quantum mechanics had unacceptable implications if taken as a complete description of physical reality. This was 1935, but in 1964 it was shown by John Bell (see Bell inequality) that - although Einstein was correct in identifying seemingly paradoxical implications of quantum mechanical nonlocality - these implications could be experimentally tested. Alain Aspect's initial experiments in 1982, and many subsequent experiments since, have definitively verified quantum entanglement.

According to the paper of J. Bell and the Copenhagen interpretation—the common interpretation of quantum mechanics by physicists since 1927 - and contrary to Einstein's ideas, quantum mechanics was *not*, at the same time a "realistic" theory and a "local" theory.

The Einstein–Podolsky–Rosen paradox shows in any case that there exist experiments by which one can measure the state of one particle and instantaneously change the state of its entangled partner - although the two particles can be an arbitrary distance apart. However, this effect does not violate causality, since no transfer of information happens. Quantum entanglement forms the basis of quantum cryptography, which is used in high-security commercial applications in banking and government.

Gravity is negligible in many areas of particle physics, so that unification between general relativity and quantum mechanics is not an urgent issue in those particular applications. However, the lack of a correct theory of quantum gravity is an important issue in cosmology and the search by physicists for an elegant "Theory of Everything" (TOE). Consequently, resolving the inconsistencies between both theories has been a major goal of 20th and 21st century physics. Many prominent physicists, including Stephen Hawking, have labored for many years in the attempt to discover a theory underlying *everything*. This TOE would combine not only the different models of subatomic physics, but also derive the four fundamental forces of nature - the strong force, electromagnetism, the weak force, and gravity - from a single force or phenomenon. While Stephen Hawking was initially a believer in the Theory of Everything, after considering Gödel's Incompleteness Theorem, he has concluded that one is not obtainable, and has stated so publicly in his lecture "Gödel and the End of Physics" (2002).[62]

11.4.4   Attempts at a unified field theory

Main article: Grand unified theory

The quest to unify the fundamental forces through quantum mechanics is still ongoing. Quantum electrodynamics (or "quantum electromagnetism"), which is currently (in the perturbative regime at least) the most accurately tested physical theory in competition with general relativity,[63][64] has been successfully merged with the weak nuclear force into the electroweak force and work is currently being done to merge the electroweak and strong force into the electrostrong force. Current predictions state that at around 10^{14} GeV the three aforementioned forces are fused into a single unified field.[65] Beyond this "grand unification", it is speculated that it may be possible to merge gravity with the other three gauge symmetries, expected to occur at roughly 10^{19} GeV. However — and while special relativity is parsimoniously incorporated into quantum electrodynamics — the expanded general relativity, currently the best theory describing the gravitation force, has not been fully incorporated into quantum theory. One of those searching for a coherent TOE is Edward Witten, a theoretical physicist who formulated the M-theory, which is an attempt at describing the supersymmetrical based string theory. M-theory/supergravity theory posits that our apparent 4-dimensional spacetime is, in reality, actually an 11-dimensional spacetime containing 10 spatial dimensions and 1 time dimension, although 7 of the spatial dimensions (hyperspace) are - at lower energies - completely 'compactified' (or infinitely curved) and not readily amenable to measurement or probing. M-theory/superstrings theory describes a 10-dimensional spacetime consisting of 6 higher dimensions + the 4 common dimensions.

Another popular theory is Loop quantum gravity (LQG), a theory first proposed by Carlo Rovelli that describes the quantum properties of gravity. It is also a theory of quantum space and quantum time, because in general relativity the geometry of spacetime is a manifestation of gravity. LQG is an attempt to merge and adapt standard quantum mechanics and standard general relativity. The main output of the theory is a physical picture of space where space is granular. The granularity is a direct consequence of the quantization. It has the same nature of the granularity of the photons in the quantum theory of electromagnetism or the discrete levels of the energy of the atoms. But here it is space itself which is discrete.

More precisely, space can be viewed as an extremely fine fabric or network "woven" of finite loops. These networks of loops are called spin networks. The evolution of a spin network over time is called a spin foam. The predicted size of this structure is the Planck length, which is approximately 1.616×10^{-35} m. According to theory, there is no meaning to length shorter than this (cf. Planck scale energy). Therefore, LQG predicts that not just matter, but also space itself, has an atomic structure.

11.5   Philosophical implications

Main article: Interpretations of quantum mechanics

Since its inception, the many counter-intuitive aspects and results of quantum mechanics have provoked strong philosophical debates and many interpretations. Even fundamental issues, such as Max Born's basic rules concerning probability amplitudes and probability distributions, took decades to be appreciated by society and many leading scientists. Richard Feynman once said, "I think I can safely say that nobody understands quantum mechanics."[66] According to Steven

Weinberg, "There is now in my opinion no entirely satisfactory interpretation of quantum mechanics."[67]

The Copenhagen interpretation - due largely to the Danish theoretical physicist Niels Bohr - remains the quantum mechanical formalism that is currently most widely accepted amongst physicists, some 75 years after its enunciation. According to this interpretation, the probabilistic nature of quantum mechanics is not a *temporary* feature which will eventually be replaced by a deterministic theory, but instead must be considered a *final* renunciation of the classical idea of "causality." It is also believed therein that any well-defined application of the quantum mechanical formalism must always make reference to the experimental arrangement, due to the conjugate nature of evidence obtained under different experimental situations.

Albert Einstein, himself one of the founders of quantum theory, rejected the quantum theoretical doctrine that the state of a system depends on the experimental arrangement for its measurement. He held that underlying quantum mechanics there should be a theory that thoroughly and directly expresses the rule against action at a distance; in other words, he insisted on the principle of locality. He inferred that the present theory was incomplete, contrary to the Copenhagen doctrine that it is complete. He therefore produced a series of objections, the most famous of which has become known as the Einstein–Podolsky–Rosen paradox.

John Bell showed that this "EPR" paradox led to experimentally testable differences between quantum mechanics and theories that rely on added hidden variables. Experiments have been performed confirming the accuracy of quantum mechanics, thereby demonstrating that quantum mechanics cannot be improved upon by addition of hidden variables.[68] The *Bohr-Einstein debates* provide a vibrant critique of the Copenhagen Interpretation from an epistemological point of view.

The Everett many-worlds interpretation, formulated in 1956, holds that *all* the possibilities described by quantum theory *simultaneously* occur in a multiverse composed of mostly independent parallel universes.[69] This is not accomplished by introducing some "new axiom" to quantum mechanics, but on the contrary, by *removing* the axiom of the collapse of the wave packet. *All* of the possible consistent states of the measured system and the measuring apparatus (including the observer) are present in a *real* physical - not just formally mathematical, as in other interpretations - quantum superposition. Such a superposition of consistent state combinations of different systems is called an entangled state. While the multiverse is deterministic, we perceive non-deterministic behavior governed by probabilities, because we can only observe the universe (i.e., the consistent state contribution to the aforementioned superposition) that we, as observers, inhabit. Everett's interpretation is perfectly consistent with John Bell's experiments and makes them intuitively understandable. However, according to the theory of quantum decoherence, these "parallel universes" will never be accessible to us. The inaccessibility can be understood as follows: once a measurement is done, the measured system becomes entangled with *both* the physicist who measured it *and* a huge number of other particles, some of which are photons flying away at the speed of light towards the other end of the universe. In order to prove that the wave function did not collapse, one would have to bring *all* these particles back and measure them again, together with the system that was originally measured. Not only is this completely impractical, but even if one *could* theoretically do this, it would have to destroy any evidence that the original measurement took place (including the physicist's memory). In light of these Bell tests, Cramer (1986) formulated his transactional interpretation.[70] Relational quantum mechanics appeared in the late 1990s as the modern derivative of the Copenhagen Interpretation.

11.6 Applications

Quantum mechanics has had enormous[71] success in explaining many of the features of our universe. Quantum mechanics is often the only tool available that can reveal the individual behaviors of the subatomic particles that make up all forms of matter (electrons, protons, neutrons, photons, and others). Quantum mechanics has strongly influenced string theories, candidates for a Theory of Everything (see reductionism).

Quantum mechanics is also critically important for understanding how individual atoms combine covalently to form molecules. The application of quantum mechanics to chemistry is known as quantum chemistry. Relativistic quantum mechanics can, in principle, mathematically describe most of chemistry. Quantum mechanics can also provide quantitative insight into ionic and covalent bonding processes by explicitly showing which molecules are energetically favorable to which others and the magnitudes of the energies involved.[72] Furthermore, most of the calculations performed in modern computational chemistry rely on quantum mechanics.

A great deal of modern technological inventions operate at a scale where quantum effects are significant. Examples include the laser, the transistor (and thus the microchip), the electron microscope, and magnetic resonance imaging (MRI). The study of semiconductors led to the invention of the diode and the transistor, which are indispensable parts of modern electronics systems and devices.

Researchers are currently seeking robust methods of directly manipulating quantum states. Efforts are being made to more fully develop quantum cryptography, which will theoretically allow guaranteed secure transmission of information. A more distant goal is the development of quantum computers, which are expected to perform certain computational tasks exponentially faster than classical computers. Instead of using classical bits, quantum computers use qubits, which can be in superpositions of states. Another active research topic is quantum teleportation, which deals with techniques to transmit quantum information over arbitrary distances.

Quantum tunneling is vital to the operation of many devices. Even in the simple light switch, the electrons in the electric current could not penetrate the potential barrier made up of a layer of oxide without quantum tunneling. Flash memory chips found in USB drives use quantum tunneling to erase their memory cells.

While quantum mechanics primarily applies to the smaller atomic regimes of matter and energy, some systems exhibit quantum mechanical effects on a large scale. Superfluidity, the frictionless flow of a liquid at temperatures near absolute zero, is one well-known example. So is the closely related phenomenon of superconductivity, the frictionless flow of an electron gas in a conducting material (an electric current) at sufficiently low temperatures.

Quantum theory also provides accurate descriptions for many previously unexplained phenomena, such as black-body radiation and the stability of the orbitals of electrons in atoms. It has also given insight into the workings of many different biological systems, including smell receptors and protein structures.[73] Recent work on photosynthesis has provided evidence that quantum correlations play an essential role in this fundamental process of plants and many other organisms.[74] Even so, classical physics can often provide good approximations to results otherwise obtained by quantum physics, typically in circumstances with large numbers of particles or large quantum numbers. Since classical formulas are much simpler and easier to compute than quantum formulas, classical approximations are used and preferred when the system is large enough to render the effects of quantum mechanics insignificant.

11.7   Examples

11.7.1   Free particle

For example, consider a free particle. In quantum mechanics, there is wave–particle duality, so the properties of the particle can be described as the properties of a wave. Therefore, its quantum state can be represented as a wave of arbitrary shape and extending over space as a wave function. The position and momentum of the particle are observables. The Uncertainty Principle states that both the position and the momentum cannot simultaneously be measured with complete precision. However, one *can* measure the position (alone) of a moving free particle, creating an eigenstate of position with a wave function that is very large (a Dirac delta) at a particular position x, and zero everywhere else. If one performs a position measurement on such a wave function, the resultant x will be obtained with 100% probability (i.e., with full certainty, or complete precision). This is called an eigenstate of position—or, stated in mathematical terms, a *generalized position eigenstate (eigendistribution)*. If the particle is in an eigenstate of position, then its momentum is completely unknown. On the other hand, if the particle is in an eigenstate of momentum, then its position is completely unknown.[75] In an eigenstate of momentum having a plane wave form, it can be shown that the wavelength is equal to h/p, where h is Planck's constant and p is the momentum of the eigenstate.[76]

11.7.2   Step potential

Main article: Solution of Schrödinger equation for a step potential
The potential in this case is given by:

$$V(x) = \begin{cases} 0, & x < 0, \\ V_0, & x \geq 0. \end{cases}$$

The solutions are superpositions of left- and right-moving waves:

$$\psi_1(x) = \frac{1}{\sqrt{k_1}} \left(A_\rightarrow e^{ik_1 x} + A_\leftarrow e^{-ik_1 x} \right) \quad x < 0$$

$$\psi_2(x) = \frac{1}{\sqrt{k_2}} \left(B_\rightarrow e^{ik_2 x} + B_\leftarrow e^{-ik_2 x} \right) \quad x > 0$$

where the wave vectors are related to the energy via

$$k_1 = \sqrt{2mE/\hbar^2}$$

$$k_2 = \sqrt{2m(E - V_0)/\hbar^2}$$

with coefficients A and B determined from the boundary conditions and by imposing a continuous derivative on the solution.

Each term of the solution can be interpreted as an incident, reflected, or transmitted component of the wave, allowing the calculation of transmission and reflection coefficients. Notably, in contrast to classical mechanics, incident particles with energies greater than the potential step are partially reflected.

11.7.3 Rectangular potential barrier

Main article: Rectangular potential barrier

This is a model for the quantum tunneling effect which plays an important role in the performance of modern technologies such as flash memory and scanning tunneling microscopy. Quantum tunneling is central to physical phenomena involved in superlattices.

11.7.4 Particle in a box

Main article: Particle in a box

The particle in a one-dimensional potential energy box is the most mathematically simple example where restraints lead to the quantization of energy levels. The box is defined as having zero potential energy everywhere *inside* a certain region, and infinite potential energy everywhere *outside* that region. For the one-dimensional case in the x direction, the time-independent Schrödinger equation may be written[77]

$$-\frac{\hbar^2}{2m} \frac{d^2 \psi}{dx^2} = E\psi.$$

With the differential operator defined by

$$\hat{p}_x = -i\hbar \frac{d}{dx}$$

the previous equation is evocative of the classic kinetic energy analogue,

$$\frac{1}{2m}\hat{p}_x^2 = E,$$

with state ψ in this case having energy E coincident with the kinetic energy of the particle.

The general solutions of the Schrödinger equation for the particle in a box are

$$\psi(x) = Ae^{ikx} + Be^{-ikx} \qquad E = \frac{\hbar^2 k^2}{2m}$$

or, from Euler's formula,

$$\psi(x) = C \sin kx + D \cos kx.$$

The infinite potential walls of the box determine the values of C, D, and k at $x = 0$ and $x = L$ where ψ must be zero. Thus, at $x = 0$,

$$\psi(0) = 0 = C \sin 0 + D \cos 0 = D$$

and $D = 0$. At $x = L$,

$$\psi(L) = 0 = C \sin kL.$$

in which C cannot be zero as this would conflict with the Born interpretation. Therefore, since $\sin(kL) = 0$, kL must be an integer multiple of π,

$$k = \frac{n\pi}{L} \qquad n = 1, 2, 3, \ldots.$$

The quantization of energy levels follows from this constraint on k, since

$$E = \frac{\hbar^2 \pi^2 n^2}{2mL^2} = \frac{n^2 h^2}{8mL^2}.$$

11.7.5 Finite potential well

Main article: Finite potential well

A finite potential well is the generalization of the infinite potential well problem to potential wells having finite depth.

The finite potential well problem is mathematically more complicated than the infinite particle-in-a-box problem as the wave function is not pinned to zero at the walls of the well. Instead, the wave function must satisfy more complicated mathematical boundary conditions as it is nonzero in regions outside the well.

11.7.6 Harmonic oscillator

Main article: Quantum harmonic oscillator
As in the classical case, the potential for the quantum harmonic oscillator is given by

$$V(x) = \frac{1}{2}m\omega^2 x^2$$

This problem can either be treated by directly solving the Schrödinger equation, which is not trivial, or by using the more elegant "ladder method" first proposed by Paul Dirac. The eigenstates are given by

$$\psi_n(x) = \sqrt{\frac{1}{2^n\, n!}} \cdot \left(\frac{m\omega}{\pi\hbar}\right)^{1/4} \cdot e^{-\frac{m\omega x^2}{2\hbar}} \cdot H_n\left(\sqrt{\frac{m\omega}{\hbar}}x\right), \qquad n = 0, 1, 2, \ldots.$$

where H_n are the Hermite polynomials,

$$H_n(x) = (-1)^n e^{x^2} \frac{d^n}{dx^n}\left(e^{-x^2}\right)$$

and the corresponding energy levels are

$$E_n = \hbar\omega\left(n + \frac{1}{2}\right)$$

This is another example illustrating the quantization of energy for bound states.

11.8 See also

- Angular momentum diagrams (quantum mechanics)

- EPR paradox

- Fractional quantum mechanics

- List of quantum-mechanical systems with analytical solutions

- Macroscopic quantum phenomena

- Phase space formulation

- Regularization (physics)

- Spherical basis

11.9 Notes

[1] Max Born & Emil Wolf, Principles of Optics, 1999, Cambridge University Press

[2] Mehra, J.; Rechenberg, H. (1982). *The historical development of quantum theory*. New York: Springer-Verlag. ISBN 0387906428.

[3] Kragh, Helge (2002). *Quantum Generations: A History of Physics in the Twentieth Century*. Princeton University Press. p. 58. ISBN 0-691-09552-3., Extract of page 58

[4] Ben-Menahem, Ari (2009). *Historical Encyclopedia of Natural and Mathematical Sciences, Volume 1*. Springer. p. 3678. ISBN 3540688315., Extract of page 3678

[5] E Arunan (2010). "Peter Debye" (PDF). *Resonance (journal)* (Indian Academy of Sciences) 15 (12).

[6] Kuhn, T. S. (1978). *Black-body theory and the quantum discontinuity 1894-1912*. Oxford: Clarendon Press. ISBN 0195023838.

[7] Kragh, Helge (1 December 2000), *Max Planck: the reluctant revolutionary*, PhysicsWorld.com

[8] Einstein, A. (1905). "Über einen die Erzeugung und Verwandlung des Lichtes betreffenden heuristischen Gesichtspunkt" [On a heuristic point of view concerning the production and transformation of light]. *Annalen der Physik* 17 (6): 132–148. Bibcode:1905AnP...322..132E. doi:10.1002/andp.19053220607. Reprinted in *The collected papers of Albert Einstein*, John Stachel, editor, Princeton University Press, 1989, Vol. 2, pp. 149-166, in German; see also *Einstein's early work on the quantum hypothesis*, ibid. pp. 134-148.

[9] van Hove, Leon (1958). "Von Neumann's contributions to quantum mechanics" (PDF). *Bulletin of the American Mathematical Society* 64: Part2:95–99. doi:10.1090/s0002-9904-1958-10206-2.

[10] *The Feynman Lectures on Physics* III 21-4 "...it was long believed that the wave function of the Schrödinger equation would never have a macroscopic representation analogous to the macroscopic representation of the amplitude for photons. On the other hand, it is now realized that the phenomena of superconductivity presents us with just this situation. accessdate=2015-11-24

[11] Richard Packard (2006) "Berkeley Experiments on Superfluid Macroscopic Quantum Effects" accessdate=2015-11-24

[12] "Quantum - Definition and More from the Free Merriam-Webster Dictionary". Merriam-webster.com. Retrieved 2012-08-18.

[13] http://mooni.fccj.org/~{}ethall/quantum/quant.htm

[14] "ysfine.com". *ysfine.com*. Retrieved 11 September 2015.

[15] Oocities.com at the Wayback Machine (archived October 26, 2009)

[16] P.A.M. Dirac, *The Principles of Quantum Mechanics*, Clarendon Press, Oxford, 1930.

[17] D. Hilbert *Lectures on Quantum Theory*, 1915–1927

[18] J. von Neumann, *Mathematische Grundlagen der Quantenmechanik*, Springer, Berlin, 1932 (English translation: *Mathematical Foundations of Quantum Mechanics*, Princeton University Press, 1955).

[19] H.Weyl "The Theory of Groups and Quantum Mechanics", 1931 (original title: "Gruppentheorie und Quantenmechanik").

[20] Dirac, P.A.M. (1958). *The Principles of Quantum Mechanics*, 4th edition, Oxford University Press, Oxford UK, p. ix: "For this reason I have chosen the symbolic method, introducing the representatives later merely as an aid to practical calculation."

[21] Greiner, Walter; Müller, Berndt (1994). *Quantum Mechanics Symmetries, Second edition*. Springer-Verlag. p. 52. ISBN 3-540-58080-8., Chapter 1, p. 52

[22] "Heisenberg - Quantum Mechanics, 1925–1927: The Uncertainty Relations". Aip.org. Retrieved 2012-08-18.

[23] Greenstein, George; Zajonc, Arthur (2006). *The Quantum Challenge: Modern Research on the Foundations of Quantum Mechanics, Second edition*. Jones and Bartlett Publishers, Inc. p. 215. ISBN 0-7637-2470-X., Chapter 8, p. 215

[24] "[Abstract] Visualization of Uncertain Particle Movement". Actapress.com. Retrieved 2012-08-18.

[25] Hirshleifer, Jack (2001). *The Dark Side of the Force: Economic Foundations of Conflict Theory*. Campbridge University Press. p. 265. ISBN 0-521-80412-4., Chapter , p.

[26] "dict.cc dictionary :: eigen :: German-English translation". *dict.cc*. Retrieved 11 September 2015.

[27] "Topics: Wave-Function Collapse". Phy.olemiss.edu. 2012-07-27. Retrieved 2012-08-18.

[28] "Collapse of the wave-function". Farside.ph.utexas.edu. Retrieved 2012-08-18.

[29] "Determinism and Naive Realism : philosophy". Reddit.com. 2009-06-01. Retrieved 2012-08-18.

[30] Michae l Trott. "Time-Evolution of a Wavepacket in a Square Well — Wolfram Demonstrations Project". Demonstrations. Retrieved 2010-10-15.

[31] Michael Trott. "Time Evolution of a Wavepacket In a Square Well". Demonstrations.wolfram.com. Retrieved 2010-10-15.

[32] Mathews, Piravonu Mathews; Venkatesan, K. (1976). *A Textbook of Quantum Mechanics*. Tata McGraw-Hill. p. 36. ISBN 0-07-096510-2., Chapter 2, p. 36

[33] "Wave Functions and the Schrödinger Equation" (PDF). Retrieved 2010-10-15.

[34]

[35] Nancy Thorndike Greenspan, "The End of the Certain World: The Life and Science of Max Born" (Basic Books, 2005), pp. 124-8 and 285-6.

[36] http://ocw.usu.edu/physics/classical-mechanics/pdf_lectures/06.pdf

[37] "The Nobel Prize in Physics 1979". Nobel Foundation. Retrieved 2010-02-16.

[38] Carl M. Bender, Daniel W. Hook, Karta Kooner (2009-12-31). "Complex Elliptic Pendulum". arXiv:1001.0131 [hep-th].

[39] See, for example, Precision tests of QED. The relativistic refinement of quantum mechanics known as quantum electrodynamics (QED) has been shown to agree with experiment to within 1 part in 10^8 for some atomic properties.

[40] Tipler, Paul; Llewellyn, Ralph (2008). *Modern Physics* (5 ed.). W. H. Freeman and Company. pp. 160–161. ISBN 978-0-7167-7550-8.

[41] "Quantum mechanics course iwhatisquantummechanics". Scribd.com. 2008-09-14. Retrieved 2012-08-18.

[42] A. Einstein, B. Podolsky, and N. Rosen, *Can quantum-mechanical description of physical reality be considered complete?* Phys. Rev. 47 777 (1935).

[43] "Between classical and quantum◆" (PDF). Retrieved 2012-08-19. replacement character in |title= at position 30 (help)

[44] (see macroscopic quantum phenomena, Bose–Einstein condensate, and Quantum machine)

[45] "Atomic Properties". Academic.brooklyn.cuny.edu. Retrieved 2012-08-18.

[46] http://assets.cambridge.org/97805218/29526/excerpt/9780521829526_excerpt.pdf

[47] Born, M., Heisenberg, W., Jordan, P. (1926). *Z. Phys.* 35: 557–615. Translated as 'On quantum mechanics II', pp. 321–385 in Van der Waerden, B.L. (1967), *Sources of Quantum Mechanics*, North-Holland, Amsterdam, "The basic difference between the theory proposed here and that used hitherto ... lies in the characteristic kinematics ...", p. 385.

[48] Dirac, P.A.M. (1930/1958). *The Principles of Quantum Mechanics*, fourth edition, Oxford University Press, Oxford UK, p. 5: "A question about what will happen to a particular photon under certain conditions is not really very precise. To make it precise one must imagine some experiment performed having a bearing on the question, and enquire what will be the result of the experiment. Only questions about the results of experiments have a real significance and it is only such questions that theoretical physics has to consider."

[49] Bohr, N. (1939). The Causality Problem in Atomic Physics, in *New Theories in Physics, Conference organized in collaboration with the International Union of Physics and the Polish Intellectual Co-operation Committee, Warsaw, May 30th – June 3rd 1938*, International Institute of Intellectual Co-operation, Paris, 1939, pp. 11–30, reprinted in *Niels Bohr, Collected Works*, volume 7 (1933 – 1958) edited by J. Kalckar, Elsevier, Amsterdam, ISBN 0-444-89892-1, pp. 303–322. "The essential lesson of the analysis of measurements in quantum theory is thus the emphasis on the necessity, in the account of the phenomena, of taking the whole experimental arrangement into consideration, in complete conformity with the fact that all unambiguous interpretation of the quantum mechanical formalism involves the fixation of the external conditions, defining the initial state of the atomic system and the character of the possible predictions as regards subsequent observable properties of that system. Any measurement in quantum theory can in fact only refer either to a fixation of the initial state or to the test of such predictions, and it is first the combination of both kinds which constitutes a well-defined phenomenon."

[50] Bohr, N. (1948). On the notions of complementarity and causality, *Dialectica* 2: 312–319. "As a more appropriate way of expression, one may advocate limitation of the use of the word *phenomenon* to refer to observations obtained under specified circumstances, including an account of the whole experiment."

[51] Ludwig, G. (1987). *An Axiomatic Basis for Quantum Mechanics*, volume 2, *Quantum Mechanics and Macrosystems*, translated by K. Just, Springer, Berlin, ISBN 978-3-642-71899-1, Chapter XIII, Special Structures in Preparation and Registration Devices, §1, Measurement chains, p. 132.

[52] Heisenberg, W. (1927). Über den anschaulichen Inhalt der quantentheoretischen Kinematik und Mechanik, *Z. Phys.* 43: 172–198. Translation as 'The actual content of quantum theoretical kinematics and mechanics' here, "But in the rigorous formulation of the law of causality, — "If we know the present precisely, we can calculate the future" — it is not the conclusion that is faulty, but the premise."

[53] Green, H.S. (1965). *Matrix Mechanics*, with a foreword by Max Born, P. Noordhoff Ltd, Groningen. "It is not possible, therefore, to provide 'initial conditions' for the prediction of the behaviour of atomic systems, in the way contemplated by classical physics. This is accepted by quantum theory, not merely as an experimental difficulty, but as a fundamental law of nature", p. 32.

[54] Rosenfeld, L. (1957). Misunderstandings about the foundations of quantum theory, pp. 41–45 in *Observation and Interpretation*, edited by S. Körner, Butterworths, London. "A phenomenon is therefore a process (endowed with the characteristic quantal wholeness) involving a definite type of interaction between the system and the apparatus."

[55] Dirac, P.A.M. (1973). Development of the physicist's conception of nature, pp. 1–55 in *The Physicist's Conception of Nature*, edited by J. Mehra, D. Reidel, Dordrecht, ISBN 90-277-0345-0, p. 5: "That led Heisenberg to his really masterful step forward, resulting in the new quantum mechanics. His idea was to build up a theory entirely in terms of quantities referring to two states."

[56] Born, M. (1927). Physical aspects of quantum mechanics, *Nature* **119**: 354–357, "These probabilities are thus dynamically determined. But what the system actually does is not determined ..."

[57] Messiah, A. (1961). *Quantum Mechanics*, volume 1, translated by G.M. Temmer from the French *Mécanique Quantique*, North-Holland, Amsterdam, p. 157.

[58] Bohr, N. (1928). The Quantum postulate and the recent development of atomic theory, *Nature* **121**: 580–590.

[59] Heisenberg, W. (1930). *The Physical Principles of the Quantum Theory*, translated by C. Eckart and F.C. Hoyt, University of Chicago Press.

[60] Goldstein, H. (1950). *Classical Mechanics*, Addison-Wesley, ISBN 0-201-02510-8.

[61] "There is as yet no logically consistent and complete relativistic quantum field theory.", p. 4. — V. B. Berestetskii, E. M. Lifshitz, L P Pitaevskii (1971). J. B. Sykes, J. S. Bell (translators). *Relativistic Quantum Theory* **4**, **part I**. *Course of Theoretical Physics (Landau and Lifshitz)* ISBN 0-08-016025-5

[62] "Stephen Hawking: Gödel and the end of physics". *cam.ac.uk*. Retrieved 11 September 2015.

[63] "The Nature of Space and Time". *google.com*. Retrieved 11 September 2015.

[64] Tatsumi Aoyama, Masashi Hayakawa, Toichiro Kinoshita, Makiko Nio (2012). "Tenth-Order QED Contribution to the Electron g-2 and an Improved Value of the Fine Structure Constant". *Physical Review Letters* **109** (11): 111807. arXiv:1205.5368v2. Bibcode:2012PhRvL.109k1807A. doi:10.1103/PhysRevLett.109.111807.

[65] Parker, B. (1993). *Overcoming some of the problems*. pp. 259–279.

[66] The Character of Physical Law (1965) Ch. 6; also quoted in The New Quantum Universe (2003), by Tony Hey and Patrick Walters

[67] Weinberg, S. "Collapse of the State Vector", Phys. Rev. A 85, 062116 (2012).

[68] "Action at a Distance in Quantum Mechanics (Stanford Encyclopedia of Philosophy)". Plato.stanford.edu. 2007-01-26. Retrieved 2012-08-18.

[69] "Everett's Relative-State Formulation of Quantum Mechanics (Stanford Encyclopedia of Philosophy)". Plato.stanford.edu. Retrieved 2012-08-18.

[70] The Transactional Interpretation of Quantum Mechanics by John Cramer. *Reviews of Modern Physics* 58, 647-688, July (1986)

[71] See, for example, the Feynman Lectures on Physics for some of the technological applications which use quantum mechanics, e.g., transistors (vol **III**, pp. 14-11 ff), integrated circuits, which are follow-on technology in solid-state physics (vol **II**, pp. 8-6), and lasers (vol **III**, pp. 9-13).

[72] *Introduction to Quantum Mechanics with Applications to Chemistry - Linus Pauling, E. Bright Wilson.* 1985-03-01. ISBN 9780486648712. Retrieved 2012-08-18.

[73] Anderson, Mark (2009-01-13). "Is Quantum Mechanics Controlling Your Thoughts? | Subatomic Particles". DISCOVER Magazine. Retrieved 2012-08-18.

[74] "Quantum mechanics boosts photosynthesis". physicsworld.com. Retrieved 2010-10-23.

[75] Davies, P. C. W.; Betts, David S. (1984). *Quantum Mechanics, Second edition*. Chapman and Hall. p. 79. ISBN 0-7487-4446-0., Chapter 6, p. 79

[76] Baofu, Peter (2007-12-31). *The Future of Complexity: Conceiving a Better Way to Understand Order and Chaos*. ISBN 9789812708991. Retrieved 2012-08-18.

[77] Derivation of particle in a box, chemistry.tidalswan.com

11.10   References

The following titles, all by working physicists, attempt to communicate quantum theory to lay people, using a minimum of technical apparatus.

- Chester, Marvin (1987) *Primer of Quantum Mechanics*. John Wiley. ISBN 0-486-42878-8

- Cox, Brian; Forshaw, Jeff (2011). *The Quantum Universe: Everything That Can Happen Does Happen:*. Allen Lane. ISBN 1-84614-432-9.

- Richard Feynman, 1985. *QED: The Strange Theory of Light and Matter*, Princeton University Press. ISBN 0-691-08388-6. Four elementary lectures on quantum electrodynamics and quantum field theory, yet containing many insights for the expert.

- Ghirardi, GianCarlo, 2004. *Sneaking a Look at God's Cards*, Gerald Malsbary, trans. Princeton Univ. Press. The most technical of the works cited here. Passages using algebra, trigonometry, and bra–ket notation can be passed over on a first reading.

- N. David Mermin, 1990, "Spooky actions at a distance: mysteries of the QT" in his *Boojums all the way through*. Cambridge University Press: 110-76.

- Victor Stenger, 2000. *Timeless Reality: Symmetry, Simplicity, and Multiple Universes*. Buffalo NY: Prometheus Books. Chpts. 5-8. Includes cosmological and philosophical considerations.

More technical:

- Bryce DeWitt, R. Neill Graham, eds., 1973. *The Many-Worlds Interpretation of Quantum Mechanics*, Princeton Series in Physics, Princeton University Press. ISBN 0-691-08131-X

- Dirac, P. A. M. (1930). *The Principles of Quantum Mechanics*. ISBN 0-19-852011-5. The beginning chapters make up a very clear and comprehensible introduction.

- Hugh Everett, 1957, "Relative State Formulation of Quantum Mechanics", *Reviews of Modern Physics* 29: 454-62.

- Feynman, Richard P.; Leighton, Robert B.; Sands, Matthew (1965). *The Feynman Lectures on Physics* 1–3. Addison-Wesley. ISBN 0-7382-0008-5.

- Griffiths, David J. (2004). *Introduction to Quantum Mechanics (2nd ed.)*. Prentice Hall. ISBN 0-13-111892-7. OCLC 40251748. A standard undergraduate text.

- Max Jammer, 1966. *The Conceptual Development of Quantum Mechanics*. McGraw Hill.

- Hagen Kleinert, 2004. *Path Integrals in Quantum Mechanics, Statistics, Polymer Physics, and Financial Markets*, 3rd ed. Singapore: World Scientific. Draft of 4th edition.

- Gunther Ludwig, 1968. *Wave Mechanics*. London: Pergamon Press. ISBN 0-08-203204-1

- George Mackey (2004). *The mathematical foundations of quantum mechanics*. Dover Publications. ISBN 0-486-43517-2.

- Albert Messiah, 1966. *Quantum Mechanics* (Vol. I), English translation from French by G. M. Temmer. North Holland, John Wiley & Sons. Cf. chpt. IV, section III.

- Omnès, Roland (1999). *Understanding Quantum Mechanics*. Princeton University Press. ISBN 0-691-00435-8. OCLC 39849482.

- Scerri, Eric R., 2006. *The Periodic Table: Its Story and Its Significance*. Oxford University Press. Considers the extent to which chemistry and the periodic system have been reduced to quantum mechanics. ISBN 0-19-530573-6

- Transnational College of Lex (1996). *What is Quantum Mechanics? A Physics Adventure*. Language Research Foundation, Boston. ISBN 0-9643504-1-6. OCLC 34661512.

- von Neumann, John (1955). *Mathematical Foundations of Quantum Mechanics*. Princeton University Press. ISBN 0-691-02893-1.

- Hermann Weyl, 1950. *The Theory of Groups and Quantum Mechanics*, Dover Publications.

- D. Greenberger, K. Hentschel, F. Weinert, eds., 2009. *Compendium of quantum physics, Concepts, experiments, history and philosophy*, Springer-Verlag, Berlin, Heidelberg.

11.11   Further reading

- Bernstein, Jeremy (2009). *Quantum Leaps*. Cambridge, Massachusetts: Belknap Press of Harvard University Press. ISBN 978-0-674-03541-6.

- Bohm, David (1989). *Quantum Theory*. Dover Publications. ISBN 0-486-65969-0.

- Eisberg, Robert; Resnick, Robert (1985). *Quantum Physics of Atoms, Molecules, Solids, Nuclei, and Particles (2nd ed.)*. Wiley. ISBN 0-471-87373-X.

- Liboff, Richard L. (2002). *Introductory Quantum Mechanics*. Addison-Wesley. ISBN 0-8053-8714-5.

- Merzbacher, Eugen (1998). *Quantum Mechanics*. Wiley, John & Sons, Inc. ISBN 0-471-88702-1.

- Sakurai, J. J. (1994). *Modern Quantum Mechanics*. Addison Wesley. ISBN 0-201-53929-2.

- Shankar, R. (1994). *Principles of Quantum Mechanics*. Springer. ISBN 0-306-44790-8.

- Stone, A. Douglas (2013). *Einstein and the Quantum*. Princeton University Press. ISBN 978-0-691-13968-5.

- Martinus J. G. Veltman, 2003 *Facts and Mysteries in Elementary Particle Physics*.

- Shushi, Tomer (2014). *The Influence of Particle Interactions on the Existence of Quantum Particles Properties* (PDF). Haifa, Israel: Journal of Physical Science and Application.

11.12   External links

- 3D animations, applications and research for basic quantum effects (animations also available in commons. (Université paris Sud))

- Quantum Cook Book by R. Shankar, Open Yale PHYS 201 material (4pp)

- The Modern Revolution in Physics - an online textbook.

- J. O'Connor and E. F. Robertson: A history of quantum mechanics.

- Introduction to Quantum Theory at Quantiki.

- Quantum Physics Made Relatively Simple: three video lectures by Hans Bethe

- H is for h-bar.

- Quantum Mechanics Books Collection: Collection of free books

Course material

- Quantum Physics Database - Fundamentals and Historical Background of Quantum Theory.

- Doron Cohen: Lecture notes in Quantum Mechanics (comprehensive, with advanced topics).

- MIT OpenCourseWare: Chemistry.

- MIT OpenCourseWare: Physics. See 8.04

- Stanford Continuing Education PHY 25: Quantum Mechanics by Leonard Susskind, see course description Fall 2007

- 5½ Examples in Quantum Mechanics

- Imperial College Quantum Mechanics Course.

- Spark Notes - Quantum Physics.

- Quantum Physics Online : interactive introduction to quantum mechanics (RS applets).

- Experiments to the foundations of quantum physics with single photons.

- AQME : Advancing Quantum Mechanics for Engineers — by T.Barzso, D.Vasileska and G.Klimeck online learning resource with simulation tools on nanohub

- Quantum Mechanics by Martin Plenio

- Quantum Mechanics by Richard Fitzpatrick

- Online course on *Quantum Transport*

FAQs

- Many-worlds or relative-state interpretation.

- Measurement in Quantum mechanics.

Media

- PHYS 201: Fundamentals of Physics II by Ramamurti Shankar, Open Yale Course

- Lectures on Quantum Mechanics by Leonard Susskind

- Everything you wanted to know about the quantum world — archive of articles from *New Scientist*.

- Quantum Physics Research from *Science Daily*

- Overbye, Dennis (December 27, 2005). "Quantum Trickery: Testing Einstein's Strangest Theory". *The New York Times*. Retrieved April 12, 2010.

- Audio: Astronomy Cast Quantum Mechanics — June 2009. Fraser Cain interviews Pamela L. Gay.

Philosophy

- "Quantum Mechanics" entry by Jenann Ismael in the *Stanford Encyclopedia of Philosophy*

- "Measurement in Quantum Theory" entry by Henry Krips in the *Stanford Encyclopedia of Philosophy*

Max Planck is considered the father of the quantum theory.

The 1927 Solvay Conference in Brussels.

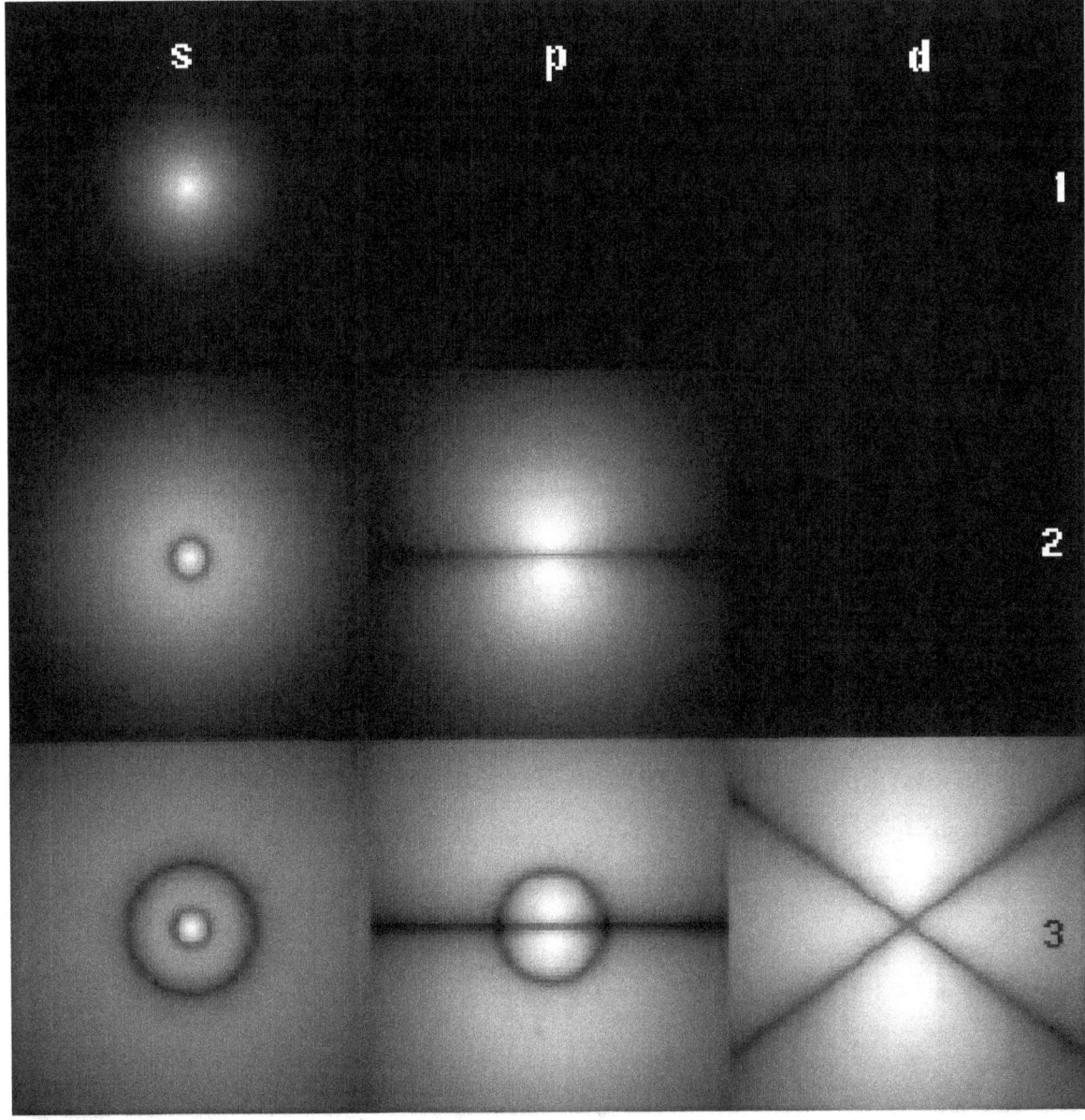

*Fig. 1: Probability densities corresponding to the wave functions of an electron in a hydrogen atom possessing definite energy levels (increasing from the top of the image to the bottom: n = 1, 2, 3, ...) and angular momenta (increasing across from left to right: s, p, d, ...). Brighter areas correspond to higher probability density in a position measurement. Such wave functions are directly comparable to Chladni's figures of acoustic modes of vibration in classical physics, and are modes of oscillation as well, possessing a sharp energy and, thus, a definite frequency. The angular momentum and energy are quantized, and take **only** discrete values like those shown (as is the case for resonant frequencies in acoustics)*

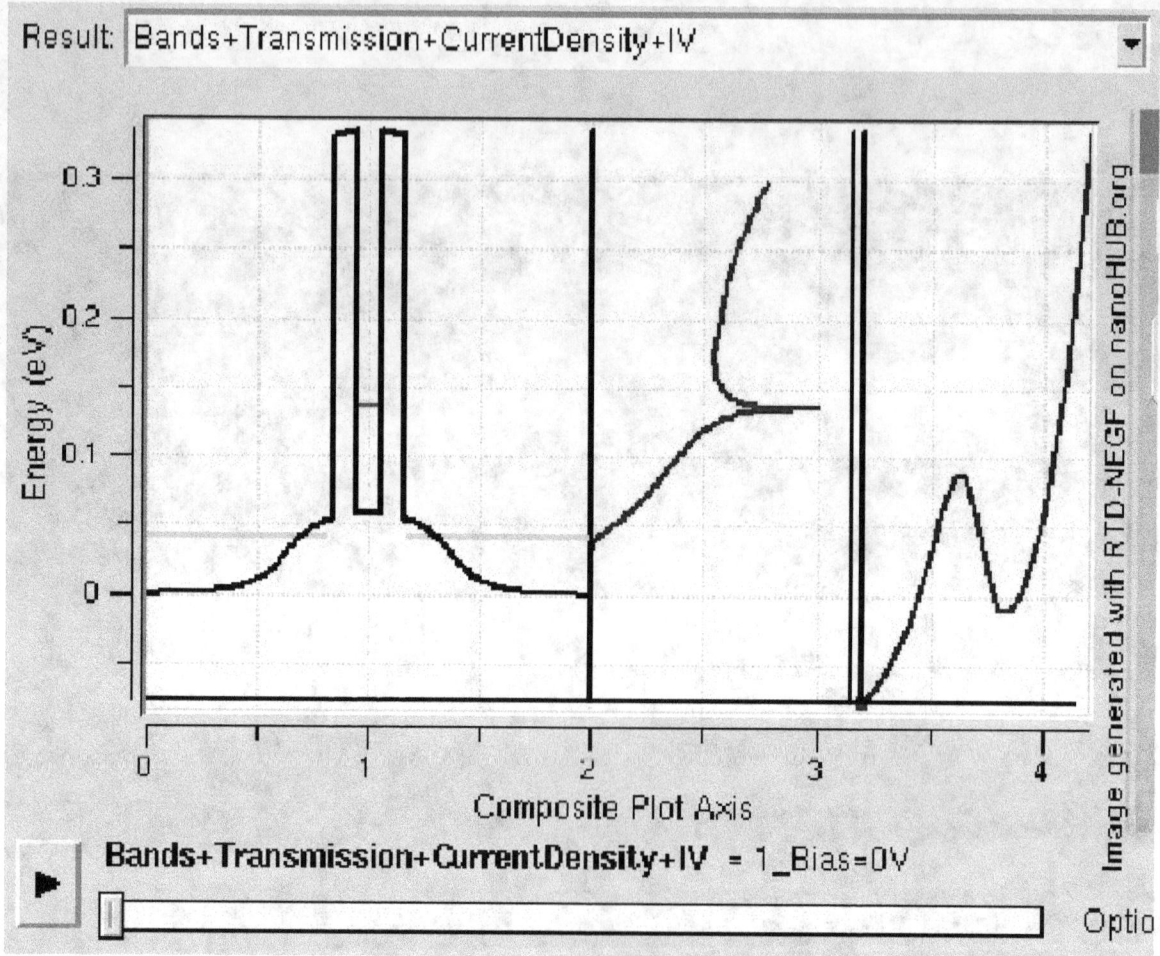

A working mechanism of a resonant tunneling diode device, based on the phenomenon of quantum tunneling through potential barriers

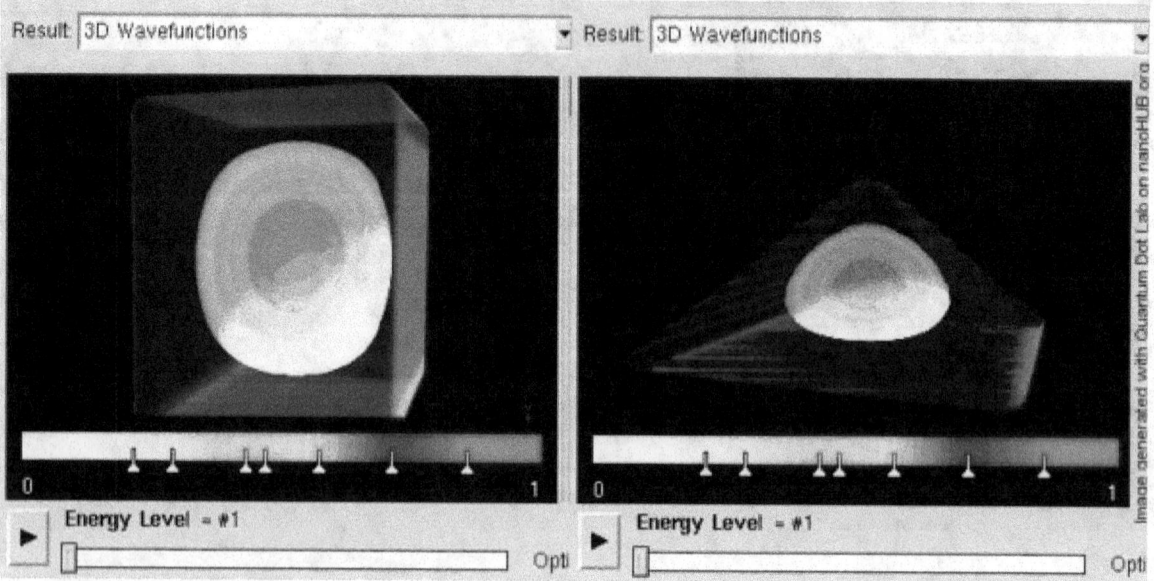

3D confined electron wave functions for each eigenstate in a Quantum Dot. Here, rectangular and triangular-shaped quantum dots are shown. Energy states in rectangular dots are more 's-type' and 'p-type'. However, in a triangular dot, the wave functions are mixed due to confinement symmetry.

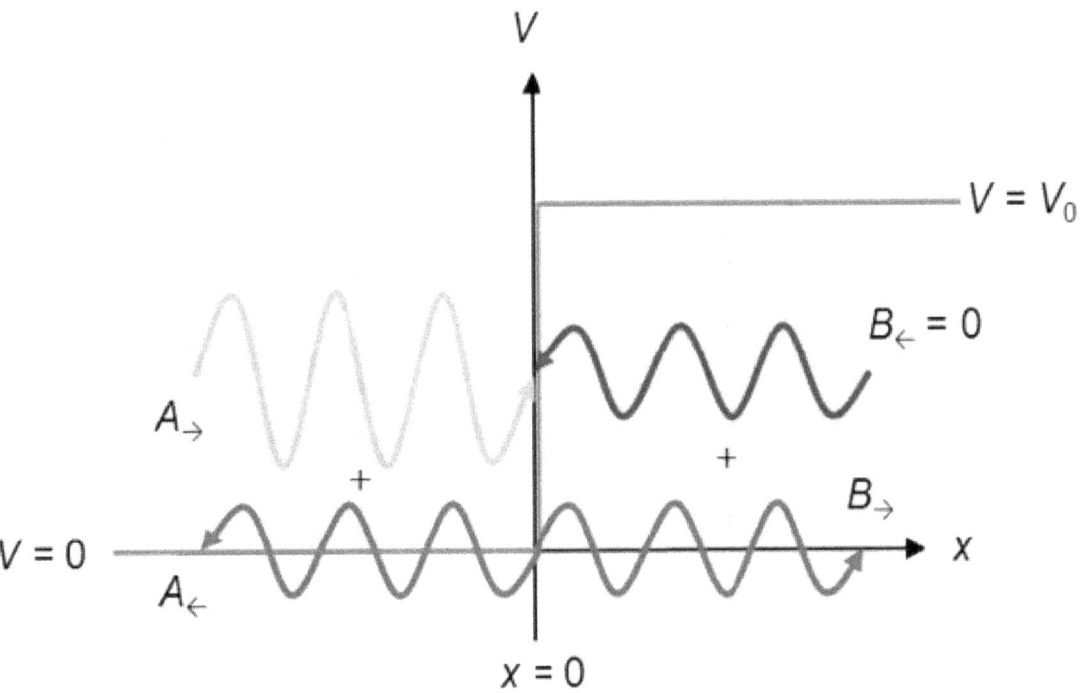

Scattering at a finite potential step of height V_0, shown in green. The amplitudes and direction of left- and right-moving waves are indicated. Yellow is the incident wave, blue are reflected and transmitted waves, red does not occur. E > V_0 for this figure.

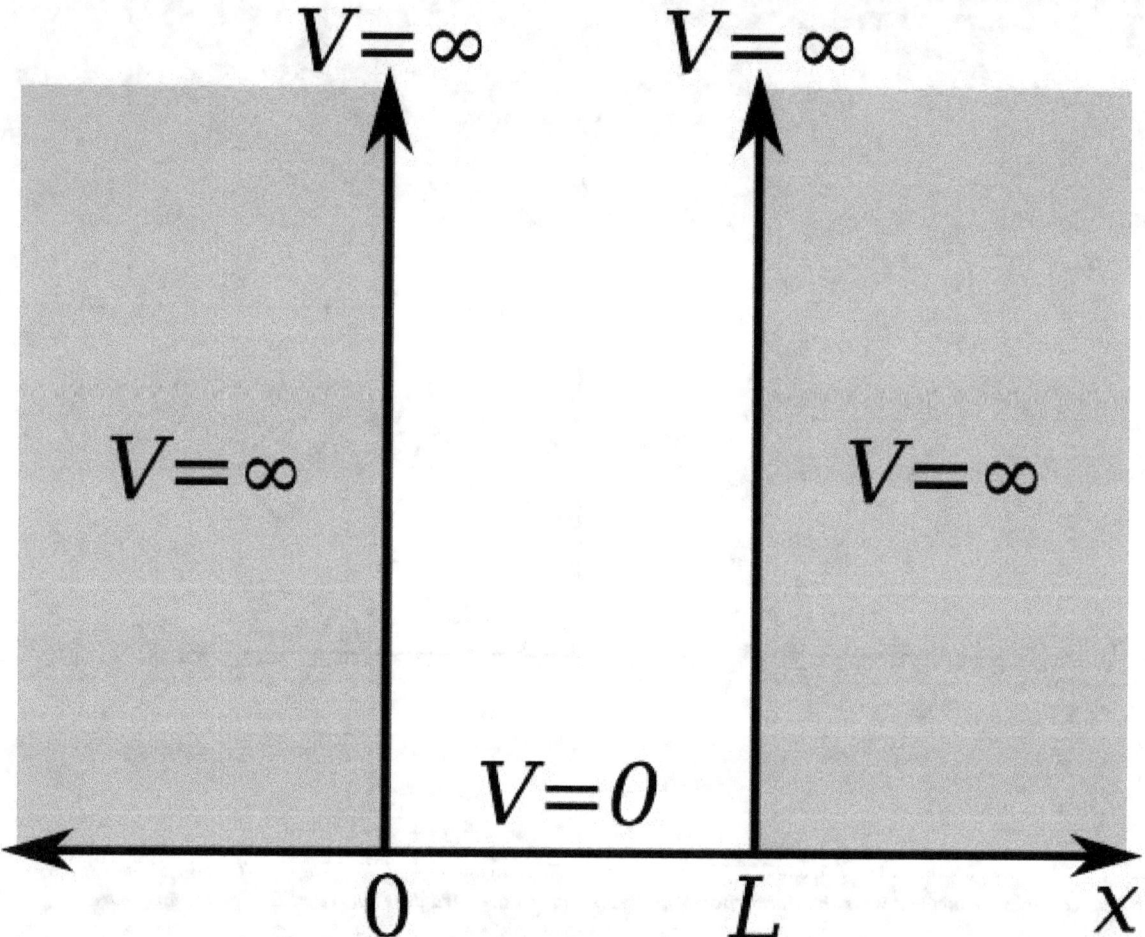

1-dimensional potential energy box (or infinite potential well)

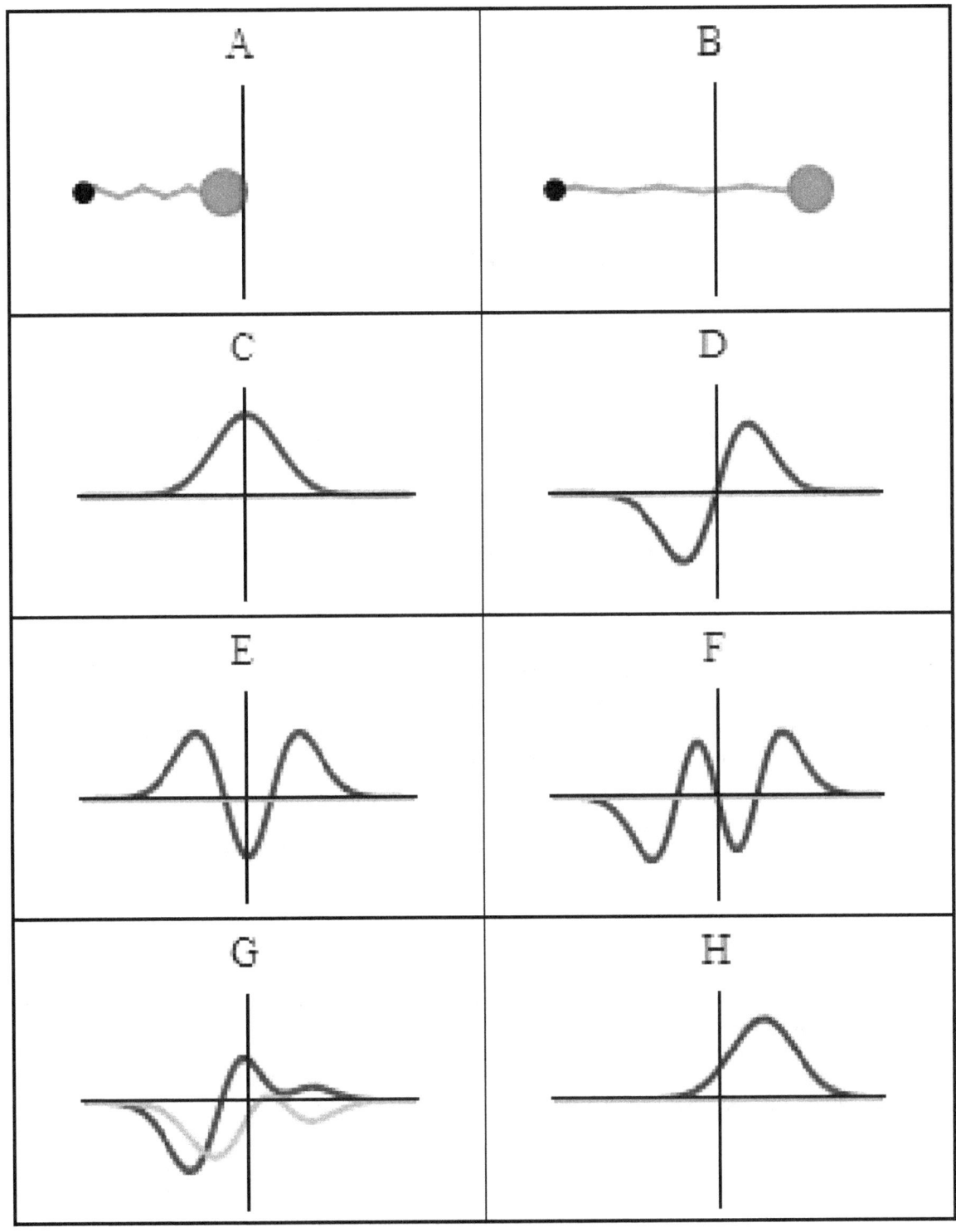

Some trajectories of a harmonic oscillator (i.e. a ball attached to a spring) in classical mechanics (A-B) and quantum mechanics (C-H).
In quantum mechanics, the position of the ball is represented by a wave (called the wave function), with the real part shown in blue
and the imaginary part shown in red. Some of the trajectories (such as C,D,E,and F) are standing waves (or "stationary states"). Each
standing-wave frequency is proportional to a possible energy level of the oscillator. This "energy quantization" does not occur in classical
physics, where the oscillator can have any energy.

Chapter 12

Formalism (mathematics)

In foundations of mathematics, philosophy of mathematics, and philosophy of logic, **formalism** is a theory that holds that statements of mathematics and logic can be considered to be statements about the consequences of certain string manipulation rules.

For example, Euclidean geometry can be considered a game whose play consists in moving around certain strings of symbols called axioms according to a set of rules called "rules of inference" to generate new strings. In playing this game one can "prove" that the Pythagorean theorem is valid because the string representing the Pythagorean theorem can be constructed using only the stated rules.

According to formalism, the truths expressed in logic and mathematics are not about numbers, sets, or triangles or any other contensive subject matter — in fact, they aren't "about" anything at all. They are syntactic forms whose shapes and locations have no meaning unless they are given an interpretation (or semantics).

Formalism is associated with rigorous method. In common use, a *formalism* means the out-turn of the effort towards formalisation of a given limited area. In other words, matters can be formally discussed once captured in a formal system, or commonly enough within something *formalisable* with claims to be one. Complete formalisation is in the domain of computer science.

Formalism stresses axiomatic proofs using theorems, specifically associated with David Hilbert. A formalist is an individual who belongs to the school of formalism, which is a certain mathematical-philosophical doctrine descending from Hilbert.

Formalists are relatively tolerant and inviting to new approaches to logic, non-standard number systems, new set theories, etc. The more games we study, the better. However, in all three of these examples, motivation is drawn from existing mathematical or philosophical concerns. The "games" are usually not arbitrary.

Recently, some formalist mathematicians have proposed that all of our *formal* mathematical knowledge should be systematically encoded in computer-readable formats, in order to facilitate automated proof checking of mathematical proofs and the use of interactive theorem proving in the development of mathematical theories and computer software. Because of their close connection with computer science, this idea is also advocated by mathematical intuitionists and constructivists in the "computability" tradition (see below).

12.1 Deductivism

Main article: Hypothetico-deductive model

Another version of formalism is often known as deductivism. In deductivism, the Pythagorean theorem is not an absolute truth, but a relative one.

This is to say, that *if* you interpret the strings in such a way that the rules of the game become true *then* you have to accept that the theorem, or, rather, the interpretation of the theorem you have given it must be a true statement. (The rules of such a game would have to include, for instance, that true statements are assigned to the axioms, and that the rules of inference are truth-preserving, etcetera.)

Under deductivism, the same view is held to be true for all other statements of formal logic and mathematics. Thus, formalism need not mean that these deductive sciences are nothing more than meaningless symbolic games. It is usually hoped that there exists some interpretation in which the rules of the game hold. Compare this position to structuralism.

Taking the deductivist view allows the working mathematician to suspend judgement on the deep philosophical questions and proceed as if solid epistemological foundations were available. Many formalists would say that in practice, the axiom systems to be studied are suggested by the demands of the particular science.

12.2 Hilbert's formalism

A major early proponent of formalism was David Hilbert, whose program was intended to be a complete and consistent axiomatization of all of mathematics. Hilbert aimed to show the consistency of mathematical systems from the assumption that the "finitary arithmetic" (a subsystem of the usual arithmetic of the positive integers, chosen to be philosophically uncontroversial) was consistent (i.e. no contradictions can be derived from the system).

The way that Hilbert tried to show that an axiomatic system was consistent was by formalizing it using a particular language (Snapper, 1979). In order to formalize an axiomatic system, you must first choose a language in which you can express and perform operations within that system. This language must include five components:

- It must include variables such as x, which can stand for some number.

- It must have quantifiers such as the symbol for the existence of an object.

- It must include equality.

- It must include connectives such as $\leftrightarrow$ for "if and only if."

- It must include certain undefined terms called parameters. For geometry, these undefined terms might be something like a point or a line, which we still choose symbols for.

Once we choose this language, Hilbert thought that we could prove all theorems within any axiomatic system using nothing more than the axioms themselves and the chosen formal language.

Gödel's conclusion in his incompleteness theorems was that you cannot prove consistency within any axiomatic system rich enough to include classical arithmetic. On the one hand, you must use only the formal language chosen to formalize this axiomatic system; on the other hand, it is impossible to prove the consistency of this language in itself (Snapper, 1979). Hilbert was originally frustrated by Gödel's work because it shattered his life's goal to completely formalize everything in number theory (Reid and Weyl, 1970). However, Gödel did not feel that he contradicted everything about Hilbert's formalist point of view. After Gödel published his work, it became apparent that proof theory still had some use, the only difference is that it could not be used to prove the consistency of all of number theory as Hilbert had hoped (Reid and Weyl, 1970). Present-day formalists use proof theory to further our understanding in mathematics, but perhaps because of Gödel's work, they make no claims about semantic meaning in the work that they do with mathematics. Proofs are simply the manipulation of symbols in our formal language starting from certain rules that we call axioms.

It is important to note that Hilbert is not considered a strict formalist as formalism is defined today. He thought there was some meaning and truth in mathematics, which is precisely why he was trying to prove the consistency of number theory. If number theory turned out to be consistent, then there had to be some sort of truth in it (Goodman, 1979). Strict formalists consider mathematics apart from its semantic meaning. They view mathematics as pure syntax: the manipulation of symbols according to certain rules. They then attempt to show that this set of rules is consistent, much like Hilbert attempted to do (Goodman, 1979). Formalists currently believe that computerized algorithms will eventually take over the task of constructing proofs. Computers will replace humans in all mathematical activities, such as checking to see if a proof is correct or not (Goodman, 1979).

David Hilbert

Hilbert was initially a deductivist, but, he considered certain metamathematical methods to yield intrinsically meaningful results and was a realist with respect to the finitary arithmetic. Later, he held the opinion that there was no other meaningful mathematics whatsoever, regardless of interpretation.

12.3   Axiomatic systems

Other formalists, such as Rudolf Carnap, Alfred Tarski and Haskell Curry, considered mathematics to be the investigation of formal axiom systems. Mathematical logicians study formal systems but are just as often realists as they are formalists.

12.4   Principia Mathematica

Perhaps the most serious attempt to formalize number theory was by the two mathematicians Bertrand Russell and Alfred North Whitehead. They created a work, Principia Mathematica, which derived number theory by the manipulation of symbols using formal logic. This work was very detailed, and they spent the better part of a decade in writing it. It was not until page 379 of the first volume that they were even able to prove that $1+1=2$.

Russell's philosophy of mathematics was not formalist, however; it is usually considered a form of logicism. He strongly criticized Hilbert's formalism.

12.5   Criticisms of formalism

Gödel indicated one of the weak points of formalism by addressing the question of consistency in axiomatic systems. More recent criticisms lie in the assertion of formalists that it is possible to computerize all of mathematics. These criticisms bring up the philosophical question of whether or not computers are able to think. Turing tests, named after Alan Turing, who created the test, are an attempt to provide criteria for judging when a computer is capable of thought. The existence of a computer which in principle could pass a Turing test would prove to formalists that computers will be able to do all of mathematics. However, there are opponents of this claim, such as John Searle, who came up with the "Chinese room" thought experiment. He presented the argument that while a computer may be able to manipulate the symbols that we give it, the machine could attach no meaning to these symbols. Since computers will not be able to deal with semantic content in mathematics (Penrose, 1989), they could not be said to "think."

Further, humans can create several ways to prove the same result, even if they might find it challenging to articulate such methods. Since creativity requires thought having a semantic foundation, a computer would not be able to create different methods of solving the same problem. Indeed, a formalist would not be able to say that these other ways of solving problems exist simply because they have not been formalized (Goodman, 1979).

Another critique of formalism is that the actual mathematical ideas that occupy mathematicians are far removed from the string manipulation games mentioned above. Formalism is thus silent to the question of which axiom systems ought to be studied, as none is more meaningful than another from a formalistic point of view.

12.6   See also

- QED project

- Mathematical formalism

- Formalized mathematics

12.7 References

- Goodman, Nicholas D. "Mathematics as an Objective Science." The American Mathematical Monthly 86.7 (1979): 540-551. Print.

- Penrose, Roger. The Emperor's New Mind: concerning Computers, Minds, and the Laws of Physics. Oxford: Oxford UP, 1989. Print.

- Reid, Constance, and Hermann Weyl. Hilbert. Berlin: Springer-Verlag, 1970. Print.

- Snapper, Ernst. "The Three Crises in Mathematics: Logicism, Intuitionism and Formalism." Mathematics Magazine 52.4 (1979): 207-16. Print.

- Weir, Alan: "Formalism in the Philosophy of Mathematics." Stanford Encyclopedia of Philosophy (2011).

Chapter 13

Wave function

Not to be confused with Wave equation.

A wave function in quantum mechanics describes the quantum state of an isolated system of one or more particles. There is *one* wave function containing all the information about the entire system, not a separate wave function for each particle in the system. Its interpretation is that of a probability amplitude. Quantities associated with measurements, such as the average momentum of a particle, can be derived from the wave function. It is a central entity in quantum mechanics and is important in all modern theories, like quantum field theory incorporating quantum mechanics, while its interpretation may differ. The most common symbols for a wave function are the Greek letters ψ or Ψ (lower-case and capital psi).

For a given system, once a representation corresponding to a maximal set of commuting observables and a suitable coordinate system is chosen, the wave function is a complex-valued function of the system's degrees of freedom corresponding to the chosen representation and coordinate system, continuous as well as discrete. Such a set of observables, by a postulate of quantum mechanics, are Hermitian linear operators on the space of states representing a set of **physical observables**, like position, momentum and spin that can, in principle, be simultaneously measured with arbitrary precision. Wave functions can be added together and multiplied by complex numbers to form new wave functions, and hence are elements of a vector space. This is the superposition principle of quantum mechanics. This vector space is endowed with an inner product such that it is a complete metric topological space with respect to the metric induced by the inner product. In this way the set of wave functions for a system form a function space that is a Hilbert space. The inner product is a measure of the overlap between physical states and is used in the foundational probabilistic interpretation of quantum mechanics, the Born rule, relating transition probabilities to inner products. The actual space depends on the system's degrees of freedom (hence on the chosen representation and coordinate system) and the exact form of the Hamiltonian entering the equation governing the dynamical behavior. In the non-relativistic case, disregarding spin, this is the Schrödinger equation.

The Schrödinger equation determines the allowed wave functions for the system and how they evolve over time. A wave function behaves qualitatively like other waves, such as water waves or waves on a string, because the Schrödinger equation is mathematically a type of wave equation. This explains the name "wave function", and gives rise to wave–particle duality. The wave of the wave function, however, is not a wave in physical space; it is a wave in an abstract mathematical "space", and in this respect it differs fundamentally from water waves or waves on a string.[1][2][3][4][5][6][7]

For a given system, the choice of which relevant degrees of freedom to use are not unique, and correspondingly the domain of the wave function is not unique. It may be taken to be a function of all the position coordinates of the particles over *position space*, or the momenta of all the particles over *momentum space*, the two are related by a Fourier transform. These descriptions are the most important, but they are not the only possibilities. Just like in classical mechanics, canonical transformations may be used in the description of a quantum system. Some particles, like electrons and photons, have nonzero spin, and the wave function must include this fundamental property as an intrinsic discrete degree of freedom. In general, for a particle with *half-integer* spin the wave function is a spinor, for a particle with *integer* spin the wave function is a tensor. Particles with spin zero are called scalar particles, those with spin 1 vector particles, and more generally for higher integer spin, tensor particles. The terminology derives from how the wave functions transform under a rotation of the coordinate system. No *elementary* particle with spin $^3/_2$ or higher is known, except for the hypothesized spin 2 graviton. Other discrete variables can be included, such as isospin. When a system has internal degrees of freedom, the

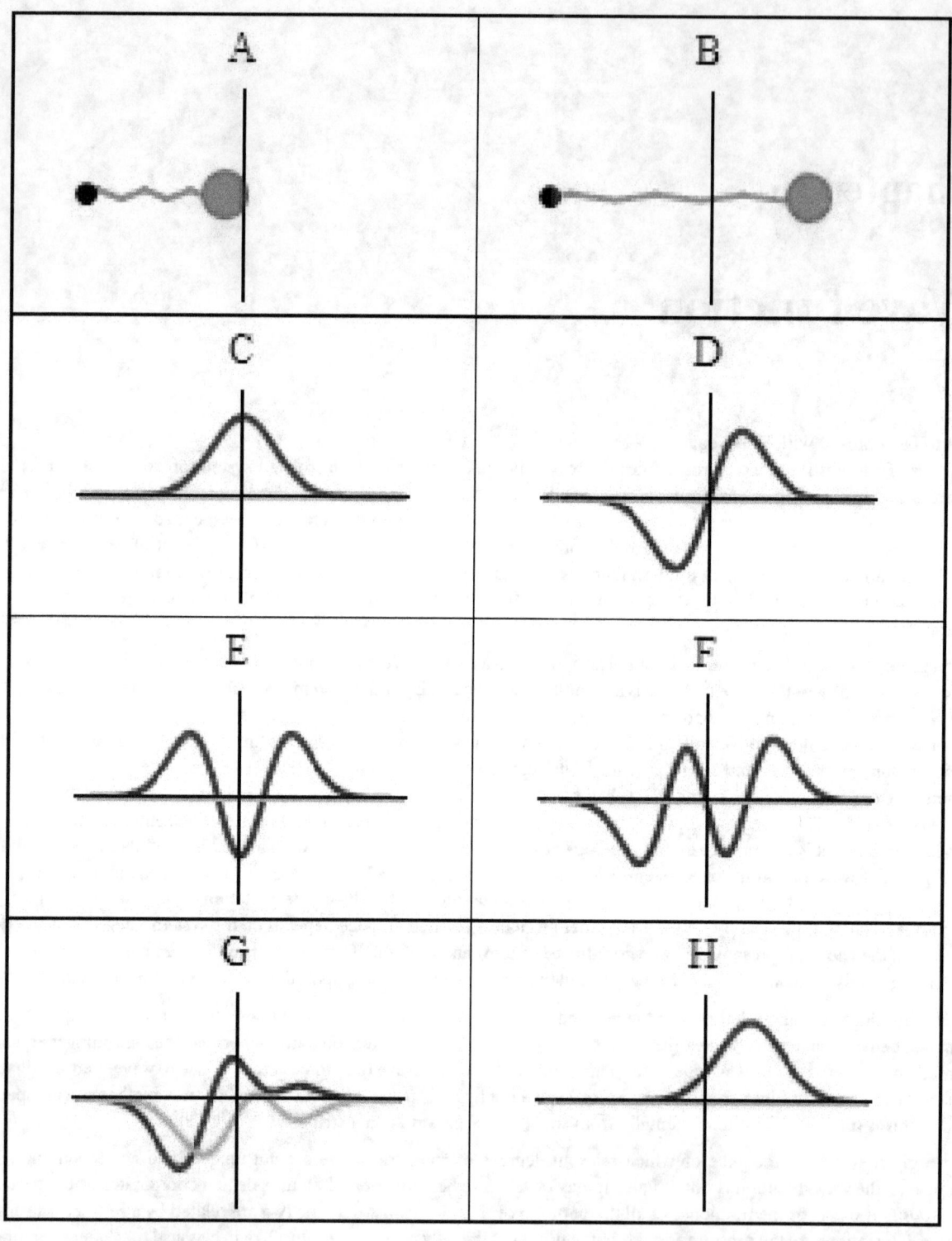

Comparison of classical and quantum harmonic oscillator conceptions for a single spinless particle. *The two processes differ greatly. The classical process (A–B) is represented as the motion of a particle along a trajectory. The quantum process (C–H) has no such trajectory. Rather, it is represented as a wave. Panels (C–F) show four different standing wave solutions of the Schrödinger equation. Panels (G–H) further show two different wave functions that are solutions of the Schrödinger equation but not standing waves.*

wave function at each point in the continuous degrees of freedom (e.g. a point in space) assigns a complex number for *each* possible value of the discrete degrees of freedom (e.g. z-component of spin). These values are often displayed in a column matrix (e.g. a 2 × 1 column vector for a non-relativistic electron with spin $\frac{1}{2}$).

In the Copenhagen interpretation, an interpretation of quantum mechanics, the squared modulus of the wave function, $|\psi|^2$, is a real number interpreted as the probability density of measuring a particle as being at a given place at a given time or having a definite momentum, and possibly having definite values for discrete degrees of freedom. The integral of this quantity, over all the system's degrees of freedom, must be 1 in accordance with the probability interpretation, this general requirement a wave function must satisfy is called the *normalization condition*. Since the wave function is complex valued, only its relative phase and relative magnitude can be measured. Its value does not in isolation tell anything about the magnitudes or directions of measurable observables; one has to apply quantum operators, whose eigenvalues correspond to sets of possible results of measurements, to the wave function ψ and calculate the statistical distributions for measurable quantities.

The unit of measurement for ψ depends on the system, and can be found by dimensional analysis of the normalization condition for the system. For one particle in three dimensions, its units are $[length]^{-3/2}$, because an integral of $|\psi|^2$ over a region of three-dimensional space is a dimensionless probability.[8]

13.1   Historical background

In 1905 Einstein postulated the proportionality between the frequency of a photon and its energy, $E = hf$,[9] and in 1916 the corresponding relation between photon momentum and wavelength, $\lambda = h/p$.[10] In 1923, De Broglie was the first to suggest that the relation $\lambda = h/p$, now called the De Broglie relation, holds for *massive* particles, the chief clue being Lorentz invariance,[11] and this can be viewed as the starting point for the modern development of quantum mechanics. The equations represent wave–particle duality for both massless and massive particles.

In the 1920s and 1930s, quantum mechanics was developed using calculus and linear algebra. Those who used the techniques of calculus included Louis de Broglie, Erwin Schrödinger, and others, developing "wave mechanics". Those who applied the methods of linear algebra included Werner Heisenberg, Max Born, and others, developing "matrix mechanics". Schrödinger subsequently showed that the two approaches were equivalent.[12]

In 1926, Schrödinger published the famous wave equation now named after him, indeed the Schrödinger equation, based on classical Conservation of energy using quantum operators and the de Broglie relations such that the solutions of the equation are the wave functions for the quantum system.[13] However, no one was clear on how to *interpret it*.[14] At first, Schrödinger and others thought that wave functions represent particles that are spread out with most of the particle being where the wave function is large.[15] This was shown to be incompatible with how elastic scattering of a wave packet representing a particle off a target appears; it spreads out in all directions.[16] While a scattered particle may scatter in any direction, it does not break up and take off in all directions. In 1926, Born provided the perspective of probability amplitude.[16][17][18] This relates calculations of quantum mechanics directly to probabilistic experimental observations. It is accepted as part of the Copenhagen interpretation of quantum mechanics. There are many other interpretations of quantum mechanics. In 1927, Hartree and Fock made the first step in an attempt to solve the N-body wave function, and developed the *self-consistency cycle*: an iterative algorithm to approximate the solution. Now it is also known as the Hartree–Fock method.[19] The Slater determinant and permanent (of a matrix) was part of the method, provided by John C. Slater.

Schrödinger did encounter an equation for the wave function that satisfied relativistic energy conservation *before* he published the non-relativistic one, but discarded it as it predicted negative probabilities and negative energies. In 1927, Klein, Gordon and Fock also found it, but incorporated the electromagnetic interaction and proved that it was Lorentz invariant. De Broglie also arrived at the same equation in 1928. This relativistic wave equation is now most commonly known as the Klein–Gordon equation.[20]

In 1927, Pauli phenomenologically found a non-relativistic equation to describe spin-1/2 particles in electromagnetic fields, now called the Pauli equation.[21] Pauli found the wave function was not described by a single complex function of space and time, but needed two complex numbers, which respectively correspond to the spin +1/2 and −1/2 states of the fermion. Soon after in 1928, Dirac found an equation from the first successful unification of special relativity and quantum mechanics applied to the electron, now called the Dirac equation. In this, the wave function is a *spinor* represented by four complex-valued components:[19] two for the electron and two for the electron's antiparticle, the positron. In the non-relativistic limit, the Dirac wave function resembles the Pauli wave function for the electron. Later, other relativistic wave equations were found.

13.1.1 Wave functions and wave equations in modern theories

All these wave equations are of enduring importance. The Schrödinger equation and the Pauli equation are under many circumstances excellent approximations of the relativistic variants. They are considerably easier to solve in practical problems than the relativistic equations. The Klein-Gordon equation and the Dirac equation, while being relativistic, do not represent full reconciliation of quantum mechanics and special relativity. The branch of quantum mechanics where these equations are studied the same way as the Schrödinger equation, often called relativistic quantum mechanics, while very successful, has its limitations (see e.g. Lamb shift) and conceptual problems (see e.g. Dirac sea).

Relativity makes it inevitable that the number of particles in a system is not constant. For full reconciliation, quantum field theory is needed.[22] In this theory, the wave equations and the wave functions have their place, but in a somewhat different guise. The main objects of interest are not the wave functions, but rather operators, so called *field operators* (or just fields where "operator" is understood) on the Hilbert space of states (to be described next section). It turns out that the original relativistic wave equations and their solutions are still needed to build the Hilbert space. Moreover, the *free fields operators*, i.e. when interactions are assumed not to exist, turn out to (formally) satisfy the same equation as do the fields (wave functions) in many cases.

Thus the Klein-Gordon equation (spin 0) and the Dirac equation (spin ½) in this guise remain in the theory. Higher spin analogues include the Proca equation (spin 1), Rarita–Schwinger equation (spin ³⁄₂), and, more generally, the Bargmann–Wigner equations. For *massless* free fields two examples are the free field Maxwell equation (spin 1) and the free field Einstein equation (spin 2) for the field operators.[23] All of them are essentially a direct consequence of the requirement of Lorentz invariance. Their solutions must transform under Lorentz transformation in a prescribed way, i.e. under a particular representation of the Lorentz group and that together with few other reasonable demands, e.g. the *cluster decomposition principle*,[24] with implications for causality is enough to fix the equations.

It should be emphasized that this applies to free field equations; interactions are not included. It should also be noted that the equations and their solutions, though needed for the theories, are not the central objects of study.

13.2 Wave functions and function spaces

The concept of Function spaces enters naturally in the discussion about wave functions. A function space is a set of functions, usually with some defining requirements on the functions, together with a topology on that set. The latter will sparsely be used here, it is only needed to obtain a precise definition of what it means for a subset of a function space to be closed. A wave function is an element of a function space partly characterized by the following concrete and abstract descriptions.

- The Schrödinger equation is linear. This means that the solutions to it, wave functions, can be added and multiplied by scalars to form a new solution.

- The superposition principle of quantum mechanics. If Ψ and Φ are two states in the abstract space of **states** of a quantum mechanical system, then $a\Psi + b\Phi$ is a valid state as well.

The first item says that the set of solutions to the Schrödinger equation is a vector space. The second item says that the set of allowable states is a vector space. This similarity is of course not accidental. Not all properties of the respective spaces have been given so far. There are also a distinctions between the spaces to keep in mind.

- Basic states are characterized by a set of quantum numbers. This is a set of eigenvalues of a maximal set of commuting observables. A choice of such a set may be called a choice of **representation**. It is a postulate of quantum mechanics that a physically observable quantity of a system, like position, momentum and spin, is represented by a linear Hermitian operator on the state space. The possible outcomes of measurement of the quantity are

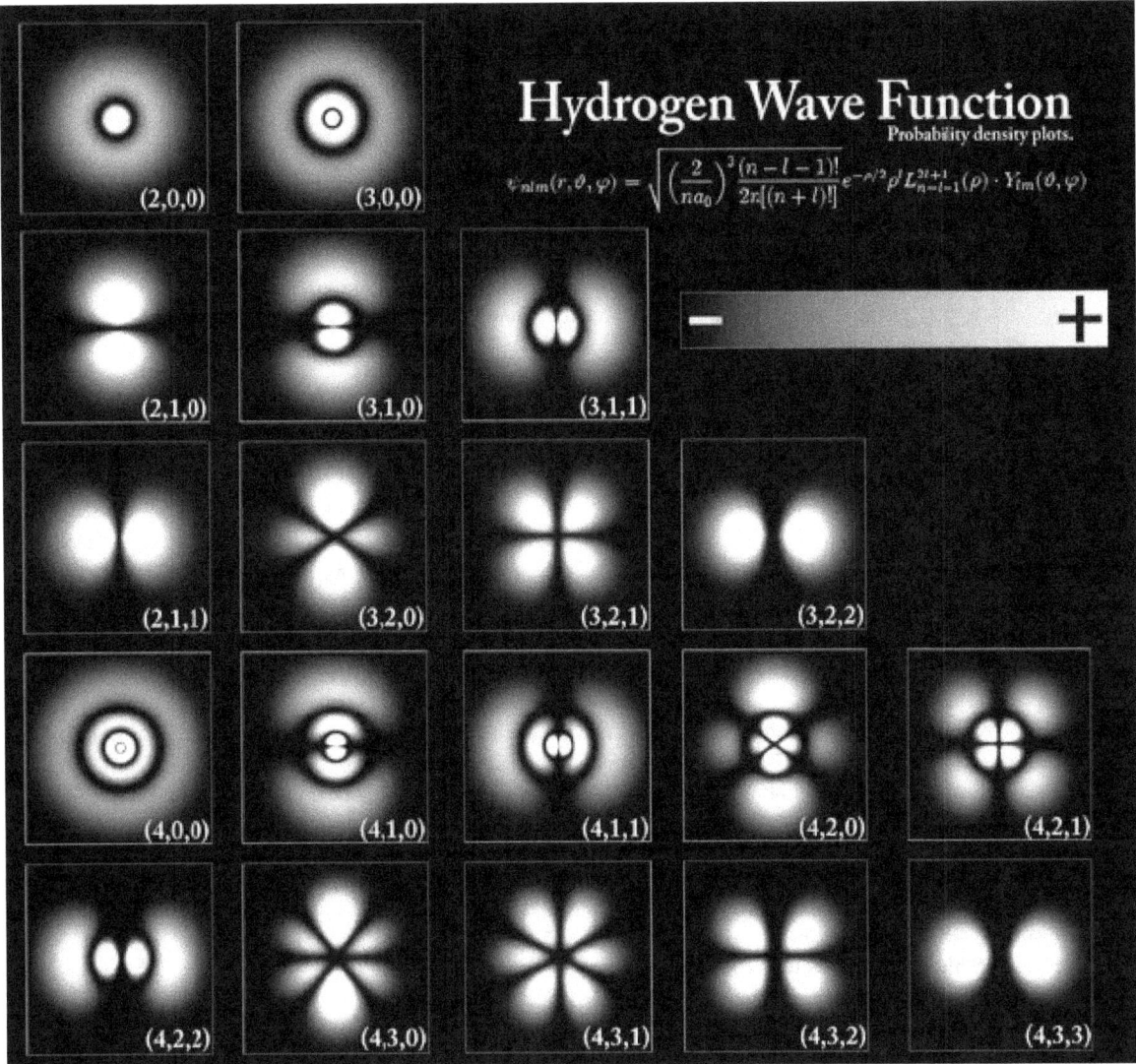

The electron probability density for the first few hydrogen atom electron orbitals shown as cross-sections. These orbitals form an orthonormal basis for the wave function of the electron. Different orbitals are depicted with different scale.

the eigenvalues of the operator.[15] Maximality refers to that no more algebraically independent linear Hermitian operator can be added to the set that commutes with the ones already present. The physical interpretation is that such a set represents what can – in theory – be simultaneously be measured with arbitrary position at a given time. The set is non-unique. It may for a one-particle system, for example, be position and spin z-projection, (x, Sz), or it may be momentum and spin y-projection, (p, Sy). At a deeper level, most observables, perhaps all, arise as generators of symmetries.[15][25][nb 1]

- Once a representation is chosen, there is still arbitrariness. It remains to choose a coordinate system. This may, for example, correspond to a choice of x-, y- and z-axis, or a choice of **curvilinear coordinates** as exemplified by the spherical coordinates used for the atomic wave functions illustrated below. This final choice also fixes a basis in abstract Hilbert space. The basic states are labeled by the quantum numbers corresponding to the maximal set of commuting observables and an appropriate coordinate system.[nb 2]

- Wave functions corresponding to a state are accordingly not unique. This has been exemplified already with momentum and position space wave functions describing the same abstract state. This non-uniqueness reflects the non-uniqueness in the choice of a maximal set of commuting observables.

- The abstract states are "abstract" only in that an arbitrary choice necessary for a particular *explicit* description of it is not given. This is the same as saying that no choice of maximal set of commuting observables has been given. This is analogous to a vector space without a specified basis.

- The wave functions of position and momenta, respectively, can be seen as a choice of representation yielding two different, but entirely equivalent, explicit descriptions of the same state for a system with no discrete degrees of freedom.

- Corresponding to the two examples in the first item, to a particular state there corresponds two wave functions, $\Psi(x, Sz)$ and $\Psi(p, Sy)$, both describing the same state. For each choice of maximal commuting sets of observables for the abstract state space, there is a corresponding representation that is associated to a function space of wave functions.

- Each choice of representation should be thought of as specifying a unique function space in which wave functions corresponding to that choice of representation lives. This distinction is best kept, even if one could argue that two such function spaces are mathematically equal, e.g. being the set of square integrable functions. One can then think of the function spaces as two distinct copies of that set.

- Between all these different function spaces and the abstract state space, there are one-to-one correspondences (here disregarding normalization and unobservable phase factors), the common denominator here being a particular abstract state. The relationship between the momentum and position space wave functions, for instance, describing the same state is the Fourier transform.

To make this concrete, in the figure to the right, the 19 sub-images are images of wave functions in position space (their norm squared). The wave functions each represent the abstract state characterized by the triple of quantum numbers (n, l, m), in the lower right of each image. These are the principal quantum number, the orbital angular momentum quantum number and the magnetic quantum number. Together with one spin-projection quantum number of the electron, this is a complete set of observables.

The figure can serve to illustrate some further properties of the function spaces of wave functions.

- In this case, the wave functions are square integrable. One can initially take the function space as the space of square integrable functions, usually denoted L^2.

- The displayed functions are solutions to the Schrödinger equation. Obviously, not every function in L^2 satisfies the Schrödinger equation for the hydrogen atom. The function space is thus a subspace of L^2.

- The displayed functions form part of a basis for the function space. To each triple (n, l, m), there corresponds a basis wave function. If spin is taken into account, there are two basis functions for each triple. The function space thus has a countable basis.

- The basis functions are mutually orthonormal. For this concept to have a meaning, there must exist an inner product. The function space is thus an inner product space. The inner product between two states intuitively measures the "overlap" between the states. The physical interpretation is that the norm squared is proportional to the transition probability between the states. That is,

$$P(\Psi \rightarrow \Phi_i) = |(\Psi, \Phi_i)|^2$$

where the i is an index composed of quantum numbers corresponding to a representation and the probabilities are the probabilities of finding the state Ψ in the definite state represented by Φi upon measurement of the physical observables corresponding to the representation, for instance, i could be the quadruple (n, l, m, Sz). This is the Born rule,[16] and is one of the fundamental postulates of quantum mechanics.

These observations encapsulate the essence of the function spaces of which wave functions are elements. Mathematically, this is expressed (in one spatial dimension, disregarding here unimportant issues of normalization) for a particle with no internal degrees of freedom as

$$\Psi = I\Psi = \int \Phi_x(\Phi_x, \Psi)dx = \int \Psi(x)\Phi_x dx = \int \Phi_p(\Phi_p, \Psi)dp = \int \Psi(p)\Phi_p dp,$$

where Ψ is any "abstract" state, Φx is an eigenfunction of the position operator representing a particle localized at x, $(\cdot,\cdot)$ represents the inner product, Φp is an eigenfunction of the momentum operator representing a particle with precise momentum p, I is the **identity operator** and the integrals (first and third) represent the completeness of momentum and position eigenstates, $\Psi(x)$ is the coordinate space wave function and $\Psi(p)$ is the wave function in momentum space. In Dirac notation, the above equation reads

$$|\Psi\rangle = I|\Psi\rangle = \int |x\rangle\langle x|\Psi\rangle dx = \int \Psi(x)|x\rangle dx = \int |p\rangle\langle p|\Psi\rangle dp = \int \Psi(p)|p\rangle dp.$$

The description is not yet complete. There is a further technical requirement on the function space, that of completeness, that allows one to take limits of sequences in the function space, and be ensured that, if the limit exists, it is an element of the function space. A complete inner product space is called a Hilbert space. The property of completeness is crucial in advanced treatments and applications of quantum mechanics. It will not be very important in the subsequent discussion of wave functions, and technical details and links may be found in footnotes like the one that follows.[nb 3] The space L^2 is a Hilbert space, with inner product presented later. The function space of the example of the figure is a subspace of L^2. A subspace of a Hilbert space is a Hilbert space if it is closed. It is here that the topology of the function space enters into its description.

It is also important to note, in order to avoid confusion, that not all functions to be discussed are elements of some Hilbert space, say L^2. The most glaring example is the set of functions $e^{2\pi i px/h}$. These are solutions of the Schrödinger equation for a free particle, but are not normalizable, hence not in L^2. But they are nonetheless fundamental for the description. One can, using them, express functions that *are* normalizable using wave packets. They are, in a sense to be made precise later, a basis (but not a Hilbert space basis) in which wave functions of interest can be expressed. There is also the artifact "normalization to a delta function" that is frequently employed for notational convenience, see further down. The delta functions themselves aren't square integrable either.

13.2.1   Physical requirements

The above description of the function space containing the wave functions is mostly mathematically motivated. The function spaces are, due to completeness, very *large* in a certain sense. Not all functions are realistic descriptions of any physical system. For instance, in the function space L^2 one can find the function that takes on the value 0 for all rational numbers and $-i$ for the irrationals in the interval [0, 1]. This *is* square integrable,[nb 4] but can hardly represent a physical state.

The following constraints on the wave function are sometimes explicitly formulated for the calculations and physical interpretation to make sense:[26][27]

- The wave function must be square integrable. This is motivated by the Copenhagen interpretation of the wave function as a probability amplitude.

- It must everywhere be everywhere continuous and everywhere continuously differentiable. This is motivated by the appearance of the Schrödinger equation.

It is possible to relax these conditions somewhat for special purposes.[nb 5] If these requirements are not met, it is not possible to interpret the wave function as a probability amplitude.[28]

This does not alter the structure of the Hilbert space that these particular wave functions inhabit, but it should be pointed out that the subspace of the square-integrable functions L^2, which is a Hilbert space, satisfying the second requirement *is not closed* in L^2, hence not a Hilbert space in itself.[nb 6] The functions that does not meet the requirements are still needed for both technical and practical reasons.[nb 7][nb 8]

13.3 Definition (one spinless particle in 1d)

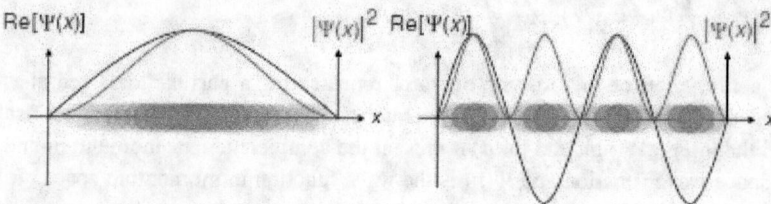

Standing waves for a particle in a box, examples of stationary states.

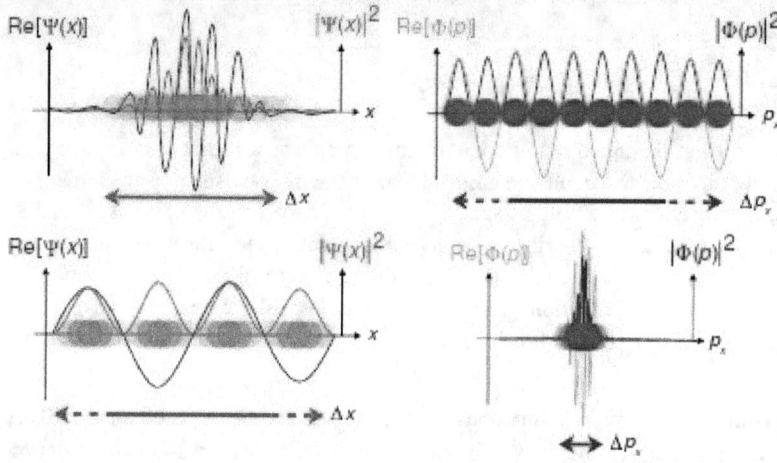

Travelling waves of a free particle.

The real parts of position wave function $\Psi(x)$ and momentum wave function $\Phi(p)$, and corresponding probability densities $|\Psi(x)|^2$ and $|\Phi(p)|^2$, for one spin-0 particle in one x or p dimension. The colour opacity of the particles corresponds to the probability density (*not* the wave function) of finding the particle at position x or momentum p.

For now, consider the simple case of a single particle, without spin, in one spatial dimension. More general cases are discussed below.

13.3.1 Position-space wave function

The state of such a particle is completely described by its wave function,

$$\Psi(x,t),$$

where x is position and t is time. This is a complex-valued function of two real variables x and t.

If interpreted as a probability amplitude, the square modulus of the wave function, the positive real number

$$|\Psi(x,t)|^2 = \Psi(x,t)^* \Psi(x,t) = \rho(x,t),$$

is interpreted as the probability density that the particle is at x. The asterisk indicates the complex conjugate. If the particle's position is measured, its location cannot be determined from the wave function, but is described by a probability distribution. The probability that its position x will be in the interval $a \leq x \leq b$ is the integral of the density over this interval:

$$P_{a \leq x \leq b}(t) = \int_a^b dx\, |\Psi(x,t)|^2$$

where t is the time at which the particle was measured. This leads to the **normalization condition**:

$$\int\limits_{-\infty}^{\infty} dx \, |\Psi(x,t)|^2 = 1 \, ,$$

because if the particle is measured, there is 100% probability that it will be *somewhere*.

Since the Schrödinger equation is linear, if any number of wave functions Ψn for $n = 1, 2, ...$ are solutions of the equation, then so is their sum, and their scalar multiples by complex numbers an. Taking scalar multiplication and addition together is known as a linear combination:

$$\sum_n a_n \Psi_n(x,t) = a_1 \Psi_1(x,t) + a_2 \Psi_2(x,t) + \cdots$$

This is the superposition principle. Multiplying a wave function Ψ by any nonzero constant complex number c to obtain $c\Psi$ does not change any information about the quantum system, because c cancels in the Schrödinger equation for $c\Psi$.

13.3.2   Momentum-space wave function

The particle also has a wave function in momentum space:

$$\Phi(p,t)$$

where p is the momentum in one dimension, which can be any value from $-\infty$ to $+\infty$, and t is time.

All the previous remarks on superposition, normalization, etc. apply similarly. In particular, if the particle's momentum is measured, the result is not deterministic, but is described by a probability distribution:

$$P_{a \leq p \leq b}(t) = \int\limits_{a}^{b} dp \, |\Phi(p,t)|^2 \, ,$$

and the normalization condition is:

$$\int\limits_{-\infty}^{\infty} dp \, |\Phi(p,t)|^2 = 1 \, .$$

13.3.3   Relation between wave functions

The position-space and momentum-space wave functions are Fourier transforms of each other, therefore both contain the same information, and either one alone is sufficient to calculate any property of the particle. As elements of **abstract physical Hilbert space**, whose elements are the possible states of the system under consideration, they represent the same object, but they are not equal when viewed as square-integrable functions. (A function and its Fourier transform are not equal.) For one dimension,[29]

$$\Phi(p,t) = \frac{1}{\sqrt{2\pi\hbar}} \int\limits_{-\infty}^{\infty} dx \, e^{-ipx/\hbar} \Psi(x,t) \quad \rightleftharpoons \quad \Psi(x,t) = \frac{1}{\sqrt{2\pi\hbar}} \int\limits_{-\infty}^{\infty} dp \, e^{ipx/\hbar} \Phi(p,t)$$

In practice, the position-space wave function is used much more often than the momentum-space wave function. The potential entering the Schrödinger equation determines in which basis the description is easiest. For the harmonic oscillator, x and p enter symmetrically, so there it doesn't matter which description one uses.

13.4 Definitions (other cases)

Following are the general forms of the wave function for systems in higher dimensions and more particles, as well as including other degrees of freedom than position coordinates or momentum components.

The position-space wave function of a single particle in three spatial dimensions is similar to the case of one spatial dimension above:

$$\Psi(\mathbf{r}, t)$$

where $\mathbf{r}$ is the position vector in three-dimensional space, and t is time. As always $\Psi(\mathbf{r}, t)$ is a complex number, for this case a complex-valued function of four real variables.

If there are many particles, in general there is only one wave function, not a separate wave function for each particle. The fact that *one* wave function describes *many* particles is what makes quantum entanglement and the EPR paradox possible. The position-space wave function for N particles is written:[19]

$$\Psi(\mathbf{r}_1, \mathbf{r}_2 \cdots \mathbf{r}_N, t)$$

where $\mathbf{r}i$ is the position of the ith particle in three-dimensional space, and t is time. Altogether, this is a complex-valued function of $3N + 1$ real variables.

In quantum mechanics there is a fundamental distinction between *identical particles* and *distinguishable* particles. For example, any two electrons are identical and fundamentally indistinguishable from each other; the laws of physics make it impossible to "stamp an identification number" on a certain electron to keep track of it.[29] This translates to a requirement on the wave function for a system of identical particles:

$$\Psi(\ldots \mathbf{r}_a, \ldots, \mathbf{r}_b, \ldots) = \pm\Psi(\ldots \mathbf{r}_b, \ldots, \mathbf{r}_a, \ldots)$$

where the $+$ sign occurs if the particles are *all bosons* and $-$ sign if they are *all fermions*. In other words, the wave function is either totally symmetric in the positions of bosons, or totally antisymmetric in the positions of fermions.[30] The physical interchange of particles corresponds to mathematically switching arguments in the wave function. The antisymmetry feature of fermionic wave functions leads to the Pauli principle. Generally, bosonic and fermionic symmetry requirements are the manifestation of particle statistics and are present in other quantum state formalisms.

For N *distinguishable* particles (no two being identical, i.e. no two having the same set of quantum numbers), there is no requirement for the wave function to be either symmetric or antisymmetric.

For a collection of particles, some identical with coordinates $\mathbf{r}_1$, $\mathbf{r}_2$, ... and others distinguishable $\mathbf{x}_1$, $\mathbf{x}_2$, ... (not identical with each other, and not identical to the aforementioned identical particles), the wave function is symmetric or antisymmetric in the identical particle coordinates $\mathbf{r}i$ only:

$$\Psi(\ldots \mathbf{r}_a, \ldots, \mathbf{r}_b, \ldots, \mathbf{x}_1, \mathbf{x}_2, \ldots) = \pm\Psi(\ldots \mathbf{r}_b, \ldots, \mathbf{r}_a, \ldots, \mathbf{x}_1, \mathbf{x}_2, \ldots)$$

Again, there is no symmetry requirement for the distinguishable particle coordinates $\mathbf{x}i$.

For a particle with spin, the wave function can be written in "position–spin space" as:

$$\Psi(\mathbf{r}, t, s_z)$$

which is a complex-valued function of position $\mathbf{r}$ in three-dimensional space, time t, and s_z, the spin projection quantum number along the z axis. (The z axis is an arbitrary choice; other axes can be used instead if the wave function is transformed appropriately, see below.) The sz parameter, unlike $\mathbf{r}$ and t, is a *discrete variable*. For example, for a spin-1/2 particle, s_z can only be $+1/2$ or $-1/2$, and not any other value. (In general, for spin s, sz can be $s, s-1, \dots, -s+1, -s$.)

Often, the complex values of the wave function for all the spin numbers are arranged into a column vector, in which there are as many entries in the column vector as there are allowed values of sz. In this case, the spin dependence is placed in indexing the entries and the wave function is a complex vector-valued function of space and time only:

$$\Psi(\mathbf{r}, t) = \begin{bmatrix} \Psi(\mathbf{r}, t, s) \\ \Psi(\mathbf{r}, t, s-1) \\ \vdots \\ \Psi(\mathbf{r}, t, -(s-1)) \\ \Psi(\mathbf{r}, t, -s) \end{bmatrix}$$

The wave function for N particles each with spin is the complex-valued function:

$$\Psi(\mathbf{r}_1, \mathbf{r}_2 \cdots \mathbf{r}_N, s_{z\,1}, s_{z\,2} \cdots s_{z\,N}, t)$$

Concerning the general case of N particles with spin in 3d, if Ψ is interpreted as a probability amplitude, the probability density is:

$$\rho(\mathbf{r}_1 \cdots \mathbf{r}_N, s_{z\,1} \cdots s_{z\,N}, t) = |\Psi(\mathbf{r}_1 \cdots \mathbf{r}_N, s_{z\,1} \cdots s_{z\,N}, t)|^2$$

and the probability that particle 1 is in region R_1 with spin $sz_1 = m_1$ *and* particle 2 is in region R_2 with spin $sz_2 = m_2$ etc. at time t is the integral of the probability density over these regions and spins:

$$P_{\mathbf{r}_1 \in R_1, s_{z\,1}=m_1, \dots, \mathbf{r}_N \in R_N, s_{z\,N}=m_N}(t) = \int_{R_1} d^3\mathbf{r}_1 \int_{R_2} d^3\mathbf{r}_2 \cdots \int_{R_N} d^3\mathbf{r}_N |\Psi(\mathbf{r}_1 \cdots \mathbf{r}_N, m_1 \cdots m_N, t)|^2$$

The multidimensional Fourier transforms of the position or position–spin space wave functions yields momentum or momentum–spin space wave functions.

13.4.1   Decompositions into products

For systems in time-independent potentials, the wave function can always be written as a function of the degrees of freedom multiplied by a time-dependent phase factor, the form of which is given by the Schrödinger equation. For the case of N particles position-spin space,

$$\Psi(\mathbf{r}_1, \mathbf{r}_2, \dots, \mathbf{r}_N, t, s_{z1}, s_{z2}, \dots, s_{zN}) = e^{-iEt/\hbar} \psi(\mathbf{r}_1, \mathbf{r}_2, \dots, \mathbf{r}_N, s_{z1}, s_{z2}, \dots, s_{zN}),$$

where E is the energy eigenvalue of the system corresponding to the eigenstate Ψ. Wave functions of this form are called stationary states.

In some situations, the wave function for a particle with spin factors into a product of a space function ψ and a spin function ξ, where each are complex-valued functions, and the time dependence can be placed in either function:

$$\Psi(\mathbf{r}, t, s_z) = \psi(\mathbf{r}, t)\xi(s_z) = \phi(\mathbf{r})\zeta(s_z, t).$$

The dynamics of each factor can be studied in isolation. This factorization is always possible when the orbital and spin angular momenta of the particle are separable in the Hamiltonian operator, that is, the Hamiltonian can be split into an orbital term and a spin term.[31] It is not possible for those interactions where an external field or any space-dependent quantity couples to the spin; examples include a particle in a magnetic field, and spin-orbit coupling. For the time-independent case this reduces to

$$\Psi(\mathbf{r}, t, s_z) = e^{-iEt/\hbar} \psi(\mathbf{r}) \xi(s_z),$$

where again E is the energy eigenvalue of the system corresponding to the eigenstate Ψ. This extends to the case of N particles:

$$\Psi(\mathbf{r}, t, s_z) = \psi(\mathbf{r}_1, \mathbf{r}_2, \ldots, \mathbf{r}_N, t) \xi(s_{z1}, s_{z2}, \ldots, s_{zN}) = \phi(\mathbf{r}_1, \mathbf{r}_2, \ldots, \mathbf{r}_N) \zeta(s_{z1}, s_{z2}, \ldots, s_{zN}, t).$$

and for the case of identical particles, each factor has to have the correct antisymmetry or symmetry, to make the overall wave function antisymmetric for fermions or symmetric for bosons.

13.5 Inner product

13.5.1 Position-space inner products

The **inner product** of two wave functions Ψ_1 and Ψ_2 is useful and important for a number of reasons given below. For the case of one spinless particle in 1d, it can be defined as the complex number (at time t)[nb 9]

$$\langle \Psi_1, \Psi_2 \rangle = \int_{-\infty}^{\infty} dx \, \Psi_1^*(x, t) \Psi_2(x, t).$$

More generally, the formulae for the inner products are integrals over all coordinates or momenta and sums over all spin quantum numbers. That is, for one spinless particle in 3d the inner product of two wave functions can be defined as the complex number:

$$\langle \Psi_1, \Psi_2 \rangle = \int_{\text{all space}} d^3\mathbf{r} \, \Psi_1^*(\mathbf{r}, t) \Psi_2(\mathbf{r}, t),$$

while for many spinless particles in 3d:

$$\langle \Psi_1, \Psi_2 \rangle = \int_{\text{all space}} d^3\mathbf{r}_1 \int_{\text{all space}} d^3\mathbf{r}_2 \cdots \int_{\text{all space}} d^3\mathbf{r}_N \, \Psi_1^*(\mathbf{r}_1 \cdots \mathbf{r}_N, t) \Psi_2(\mathbf{r}_1 \cdots \mathbf{r}_N, t)$$

(altogether, this is N three-dimensional volume integrals with differential volume elements d^3ri, also written "dVi" or "$dxi\,dyi\,dzi$"). For one particle with spin in 3d:

$$\langle \Psi_1, \Psi_2 \rangle = \sum_{\text{all } s_z} \int_{\text{all space}} d^3\mathbf{r} \Psi_1^*(\mathbf{r}, t, s_z) \Psi_2(\mathbf{r}, t, s_z),$$

and for the general case of N particles with spin in 3d:

$$\langle \Psi_1, \Psi_2 \rangle = \sum_{s_z N} \cdots \sum_{s_z 2} \sum_{s_z 1} \int_{\text{all space}} d^3 \mathbf{r}_1 \int_{\text{all space}} d^3 \mathbf{r}_2 \cdots \int_{\text{all space}} d^3 \mathbf{r}_N \Psi_1^* (\mathbf{r}_1 \cdots \mathbf{r}_N, s_{z1} \cdots s_z N, t) \Psi_2 (\mathbf{r}_1 \cdots \mathbf{r}_N, s_{z1} \cdots s_z N, t)$$

(altogether, N three-dimensional volume integrals followed by N sums over the spins).

In the Copenhagen interpretation, the modulus squared of the inner product (a complex number) gives a real number

$$|\langle \Psi_1, \Psi_2 \rangle|^2 = P(\Psi_2 \rightarrow \Psi_1) ,$$

which is interpreted as the probability of the wave function Ψ_2 "collapsing" to the new wave function Ψ_1 upon measurement of an observable, whose eigenvalues are the possible results of the measurement, with Ψ_1 being an eigenvector of the resulting eigenvalue.

Although the inner product of two wave functions is a complex number, the inner product of a wave function Ψ with itself,

$$\langle \Psi, \Psi \rangle = \|\Psi\|^2 ,$$

is *always* a positive real number. The number $\|\Psi\|$ (not $\|\Psi\|^2$) is called the **norm** of the wave function Ψ, and is not the same as the modulus $|\Psi|$.

A wave function is normalized if:

$$\langle \Psi, \Psi \rangle = 1 .$$

If Ψ is not normalized, then dividing by its norm gives the normalized function $\Psi/\|\Psi\|$.

Two wave functions Ψ_1 and Ψ_2 are orthogonal if their inner product is zero:

$$\langle \Psi_1, \Psi_2 \rangle = 0 .$$

A set of wave functions Ψ_1, Ψ_2, ... are orthonormal if they are each normalized and are all orthogonal to each other:

$$\langle \Psi_m, \Psi_n \rangle = \delta_{mn} ,$$

where m and n each take values 1, 2, ..., and δmn is the Kronecker delta (+1 for $m = n$ and 0 for $m \neq n$). Orthonormality of wave functions is instructive to consider since this guarantees linear independence of the functions. (However, the wave functions do not have to be orthonormal and can still be linearly independent, but the inner product of Ψm and Ψn is more complicated than the mere δmn).

Returning to the superposition above:

$$\Psi = \sum_n a_n \psi_n$$

if the basis wave functions ψn are orthonormal, then the coefficients have a particularly simple form:

$$a_n = \langle \psi_n, \Psi \rangle$$

If the basis wave functions were not orthonormal, then the coefficients would be different.

13.5.2 Momentum-space inner products

Analogous to the position case, the inner product of two wave functions $\Phi_1(p, t)$ and $\Phi_2(p, t)$ can be defined as:

$$\langle \Phi_1, \Phi_2 \rangle = \int_{-\infty}^{\infty} dp\, \Phi_1^*(p, t) \Phi_2(p, t),$$

and similarly for more particles in higher dimensions.

One particular solution to the time-independent Schrödinger equation is

$$\Psi_p(x) = e^{ipx/\hbar},$$

a plane wave, which can be used in the description of a particle with momentum exactly p, since it is an eigenfunction of the momentum operator. These functions are not normalizable to unity (they aren't square-integrable), so they are not really elements of physical Hilbert space. The set

$$\{\Psi_p(x, t), -\infty \le p \le \infty\}$$

forms what is called the **momentum basis**. This "basis" is not a basis in the usual mathematical sense. For one thing, since the functions aren't normalizable, they are instead **normalized to a delta function**,

$$\langle \Psi_p, \Psi_{p'} \rangle = \delta(p - p').$$

For another thing, though they are linearly independent, there are too many of them (they form an uncountable set) for a basis for physical Hilbert space. They can still be used to express all functions in it using Fourier transforms as described above.

13.6 Units of the wave function

Although wave functions are complex numbers, both the real and imaginary parts each have the same units (the imaginary unit i is a pure number without physical units). The units of ψ depend on the number of particles N the wave function describes, and the number of spatial or momentum dimensions n of the system.

When integrating $|\psi|^2$ over all the coordinates, the volume element $d^n \mathbf{r}_1 d^n \mathbf{r}_2 ... d^n \mathbf{r} N$ has units of [length]^{Nn}. Since the normalization conditions require the integral to be the unitless number 1, $|\psi|^2$ must have units of [length]$^{-Nn}$, thus the units of $|\psi|$ and hence ψ are [length]$^{-Nn/2}$. Likewise, in momentum space, length is replaced by momentum, and the units are [momentum]$^{-Nn/2}$. These results are true for particles of any spin, since for particles with spin, the summations are over dimensionless spin quantum numbers.

13.7 More on wave functions and abstract state space

Main article: Quantum state

As has been demonstrated, the set of all possible normalizable wave functions for a system with a particular choice of basis constitute a Hilbert space. This vector space is in general infinite-dimensional. Due to the multiple possible choices of basis, these Hilbert spaces are not unique. One therefore talks about an abstract Hilbert space, **state space**, where the choice of basis is left undetermined. The choice of basis corresponds to a choice of a maximal set of quantum numbers, each quantum number corresponding to an observable. Two observables corresponding to quantum numbers in the maximal set must commute, therefore, the basis isn't entirely arbitrary, but nonetheless, there are always several choices.

Specifically, each state is represented as an abstract vector in state space[32]

$$|\Psi\rangle$$

where $|\Psi\rangle$ is a "ket" (a vector) written in Dirac's bra–ket notation.[33] Kets that differ by multiplication by a scalar represent the same state. A ray in Hilbert space is a set of normalized vectors differing by a complex number of modulus 1. If $|\psi\rangle$ and $|\phi\rangle$ are two states in the vector space, and a and b are two complex numbers, then the linear combination

$$|\Psi\rangle = a|\psi\rangle + b|\phi\rangle$$

is also in the same vector space. The state space is postulated to have an inner product, denoted by

$$\langle\Psi_1|\Psi_2\rangle,$$

that is (usually, this differs) linear in the first argument and antilinear in the second argument. The dual vectors are denoted as "bras", $\langle\Psi|$. These are linear functionals, elements of the dual space to the state space. The inner product, once chosen, can be used to define a unique map from state space to its dual, see Riesz representation theorem. this map is antilinear. One has

$$\langle\Psi| = a^*\langle\psi| + b^*\langle\phi| \leftrightarrow a|\psi\rangle + b|\phi\rangle = |\Psi\rangle,$$

where the asterisk denotes the complex conjugate. For this reason one has under this map

$$\langle\Phi|\Psi\rangle = \langle\Phi|(|\Psi\rangle),$$

and one may, as a practical consequence, at least notation-wise in this formalism, ignore that bra's are dual vectors.

The state vector for the system evolves in time according to the Schrödinger equation, or other dynamical pictures of quantum mechanics- In bra-ket notation this reads,

$$i\hbar\frac{d}{dt}|\Psi\rangle = \hat{H}|\Psi\rangle$$

Abstract state space is also, by definition, required to be a Hilbert space. The only requirement missing for this in the description so far is completeness. See the quantum state article for more explanation of the Hilbert space formalism and its consequences to quantum physics.

The connection to the Hilbert spaces of wave functions is made as follows. If (a, b, ... l, m, ...) is a maximal set of quantum numbers, denote the state corresponding to *fixed choices* of these quantum numbers by

$$|a, b, \ldots, l, m, \ldots\rangle.$$

The wave function corresponding to an arbitrary state $|\Psi\rangle$ is denoted

$$\langle a, b, \ldots, l, m, \ldots |\Psi\rangle,$$

for a concrete example,

$$\Psi(x) = \langle x | \Psi \rangle.$$

There are several advantages to understanding wave functions as representing elements of an abstract vector space:

- All the powerful tools of linear algebra can be used to manipulate and understand wave functions. For example:
  - Linear algebra explains how a vector space can be given a basis, and then any vector in the vector space can be expressed in this basis. This explains the relationship between a wave function in position space and a wave function in momentum space, and suggests that there are other possibilities too.
  - Bra–ket notation can be used to manipulate wave functions.
- The idea that quantum states are vectors in an abstract vector space (technically, a complex projective Hilbert space) is completely general in all aspects of quantum mechanics and quantum field theory, whereas the idea that quantum states are complex-valued "wave" functions of space is only true in certain situations.

Following is a summary of the bra–ket formalism applied to wave functions, with general discrete or continuous bases.

13.7.1 Discrete and continuous bases

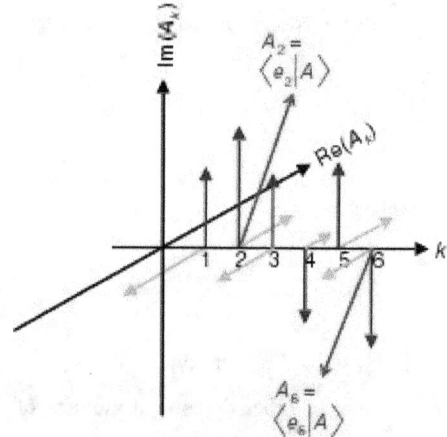

Discrete components A_k of a complex vector $|A\rangle = \sum_k A_k|e_k\rangle$, which belongs to a *countably infinite*-dimensional Hilbert space; there are countably infinitely many k values and basis vectors $|e_k\rangle$.

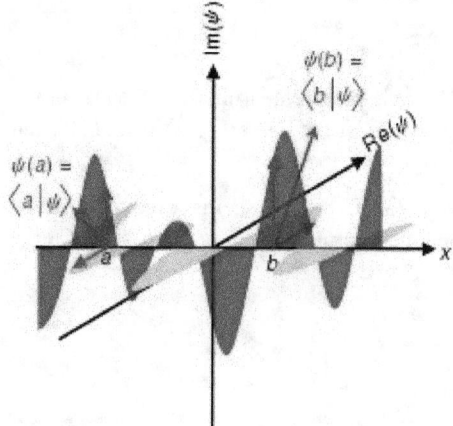

Continuous components $\psi(x)$ of a complex vector $|\psi\rangle = \int dx\ \psi(x)|x\rangle$, which belongs to an *uncountably infinite*-dimensional Hilbert space; there are uncountably infinitely many x values and basis vectors $|x\rangle$.

Components of complex vectors plotted against index number; discrete k and continuous x. Two probability amplitudes out of infinitely many are highlighted.

A Hilbert space with a discrete basis $|\varepsilon i\rangle$ for $i = 1, 2...n$ is orthonormal if the inner product of all pairs of basis kets are given by the Kronecker delta:

$$\langle \varepsilon_i | \varepsilon_j \rangle = \delta_{ij} .$$

Orthonormal bases are convenient to work with because the inner product of two vectors have simple expressions. A wave function $|\Psi\rangle$ expressed in this discrete basis of the Hilbert space, and the corresponding bra in the dual space, are respectively given by:

$$|\Psi\rangle = \sum_{i=1}^{n} c_i |\varepsilon_i\rangle = \begin{bmatrix} c_1 \\ \vdots \\ c_n \end{bmatrix} \quad \langle \Psi| = |\Psi\rangle^\dagger = \sum_{i=1}^{n} c_i^* \langle \varepsilon_i| = \begin{bmatrix} c_1^* & \cdots & c_n^* \end{bmatrix} ,$$

where the complex numbers

$$c_i = \langle \varepsilon_i | \Psi \rangle$$

are the components of the vector. The column vector is a useful way to list the numbers, and operations on the entire vector can be done according to matrix addition and multiplication. The entire vector $|\Psi\rangle$ is independent of the basis, but the components depend on the basis. If a change of basis is made, the components of the vector must also change to compensate.

A Hilbert space with a continuous basis $\{ |\varepsilon\rangle \}$ is orthonormal if the inner product of all pairs of basis kets are given by the Dirac delta function:

$$\langle \varepsilon | \varepsilon' \rangle = \delta(\varepsilon - \varepsilon') .$$

As with the discrete bases, a symbol ε is used in the basis states, two common notations are $|\varepsilon\rangle$ and sometimes $|\Psi\varepsilon\rangle$. A particular basis ket may be subscripted $|\varepsilon_0\rangle \equiv |\Psi\varepsilon_0\rangle$ or primed $|\varepsilon'\rangle \equiv |\Psi\varepsilon'\rangle$, or simply given another symbol in place of ε.

While discrete basis vectors are summed over a discrete index, continuous basis vectors are integrated over a continuous index (a variable of a function). In what follows, all integrals are with respect to the real-valued basis variable ε (not complex-valued), over the required range. Usually this is just the real line or subsets of it. The state $|\Psi\rangle$ in the continuous basis of the Hilbert space, with the corresponding bra in the dual space, are respectively given by:[34]

$$|\Psi\rangle = \int d\varepsilon |\varepsilon\rangle \Psi(\varepsilon) , \quad \langle \Psi| = \int d\varepsilon \langle \varepsilon| \Psi(\varepsilon)^* ,$$

where the components are the complex-valued functions

$$\Psi(\varepsilon) = \langle \varepsilon | \Psi \rangle$$

of a real variable ε.

13.7.2 Completeness conditions

The **completeness conditions** (also called **closure relations**) are

$$\sum_{i=1}^{n} |\varepsilon_i\rangle\langle\varepsilon_i| = 1, \quad \int d\varepsilon\, |\varepsilon\rangle\langle\varepsilon| = 1$$

for discrete and continuous orthonormal bases, respectively. An orthonormal set of kets form bases if and only if they satisfy these relations.[34] In each case, the equality to unity means this is an identity operator; its action on any state leaves it unchanged. Multiplying any state on the right of these gives the representation of the state $|\Psi\rangle$ in the basis. The inner product of a first state $|\Psi_1\rangle$ with a second $|\Psi_2\rangle$ can also be obtained by multiplying $|\Psi_1\rangle$ on the left and $|\Psi_2\rangle$ on the right of the relevant completeness condition.

13.7.3 Inner product

Physically, the nature of the inner product is dependent on the basis in use, because the basis is chosen to reflect the quantum state of the system.

If $|\Psi_1\rangle$ is a state in the above basis with components $c_1, c_2, ..., cn$ and $|\Psi_2\rangle$ is another state in the same basis with components $z_1, z_2, ..., zn$, the inner product is the complex number:

$$\langle\Psi_1|\Psi_2\rangle = \left(\sum_i z_i^*\langle\varepsilon_i|\right)\left(\sum_j c_j|\varepsilon_j\rangle\right) = \sum_{ij} z_i^* c_j \langle\varepsilon_i|\varepsilon_j\rangle = \sum_i z_i^* c_i .$$

If $|\Psi_1\rangle$ is a state in the above continuous basis with components $\Psi_1(\varepsilon')$, and $|\Psi_2\rangle$ is another state in the same basis with components $\Psi_2(\varepsilon)$, the inner product is the complex number:

$$\langle\Psi_1|\Psi_2\rangle = \left(\int d\varepsilon'\, \Psi_1(\varepsilon')^*\langle\varepsilon'|\right)\left(\int d\varepsilon\, \Psi_2(\varepsilon)|\varepsilon\rangle\right) = \int d\varepsilon' \int d\varepsilon\, \Psi_1(\varepsilon')^* \Psi_2(\varepsilon)\langle\varepsilon'|\varepsilon\rangle = \int d\varepsilon\, \Psi_1(\varepsilon)^* \Psi_2(\varepsilon) .$$

where the integrals are taken over all ε and ε'.

The square of the **norm (magnitude)** of the state vector $|\Psi\rangle$ is given by the inner product of $|\Psi\rangle$ with itself, a real number:

$$\|\Psi\|^2 = \langle\Psi|\Psi\rangle = \sum_{j=1}^{n} |c_j|^2, \quad \|\Psi\|^2 = \langle\Psi|\Psi\rangle = \int d\varepsilon\, |\Psi(\varepsilon)|^2$$

for the discrete and continuous bases, respectively. Each say the projection of a complex probability amplitude onto itself is real. If $|\Psi\rangle$ is normalized, these expressions would be each separately equal to 1. If the state is not normalized, then dividing by its magnitude normalizes the state:

$$|\Psi_N\rangle = \frac{1}{\|\Psi\|}|\Psi\rangle$$

13.7.4 Normalized components and probabilities

In the literature, the following results are often presented with normalized wave functions. Here, we keep the normalization factors to show where they appear if the wave function is not already normalized.

For the discrete basis, projecting the normalized state $|\Psi N\rangle$ onto a particular state the system may collapse to, $|eq\rangle$, gives the complex number;

$$\langle \varepsilon_q | \Psi_N \rangle = \langle \varepsilon_q | \frac{1}{\|\Psi\|} \left(\sum_{i=1}^{n} c_i | \varepsilon_i \rangle \right) = \frac{c_q}{\|\Psi\|} \, ,$$

so the modulus squared of this gives a real number;

$$P(\varepsilon_q) = |\langle \varepsilon_q | \Psi_N \rangle|^2 = \frac{|c_q|^2}{\|\Psi\|^2} \, ,$$

In the Copenhagen interpretation, this is the probability of state $|eq\rangle$ occurring.

In the continuous basis, the projection of the normalized state onto some particular basis $|\varepsilon'\rangle$ is a complex-valued function;

$$\langle \varepsilon' | \Psi_N \rangle = \langle \varepsilon' | \left(\frac{1}{\|\Psi\|} \int d\varepsilon | \varepsilon \rangle \Psi(\varepsilon) \right) = \frac{1}{\|\Psi\|} \int d\varepsilon \langle \varepsilon' | \varepsilon \rangle \Psi(\varepsilon) = \frac{1}{\|\Psi\|} \int d\varepsilon \delta(\varepsilon' - \varepsilon) \Psi(\varepsilon) = \frac{\Psi(\varepsilon')}{\|\Psi\|} \, ,$$

so the squared modulus is a real-valued function

$$\rho(\varepsilon') = |\langle \varepsilon' | \Psi_N \rangle|^2 = \frac{|\Psi(\varepsilon')|^2}{\|\Psi\|^2}$$

In the Copenhagen interpretation, this function is the *probability density function* of measuring the observable ε', so integrating this with respect to ε' between $a \leq \varepsilon' \leq b$ gives:

$$P_{a \leq \varepsilon \leq b} = \frac{1}{\|\Psi\|^2} \int_a^b d\varepsilon' |\Psi(\varepsilon')|^2 = \frac{1}{\|\Psi\|^2} \int_a^b d\varepsilon' |\langle \varepsilon' | \Psi \rangle|^2 \, ,$$

the probability of finding the system with ε' between $\varepsilon' = a$ and $\varepsilon' = b$.

13.7.5   Wave function collapse

The physical meaning of the components of $|\Psi\rangle$ is given by the *wave function collapse postulate*, also known as wave function collapse. If the observable(s) ε (momentum and/or spin, position and/or spin, etc.) corresponding to states $|ei\rangle$ has distinct and definite values, λi, and a measurement of that variable is performed on a system in the state $|\Psi\rangle$ then the probability of measuring λi is $|\langle ei|\Psi\rangle|^2$. If the measurement yields λi, the system "collapses" to the state $|ei\rangle$ irreversibly and instantaneously.

13.7.6   Time dependence

Main article: Dynamical pictures (quantum mechanics)

In the Schrödinger picture, the states evolve in time, so the time dependence is placed in $|\Psi\rangle$ according to[35]

$$|\Psi(t)\rangle = \sum_i |\varepsilon_i\rangle \langle \varepsilon_i | \Psi(t) \rangle = \sum_i c_i(t) |\varepsilon_i\rangle$$

for discrete bases, or

$$|\Psi(t)\rangle = \int d\varepsilon\, |\varepsilon\rangle\langle\varepsilon|\Psi(t)\rangle = \int d\varepsilon\, \Psi(\varepsilon,t)|\varepsilon\rangle$$

for continuous bases. However, in the Heisenberg picture the states $|\Psi\rangle$ are constant in time and time dependence is placed in the Heisenberg operators, so $|\Psi\rangle$ is not written as $|\Psi(t)\rangle$. The Heisenberg picture wave function is a snapshot of a Schrödinger picture wave function, representing the whole spacetime history of the system. In the interaction picture (also called Dirac picture), the time dependence is placed in both the states and operators, the subdivision depending on the interaction term in the Hamiltonian, and can be viewed as intermediate between the Heisenberg and Schrödinger pictures. It is useful primarily in computing S-matrix elements.[36]

13.7.7 Tensor product

Further information: Bra-ket notation § Composite bras and kets

It is useful to introduce another operation with the physical interpretation of forming composite states from a collection of other states. This is the tensor product. Given two systems described by states $|\Psi\rangle$ and $|\Phi\rangle$, the tensor product of the states forms the composite state denoted by $|\Psi\rangle\otimes|\Phi\rangle$ or simply without any operation symbol $|\Psi\rangle|\Phi\rangle$, and the new system includes both of the original systems together. The tensor product state $|\Psi\rangle|\Phi\rangle$ lives in a new space; the tensor product of the original Hilbert spaces. The bases spanning this space are the tensor products of the original bases. The product is not commutative in general, so $|\Psi\rangle|\Phi\rangle \neq |\Phi\rangle|\Psi\rangle$. If $|\Psi\rangle$ has components c_i and $|\Phi\rangle$ has components z_j, each in a discrete orthonormal basis $|\varepsilon_k\rangle$, then:

$$|\Psi\rangle|\Phi\rangle = \left(\sum_i c_i|\varepsilon_i\rangle\right)\left(\sum_j z_j|\varepsilon_j\rangle\right) = \sum_{i,j} c_i z_j|\varepsilon_i\rangle|\varepsilon_j\rangle$$

and the notation can be simplified by abbreviating $|A\rangle = |\Psi\rangle|\Phi\rangle$, $A_{ij} = c_i z_j$, and $|E_{ij}\rangle = |\varepsilon_i\rangle|\varepsilon_j\rangle$, so that

$$|A\rangle = \sum_{i,j} A_{ij}|E_{ij}\rangle$$

The same procedure follows for continuous bases using integration. This can also be extended to any number of states, however taking tensor products for fermions and bosons is complicated by the symmetry requirements, see identical particles for general results.

13.8 Position representations

This section applies mostly to non-relativistic quantum mechanics. In relativistic quantum mechanics, eigenstates of the position operator are problematic due to a relativistic extension of Heisenberg's uncertainty principle. In relativistic quantum field theory, they are not used at all to label physical states. Associated to a particle perfectly localized to a point in space is an infinite uncertainty in energy. This leads to pair production in the relativistic regime. Thus such a particle automatically has companions, leading to a breakdown of the description.

13.8.1 State space for one spin-0 particle in 1d

For a spinless particle in one spatial dimension (the x-axis or real line), the state $|\Psi\rangle$ can be expanded in terms of a continuum of basis states; $|x\rangle$, also written $|\Psi x\rangle$, corresponding to the set of all position coordinates x. The completeness condition for this basis is

$$1 = \int\limits_{-\infty}^{\infty} dx\, |x\rangle\langle x|$$

and the orthogonality relation is

$$\langle x'|x\rangle = \delta(x' - x)$$

The state $|\Psi\rangle$ is expressed by:

$$|\Psi\rangle = \left(\int\limits_{-\infty}^{\infty} dx\, |x\rangle\langle x|\right)|\Psi\rangle = \int\limits_{-\infty}^{\infty} dx\, |x\rangle\langle x|\Psi\rangle = \int\limits_{-\infty}^{\infty} dx\, \Psi(x)|x\rangle$$

in which the "wave function" described as a function is a component of the complex state vector.

$$\Psi(x) = \langle x|\Psi\rangle$$

The inner product as stated at the beginning of this article is:

$$\langle \Psi_1|\Psi_2\rangle = \langle \Psi_1|\left(\int\limits_{-\infty}^{\infty} dx\, |x\rangle\langle x|\right)|\Psi_2\rangle = \int\limits_{-\infty}^{\infty} dx\, \langle \Psi_1|x\rangle\langle x|\Psi_2\rangle = \int\limits_{-\infty}^{\infty} dx\, \Psi_1(x)^* \Psi_2(x)\,.$$

If the particle is confined to a region R (a subset of the x-axis), the integrals in the inner product and completeness condition would be integrals over R.

13.8.2 State space (other cases)

The previous example can be extended to more particles in higher dimensions, and include spin.

For one spinless particle in 3d, the basis states are $|r\rangle$ and any state vector $|\Psi\rangle$ in this space is expressed in terms of the basis vectors as $|r\rangle$:

$$|\Psi\rangle = \int\limits_{\text{all space}} d^3\mathbf{r}\,|\mathbf{r}\rangle\langle\mathbf{r}|\Psi\rangle$$

with components:

$$\langle\mathbf{r}|\Psi\rangle = \Psi(\mathbf{r})$$

For N spinless particles in 3d, the basis states are $|r_1, ..., rN\rangle$. This is the tensor product of the one-particle position bases $|r_1\rangle, |r_2\rangle, ..., |rN\rangle$, each of which spans the separate one-particle Hilbert spaces, so $|r_1, ..., rN\rangle$ are the basis states for the tensor product of the one-particle Hilbert spaces (the Hilbert space for the composite many particle system). Any state vector $|\Psi\rangle$ in this space is

$$|\Psi\rangle = \int\limits_{\text{all space}} d^3\mathbf{r}_N \cdots \int\limits_{\text{all space}} d^3\mathbf{r}_2 \int\limits_{\text{all space}} d^3\mathbf{r}_1 |\mathbf{r}_1, \mathbf{r}_2, \ldots, \mathbf{r}_N\rangle\langle\mathbf{r}_1, \mathbf{r}_2, \ldots, \mathbf{r}_N|\Psi\rangle$$

with components:

$$\langle\mathbf{r}_1, \mathbf{r}_2, \ldots, \mathbf{r}_N|\Psi\rangle = \Psi(\mathbf{r}_1, \mathbf{r}_2, \ldots, \mathbf{r}_N)$$

For one particle with spin in 3d, the basis states are $|\mathbf{r}, s_z\rangle$, the tensor product of the position basis $|\mathbf{r}\rangle$ and spin basis $|s_z\rangle$, which exists in a new space from the spin space and position space alone. Any state $|\Psi\rangle$ in this space is:

$$|\Psi\rangle = \sum_{s_z} \int\limits_{\text{all space}} d^3\mathbf{r} |\mathbf{r}, s_z\rangle\langle\mathbf{r}, s_z|\Psi\rangle$$

with components:

$$\langle\mathbf{r}, s_z|\Psi\rangle = \Psi(\mathbf{r}, s_z)$$

For N particles with spin in 3d, the basis states are $|\mathbf{r}_1, \ldots, \mathbf{r}N, s_{z\,1}, \ldots, s_{z\,N}\rangle$, the tensor product of the position basis $|\mathbf{r}_1, \ldots, \mathbf{r}N\rangle$ and spin basis $|s_{z\,1}, \ldots, s_{z\,N}\rangle$, which exists in a new space from the spin space and position space alone. Any state in this space is:

$$|\Psi\rangle = \sum_{s_{z\,1}, \ldots, s_{z\,N}} \int\limits_{\text{all space}} d^3\mathbf{r}_N \cdots \int\limits_{\text{all space}} d^3\mathbf{r}_1 |\mathbf{r}_1, \ldots, \mathbf{r}_N, s_{z\,1}, \ldots, s_{z\,N}\rangle\langle\mathbf{r}_1, \ldots, \mathbf{r}_N, s_{z\,1}, \ldots, s_{z\,N}|\Psi\rangle$$

with components:

$$\langle\mathbf{r}_1, \ldots, \mathbf{r}_N, s_{z\,1}, \ldots, s_{z\,N}|\Psi\rangle = \Psi(\mathbf{r}_1, \ldots, \mathbf{r}_N, s_{z\,1}, \ldots, s_{z\,N})$$

If the particles are restricted to regions of position space, then the integrals in the completeness relations are taken over those regions, rather than the entire coordinate space. For the general case of many particles with spin in 3d, if particle 1 is in region R_1, particle 2 is in region R_2, and so on, the state in this position–spin representation is:

$$|\Psi\rangle = \sum_{s_{z\,1}, \ldots, s_{z\,N}} \int\limits_{R_N} d^3\mathbf{r}_N \cdots \int\limits_{R_1} d^3\mathbf{r}_1\, \Psi(\mathbf{r}_1, \ldots, \mathbf{r}_N, s_{z\,1}, \ldots, s_{z\,N})|\mathbf{r}_1, \ldots, \mathbf{r}_N, s_{z\,1}, \ldots, s_{z\,N}\rangle$$

The orthogonality relation for this basis is:

$$\langle\mathbf{x}_1, \ldots, \mathbf{x}_N, m_1, \ldots, m_N|\mathbf{r}_1, \ldots, \mathbf{r}_N, s_{z\,1}, \ldots, s_{z\,N}\rangle = \delta_{m_1\,s_{z\,1}} \cdots \delta_{m_N\,s_{z\,N}} \delta(\mathbf{x}_1 - \mathbf{r}_1) \cdots \delta(\mathbf{x}_N - \mathbf{r}_N)$$

and the inner product of $|\Psi_1\rangle$ and $|\Psi_2\rangle$ is:

$$\langle\Psi_1|\Psi_2\rangle = \sum_{s_{z\,1}, \ldots, s_{z\,N}} \int\limits_{R_N} d^3\mathbf{r}_N \cdots \int\limits_{R_1} d^3\mathbf{r}_1\, \Psi_1(\mathbf{r}_1, \ldots, \mathbf{r}_N, s_{z\,1}, \ldots, s_{z\,N})^* \Psi_2(\mathbf{r}_1, \ldots, \mathbf{r}_N, s_{z\,1}, \ldots, s_{z\,N}).$$

Momentum space wave functions are similar, using the momentum vectors of the particles as continuous bases, namely $|\mathbf{p}\rangle$, $|\mathbf{p}_1, \mathbf{p}_2, \ldots, \mathbf{p}N\rangle$, etc.

13.9 Ontology

Main article: Interpretations of quantum mechanics

Whether the wave function really exists, and what it represents, are major questions in the interpretation of quantum mechanics. Many famous physicists of a previous generation puzzled over this problem, such as Schrödinger, Einstein and Bohr. Some advocate formulations or variants of the Copenhagen interpretation (e.g. Bohr, Wigner and von Neumann) while others, such as Wheeler or Jaynes, take the more classical approach[37] and regard the wave function as representing information in the mind of the observer, i.e. a measure of our knowledge of reality. Some, including Schrödinger, Bohm and Everett and others, argued that the wave function must have an objective, physical existence. Einstein thought that a complete description of physical reality should refer directly to physical space and time, as distinct from the wave function, which refers to an abstract mathematical space.[38]

13.10 Examples

13.10.1 Free particle

Main article: Free particle

A free particle in 3d with wave vector $\mathbf{k}$ and angular frequency ω has a wave function

$$\Psi(\mathbf{r}, t) = Ae^{i(\mathbf{k}\cdot\mathbf{r} - \omega t)}.$$

13.10.2 Particle in a box

Main article: Particle in a box

A particle is restricted to a 1D region between $x = 0$ and $x = L$; its wave function is:

$$\Psi(x, t) = Ae^{i(kx - \omega t)}, \qquad 0 \le x \le L$$
$$\Psi(x, t) = 0, \qquad\qquad x < 0, x > L$$

To normalize the wave function we need to find the value of the arbitrary constant A; solved from

$$\int_{-\infty}^{\infty} dx\, |\Psi|^2 = 1.$$

From Ψ, we have $|\Psi|^2 = A^2$, so the integral becomes;

$$\int_{-\infty}^{0} dx \cdot 0 + \int_{0}^{L} dx\, A^2 + \int_{L}^{\infty} dx \cdot 0 = 1,$$

Solving this equation gives $A = 1/\sqrt{L}$, so the normalized wave function in the box is;

$$\Psi(x, t) = \frac{1}{\sqrt{L}} e^{i(kx - \omega t)}, \quad 0 \le x \le L.$$

13.10.3 One-dimensional quantum tunnelling

Main articles: Finite potential barrier and Quantum tunnelling

One of most prominent features of the wave mechanics is a possibility for a particle to reach a location with a prohibitive (in classical mechanics) force potential. In the one-dimensional case of particles with energy less than V_0 in the square potential

$$V(x) = \begin{cases} V_0 & |x| < a \\ 0 & \text{otherwise,} \end{cases}$$

the steady-state solutions to the wave equation have the form (for some constants k, κ)

$$\psi(x) = \begin{cases} A_r \exp(ikx) + A_l \exp(-ikx) & x < -a, \\ B_r \exp(\kappa x) + B_l \exp(-\kappa x) & |x| \leq a, \\ C_r \exp(ikx) + C_l \exp(-ikx) & x > a. \end{cases}$$

Note that these wave functions are not normalized; see scattering theory for discussion.

The standard interpretation of this is as a stream of particles being fired at the step from the left (the direction of negative x): setting $A_r = 1$ corresponds to firing particles singly; the terms containing A_r and C_r signify motion to the right, while A_l and C_l – to the left. Under this beam interpretation, put $C_l = 0$ since no particles are coming from the right. By applying the continuity of wave functions and their derivatives at the boundaries, it is hence possible to determine the constants above.

13.10.4 Quantum Dots

In a semiconductor crystallite whose radius is smaller than the size of its exciton Bohr radius, the excitons are squeezed, leading to quantum confinement. The energy levels can then be modeled using the particle in a box model in which the energy of different states is dependent on the length of the box.

13.10.5 Other

Some examples of wave functions for specific applications include:

- Finite square well

- Delta potential

- Quantum harmonic oscillator

- Hydrogen atom and Hydrogen-like atom

13.11   See also

- Boson

- de Broglie–Bohm theory

- Double-slit experiment

- Faraday wave

- Fermion

- Schrödinger equation

- Wave function collapse

- Wave packet

- Phase space formulation of quantum mechanics, wave functions are replaced by quasi-probability distributions that place the position and momenta variables on equal footing.

13.12   Remarks

[1] For this statement to make sense, the observables need to be elements of a maximal commuting set. To see this, it is a simple matter to note that, for example, the momentum operator of the i'th particle in an n-particle system is *not* a generator of any symmetry in nature. On the other hand, the *total* angular momentum *is* a generator of a symmetry in nature; the translational symmetry.

[2] The resulting basis may or may not technically be a basis in the mathematical sense of Hilbert spaces. For instance, states of definite position and definite momentum are not square integrable. This may be overcome with the use of wave packets or by enclosing the system in a "box". See further remarks below.

[3] In technical terms, this is formulated the following way. The inner product yields a norm. This norm in turn induces a metric. If this metric is complete, then the aforementioned limits will be in the function space. The inner product space is then called complete. A complete inner product space is a Hilbert space. The abstract state space is always taken as a Hilbert space. The matching requirement for the function spaces is a natural one. The Hilbert space property of the abstract state space was originally extracted from the observation that the function spaces forming normalizable solutions to the Schrödinger equation are Hilbert spaces.

[4] As is explained in a later footnote, the integral must be taken to be the Lebesgue integral, the Riemann integral is not sufficient.

[5] One such relaxation is that the wave function must belong to the Sobolev space $W^{1,2}$. It means that it is differentiable in the sense of distributions, and its gradient is square-integrable. This relaxation is necessary for potentials that are not functions but are distributions, such as the Dirac delta function.

[6] It is easy to visualize a sequence of functions meeting the requirement that converges to a *discontinuous* function. For this, modify an example given in Inner product space#Examples. This element though *is* an element of L^2.

[7] For instance, in perturbation theory one may construct a sequence of functions approximating the true wave function. This sequence will be guaranteed to converge in a larger space, but without the assumption of a full-fledged Hilbert space, it will not be guaranteed that the convergence is to a function in the relevant space and hence solving the original problem.

[8] Some functions not being square-integrable, like the plane-wave free particle solutions are necessary for the description as outlined in a previous note and also further below.

[9] The functions are here assumed to be elements of L^2, the space of square integrable functions. The elements of this space are more precisely equivalence classes of square integrable functions, two functions declared equivalent if they differ on a set of Lebesgue measure 0. This is necessary to obtain an inner product (that is, $(\Psi, \Psi) = 0 \Rightarrow \Psi \equiv 0$) as opposed to a **semi-inner product**. The integral is taken to be the Lebesgue integral. This is essential for completeness of the space, thus yielding a complete inner product space = Hilbert space.

13.13   Notes

[1] Born 1927, pp. 354–357

[2] Heisenberg 1958, p. 143

[3] Heisenberg, W. (1927/1985/2009). Heisenberg is translated by Camilleri 2009, p. 71, (from Bohr 1985, p. 142).

[4] Murdoch 1987, p. 43

[5] de Broglie 1960, p. 48

[6] Landau Lifshitz, p. 6

[7] Newton 2002, pp. 19–21

[8] Lerner & Trigg 1991, pp. 1223–1229

[9] Einstein 1905, pp. 132–148 (in German), Arons & Peppard 1965, p. 367 (in English)

[10] Einstein 1916, pp. 47–62 and a nearly identical version Einstein 1917, pp. 121–128 translated in ter Haar 1967, pp. 167–183.

[11] de Broglie 1923, pp. 507–510,548,630

[12] Hanle 1977, pp. 606–609

[13] Schrödinger 1926, pp. 1049–1070

[14] Tipler, Mosca & Freeman 2008

[15] Weinberg 2013

[16] Born 1926a, translated in Wheeler & Zurek 1983 at pages 52–55.

[17] Born 1926b, translated in Ludwig 1968, pp. 206–225. Also here.

[18] Young & Freedman 2008, p. 1333

[19] Atkins 1974

[20] Martin & Shaw 2008

[21] Pauli 1927, pp. 601–623.

[22] Weinberg (2002) takes the standpoint that quantum field theory appears the way it does because it is the *only* way to reconcile quantum mechanics with special relativity.

[23] Weinberg (2002) See especially chapter 5, where some of these results are derived.

[24] Weinberg 2002 Chapter 4.

[25] Weinberg 2002

[26] Eisberg & Resnick 1985

[27] Rae 2008

[28] Atkins 1974, p. 258

[29] Griffiths 2004

[30] Zettili 2009, p. 463

[31] Shankar 1994, p. 378–379

[32] Dirac 1982

[33] Dirac 1939

[34] (Peleg et al. 2010) pp. 64–65.

[35] (Peleg et al. 2010, pp. 68–69)

[36] Weinberg 2002 Chapter 3, Scattering matrix.

[37] Jaynes 2003

[38] Einstein 1998, p. 682

13.14   References

- Atkins, P. W. (1974). *Quanta: A Handbook of Concepts*. ISBN 0-19-855494-X.

- Arons, A. B.; Peppard, M. B. (1965). "Einstein's proposal of the photon concept: A translation of the *Annalen der Physik* paper of 1905" (PDF). *American Journal of Physics* **33** (5): 367. Bibcode:1965AmJPh..33..367A. doi:10.1119/1.1971542.

- Bohr, N. (1985). J. Kalckar, ed. *Niels Bohr - Collected Works: Foundations of Quantum Physics I (1926 - 1932)* **6**. Amsterdam: North Holland. ISBN 9780444532893.

- Born, M. (1926a). "Zur Quantenmechanik der Stossvorgange". *Z. f . Physik* **37**: 863–867. Bibcode:1926ZPhy...37 doi:10.1007/bf01397477.

- Born, M. (1926b). "Quantenmechanik der Stossvorgange". *Z. f . Physik* **38**: 803–827. Bibcode:1926ZPhy...38..8   doi:10.1007/bf01397184.

- Born, M. (1927). "Physical aspects of quantum mechanics". *Nature* **119**: 354–357. Bibcode:1927Natur.119..354B. doi:10.1038/119354a0.

- de Broglie, L. (1923). "Radiations—Ondes et quanta" [Radiation—Waves and quanta]. *Comptes Rendus* (in French) **177**: 507–510, 548, 630. Online copy (French) Online copy (English)

- de Broglie, L. (1960). *Non-linear Wave Mechanics: a Causal Interpretation*. Amsterdam: Elsevier.

- Camilleri, K. (2009). *Heisenberg and the Interpretation of Quantum Mechanics: the Physicist as Philosopher*. Cambridge UK: Cambridge University Press. ISBN 978-0-521-88484-6.

- Dirac, P. A. M. (1982). *The principles of quantum mechanics*. The international series on monographs on physics (4th ed.). Oxford University Press. ISBN 0 19 852011 5.

- Dirac, P. A. M. (1939). "A new notation for quantum mechanics". *Mathematical Proceedings of the Cambridge Philosophical Society* **35** (3): 416–418. Bibcode:1939PCPS...35..416D. doi:10.1017/S0305004100021162.

- Einstein, A. (1905). "Über einen die Erzeugung und Verwandlung des Lichtes betreffenden heuristischen Gesichtspunkt". *Annalen der Physik* (in German) **17** (6): 132–148. Bibcode:1905AnP...322..132E. doi:10.1002/

- Einstein, A. (1916). "Zur Quantentheorie der Strahlung". *Mitteilungen der Physikalischen Gesellschaft Zürich* **18**: 47–62.

- Einstein, A. (1917). "Zur Quantentheorie der Strahlung". *Physikalische Zeitschrift* (in German) **18**: 121–128. Bibcode:1917PhyZ...18..121E.

- Einstein, A. (1998). P. A. Schlipp, ed. *Albert Einstein: Philosopher-Scientist*. The Library of Living Philosophers **VII** (3rd ed.). La Salle Publishing Company, Illinois: Open Court. ISBN 0-87548-133-7.

- Eisberg, R.; Resnick, R. (1985). *Quantum Physics of Atoms, Molecules, Solids, Nuclei and Particles* (2nd ed.). John Wiley & Sons. ISBN 978-0-471-87373-0.

- Griffiths, D. J. (2004). *Introduction to Quantum Mechanics* (2nd ed.). Essex England: Pearson Education Ltd. ISBN 978-0131118928.

- Heisenberg, W. (1958). *Physics and Philosophy: the Revolution in Modern Science.* New York: Harper & Row.

- Hanle, P.A. (1977), "Erwin Schrodinger's Reaction to Louis de Broglie's Thesis on the Quantum Theory.", *Isis* **68** (4), doi:10.1086/351880

- Jaynes, E. T. (2003). G. Larry Bretthorst, ed. *Probability Theory: The Logic of Science.* Cambridge University Press. ISBN 978-0-521 59271-0.

- Landau, L.D.; Lifshitz, E. M. (1977). *Quantum Mechanics: Non-Relativistic Theory.* Vol. 3 (3rd ed.). Pergamon Press. ISBN 978-0-08-020940-1. Online copy

- Lerner, R.G.; Trigg, G.L. (1991). *Encyclopaedia of Physics* (2nd ed.). VHC Publishers. ISBN 0-89573-752-3.

- Ludwig, G. (1968). *Wave Mechanics.* Oxford UK: Pergamon Press. ISBN 0-08-203204-1. LCCN 66-30631.

- Murdoch, D. (1987). *Niels Bohr's Philosophy of Physics.* Cambridge UK: Cambridge University Press. ISBN 0-521-33320-2.

- Newton, R.G. (2002). *Quantum Physics: a Text for Graduate Student.* New York: Springer. ISBN 0-387-95473-2.

- Pauli, Wolfgang (1927). "Zur Quantenmechanik des magnetischen Elektrons". *Zeitschrift für Physik* (in German) **43**. Bibcode:1927ZPhy...43..601P. doi:10.1007/bf01397326.

- Peleg, Y.; Pnini, R.; Zaarur, E.; Hecht, E. (2010). *Quantum mechanics.* Schaum's outlines (2nd ed.). McGraw Hill. ISBN 978-0-07-162358-2.

- Rae, A.I.M. (2008). *Quantum Mechanics* **2** (5th ed.). Taylor & Francis Group. ISBN 1-5848-89705.

- Schrödinger, E. (1926). "An Undulatory Theory of the Mechanics of Atoms and Molecules" (PDF). *Physical Review* **28** (6): 1049–1070. Bibcode:1926PhRv...28.1049S. doi:10.1103/PhysRev.28.1049. Archived from the original (PDF) on 17 December 2008.

- Shankar, R. (1994). *Principles of Quantum Mechanics* (2nd ed.). ISBN 0306447908.

- Martin, B.R.; Shaw, G. (2008). *Particle Physics.* Manchester Physics Series (3rd ed.). John Wiley & Sons. ISBN 978-0-470-03294-7.

- ter Haar, D. (1967). *The Old Quantum Theory.* Pergamon Press. pp. 167–183. LCCN 66029628.

- Tipler, P. A.; Mosca, G.; Freeman (2008). *Physics for Scientists and Engineers – with Modern Physics* (6th ed.). ISBN 0-7167-8964-7.

- Weinberg, S. (2013), *Lectures in Quantum Mechanics*, Cambridge University Press, ISBN 978-1-107-02872-2

- Weinberg, S. (2002), *The Quantum Theory of Fields* **1**, Cambridge University Press, ISBN 0-521-55001-7

- Young, H. D.; Freedman, R. A. (2008). Pearson, ed. *Sears' and Zemansky's University Physics* (12th ed.). Addison-Wesley. ISBN 978-0-321-50130-1.

- Wheeler, J.A.; Zurek, W.H. (1983). *Quantum Theory and Measurement.* Princeton NJ: Princeton University Press.

- Zettili, N. (2009). *Quantum Mechanics: Concepts and Applications* (2nd ed.). ISBN 978-0-470-02679-3.

13.15   Further reading

- Yong-Ki Kim (September 2, 2000). "Practical Atomic Physics" (PDF). *National Institute of Standards and Technology* (Maryland): 1 (55 pages). Retrieved 2010-08-17.

- Polkinghorne, John (2002). *Quantum Theory, A Very Short Introduction.* Oxford University Press. ISBN 0-19-280252-6.

13.16   External links

- . . .

- Normalization.

- Quantum Mechanics and Quantum Computation at BerkeleyX

- Einstein, *The quantum theory of radiation*

Continuously differentiable

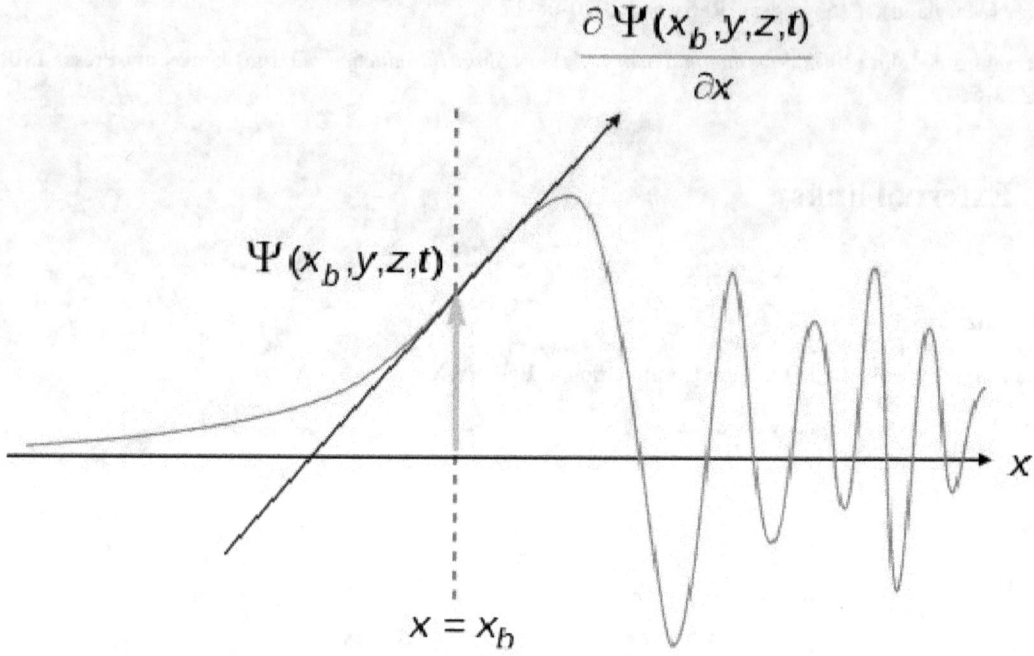

Discontinuous

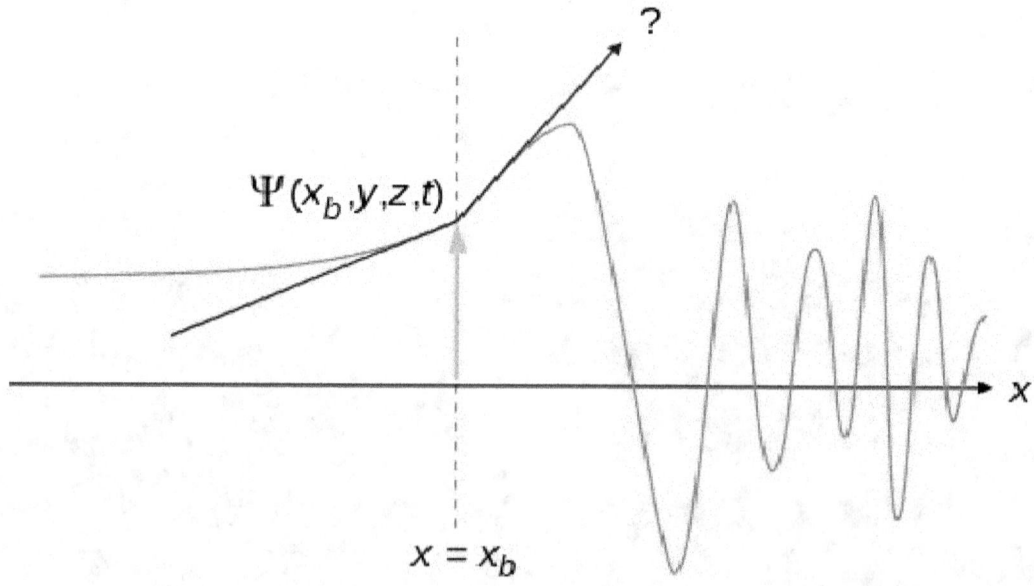

Continuity of the wave function and its first spatial derivative (in the x direction, y and z coordinates not shown), at some time t

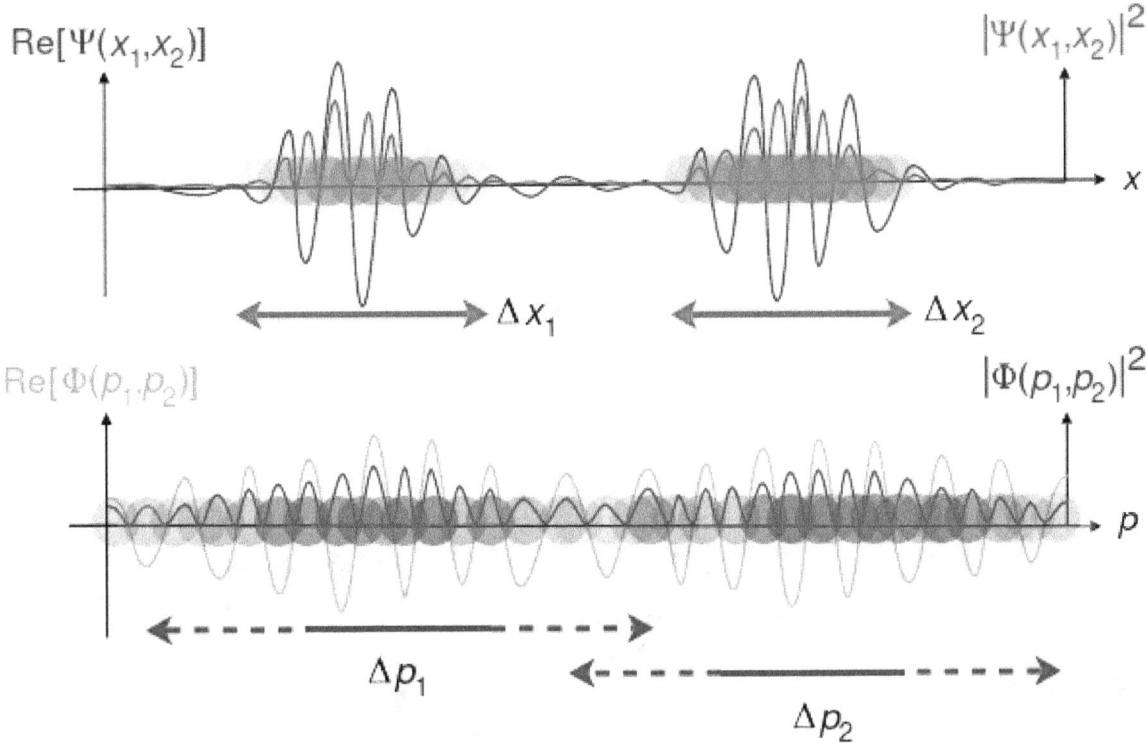

Traveling waves of two free particles, with two of three dimensions suppressed. Top is position space wave function, bottom is momentum space wave function, with corresponding probability densities.

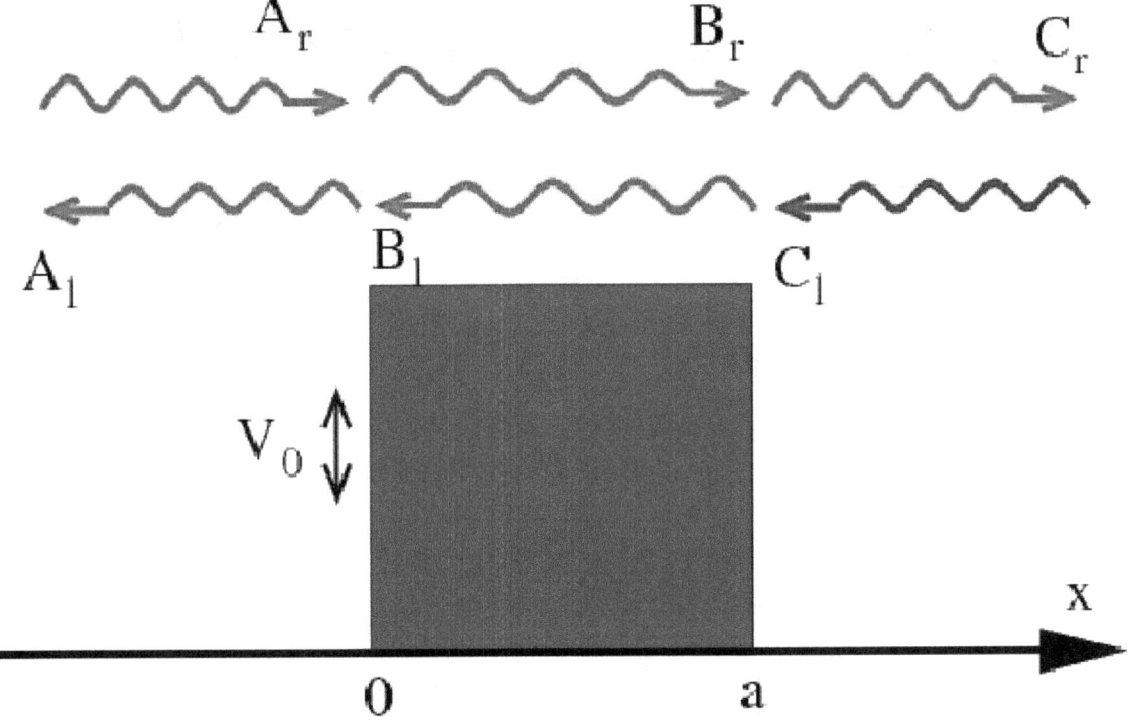

Scattering at a finite potential barrier of height V_0 The amplitudes and direction of left and right moving waves are indicated. In red, those waves used for the derivation of the reflection and transmission amplitude. $E > V_0$ for this illustration.

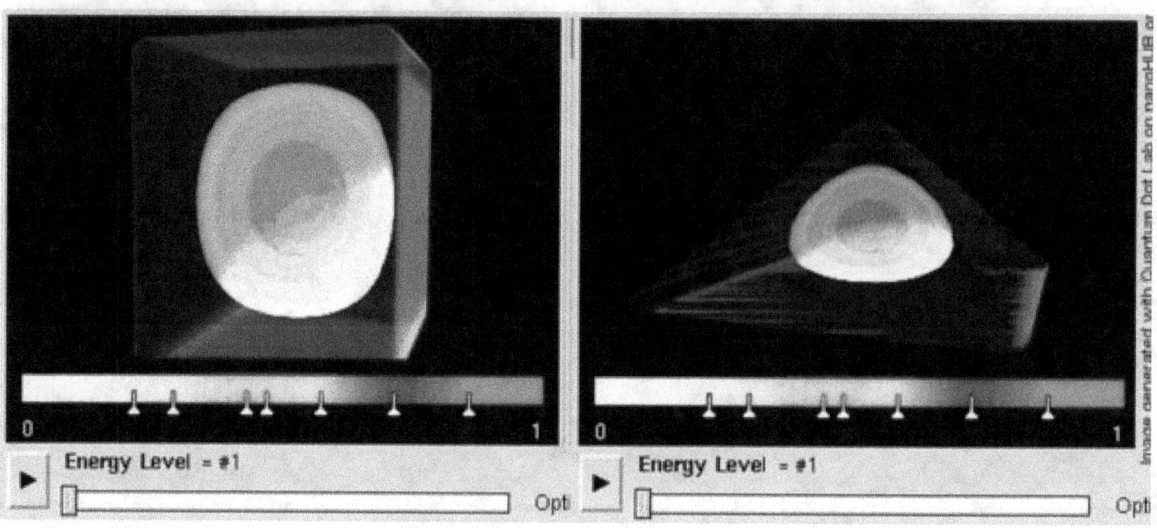

3D confined electron wave functions in a quantum dot. Here, rectangular and triangular-shaped quantum dots are shown. Energy states in rectangular dots are more s-type and p-type. However, in a triangular dot the wave functions are mixed due to confinement symmetry. (Click for animation)

Chapter 14

Schrödinger equation

For a more general introduction to the topic, see Introduction to quantum mechanics.

In quantum mechanics, the **Schrödinger equation** is a partial differential equation that describes how the quantum state of a quantum system changes with time. It was formulated in late 1925, and published in 1926, by the Austrian physicist Erwin Schrödinger.[1]

In classical mechanics Newton's second law, ($\mathbf{F} = m\mathbf{a}$), is used to mathematically predict what a given system will do at any time after a known initial condition. In quantum mechanics, the analogue of Newton's law is Schrödinger's equation for a quantum system (usually atoms, molecules, and subatomic particles whether free, bound, or localized). It is not a simple algebraic equation, but in general a linear partial differential equation, describing the time-evolution of the system's wave function (also called a "state function").[2]:1–2

The concept of a wavefunction is a fundamental postulate of quantum mechanics. Although Schrödinger's equation is often presented as a separate postulate, some authors[3]:Chapter 3 show that some properties resulting from Schrödinger's equation may be deduced just from symmetry principles alone, for example the commutation relations. Generally, "derivations" of the Schrödinger equation demonstrate its mathematical plausibility for describing wave-particle duality, but to date there are no universally accepted derivations of Schrödinger's equation from appropriate axioms.

In the Copenhagen interpretation of quantum mechanics, the wave function is the most complete description that can be given of a physical system. Solutions to Schrödinger's equation describe not only molecular, atomic, and subatomic systems, but also macroscopic systems, possibly even the whole universe.[4]:292ff The Schrödinger equation, in its most general form, is consistent with both classical mechanics and special relativity, but the original formulation by Schrödinger himself was non-relativistic.

The Schrödinger equation is not the only way to make predictions in quantum mechanics—other formulations can be used, such as Werner Heisenberg's matrix mechanics, and Richard Feynman's path integral formulation.

14.1   Equation

14.1.1   Time-dependent equation

The form of the Schrödinger equation depends on the physical situation (see below for special cases). The most general form is the time-dependent Schrödinger equation, which gives a description of a system evolving with time:[5]:143

where i is the imaginary unit, $\hbar$ is the Planck constant divided by 2π, the symbol $\partial/\partial t$ indicates a partial derivative with respect to time t, Ψ (the Greek letter Psi) is the wave function of the quantum system, and $\hat{H}$ is the Hamiltonian operator (which characterizes the total energy of any given wave function and takes different forms depending on the situation).

The most famous example is the non-relativistic Schrödinger equation for a single particle moving in an electric field (but not a magnetic field; see the Pauli equation):[6]

where μ is the particle's "reduced mass", V is its potential energy, ∇^2 is the Laplacian (a differential operator), and Ψ is the wave function (more precisely, in this context, it is called the "position-space wave function"). In plain language, it means "total energy equals kinetic energy plus potential energy", but the terms take unfamiliar forms for reasons explained below.

Given the particular differential operators involved, this is a linear partial differential equation. It is also a diffusion equation, but unlike the heat equation, this one is also a wave equation given the imaginary unit present in the transient term.

The term *"Schrödinger equation"* can refer to both the general equation (first box above), or the specific nonrelativistic version (second box above and variations thereof). The general equation is indeed quite general, used throughout quantum mechanics, for everything from the Dirac equation to quantum field theory, by plugging in various complicated expressions for the Hamiltonian. The specific nonrelativistic version is a simplified approximation to reality, which is quite accurate in many situations, but very inaccurate in others (see relativistic quantum mechanics and relativistic quantum field theory).

To apply the Schrödinger equation, the Hamiltonian operator is set up for the system, accounting for the kinetic and potential energy of the particles constituting the system, then inserted into the Schrödinger equation. The resulting partial differential equation is solved for the wave function, which contains information about the system.

14.1.2 Time-independent equation

The time-independent Schrödinger equation predicts that wave functions can form standing waves, called stationary states (also called "orbitals", as in atomic orbitals or molecular orbitals). These states are important in their own right, and if the stationary states are classified and understood, then it becomes easier to solve the time-dependent Schrödinger equation for *any* state. The *time-independent Schrödinger equation* is the equation describing stationary states. (It is only used when the Hamiltonian itself is not dependent on time. However, even in this case the total wave function still has a time dependency.)

In words, the equation states:

> *When the Hamiltonian operator acts on a certain wave function Ψ, and the result is proportional to the same wave function Ψ, then Ψ is a stationary state, and the proportionality constant, E, is the energy of the state Ψ.*

The time-independent Schrödinger equation is discussed further below. In linear algebra terminology, this equation is an eigenvalue equation.

As before, the most famous manifestation is the non-relativistic Schrödinger equation for a single particle moving in an electric field (but not a magnetic field):

with definitions as above.

14.2 Implications

The Schrödinger equation and its solutions introduced a breakthrough in thinking about physics. Schrödinger's equation was the first of its type, and solutions led to consequences that were very unusual and unexpected for the time.

14.2.1   Total, kinetic, and potential energy

The *overall* form of the equation is *not* unusual or unexpected as it uses the principle of the conservation of energy. The terms of the nonrelativistic Schrödinger equation can be interpreted as total energy of the system, equal to the system kinetic energy plus the system potential energy. In this respect, it is just the same as in classical physics.

14.2.2   Quantization

The Schrödinger equation predicts that if certain properties of a system are measured, the result may be *quantized*, meaning that only specific discrete values can occur. One example is *energy quantization*: the energy of an electron in an atom is always one of the quantized energy levels, a fact discovered via atomic spectroscopy. (Energy quantization is discussed below.) Another example is quantization of angular momentum. This was an *assumption* in the earlier Bohr model of the atom, but it is a *prediction* of the Schrödinger equation.

Another result of the Schrödinger equation is that not every measurement gives a quantized result in quantum mechanics. For example, position, momentum, time, and (in some situations) energy can have any value across a continuous range.[7]:165–167

14.2.3   Measurement and uncertainty

Main articles: Measurement in quantum mechanics, Heisenberg uncertainty principle and Interpretations of quantum mechanics

In classical mechanics, a particle has, at every moment, an exact position and an exact momentum. These values change deterministically as the particle moves according to Newton's laws. Under the Copenhagen interpretation of quantum mechanics, particles do not have exactly determined properties, and when they are measured, the result is randomly drawn from a probability distribution. The Schrödinger equation predicts what the probability distributions are, but fundamentally cannot predict the exact result of each measurement.

The Heisenberg uncertainty principle is the statement of the inherent measurement uncertainty in quantum mechanics. It states that the more precisely a particle's position is known, the less precisely its momentum is known, and vice versa.

The Schrödinger equation describes the (deterministic) evolution of the wave function of a particle. However, even if the wave function is known exactly, the result of a specific measurement on the wave function is uncertain.

14.2.4   Quantum tunneling

Main article: Quantum tunneling
In classical physics, when a ball is rolled slowly up a large hill, it will come to a stop and roll back, because it doesn't have enough energy to get over the top of the hill to the other side. However, the Schrödinger equation predicts that there is a small probability that the ball will get to the other side of the hill, even if it has too little energy to reach the top. This is called quantum tunneling. It is related to the distribution of energy: although the ball's assumed position seems to be on one side of the hill, there is a chance of finding it on the other side.

14.2.5   Particles as waves

Main articles: Matter wave, Wave–particle duality and Double-slit experiment
The nonrelativistic Schrödinger equation is a type of partial differential equation called a wave equation. Therefore, it is often said particles can exhibit behavior usually attributed to waves. In some modern interpretations this description is reversed – the quantum state, i.e. wave, is the only genuine physical reality, and under the appropriate conditions it can show features of particle-like behavior. However, Ballentine [3]:Chapter 4, p.99 shows that such an interpretation has

problems. Ballentine points out that whilst it is arguable to associate a physical wave with a single particle, there is still only *one* Schrödinger wave equation for many particles. He points out:

> *"If a physical wave field were associated with a particle, or if a particle were identified with a wave packet, then corresponding to N interacting particles there should be N interacting waves in ordinary three-dimensional space. But according to (4.6) that is not the case; instead there is one "wave" function in an abstract 3N-dimensional configuration space. The misinterpretation of psi as a physical wave in ordinary space is possible only because the most common applications of quantum mechanics are to one-particle states, for which configuration space and ordinary space are isomorphic."*

Two-slit diffraction is a famous example of the strange behaviors that waves regularly display, that are not intuitively associated with particles. The overlapping waves from the two slits cancel each other out in some locations, and reinforce each other in other locations, causing a complex pattern to emerge. Intuitively, one would not expect this pattern from firing a single particle at the slits, because the particle should pass through one slit or the other, not a complex overlap of both.

However, since the Schrödinger equation is a wave equation, a single particle fired through a double-slit *does* show this same pattern (figure on right). Note: The experiment must be repeated many times for the complex pattern to emerge. Although this is counterintuitive, the prediction is correct; in particular, electron diffraction and neutron diffraction are well understood and widely used in science and engineering.

Related to diffraction, particles also display superposition and interference.

The superposition property allows the particle to be in a quantum superposition of two or more quantum states at the same time. However, it is noted that a "quantum state" in QM means the *probability* that a system will be, for example at a position x, not that the system will actually be at position x. It does not infer that the particle itself may be in two classical states at once. Indeed, QM is generally unable to assign values for properties prior to measurement at all.

14.3 Interpretation of the wave function

Main article: Interpretations of quantum mechanics

The Schrödinger equation provides a way to calculate the wave function of a system and how it changes dynamically in time. However, the Schrödinger equation does not directly say *what*, exactly, the wave function is. Interpretations of quantum mechanics address questions such as what the relation is between the wave function, the underlying reality, and the results of experimental measurements.

An important aspect is the relationship between the Schrödinger equation and wavefunction collapse. In the oldest Copenhagen interpretation, particles follow the Schrödinger equation *except* during wavefunction collapse, during which they behave entirely differently. The advent of quantum decoherence theory allowed alternative approaches (such as the Everett many-worlds interpretation and consistent histories), wherein the Schrödinger equation is *always* satisfied, and wavefunction collapse should be explained as a consequence of the Schrödinger equation.

14.4 Historical background and development

Main article: Theoretical and experimental justification for the Schrödinger equation

Following Max Planck's quantization of light (see black body radiation), Albert Einstein interpreted Planck's quanta to be photons, particles of light, and proposed that the energy of a photon is proportional to its frequency, one of the first signs of wave–particle duality. Since energy and momentum are related in the same way as frequency and wavenumber in special relativity, it followed that the momentum p of a photon is inversely proportional to its wavelength λ, or proportional to its wavenumber k.

$$p = \frac{h}{\lambda} = \hbar k$$

where h is Planck's constant. Louis de Broglie hypothesized that this is true for all particles, even particles which have mass such as electrons. He showed that, assuming that the matter waves propagate along with their particle counterparts, electrons form standing waves, meaning that only certain discrete rotational frequencies about the nucleus of an atom are allowed.[8] These quantized orbits correspond to discrete energy levels, and de Broglie reproduced the Bohr model formula for the energy levels. The Bohr model was based on the assumed quantization of angular momentum L according to:

$$L = n\frac{h}{2\pi} = n\hbar.$$

According to de Broglie the electron is described by a wave and a whole number of wavelengths must fit along the circumference of the electron's orbit:

$$n\lambda = 2\pi r.$$

This approach essentially confined the electron wave in one dimension, along a circular orbit of radius r.

In 1921, prior to de Broglie, Arthur C. Lunn at the University of Chicago had used the same argument based on the completion of the relativistic energy–momentum 4-vector to derive what we now call the de Broglie relation.[9] Unlike de Broglie, Lunn went on to formulate the differential equation now known as the Schrödinger equation, and solve for its energy eigenvalues for the hydrogen atom. Unfortunately the paper was rejected by the Physical Review, as recounted by Kamen.[10]

Following up on de Broglie's ideas, physicist Peter Debye made an offhand comment that if particles behaved as waves, they should satisfy some sort of wave equation. Inspired by Debye's remark, Schrödinger decided to find a proper 3-dimensional wave equation for the electron. He was guided by William R. Hamilton's analogy between mechanics and optics, encoded in the observation that the zero-wavelength limit of optics resembles a mechanical system — the trajectories of light rays become sharp tracks that obey Fermat's principle, an analog of the principle of least action.[11] A modern version of his reasoning is reproduced below. The equation he found is:[12]

$$i\hbar\frac{\partial}{\partial t}\Psi(\mathbf{r},\,t) = -\frac{\hbar^2}{2m}\nabla^2\Psi(\mathbf{r},\,t) + V(\mathbf{r})\Psi(\mathbf{r},\,t).$$

However, by that time, Arnold Sommerfeld had refined the Bohr model with relativistic corrections.[13][14] Schrödinger used the relativistic energy momentum relation to find what is now known as the Klein–Gordon equation in a Coulomb potential (in natural units):

$$\left(E + \frac{e^2}{r}\right)^2 \psi(x) = -\nabla^2\psi(x) + m^2\psi(x).$$

He found the standing waves of this relativistic equation, but the relativistic corrections disagreed with Sommerfeld's formula. Discouraged, he put away his calculations and secluded himself in an isolated mountain cabin in December 1925.[15]

While at the cabin, Schrödinger decided that his earlier non-relativistic calculations were novel enough to publish, and decided to leave off the problem of relativistic corrections for the future. Despite the difficulties in solving the differential equation for hydrogen (he had sought help from his friend the mathematician Hermann Weyl[16]:3) Schrödinger showed that his non-relativistic version of the wave equation produced the correct spectral energies of hydrogen in a paper published in 1926.[16]:1[17] In the equation, Schrödinger computed the hydrogen spectral series by treating a hydrogen atom's electron as a wave $\Psi(x, t)$, moving in a potential well V, created by the proton. This computation accurately reproduced the energy levels of the Bohr model. In a paper, Schrödinger himself explained this equation as follows:

This 1926 paper was enthusiastically endorsed by Einstein, who saw the matter-waves as an intuitive depiction of nature, as opposed to Heisenberg's matrix mechanics, which he considered overly formal.[19]

The Schrödinger equation details the behavior of Ψ but says nothing of its *nature*. Schrödinger tried to interpret it as a charge density in his fourth paper, but he was unsuccessful.[20]:219 In 1926, just a few days after Schrödinger's fourth and final paper was published, Max Born successfully interpreted Ψ as the probability amplitude, whose absolute square is equal to probability density.[20]:220 Schrödinger, though, always opposed a statistical or probabilistic approach, with its associated discontinuities—much like Einstein, who believed that quantum mechanics was a statistical approximation to an underlying deterministic theory— and never reconciled with the Copenhagen interpretation.[21]

Louis de Broglie in his later years proposed a real valued wave function connected to the complex wave function by a proportionality constant and developed the De Broglie–Bohm theory.

14.5 The wave equation for particles

Main article: Wave–particle duality

The Schrödinger equation is a wave equation, since the *solutions* are functions which describe wave-like motions. Wave equations in physics can normally be derived from other physical laws – the wave equation for mechanical vibrations on strings and in matter can be derived from Newton's laws – where the wave function represents the displacement of matter, and electromagnetic waves from Maxwell's equations, where the wave functions are electric and magnetic fields. The basis for Schrödinger's equation, on the other hand, is the energy of the system and a separate postulate of quantum mechanics: the wave function is a description of the system.[22] The Schrödinger equation is therefore a new concept in itself; as Feynman put it:

The foundation of the equation is structured to be a linear differential equation based on classical energy conservation, and consistent with the De Broglie relations. The solution is the wave function ψ, which contains all the information that can be known about the system. In the Copenhagen interpretation, the modulus of ψ is related to the probability the particles are in some spatial configuration at some instant of time. Solving the equation for ψ can be used to predict how the particles will behave under the influence of the specified potential and with each other.

The Schrödinger equation was developed principally from the De Broglie hypothesis, a wave equation that would describe particles,[24] and can be constructed as shown informally in the following sections.[25] For a more rigorous description of Schrödinger's equation, see also.[26]

14.5.1 Consistency with energy conservation

The total energy E of a particle is the sum of kinetic energy T and potential energy V, this sum is also the frequent expression for the Hamiltonian H in classical mechanics:

$$E = T + V = H$$

Explicitly, for a particle in one dimension with position x, mass m and momentum p, and potential energy V which generally varies with position and time t:

$$E = \frac{p^2}{2m} + V(x, t) = H.$$

For three dimensions, the position vector $\mathbf{r}$ and momentum vector $\mathbf{p}$ must be used:

$$E = \frac{\mathbf{p} \cdot \mathbf{p}}{2m} + V(\mathbf{r}, t) = H$$

This formalism can be extended to any fixed number of particles: the total energy of the system is then the total kinetic energies of the particles, plus the total potential energy, again the Hamiltonian. However, there can be interactions between the particles (an N-body problem), so the potential energy V can change as the spatial configuration of particles changes, and possibly with time. The potential energy, in general, is *not* the sum of the separate potential energies for each particle, it is a function of all the spatial positions of the particles. Explicitly:

$$E = \sum_{n=1}^{N} \frac{\mathbf{p}_n \cdot \mathbf{p}_n}{2m_n} + V(\mathbf{r}_1, \mathbf{r}_2 \cdots \mathbf{r}_N, t) = H$$

14.5.2 Linearity

The simplest wavefunction is a plane wave of the form:

$$\Psi(\mathbf{r}, t) = A e^{i(\mathbf{k} \cdot \mathbf{r} - \omega t)}$$

where the A is the amplitude, $\mathbf{k}$ the wavevector, and ω the angular frequency, of the plane wave. In general, physical situations are not purely described by plane waves, so for generality the superposition principle is required; any wave can be made by superposition of sinusoidal plane waves. So if the equation is linear, a linear combination of plane waves is also an allowed solution. Hence a necessary and separate requirement is that the Schrödinger equation is a linear differential equation.

For discrete $\mathbf{k}$ the sum is a superposition of plane waves:

$$\Psi(\mathbf{r}, t) = \sum_{n=1}^{\infty} A_n e^{i(\mathbf{k}_n \cdot \mathbf{r} - \omega_n t)}$$

for some real amplitude coefficients A_n, and for continuous $\mathbf{k}$ the sum becomes an integral, the Fourier transform of a momentum space wavefunction:[27]

$$\Psi(\mathbf{r}, t) = \frac{1}{(\sqrt{2\pi})^3} \int \Phi(\mathbf{k}) e^{i(\mathbf{k} \cdot \mathbf{r} - \omega t)} d^3\mathbf{k}$$

where $d^3\mathbf{k} = dk_x dk_y dk_z$ is the differential volume element in $\mathbf{k}$-space, and the integrals are taken over all $\mathbf{k}$-space. The momentum wavefunction $\Phi(\mathbf{k})$ arises in the integrand since the position and momentum space wavefunctions are Fourier transforms of each other.

14.5.3 Consistency with the De Broglie relations

Einstein's light quanta hypothesis (1905) states that the energy E of a photon is proportional to the frequency ν (or angular frequency, $\omega = 2\pi\nu$) of the corresponding quantum wavepacket of light:

$$E = h\nu = \hbar\omega$$

Likewise De Broglie's hypothesis (1924) states that any particle can be associated with a wave, and that the momentum p of the particle is inversely proportional to the wavelength λ of such a wave (or proportional to the wavenumber, $k = 2\pi/\lambda$), in one dimension, by:

$$p = \frac{h}{\lambda} = \hbar k \ ,$$

while in three dimensions, wavelength λ is related to the magnitude of the wavevector $\mathbf{k}$:

$$\mathbf{p} = \hbar\mathbf{k} \ , \quad |\mathbf{k}| = \frac{2\pi}{\lambda} \ .$$

The Planck–Einstein and de Broglie relations illuminate the deep connections between energy with time, and space with momentum, and express wave–particle duality. In practice, natural units comprising $\hbar = 1$ are used, as the De Broglie *equations* reduce to *identities*: allowing momentum, wavenumber, energy and frequency to be used interchangeably, to prevent duplication of quantities, and reduce the number of dimensions of related quantities. For familiarity SI units are still used in this article.

Schrödinger's insight, late in 1925, was to express the phase of a plane wave as a complex phase factor using these relations:

$$\Psi = Ae^{i(\mathbf{k}\cdot\mathbf{r} - \omega t)} = Ae^{i(\mathbf{p}\cdot\mathbf{r} - Et)/\hbar}$$

and to realize that the first order partial derivatives were:

with respect to space:

$$\nabla\Psi = \frac{i}{\hbar}\mathbf{p}Ae^{i(\mathbf{p}\cdot\mathbf{r} - Et)/\hbar} = \frac{i}{\hbar}\mathbf{p}\Psi$$

with respect to time:

$$\frac{\partial\Psi}{\partial t} = -\frac{iE}{\hbar}Ae^{i(\mathbf{p}\cdot\mathbf{r} - Et)/\hbar} = -\frac{iE}{\hbar}\Psi$$

Another postulate of quantum mechanics is that all observables are represented by linear Hermitian operators which act on the wavefunction, and the eigenvalues of the operator are the values the observable takes. The previous derivatives are consistent with the energy operator, corresponding to the time derivative,

$$\hat{E}\Psi = i\hbar\frac{\partial}{\partial t}\Psi = E\Psi$$

where E are the energy eigenvalues, and the momentum operator, corresponding to the spatial derivatives (the gradient ∇),

$$\hat{\mathbf{p}}\Psi = -i\hbar\nabla\Psi = \mathbf{p}\Psi$$

where $\mathbf{p}$ is a vector of the momentum eigenvalues. In the above, the "hats" ($\wedge$) indicate these observables are operators, not simply ordinary numbers or vectors. The energy and momentum operators are *differential operators*, while the potential energy function V is just a multiplicative factor.

Substituting the energy and momentum operators into the classical energy conservation equation obtains the operator:

$$E = \frac{\mathbf{p}\cdot\mathbf{p}}{2m} + V \quad \rightarrow \quad \hat{E} = \frac{\hat{\mathbf{p}}\cdot\hat{\mathbf{p}}}{2m} + V$$

so in terms of derivatives with respect to time and space, acting this operator on the wavefunction Ψ immediately led Schrödinger to his equation:

$$i\hbar\frac{\partial \Psi}{\partial t} = -\frac{\hbar^2}{2m}\nabla^2\Psi + V\Psi$$

Wave–particle duality can be assessed from these equations as follows. The kinetic energy T is related to the square of momentum $\mathbf{p}$. As the particle's momentum increases, the kinetic energy increases more rapidly, but since the wavenumber $|\mathbf{k}|$ increases the wavelength λ decreases. In terms of ordinary scalar and vector quantities (not operators):

$$\mathbf{p}\cdot\mathbf{p} \propto \mathbf{k}\cdot\mathbf{k} \propto T \propto \frac{1}{\lambda^2}$$

The kinetic energy is also proportional to the second spatial derivatives, so it is also proportional to the magnitude of the *curvature* of the wave, in terms of operators:

$$\hat{T}\Psi = \frac{-\hbar^2}{2m}\nabla\cdot\nabla\Psi \propto \nabla^2\Psi .$$

As the curvature increases, the amplitude of the wave alternates between positive and negative more rapidly, and also shortens the wavelength. So the inverse relation between momentum and wavelength is consistent with the energy the particle has, and so the energy of the particle has a connection to a wave, all in the same mathematical formulation.[24]

14.5.4 Wave and particle motion

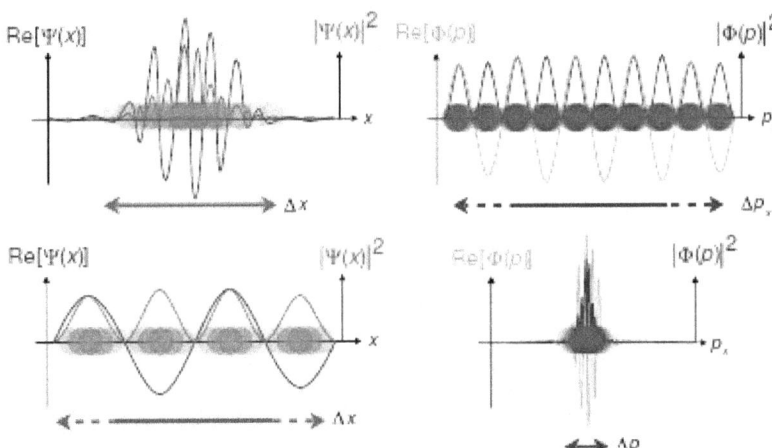

Increasing levels of wavepacket localization, meaning the particle has a more localized position.

Perfect localization

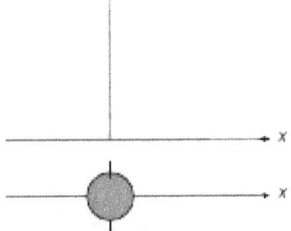

In the limit $\hbar \to 0$, the particle's position and momentum become known exactly. This is equivalent to the classical particle.

Schrödinger required that a wave packet solution near position **r** with wavevector near **k** will move along the trajectory determined by classical mechanics for times short enough for the spread in **k** (and hence in velocity) not to substantially increase the spread in **r**. Since, for a given spread in **k**, the spread in velocity is proportional to Planck's constant $\hbar$, it is sometimes said that in the limit as $\hbar$ approaches zero, the equations of classical mechanics are restored from quantum mechanics.[28] Great care is required in how that limit is taken, and in what cases.

The limiting short-wavelength is equivalent to $\hbar$ tending to zero because this is limiting case of increasing the wave packet localization to the definite position of the particle (see images right). Using the Heisenberg uncertainty principle for position and momentum, the products of uncertainty in position and momentum become zero as $\hbar \to 0$:

$$\sigma(x)\sigma(p_x) \geqslant \frac{\hbar}{2} \quad \to \quad \sigma(x)\sigma(p_x) \geqslant 0$$

where σ denotes the (root mean square) measurement uncertainty in x and px (and similarly for the y and z directions) which implies the position and momentum can only be known to arbitrary precision in this limit.

The Schrödinger equation in its general form

$$i\hbar \frac{\partial}{\partial t} \Psi(\mathbf{r}, t) = \hat{H}\Psi(\mathbf{r}, t)$$

is closely related to the Hamilton–Jacobi equation (HJE)

$$\frac{\partial}{\partial t} S(q_i, t) = H\left(q_i, \frac{\partial S}{\partial q_i}, t\right)$$

where S is action and H is the Hamiltonian function (not operator). Here the generalized coordinates qi for $i = 1, 2, 3$ (used in the context of the HJE) can be set to the position in Cartesian coordinates as $\mathbf{r} = (q_1, q_2, q_3) = (x, y, z)$.[28]

Substituting

$$\Psi = \sqrt{\rho(\mathbf{r}, t)}e^{iS(\mathbf{r},t)/\hbar}$$

where ϱ is the probability density, into the Schrödinger equation and then taking the limit $\hbar \to 0$ in the resulting equation, yields the Hamilton–Jacobi equation.

The implications are:

- The motion of a particle, described by a (short-wavelength) wave packet solution to the Schrödinger equation, is also described by the Hamilton–Jacobi equation of motion.

- The Schrödinger equation includes the wavefunction, so its wave packet solution implies the position of a (quantum) particle is fuzzily spread out in wave fronts. On the contrary, the Hamilton–Jacobi equation applies to a (classical) particle of definite position and momentum, instead the position and momentum at all times (the trajectory) are deterministic and can be simultaneously known.

14.6 Non-relativistic quantum mechanics

The quantum mechanics of particles without accounting for the effects of special relativity, for example particles propagating at speeds much less than light, is known as **non-relativistic quantum mechanics**. Following are several forms of Schrödinger's equation in this context for different situations: time independence and dependence, one and three spatial dimensions, and one and N particles.

In actuality, the particles constituting the system do not have the numerical labels used in theory. The language of mathematics forces us to label the positions of particles one way or another, otherwise there would be confusion between symbols representing which variables are for which particle.[26]

14.6.1   Time independent

If the Hamiltonian is not an explicit function of time, the equation is separable into a product of spatial and temporal parts. In general, the wavefunction takes the form:

$$\Psi(\text{coords space}, t) = \psi(\text{coords space})\tau(t) \,.$$

where ψ(space coords) is a function of all the spatial coordinate(s) of the particle(s) constituting the system only, and $\tau(t)$ is a function of time only.

Substituting for ψ into the Schrödinger equation for the relevant number of particles in the relevant number of dimensions, solving by separation of variables implies the general solution of the time-dependent equation has the form:[12]

$$\Psi(\text{coords space}, t) = \psi(\text{coords space})e^{-iEt/\hbar} \,.$$

Since the time dependent phase factor is always the same, only the spatial part needs to be solved for in time independent problems. Additionally, the energy operator $\hat{E} = i\hbar\partial/\partial t$ can always be replaced by the energy eigenvalue E, thus the time independent Schrödinger equation is an eigenvalue equation for the Hamiltonian operator:[5]:143ff

$$\hat{H}\psi = E\psi$$

This is true for any number of particles in any number of dimensions (in a time independent potential). This case describes the standing wave solutions of the time-dependent equation, which are the states with definite energy (instead of a probability distribution of different energies). In physics, these standing waves are called "stationary states" or "energy eigenstates"; in chemistry they are called "atomic orbitals" or "molecular orbitals". Superpositions of energy eigenstates change their properties according to the relative phases between the energy levels.

The energy eigenvalues from this equation form a discrete spectrum of values, so mathematically energy must be quantized. More specifically, the energy eigenstates form a basis – any wavefunction may be written as a sum over the discrete energy states or an integral over continuous energy states, or more generally as an integral over a measure. This is the spectral theorem in mathematics, and in a finite state space it is just a statement of the completeness of the eigenvectors of a Hermitian matrix.

14.6.2   One-dimensional examples

For a particle in one dimension, the Hamiltonian is:

$$\hat{H} = \frac{\hat{p}^2}{2m} + V(x), \quad \hat{p} = -i\hbar\frac{d}{dx}$$

and substituting this into the general Schrödinger equation gives:

$$-\frac{\hbar^2}{2m}\frac{d^2}{dx^2}\psi(x) + V(x)\psi(x) = E\psi(x)$$

This is the only case the Schrödinger equation is an ordinary differential equation, rather than a partial differential equation. The general solutions are always of the form:

$$\Psi(x, t) = \psi(x)e^{-iEt/\hbar} \,.$$

For N particles in one dimension, the Hamiltonian is:

$$\hat{H} = \sum_{n=1}^{N} \frac{\hat{p}_n^2}{2m_n} + V(x_1, x_2, \cdots x_N), \quad \hat{p}_n = -i\hbar \frac{\partial}{\partial x_n}$$

where the position of particle n is xn. The corresponding Schrödinger equation is:

$$-\frac{\hbar^2}{2} \sum_{n=1}^{N} \frac{1}{m_n} \frac{\partial^2}{\partial x_n^2} \psi(x_1, x_2, \cdots x_N) + V(x_1, x_2, \cdots x_N)\psi(x_1, x_2, \cdots x_N) = E\psi(x_1, x_2, \cdots x_N).$$

so the general solutions have the form:

$$\Psi(x_1, x_2, \cdots x_N, t) = e^{-iEt/\hbar} \psi(x_1, x_2 \cdots x_N)$$

For non-interacting distinguishable particles,[29] the potential of the system only influences each particle separately, so the total potential energy is the sum of potential energies for each particle:

$$V(x_1, x_2, \cdots x_N) = \sum_{n=1}^{N} V(x_n).$$

and the wavefunction can be written as a product of the wavefunctions for each particle:

$$\Psi(x_1, x_2, \cdots x_N, t) = e^{-iEt/\hbar} \prod_{n=1}^{N} \psi(x_n),$$

For non-interacting identical particles, the potential is still a sum, but wavefunction is a bit more complicated - it is a sum over the permutations of products of the separate wavefunctions to account for particle exchange. In general for interacting particles, the above decompositions are *not* possible.

Free particle

For no potential, $V = 0$, so the particle is free and the equation reads:[5]:151ff

$$-E\psi = \frac{\hbar^2}{2m} \frac{d^2\psi}{dx^2}$$

which has oscillatory solutions for $E > 0$ (the C_n are arbitrary constants):

$$\psi_E(x) = C_1 e^{i\sqrt{2mE/\hbar^2}\, x} + C_2 e^{-i\sqrt{2mE/\hbar^2}\, x}$$

and exponential solutions for $E < 0$

$$\psi_{-|E|}(x) = C_1 e^{\sqrt{2m|E|/\hbar^2}\, x} + C_2 e^{-\sqrt{2m|E|/\hbar^2}\, x}.$$

The exponentially growing solutions have an infinite norm, and are not physical. They are not allowed in a finite volume with periodic or fixed boundary conditions.

See also free particle and wavepacket for more discussion on the free particle.

Constant potential

For a constant potential, $V = V_0$, the solution is oscillatory for $E > V_0$ and exponential for $E < V_0$, corresponding to energies that are allowed or disallowed in classical mechanics. Oscillatory solutions have a classically allowed energy and correspond to actual classical motions, while the exponential solutions have a disallowed energy and describe a small amount of quantum bleeding into the classically disallowed region, due to quantum tunneling. If the potential V_0 grows to infinity, the motion is classically confined to a finite region. Viewed far enough away, every solution is reduced an exponential; the condition that the exponential is decreasing restricts the energy levels to a discrete set, called the allowed energies.[27]

Harmonic oscillator

Main article: Quantum harmonic oscillator

The Schrödinger equation for this situation is

$$E\psi = -\frac{\hbar^2}{2m}\frac{d^2}{dx^2}\psi + \frac{1}{2}m\omega^2 x^2 \psi$$

It is a notable quantum system to solve for; since the solutions are exact (but complicated – in terms of Hermite polynomials), and it can describe or at least approximate a wide variety of other systems, including vibrating atoms, molecules,[30] and atoms or ions in lattices,[31] and approximating other potentials near equilibrium points. It is also the basis of perturbation methods in quantum mechanics.

There is a family of solutions – in the position basis they are

$$\psi_n(x) = \sqrt{\frac{1}{2^n\,n!}} \cdot \left(\frac{m\omega}{\pi\hbar}\right)^{1/4} \cdot e^{-\frac{m\omega x^2}{2\hbar}} \cdot H_n\left(\sqrt{\frac{m\omega}{\hbar}}x\right)$$

where $n = 0,1,2,...$, and the functions Hn are the Hermite polynomials.

14.6.3   Three-dimensional examples

The extension from one dimension to three dimensions is straightforward, all position and momentum operators are replaced by their three-dimensional expressions and the partial derivative with respect to space is replaced by the gradient operator.

The Hamiltonian for one particle in three dimensions is:

$$\hat{H} = \frac{\hat{\mathbf{p}} \cdot \hat{\mathbf{p}}}{2m} + V(\mathbf{r}), \quad \hat{\mathbf{p}} = -i\hbar\nabla$$

generating the equation:

$$-\frac{\hbar^2}{2m}\nabla^2\psi(\mathbf{r}) + V(\mathbf{r})\psi(\mathbf{r}) = E\psi(\mathbf{r})$$

with stationary state solutions of the form:

$$\Psi(\mathbf{r},t) = \psi(\mathbf{r})e^{-iEt/\hbar}$$

where the position of the particle is $\mathbf{r}$. Two useful coordinate systems for solving the Schrödinger equation are Cartesian coordinates so that $\mathbf{r} = (x, y, z)$ and spherical polar coordinates so that $\mathbf{r} = (r, \theta, \varphi)$, although other orthogonal coordinates are useful for solving the equation for systems with certain geometric symmetries.

For N particles in three dimensions, the Hamiltonian is:

$$\hat{H} = \sum_{n=1}^{N} \frac{\hat{\mathbf{p}}_n \cdot \hat{\mathbf{p}}_n}{2m_n} + V(\mathbf{r}_1, \mathbf{r}_2, \cdots \mathbf{r}_N), \quad \hat{\mathbf{p}}_n = -i\hbar\nabla_n$$

where the position of particle n is $\mathbf{r}n$ and the gradient operators are partial derivatives with respect to the particle's position coordinates. In Cartesian coordinates, for particle n, the position vector is $\mathbf{r}n = (xn, yn, zn)$ while the gradient and Laplacian operator are respectively:

$$\nabla_n = \mathbf{e}_x \frac{\partial}{\partial x_n} + \mathbf{e}_y \frac{\partial}{\partial y_n} + \mathbf{e}_z \frac{\partial}{\partial z_n}, \quad \nabla_n^2 = \nabla_n \cdot \nabla_n = \frac{\partial^2}{\partial x_n{}^2} + \frac{\partial^2}{\partial y_n{}^2} + \frac{\partial^2}{\partial z_n{}^2}$$

The Schrödinger equation is:

$$-\frac{\hbar^2}{2} \sum_{n=1}^{N} \frac{1}{m_n} \nabla_n^2 \Psi(\mathbf{r}_1, \mathbf{r}_2, \cdots \mathbf{r}_N) + V(\mathbf{r}_1, \mathbf{r}_2, \cdots \mathbf{r}_N)\Psi(\mathbf{r}_1, \mathbf{r}_2, \cdots \mathbf{r}_N) = E\Psi(\mathbf{r}_1, \mathbf{r}_2, \cdots \mathbf{r}_N)$$

with stationary state solutions:

$$\Psi(\mathbf{r}_1, \mathbf{r}_2 \cdots \mathbf{r}_N, t) = e^{-iEt/\hbar}\psi(\mathbf{r}_1, \mathbf{r}_2 \cdots \mathbf{r}_N)$$

Again, for non-interacting distinguishable particles the potential is the sum of particle potentials

$$V(\mathbf{r}_1, \mathbf{r}_2, \cdots \mathbf{r}_N) = \sum_{n=1}^{N} V(\mathbf{r}_n)$$

and the wavefunction is a product of the particle wavefuntions

$$\Psi(\mathbf{r}_1, \mathbf{r}_2 \cdots \mathbf{r}_N, t) = e^{-iEt/\hbar} \prod_{n=1}^{N} \psi(\mathbf{r}_n)\,.$$

For non-interacting identical particles, the potential is a sum but the wavefunction is a sum over permutations of products. The previous two equations do not apply to interacting particles.

Following are examples where exact solutions are known. See the main articles for further details.

Hydrogen atom

This form of the Schrödinger equation can be applied to the hydrogen atom:[22][24]

$$E\psi = -\frac{\hbar^2}{2\mu}\nabla^2\psi - \frac{e^2}{4\pi\varepsilon_0 r}\psi$$

where e is the electron charge, $\mathbf{r}$ is the position of the electron ($r = |\mathbf{r}|$ is the magnitude of the position), the potential term is due to the Coulomb interaction, wherein ε_0 is the electric constant (permittivity of free space) and

$$\mu = \frac{m_e m_p}{m_e + m_p}$$

is the 2-body reduced mass of the hydrogen nucleus (just a proton) of mass m_p and the electron of mass m_e. The negative sign arises in the potential term since the proton and electron are oppositely charged. The reduced mass in place of the electron mass is used since the electron and proton together orbit each other about a common centre of mass, and constitute a two-body problem to solve. The motion of the electron is of principle interest here, so the equivalent one-body problem is the motion of the electron using the reduced mass.

The wavefunction for hydrogen is a function of the electron's coordinates, and in fact can be separated into functions of each coordinate.[32] Usually this is done in spherical polar coordinates:

$$\psi(r, \theta, \phi) = R(r)Y_\ell^m(\theta, \phi) = R(r)\Theta(\theta)\Phi(\phi)$$

where R are radial functions and $Y_\ell^m(\theta, \phi)$ are spherical harmonics of degree ℓ and order m. This is the only atom for which the Schrödinger equation has been solved for exactly. Multi-electron atoms require approximative methods. The family of solutions are:[33]

$$\psi_{n\ell m}(r, \theta, \phi) = \sqrt{\left(\frac{2}{na_0}\right)^3 \frac{(n - \ell - 1)!}{2n[(n + \ell)!]}} e^{-r/na_0} \left(\frac{2r}{na_0}\right)^\ell L_{n-\ell-1}^{2\ell+1}\left(\frac{2r}{na_0}\right) \cdot Y_\ell^m(\theta, \phi)$$

where:

- $a_0 = \frac{4\pi\varepsilon_0\hbar^2}{m_e e^2}$ is the Bohr radius,

- $L_{n-\ell-1}^{2\ell+1}(\cdots)$ are the generalized Laguerre polynomials of degree $n - \ell - 1$.

- n, ℓ, m are the principal, azimuthal, and magnetic quantum numbers respectively: which take the values:

$$n = 1, 2, 3, \ldots$$
$$\ell = 0, 1, 2, \ldots, n - 1$$
$$m = -\ell, \ldots, \ell$$

NB: generalized Laguerre polynomials are defined differently by different authors—see main article on them and the hydrogen atom.

Two-electron atoms or ions

The equation for any two-electron system, such as the neutral helium atom (He, $Z = 2$), the negative hydrogen ion (H$^-$, $Z = 1$), or the positive lithium ion (Li$^+$, $Z = 3$) is:[25]

$$E\psi = -\hbar^2 \left[\frac{1}{2\mu}\left(\nabla_1^2 + \nabla_2^2\right) + \frac{1}{M}\nabla_1 \cdot \nabla_2\right]\psi + \frac{e^2}{4\pi\varepsilon_0}\left[\frac{1}{r_{12}} - Z\left(\frac{1}{r_1} + \frac{1}{r_2}\right)\right]\psi$$

where $\mathbf{r}_1$ is the position of one electron ($r_1 = |\mathbf{r}_1|$ is its magnitude), $\mathbf{r}_2$ is the position of the other electron ($r_2 = |\mathbf{r}_2|$ is the magnitude), $r_{12} = |\mathbf{r}_{12}|$ is the magnitude of the separation between them given by

$$|\mathbf{r}_{12}| = |\mathbf{r}_2 - \mathbf{r}_1|$$

μ is again the two-body reduced mass of an electron with respect to the nucleus of mass M, so this time

$$\mu = \frac{m_e M}{m_e + M}$$

and Z is the atomic number for the element (not a quantum number).

The cross-term of two laplacians

$$\frac{1}{M} \nabla_1 \cdot \nabla_2$$

is known as the *mass polarization term*, which arises due to the motion of atomic nuclei. The wavefunction is a function of the two electron's positions:

$$\psi = \psi(\mathbf{r}_1, \mathbf{r}_2).$$

There is no closed form solution for this equation.

14.6.4 Time dependent

This is the equation of motion for the quantum state. In the most general form, it is written:[1]·143ff

$$i\hbar \frac{\partial}{\partial t} \Psi = \hat{H} \Psi.$$

and the solution, the wavefunction, is a function of all the particle coordinates of the system and time. Following are specific cases.

For one particle in one dimension, the Hamiltonian

$$\hat{H} = \frac{\hat{p}^2}{2m} + V(x, t), \quad \hat{p} = -i\hbar \frac{\partial}{\partial x}$$

generates the equation:

$$i\hbar \frac{\partial}{\partial t} \Psi(x, t) = -\frac{\hbar^2}{2m} \frac{\partial^2}{\partial x^2} \Psi(x, t) + V(x, t) \Psi(x, t)$$

For N particles in one dimension, the Hamiltonian is:

$$\hat{H} = \sum_{n=1}^{N} \frac{\hat{p}_n^2}{2m_n} + V(x_1, x_2, \cdots x_N, t), \quad \hat{p}_n = -i\hbar \frac{\partial}{\partial x_n}$$

where the position of particle n is x_n, generating the equation:

$$i\hbar \frac{\partial}{\partial t} \Psi(x_1, x_2 \cdots x_N, t) = -\frac{\hbar^2}{2} \sum_{n=1}^{N} \frac{1}{m_n} \frac{\partial^2}{\partial x_n^2} \Psi(x_1, x_2 \cdots x_N, t) + V(x_1, x_2 \cdots x_N, t) \Psi(x_1, x_2 \cdots x_N, t).$$

For one particle in three dimensions, the Hamiltonian is:

$$\hat{H} = \frac{\hat{\mathbf{p}} \cdot \hat{\mathbf{p}}}{2m} + V(\mathbf{r}, t), \quad \hat{\mathbf{p}} = -i\hbar\nabla$$

generating the equation:

$$i\hbar\frac{\partial}{\partial t}\Psi(\mathbf{r}, t) = -\frac{\hbar^2}{2m}\nabla^2\Psi(\mathbf{r}, t) + V(\mathbf{r}, t)\Psi(\mathbf{r}, t)$$

For N particles in three dimensions, the Hamiltonian is:

$$\hat{H} = \sum_{n=1}^{N}\frac{\hat{\mathbf{p}}_n \cdot \hat{\mathbf{p}}_n}{2m_n} + V(\mathbf{r}_1, \mathbf{r}_2, \cdots \mathbf{r}_N, t), \quad \hat{\mathbf{p}}_n = -i\hbar\nabla_n$$

where the position of particle n is $\mathbf{r}n$, generating the equation:[5]:141

$$i\hbar\frac{\partial}{\partial t}\Psi(\mathbf{r}_1, \mathbf{r}_2, \cdots \mathbf{r}_N, t) = -\frac{\hbar^2}{2}\sum_{n=1}^{N}\frac{1}{m_n}\nabla_n^2\Psi(\mathbf{r}_1, \mathbf{r}_2, \cdots \mathbf{r}_N, t) + V(\mathbf{r}_1, \mathbf{r}_2, \cdots \mathbf{r}_N, t)\Psi(\mathbf{r}_1, \mathbf{r}_2, \cdots \mathbf{r}_N, t)$$

This last equation is in a very high dimension, so the solutions are not easy to visualize.

14.7   Solution methods

14.8   Properties

The Schrödinger equation has the following properties: some are useful, but there are shortcomings. Ultimately, these properties arise from the Hamiltonian used, and solutions to the equation.

14.8.1   Linearity

See also: Linear differential equation

In the development above, the Schrödinger equation was made to be linear for generality, though this has other implications. If two wave functions ψ_1 and ψ_2 are solutions, then so is any linear combination of the two:

$$\psi = a\psi_1 + b\psi_2$$

where a and b are any complex numbers (the sum can be extended for any number of wavefunctions). This property allows superpositions of quantum states to be solutions of the Schrödinger equation. Even more generally, it holds that a general solution to the Schrödinger equation can be found by taking a weighted sum over all single state solutions achievable. For example, consider a wave function $\Psi(x, t)$ such that the wave function is a product of two functions: one time independent, and one time dependent. If states of definite energy found using the time independent Schrödinger equation are given by $\psi E(x)$ with amplitude An and time dependent phase factor is given by

$$e^{-iE_n t/\hbar},$$

then a valid general solution is

$$\Psi(x,t) = \sum_n A_n \psi_{E_n}(x) e^{-iE_n t/\hbar}.$$

Additionally, the ability to scale solutions allows one to solve for a wave function without normalizing it first. If one has a set of normalized solutions ψn, then

$$\Psi = \sum_n A_n \psi_n$$

can be normalized by ensuring that

$$\sum_n |A_n|^2 = 1.$$

This is much more convenient than having to verify that

$$\int_{-\infty}^{\infty} |\Psi(x)|^2 \, dx = \int_{-\infty}^{\infty} \Psi(x)\Psi^*(x) \, dx = 1.$$

14.8.2 Real energy eigenstates

For the time-independent equation, an additional feature of linearity follows: if two wave functions ψ_1 and ψ_2 are solutions to the time-independent equation with the same energy E, then so is any linear combination:

$$\hat{H}(a\psi_1 + b\psi_2) = a\hat{H}\psi_1 + b\hat{H}\psi_2 = E(a\psi_1 + b\psi_2).$$

Two different solutions with the same energy are called *degenerate*.[27]

In an arbitrary potential, if a wave function ψ solves the time-independent equation, so does its complex conjugate, denoted ψ^*. By taking linear combinations, the real and imaginary parts of ψ are each solutions. If there is no degeneracy they can only differ by a factor.

In the time-dependent equation, complex conjugate waves move in opposite directions. If $\Psi(x, t)$ is one solution, then so is $\Psi(x, -t)$. The symmetry of complex conjugation is called time-reversal symmetry.

14.8.3 Space and time derivatives

The Schrödinger equation is first order in time and second in space, which describes the time evolution of a quantum state (meaning it determines the future amplitude from the present).

Explicitly for one particle in 3-dimensional Cartesian coordinates – the equation is

$$i\hbar \frac{\partial \Psi}{\partial t} = -\frac{\hbar^2}{2m} \left(\frac{\partial^2 \Psi}{\partial x^2} + \frac{\partial^2 \Psi}{\partial y^2} + \frac{\partial^2 \Psi}{\partial z^2} \right) + V(x, y, z, t)\Psi.$$

The first time partial derivative implies the initial value (at $t = 0$) of the wavefunction

$\Psi(x, y, z, 0)$

is an arbitrary constant. Likewise – the second order derivatives with respect to space implies the wavefunction *and* its first order spatial derivatives

$\Psi(x_b, y_b, z_b, t)$

$$\frac{\partial}{\partial x}\Psi(x_b, y_b, z_b, t) \quad \frac{\partial}{\partial y}\Psi(x_b, y_b, z_b, t) \quad \frac{\partial}{\partial z}\Psi(x_b, y_b, z_b, t)$$

are all arbitrary constants at a given set of points, where *xb, yb, zb* are a set of points describing boundary *b* (derivatives are evaluated at the boundaries). Typically there are one or two boundaries, such as the step potential and particle in a box respectively.

As the first order derivatives are arbitrary, the wavefunction can be a continuously differentiable function of space, since at any boundary the gradient of the wavefunction can be matched.

On the contrary, wave equations in physics are usually *second order in time*, notable are the family of classical wave equations and the quantum Klein–Gordon equation.

14.8.4   Local conservation of probability

Main articles: Probability current and Continuity equation

The Schrödinger equation is consistent with probability conservation. Multiplying the Schrödinger equation on the right by the complex conjugate wavefunction, and multiplying the wavefunction to the left of the complex conjugate of the Schrödinger equation, and subtracting, gives the continuity equation for probability:[34]

$$\frac{\partial}{\partial t}\rho(\mathbf{r}, t) + \nabla \cdot \mathbf{j} = 0,$$

where

$$\rho = |\Psi|^2 = \Psi^*(\mathbf{r}, t)\Psi(\mathbf{r}, t)$$

is the probability density (probability per unit volume, * denotes complex conjugate), and

$$\mathbf{j} = \frac{1}{2m}\left(\Psi^*\hat{\mathbf{p}}\Psi - \Psi\hat{\mathbf{p}}\Psi^*\right)$$

is the probability current (flow per unit area).

Hence predictions from the Schrödinger equation do not violate probability conservation.

14.8.5   Positive energy

If the potential is bounded from below, meaning there is a minimum value of potential energy, the eigenfunctions of the Schrödinger equation have energy which is also bounded from below. This can be seen most easily by using the variational principle, as follows. (See also below).

For any linear operator $\hat{A}$ bounded from below, the eigenvector with the smallest eigenvalue is the vector ψ that minimizes the quantity

$\langle \psi | \hat{A} | \psi \rangle$

over all ψ which are normalized.[34] In this way, the smallest eigenvalue is expressed through the variational principle. For the Schrödinger Hamiltonian $\hat{H}$ bounded from below, the smallest eigenvalue is called the ground state energy. That energy is the minimum value of

$$\langle \psi | \hat{H} | \psi \rangle = \int \psi^*(\mathbf{r}) \left[-\frac{\hbar^2}{2m} \nabla^2 \psi(\mathbf{r}) + V(\mathbf{r}) \psi(\mathbf{r}) \right] d^3\mathbf{r} = \int \left[\frac{\hbar^2}{2m} |\nabla \psi|^2 + V(\mathbf{r}) |\psi|^2 \right] d^3\mathbf{r} = \langle \hat{H} \rangle$$

(using integration by parts). Due to the complex modulus of ψ squared (which is positive definite), the right hand side always greater than the lowest value of $V(x)$. In particular, the ground state energy is positive when $V(x)$ is everywhere positive.

For potentials which are bounded below and are not infinite over a region, there is a ground state which minimizes the integral above. This lowest energy wavefunction is real and positive definite – meaning the wavefunction can increase and decrease, but is positive for all positions. It physically cannot be negative: if it were, smoothing out the bends at the sign change (to minimize the wavefunction) rapidly reduces the gradient contribution to the integral and hence the kinetic energy, while the potential energy changes linearly and less quickly. The kinetic and potential energy are both changing at different rates, so the total energy is not constant, which can't happen (conservation). The solutions are consistent with Schrödinger equation if this wavefunction is positive definite.

The lack of sign changes also shows that the ground state is nondegenerate, since if there were two ground states with common energy E, not proportional to each other, there would be a linear combination of the two that would also be a ground state resulting in a zero solution.

14.8.6 Analytic continuation to diffusion

See also: Path integral formulation (The Schrödinger equation)

The above properties (positive definiteness of energy) allow the analytic continuation of the Schrödinger equation to be identified as a stochastic process. This can be interpreted as the Huygens–Fresnel principle applied to De Broglie waves; the spreading wavefronts are diffusive probability amplitudes.[34]

For a free particle (not subject to a potential) in a random walk, substituting $\tau = it$ into the time-dependent Schrödinger equation gives:[35]

$$\frac{\partial}{\partial \tau} X(\mathbf{r}, \tau) = \frac{\hbar}{2m} \nabla^2 X(\mathbf{r}, \tau), \quad X(\mathbf{r}, \tau) = \Psi(\mathbf{r}, \tau/i)$$

which has the same form as the diffusion equation, with diffusion coefficient $\hbar/2m$.

14.9 Relativistic quantum mechanics

Relativistic quantum mechanics is obtained where quantum mechanics and special relativity simultaneously apply. In general, one wishes to build relativistic wave equations from the relativistic energy–momentum relation

$$E^2 = (pc)^2 + (m_0 c^2)^2,$$

instead of classical energy equations. The Klein–Gordon equation and the Dirac equation are two such equations. The Klein–Gordon equation,

$$\frac{1}{c^2}\frac{\partial^2}{\partial t^2}\psi - \nabla^2\psi + \frac{m^2 c^2}{\hbar^2}\psi = 0.$$

was the first such equation to be obtained, even before the non-relativistic one, and applies to massive spinless particles. The Dirac equation arose from taking the "square root" of the Klein–Gordon equation by factorizing the entire relativistic wave operator into a product of two operators – one of these is the operator for the entire Dirac equation.

The general form of the Schrödinger equation remains true in relativity, but the Hamiltonian is less obvious. For example, the Dirac Hamiltonian for a particle of mass m and electric charge q in an electromagnetic field (described by the electromagnetic potentials φ and $\mathbf{A}$) is:

$$\hat{H}_{\text{Dirac}} = \gamma^0 \left[c\boldsymbol{\gamma} \cdot (\hat{\mathbf{p}} - q\mathbf{A}) + mc^2 + \gamma^0 q\phi \right],$$

in which the $\boldsymbol{\gamma} = (\gamma^1, \gamma^2, \gamma^3)$ and γ^0 are the Dirac gamma matrices related to the spin of the particle. The Dirac equation is true for all spin-1/2 particles, and the solutions to the equation are 4-component spinor fields with two components corresponding to the particle and the other two for the antiparticle.

For the Klein–Gordon equation, the general form of the Schrödinger equation is inconvenient to use, and in practice the Hamiltonian is not expressed in an analogous way to the Dirac Hamiltonian. The equations for relativistic quantum fields can be obtained in other ways, such as starting from a Lagrangian density and using the Euler-Lagrange equations for fields, or use the representation theory of the Lorentz group in which certain representations can be used to fix the equation for a free particle of given spin (and mass).

In general, the Hamiltonian to be substituted in the general Schrödinger equation is not just a function of the position and momentum operators (and possibly time), but also of spin matrices. Also, the solutions to a relativistic wave equation, for a massive particle of spin s, are complex-valued $2(2s+1)$-component spinor fields.

14.10   Quantum field theory

The general equation is also valid and used in quantum field theory, both in relativistic and non-relativistic situations. However, the solution ψ is no longer interpreted as a "wave", but should be interpreted as an operator acting on states existing in a Fock space.

14.11   See also

- Fractional Schrödinger equation

- Nonlinear Schrödinger equation

- Quantum carpet

- Quantum revival

- Relation between Schrödinger's equation and the path integral formulation of quantum mechanics

- Schrödinger field

- Schrödinger picture

- Schrödinger's cat

- Theoretical and experimental justification for the Schrödinger equation

14.12 Notes

[1] Schrödinger, E. (1926). "An Undulatory Theory of the Mechanics of Atoms and Molecules" (PDF). *Physical Review* **28** (6): 1049–1070. Bibcode:1926PhRv...28.1049S. doi:10.1103/PhysRev.28.1049. Archived from the original (PDF) on 17 December 2008.

[2] Griffiths, David J. (2004). *Introduction to Quantum Mechanics (2nd ed.)*, Prentice Hall, ISBN 0-13-111892-7

[3] Ballentine, Leslie (1998), *Quantum Mechanics: A Modern Development*, World Scientific Publishing Co., ISBN 9810241054

[4] Laloe, Franck (2012), *Do We Really Understand Quantum Mechanics*, Cambridge University Press, ISBN 978-1-107-02501-1

[5] Shankar, R. (1994). *Principles of Quantum Mechanics* (2nd ed.). Kluwer Academic/Plenum Publishers. ISBN 978-0-306-44790-7.

[6] http://hyperphysics.phy-astr.gsu.edu/hbase/quantum/scheq.html

[7] Nouredine Zettili (17 February 2009). *Quantum Mechanics: Concepts and Applications*. John Wiley & Sons. ISBN 978-0-470-02678-6.

[8] de Broglie, L. (1925). "Recherches sur la théorie des quanta" [On the Theory of Quanta] (PDF). *Annales de Physique* **10** (3): 22–128. Translated version at the Wayback Machine (archived May 9, 2009).

[9] Weissman, M.B.; V. V. Iliev; I. Gutman (2008). "A pioneer remembered: biographical notes about Arthur Constant Lunn". *Communications in Mathematical and in Computer Chemistry* **59** (3): 687–708.

[10] Kamen, Martin D. (1985). *Radiant Science, Dark Politics*. Berkeley and Los Angeles, CA: University of California Press. pp. 29–32. ISBN 0-520-04929-2.

[11] Schrodinger, E. (1984). *Collected papers*. Friedrich Vieweg und Sohn. ISBN 3-7001-0573-8. See introduction to first 1926 paper.

[12] Encyclopaedia of Physics (2nd Edition), R.G. Lerner, G.L. Trigg, VHC publishers, 1991, (Verlagsgesellschaft) 3-527-26954-1, (VHC Inc.) ISBN 0-89573-752-3

[13] Sommerfeld, A. (1919). *Atombau und Spektrallinien*. Braunschweig: Friedrich Vieweg und Sohn. ISBN 3-87144-484-7.

[14] For an English source, see Haar, T. "The Old Quantum Theory".

[15] Rhodes, R. (1986). *Making of the Atomic Bomb*. Touchstone. ISBN 0-671-44133-7.

[16] Erwin Schrödinger (1982). *Collected Papers on Wave Mechanics: Third Edition*. American Mathematical Soc. ISBN 978-0-8218-3524-1.

[17] Schrödinger, E. (1926). "Quantisierung als Eigenwertproblem; von Erwin Schrödinger". *Annalen der Physik* **384**: 361–377. doi:10.1002/andp.19263840404.

[18] Erwin Schrödinger, "The Present situation in Quantum Mechanics," p. 9 of 22. The English version was translated by John D. Trimmer. The translation first appeared first in **Proceedings of the American Philosophical Society**, 124, 323–38. It later appeared as Section I.11 of Part I of **Quantum Theory and Measurement** by J.A. Wheeler and W.H. Zurek, eds., Princeton University Press, New Jersey 1983).

[19] Einstein, A.; *et. al.* "Letters on Wave Mechanics: Schrodinger–Planck–Einstein–Lorentz".

[20] Moore, W.J. (1992). *Schrödinger: Life and Thought*. Cambridge University Press. ISBN 0-521-43767-9.

[21] It is clear that even in his last year of life, as shown in a letter to Max Born, that Schrödinger never accepted the Copenhagen interpretation.[20]:220

[22] Molecular Quantum Mechanics Parts I and II: An Introduction to Quantum Chemistry (Volume 1), P.W. Atkins, Oxford University Press, 1977, ISBN 0-19-855129-0

[23] The New Quantum Universe, T.Hey, P.Walters, Cambridge University Press, 2009, ISBN 978-0-521-56457-1

[24] Quanta: A handbook of concepts, P.W. Atkins, Oxford University Press, 1974, ISBN 0-19-855493-1

[25] Physics of Atoms and Molecules, B.H. Bransden, C.J.Joachain, Longman, 1983, ISBN 0-582-44401-2

[26] Quantum Physics of Atoms, Molecules, Solids, Nuclei and Particles (2nd Edition), R. Resnick, R. Eisberg, John Wiley & Sons, 1985, ISBN 978-0-471-87373-0

[27] Quantum Mechanics Demystified, D. McMahon, Mc Graw Hill (USA), 2006, ISBN(10) 0 07 145546 9

[28] Analytical Mechanics, L.N. Hand, J.D. Finch, Cambridge University Press, 2008, ISBN 978-0-521-57572-0

[29] N. Zettili. *Quantum Mechanics: Concepts and Applications* (2nd ed.). p. 458. ISBN 978-0-470-02679-3.

[30] Physical chemistry, P.W. Atkins, Oxford University Press, 1978, ISBN 0-19-855148-7

[31] Solid State Physics (2nd Edition), J.R. Hook, H.E. Hall, Manchester Physics Series, John Wiley & Sons, 2010, ISBN 978-0-471-92804-1

[32] Physics for Scientists and Engineers – with Modern Physics (6th Edition), P. A. Tipler, G. Mosca, Freeman, 2008, ISBN 0-7167-8964-7

[33] David Griffiths (2008). *Introduction to elementary particles*. Wiley-VCH. pp. 162–. ISBN 978-3-527-40601-2. Retrieved 27 June 2011.

[34] Quantum Mechanics, E. Abers, Pearson Ed., Addison Wesley, Prentice Hall Inc, 2004, ISBN 978-0-13-146100-0

[35] http://www.stt.msu.edu/~{}mcubed/Relativistic.pdf

14.13   References

- P. A. M. Dirac (1958). *The Principles of Quantum Mechanics* (4th ed.). Oxford University Press.

- B.H. Bransden and C.J. Joachain (2000). *Quantum Mechanics* (2nd ed.). Prentice Hall PTR. ISBN 0-582-35691-1.

- David J. Griffiths (2004). *Introduction to Quantum Mechanics* (2nd ed.). Benjamin Cummings. ISBN 0-13-124405-1.

- Richard Liboff (2002). *Introductory Quantum Mechanics* (4th ed.). Addison Wesley. ISBN 0-8053-8714-5.

- David Halliday (2007). *Fundamentals of Physics* (8th ed.). Wiley. ISBN 0-471-15950-6.

- Serway, Moses, and Moyer (2004). *Modern Physics* (3rd ed.). Brooks Cole. ISBN 0-534-49340-8.

- Schrödinger, Erwin (December 1926). "An Undulatory Theory of the Mechanics of Atoms and Molecules". *Phys. Rev.* 28 (6): 1049–1070. Bibcode:1926PhRv...28.1049S. doi:10.1103/PhysRev.28.1049.

- Teschl, Gerald (2009). *Mathematical Methods in Quantum Mechanics; With Applications to Schrödinger Operators*. Providence: American Mathematical Society. ISBN 978-0-8218-4660-5.

14.14   External links

- Hazewinkel, Michiel, ed. (2001), "Schrödinger equation", *Encyclopedia of Mathematics*, Springer, ISBN 978-1-55608-010-4

- Quantum Physics — textbook by Benjamin Crowell with a treatment of the time-independent Schrödinger equation

- Linear Schrödinger Equation at EqWorld: The World of Mathematical Equations.

- Nonlinear Schrödinger Equation at EqWorld: The World of Mathematical Equations.

- The Schrödinger Equation in One Dimension as well as the directory of the book.

- All about 3D Schrödinger Equation

- Mathematical aspects of Schrödinger equations are discussed on the Dispersive PDE Wiki.

- Web-Schrödinger: Interactive solution of the 2D time-dependent and stationary Schrödinger equation

- An alternate reasoning behind the Schrödinger Equation

- Online software-Periodic Potential Lab Solves the time-independent Schrödinger equation for arbitrary periodic potentials.

- What Do You Do With a Wavefunction?

- The Young Double-Slit Experiment

Schrödinger equation as part of a monument in front of Warsaw University's Centre of New Technologies

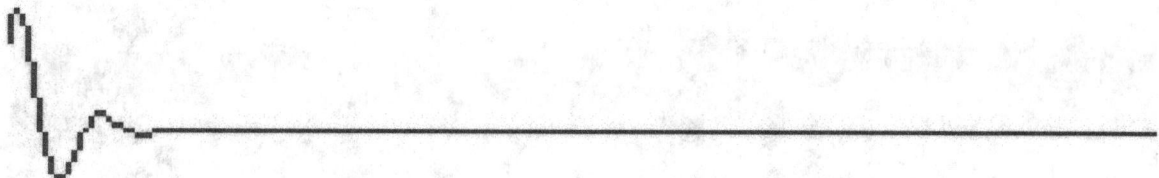

A wave function that satisfies the non-relativistic Schrödinger equation with $V = 0$. *In other words, this corresponds to a particle traveling freely through empty space. The real part of the wave function is plotted here.*

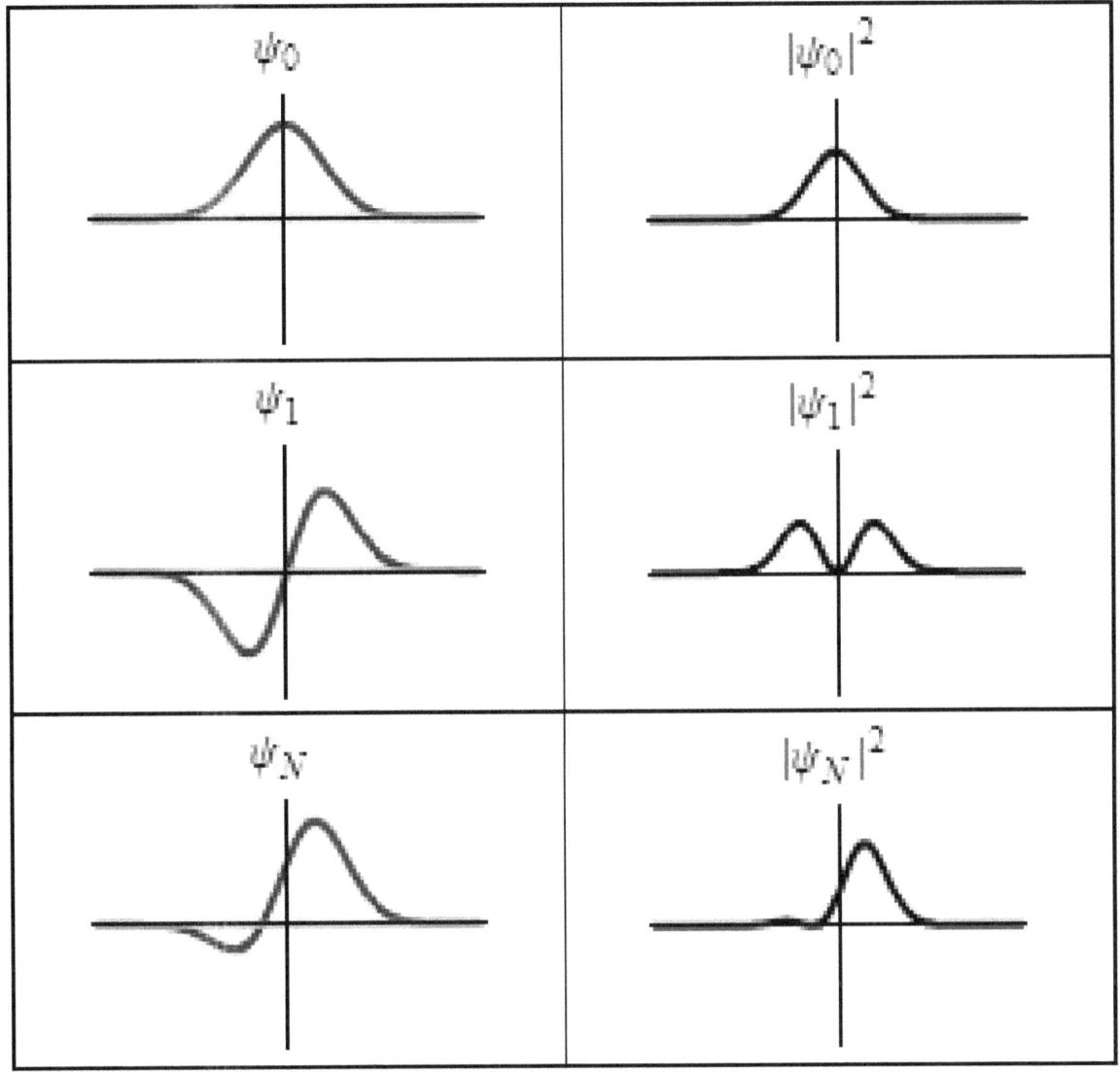

Each of these three rows is a wave function which satisfies the time-dependent Schrödinger equation for a harmonic oscillator. Left: The real part (blue) and imaginary part (red) of the wave function. Right: The probability distribution of finding the particle with this wave function at a given position. The top two rows are examples of **stationary states**, which correspond to standing waves. The bottom row is an example of a state which is not a stationary state. The right column illustrates why stationary states are called "stationary".

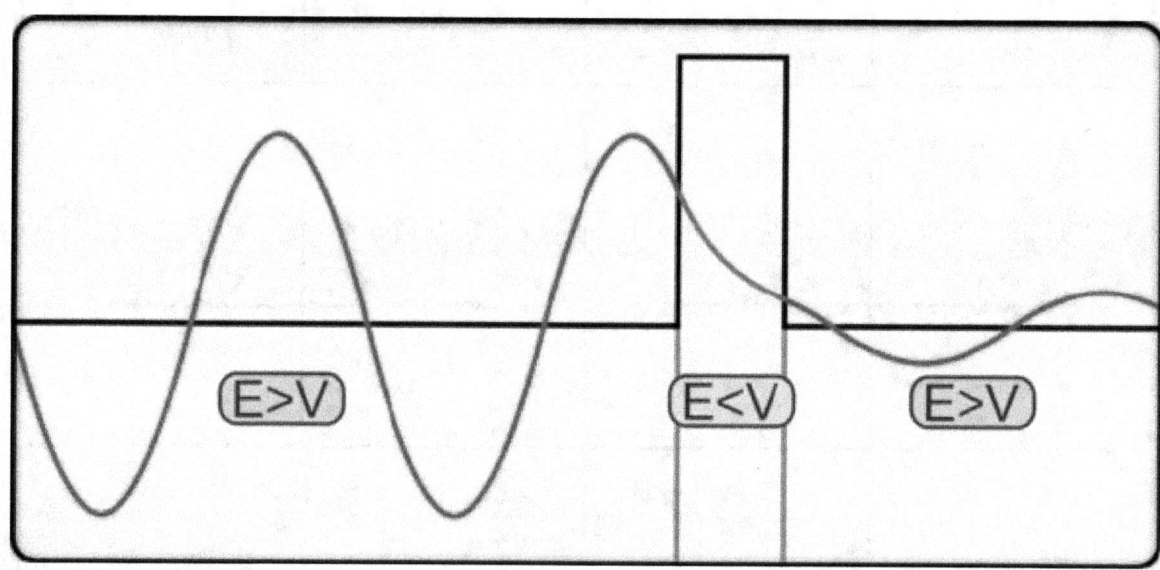

Quantum tunneling through a barrier. A particle coming from the left does not have enough energy to climb the barrier. However, it can sometimes "tunnel" to the other side.

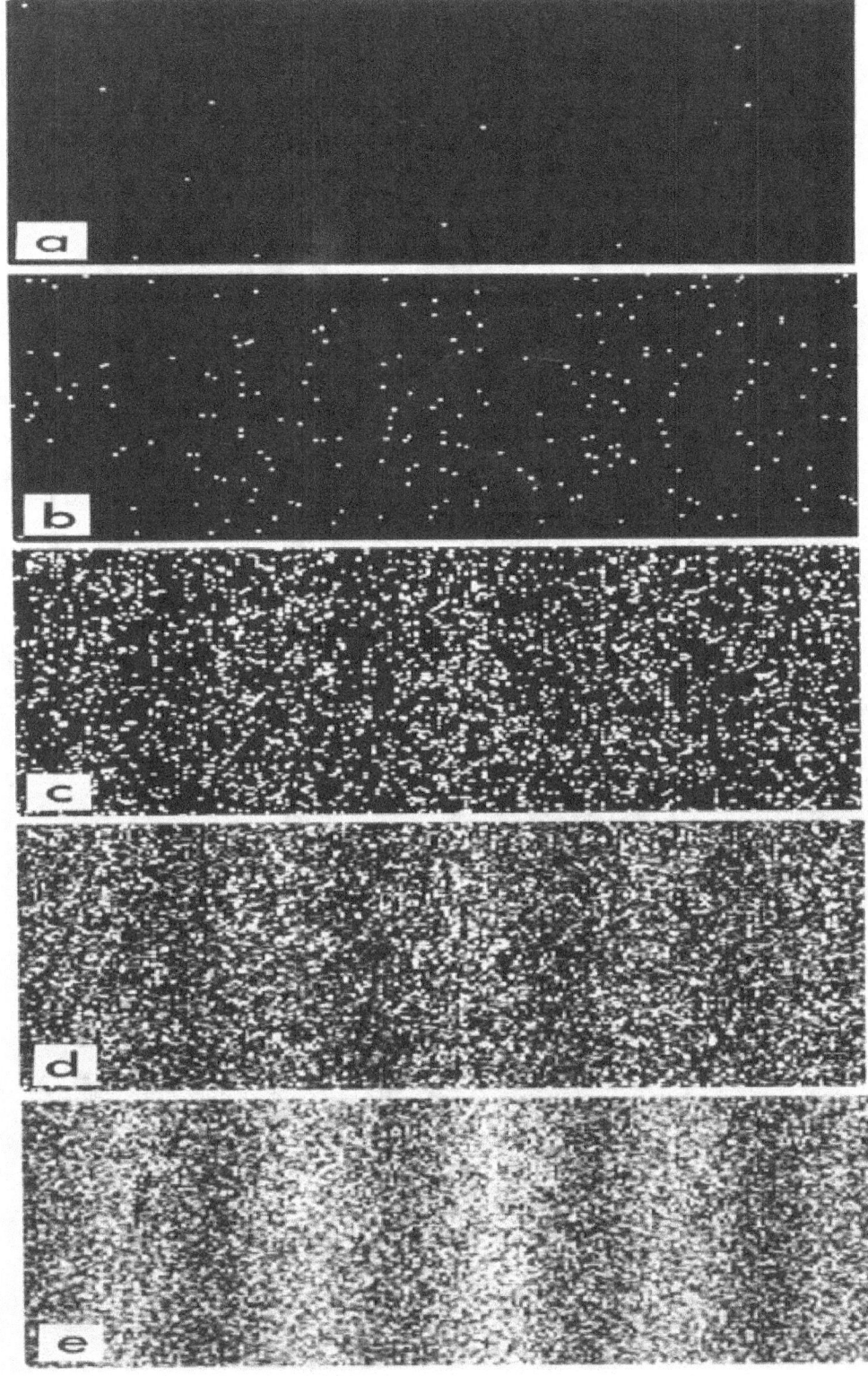

A double slit experiment showing the accumulation of electrons on a screen as time passes.

Erwin Schrödinger

$$-\frac{\hbar^2}{2m}\frac{\partial^2\Psi}{\partial x^2}+V\Psi=E\Psi$$

	$E-V(x)<0$	$E-V(x)>0$
$\Psi>0$	$\left[E-V(x)\right]\Psi<0$ $\dfrac{\partial^2\Psi}{\partial x^2}>0 \quad T<0$	$\left[E-V(x)\right]\Psi>0$ $\dfrac{\partial^2\Psi}{\partial x^2}<0 \quad T>0$
$\Psi<0$	$\left[E-V(x)\right]\Psi>0$ $\dfrac{\partial^2\Psi}{\partial x^2}<0 \quad T>0$	$\left[E-V(x)\right]\Psi<0$ $\dfrac{\partial^2\Psi}{\partial x^2}>0 \quad T<0$

Diagrammatic summary of the quantities related to the wavefunction, as used in De broglie's hypothesis and development of the Schrödinger equation.[24]

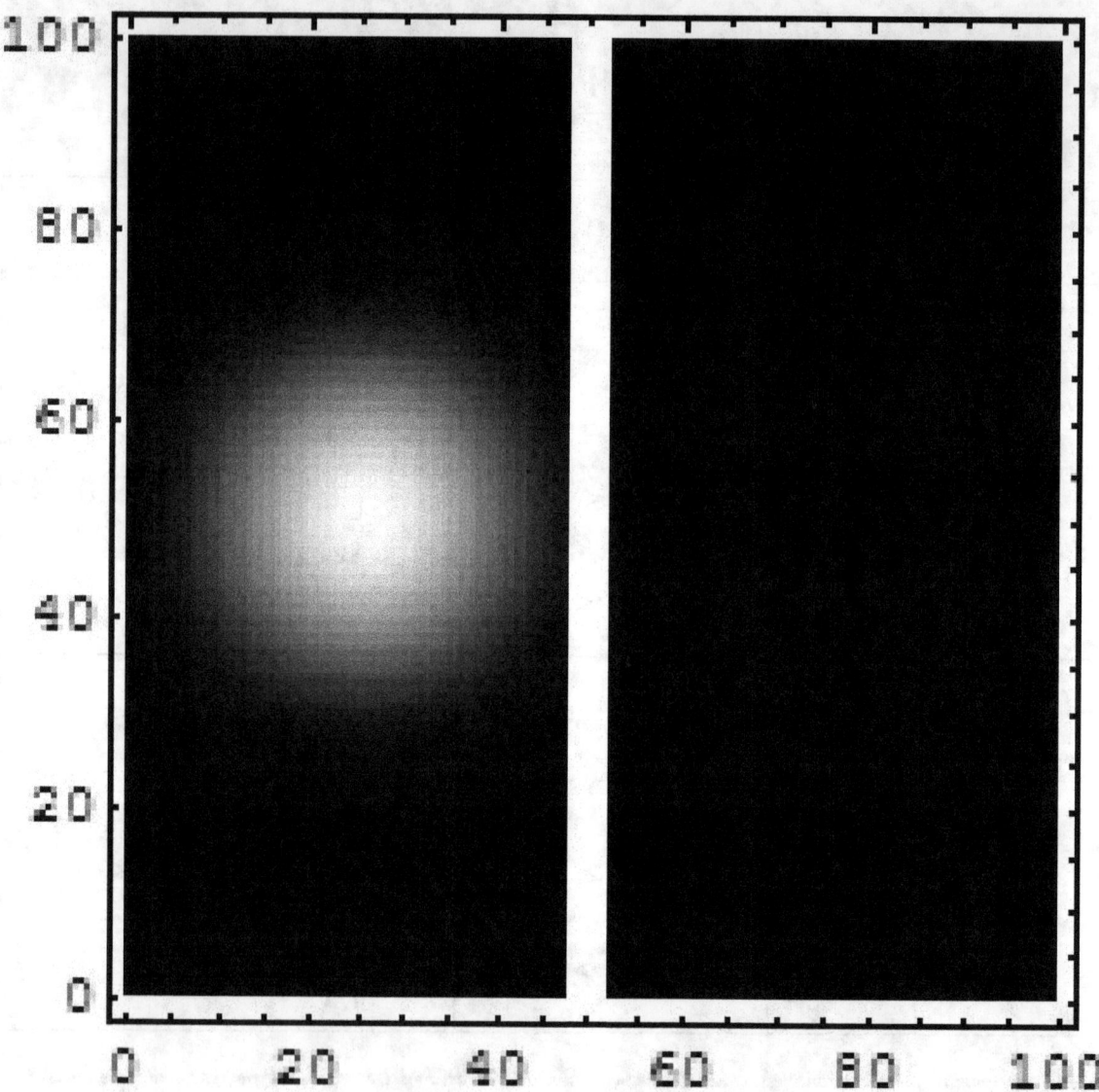

Animation of a de Broglie wave incident on a barrier.

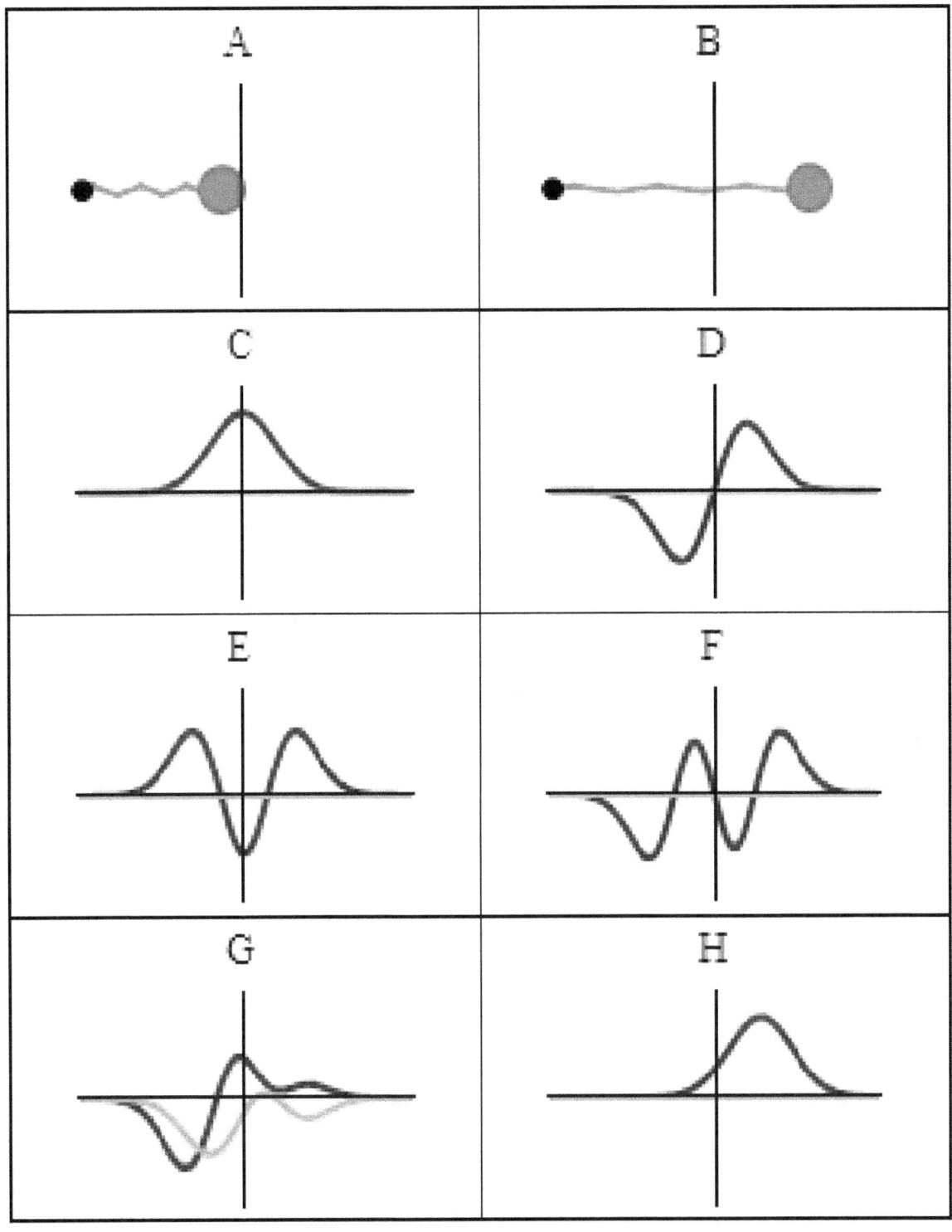

A harmonic oscillator in classical mechanics (A–B) and quantum mechanics (C–H). In (A–B), a ball, attached to a spring, oscillates back and forth. (C–H) are six solutions to the Schrödinger Equation for this situation. The horizontal axis is position, the vertical axis is the real part (blue) or imaginary part (red) of the wavefunction. Stationary states, or energy eigenstates, which are solutions to the time-independent Schrödinger Equation, are shown in C,D,E,F, but not G or H.

Continuously differentiable

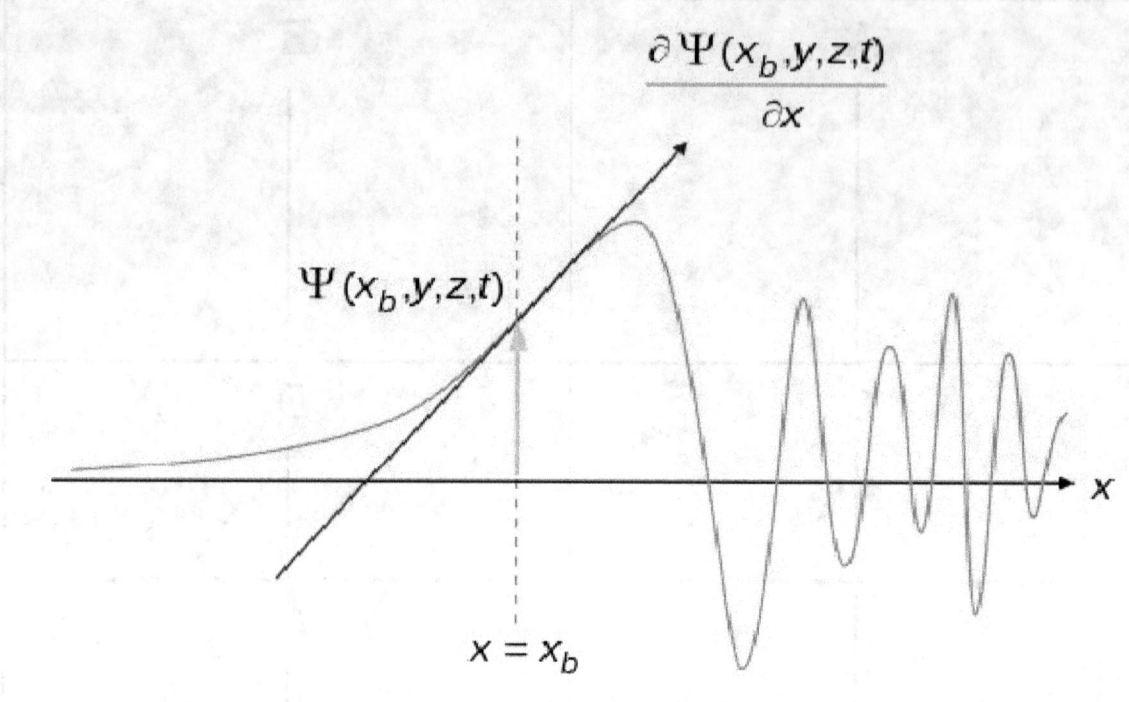

Discontinuous

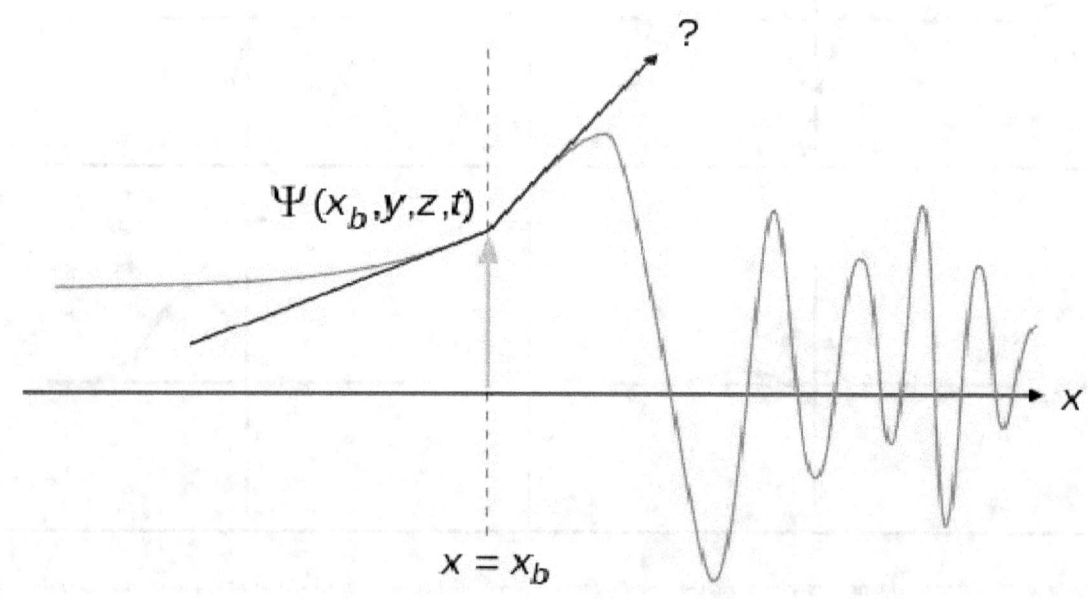

Continuity of the wavefunction and its first spatial derivative (in the x direction, y and z coordinates not shown), at some time t.

Chapter 15

Quantum field theory

"Relativistic quantum field theory" redirects here. For other uses, see Relativity.

In theoretical physics, **quantum field theory** (**QFT**) is a theoretical framework for constructing quantum mechanical models of subatomic particles in particle physics and quasiparticles in condensed matter physics. A QFT treats particles as excited states of an underlying physical field, so these are called field quanta.

In quantum field theory, quantum mechanical interactions between particles are described by interaction terms between the corresponding underlying quantum fields.

15.1 Definition

Quantum electrodynamics (QED) has one electron field and one photon field; quantum chromodynamics (QCD) has one field for each type of quark; and, in condensed matter, there is an atomic displacement field that gives rise to phonon particles. Edward Witten describes QFT as "by far" the most difficult theory in modern physics.[1]

15.1.1 Dynamics

See also: Relativistic dynamics

Ordinary quantum mechanical systems have a fixed number of particles, with each particle having a finite number of degrees of freedom. In contrast, the excited states of a QFT can represent any number of particles. This makes quantum field theories especially useful for describing systems where the particle count/number may change over time, a crucial feature of relativistic dynamics.

15.1.2 States

QFT interaction terms are similar in spirit to those between charges with electric and magnetic fields in Maxwell's equations. However, unlike the classical fields of Maxwell's theory, fields in QFT generally exist in quantum superpositions of states and are subject to the laws of quantum mechanics.

Because the fields are continuous quantities over space, there exist excited states with arbitrarily large numbers of particles in them, providing QFT systems with an effectively infinite number of degrees of freedom. Infinite degrees of freedom can easily lead to divergences of calculated quantities (e.g., the quantities become infinite). Techniques such as renormalization of QFT parameters or discretization of spacetime, as in lattice QCD, are often used to avoid such infinities so as to yield physically meaningful results.

15.1.3 Fields and radiation

The gravitational field and the electromagnetic field are the only two fundamental fields in nature that have infinite range and a corresponding classical low-energy limit, which greatly diminishes and hides their "particle-like" excitations. Albert Einstein in 1905, attributed "particle-like" and discrete exchanges of momenta and energy, characteristic of "field quanta", to the electromagnetic field. Originally, his principal motivation was to explain the thermodynamics of radiation. Although the photoelectric effect and Compton scattering strongly suggest the existence of the photon, it might alternately be explained by a mere quantization of emission; more definitive evidence of the quantum nature of radiation is now taken up into modern quantum optics as in the antibunching effect.[2]

15.2 Theories

There is currently no complete quantum theory of the remaining fundamental force, gravity. Many of the proposed theories to describe gravity as a QFT postulate the existence of a graviton particle that mediates the gravitational force. Presumably, the as yet unknown correct quantum field-theoretic treatment of the gravitational field will behave like Einstein's general theory of relativity in the low-energy limit. Quantum field theory of the fundamental forces itself has been postulated to be the low-energy effective field theory limit of a more fundamental theory such as superstring theory.

Most theories in standard particle physics are formulated as **relativistic quantum field theories**, such as QED, QCD, and the Standard Model. QED, the quantum field-theoretic description of the electromagnetic field, approximately reproduces Maxwell's theory of electrodynamics in the low-energy limit, with small non-linear corrections to the Maxwell equations required due to virtual electron–positron pairs.

In the perturbative approach to quantum field theory, the full field interaction terms are approximated as a perturbative expansion in the number of particles involved. Each term in the expansion can be thought of as forces between particles being mediated by other particles. In QED, the electromagnetic force between two electrons is caused by an exchange of photons. Similarly, intermediate vector bosons mediate the weak force and gluons mediate the strong force in QCD. The notion of a force-mediating particle comes from perturbation theory, and does not make sense in the context of non-perturbative approaches to QFT, such as with bound states.

15.3 History

Main article: History of quantum field theory

15.3.1 Foundations

The early development of the field involved Dirac, Fock, Pauli, Heisenberg and Bogolyubov. This phase of development culminated with the construction of the theory of quantum electrodynamics in the 1950s.

15.3.2 Gauge theory

Gauge theory was formulated and quantized, leading to the **unification of forces** embodied in the standard model of particle physics. This effort started in the 1950s with the work of Yang and Mills, was carried on by Martinus Veltman and a host of others during the 1960s and completed by the 1970s through the work of Gerard 't Hooft, Frank Wilczek, David Gross and David Politzer.

15.3.3   Grand synthesis

Parallel developments in the understanding of phase transitions in condensed matter physics led to the study of the renormalization group. This in turn led to the grand synthesis of theoretical physics, which unified theories of particle and condensed matter physics through quantum field theory. This involved the work of Michael Fisher and Leo Kadanoff in the 1970s, which led to the seminal reformulation of quantum field theory by Kenneth G. Wilson in 1975.

15.4   Principles

15.4.1   Classical and quantum fields

Main article: Classical field theory

A classical field is a function defined over some region of space and time.[3] Two physical phenomena which are described by classical fields are Newtonian gravitation, described by Newtonian gravitational field $g(x, t)$, and classical electromagnetism, described by the electric and magnetic fields $E(x, t)$ and $B(x, t)$. Because such fields can in principle take on distinct values at each point in space, they are said to have infinite degrees of freedom.[3]

Classical field theory does not, however, account for the quantum-mechanical aspects of such physical phenomena. For instance, it is known from quantum mechanics that certain aspects of electromagnetism involve discrete particles—photons—rather than continuous fields. The business of *quantum* field theory is to write down a field that is, like a classical field, a function defined over space and time, but which also accommodates the observations of quantum mechanics. This is a *quantum field*.

It is not immediately clear *how* to write down such a quantum field, since quantum mechanics has a structure very unlike a field theory. In its most general formulation, quantum mechanics is a theory of abstract operators (observables) acting on an abstract state space (Hilbert space), where the observables represent physically observable quantities and the state space represents the possible states of the system under study.[4] For instance, the fundamental observables associated with the motion of a single quantum mechanical particle are the position and momentum operators $\hat{x}$ and $\hat{p}$. Field theory, in contrast, treats x as a way to index the field rather than as an operator.[5]

There are two common ways of developing a quantum field: the path integral formalism and canonical quantization.[6] The latter of these is pursued in this article.

Lagrangian formalism

Quantum field theory frequently makes use of the Lagrangian formalism from classical field theory. This formalism is analogous to the Lagrangian formalism used in classical mechanics to solve for the motion of a particle under the influence of a field. In classical field theory, one writes down a Lagrangian density, $\mathcal{L}$, involving a field, $\varphi(x,t)$, and possibly its first derivatives ($\partial\varphi/\partial t$ and $\nabla\varphi$), and then applies a field-theoretic form of the Euler–Lagrange equation. Writing coordinates $(t, x) = (x^0, x^1, x^2, x^3) = x^\mu$, this form of the Euler–Lagrange equation is[3]

$$\frac{\partial}{\partial x^\mu}\left[\frac{\partial\mathcal{L}}{\partial(\partial\phi/\partial x^\mu)}\right] - \frac{\partial\mathcal{L}}{\partial\phi} = 0,$$

where a sum over μ is performed according to the rules of Einstein notation.

By solving this equation, one arrives at the "equations of motion" of the field.[3] For example, if one begins with the Lagrangian density

$$\mathcal{L}(\phi, \nabla\phi) = -\rho(t, x)\,\phi(t, x) - \frac{1}{8\pi G}|\nabla\phi|^2,$$

and then applies the Euler–Lagrange equation, one obtains the equation of motion

$$4\pi G\rho(t,\mathbf{x}) = \nabla^2\phi.$$

This equation is Newton's law of universal gravitation, expressed in differential form in terms of the gravitational potential $\varphi(t,\mathbf{x})$ and the mass density $\rho(t,\mathbf{x})$. Despite the nomenclature, the "field" under study is the gravitational potential, φ, rather than the gravitational field, $\mathbf{g}$. Similarly, when classical field theory is used to study electromagnetism, the "field" of interest is the electromagnetic four-potential (V/c, $\mathbf{A}$), rather than the electric and magnetic fields $\mathbf{E}$ and $\mathbf{B}$.

Quantum field theory uses this same Lagrangian procedure to determine the equations of motion for quantum fields. These equations of motion are then supplemented by commutation relations derived from the canonical quantization procedure described below, thereby incorporating quantum mechanical effects into the behavior of the field.

15.4.2 Single- and many-particle quantum mechanics

Main articles: Quantum mechanics and First quantization

In quantum mechanics, a particle (such as an electron or proton) is described by a complex wavefunction, $\psi(x,t)$, whose time-evolution is governed by the Schrödinger equation:

$$-\frac{\hbar^2}{2m}\frac{\partial^2}{\partial x^2}\psi(x,t) + V(x)\psi(x,t) = i\hbar\frac{\partial}{\partial t}\psi(x,t).$$

Here m is the particle's mass and $V(x)$ is the applied potential. Physical information about the behavior of the particle is extracted from the wavefunction by constructing expected values for various quantities; for example, the expected value of the particle's position is given by integrating $\psi^*(x)\, x\, \psi(x)$ over all space, and the expected value of the particle's momentum is found by integrating $-i\hbar\psi^*(x)d\psi/dx$. The quantity $\psi^*(x)\psi(x)$ is itself in the Copenhagen interpretation of quantum mechanics interpreted as a probability density function. This treatment of quantum mechanics, where a particle's wavefunction evolves against a classical background potential $V(x)$, is sometimes called *first quantization*.

This description of quantum mechanics can be extended to describe the behavior of multiple particles, so long as the number and the type of particles remain fixed. The particles are described by a wavefunction $\psi(x_1, x_2, \ldots, xN, t)$, which is governed by an extended version of the Schrödinger equation.

Often one is interested in the case where N particles are all of the same type (for example, the 18 electrons orbiting a neutral argon nucleus). As described in the article on identical particles, this implies that the state of the entire system must be either symmetric (bosons) or antisymmetric (fermions) when the coordinates of its constituent particles are exchanged. This is achieved by using a Slater determinant as the wavefunction of a fermionic system (and a Slater permanent for a bosonic system), which is equivalent to an element of the symmetric or antisymmetric subspace of a tensor product.

For example, the general quantum state of a system of N bosons is written as

$$|\phi_1 \cdots \phi_N\rangle = \sqrt{\frac{\prod_j N_j!}{N!}} \sum_{p \in S_N} |\phi_{p(1)}\rangle \otimes \cdots \otimes |\phi_{p(N)}\rangle,$$

where $|\phi_i\rangle$ are the single-particle states, Nj is the number of particles occupying state j, and the sum is taken over all possible permutations p acting on N elements. In general, this is a sum of $N!$ (N factorial) distinct terms. $\sqrt{\frac{\prod_j N_j!}{N!}}$ is a normalizing factor.

There are several shortcomings to the above description of quantum mechanics, which are addressed by quantum field theory. First, it is unclear how to extend quantum mechanics to include the effects of special relativity.[7] Attempted replacements for the Schrödinger equation, such as the Klein–Gordon equation or the Dirac equation, have many unsatisfactory qualities; for instance, they possess energy eigenvalues that extend to $-\infty$, so that there seems to be no easy definition of a ground state. It turns out that such inconsistencies arise from relativistic wavefunctions not having a well-defined probabilistic interpretation in position space, as probability conservation is not a relativistically covariant concept. The second shortcoming, related to the first, is that in quantum mechanics there is no mechanism to describe particle creation and annihilation;[8] this is crucial for describing phenomena such as pair production, which result from the conversion between mass and energy according to the relativistic relation $E = mc^2$.

15.4.3 Second quantization

Main article: Second quantization

In this section, we will describe a method for constructing a quantum field theory called **second quantization**. This basically involves choosing a way to index the quantum mechanical degrees of freedom in the space of multiple identical-particle states. It is based on the Hamiltonian formulation of quantum mechanics.

Several other approaches exist, such as the Feynman path integral,[9] which uses a Lagrangian formulation. For an overview of some of these approaches, see the article on quantization.

Bosons

For simplicity, we will first discuss second quantization for bosons, which form perfectly symmetric quantum states. Let us denote the mutually orthogonal single-particle states which are possible in the system by $|\phi_1\rangle, |\phi_2\rangle, |\phi_3\rangle$, and so on. For example, the 3-particle state with one particle in state $|\phi_1\rangle$ and two in state $|\phi_2\rangle$ is

$$\frac{1}{\sqrt{3}} \left[|\phi_1\rangle|\phi_2\rangle|\phi_2\rangle + |\phi_2\rangle|\phi_1\rangle|\phi_2\rangle + |\phi_2\rangle|\phi_2\rangle|\phi_1\rangle \right].$$

The first step in second quantization is to express such quantum states in terms of **occupation numbers**, by listing the number of particles occupying each of the single-particle states $|\phi_1\rangle, |\phi_2\rangle$, etc. This is simply another way of labelling the states. For instance, the above 3-particle state is denoted as

$$|1, 2, 0, 0, 0, \ldots\rangle.$$

An N-particle state belongs to a space of states describing systems of N particles. The next step is to combine the individual N-particle state spaces into an extended state space, known as Fock space, which can describe systems of any number of particles. This is composed of the state space of a system with no particles (the so-called vacuum state, written as $|0\rangle$), plus the state space of a 1-particle system, plus the state space of a 2-particle system, and so forth. States describing a definite number of particles are known as Fock states: a general element of Fock space will be a linear combination of Fock states. There is a one-to-one correspondence between the occupation number representation and valid boson states in the Fock space.

At this point, the quantum mechanical system has become a quantum field in the sense we described above. The field's elementary degrees of freedom are the occupation numbers, and each occupation number is indexed by a number j indicating which of the single-particle states $|\phi_1\rangle, |\phi_2\rangle, \ldots, |\phi_j\rangle, \ldots$ it refers to:

$$|N_1, N_2, N_3, \ldots, N_j, \ldots\rangle.$$

The properties of this quantum field can be explored by defining creation and annihilation operators, which add and subtract particles. They are analogous to ladder operators in the quantum harmonic oscillator problem, which added and subtracted energy quanta. However, these operators literally create and annihilate particles of a given quantum state. The bosonic annihilation operator a_2 and creation operator $a_2^\dagger$ are easily defined in the occupation number representation as having the following effects:

$$a_2 | N_1, N_2, N_3, \ldots \rangle = \sqrt{N_2} \, | \, N_1, (N_2 - 1), N_3, \ldots \rangle,$$

$$a_2^\dagger | N_1, N_2, N_3, \ldots \rangle = \sqrt{N_2 + 1} \, | \, N_1, (N_2 + 1), N_3, \ldots \rangle.$$

It can be shown that these are operators in the usual quantum mechanical sense, i.e. linear operators acting on the Fock space. Furthermore, they are indeed Hermitian conjugates, which justifies the way we have written them. They can be shown to obey the commutation relation

$$\left[a_i, a_j \right] = 0 \quad , \quad \left[a_i^\dagger, a_j^\dagger \right] = 0 \quad , \quad \left[a_i, a_j^\dagger \right] = \delta_{ij},$$

where δ stands for the Kronecker delta. These are precisely the relations obeyed by the ladder operators for an infinite set of independent quantum harmonic oscillators, one for each single-particle state. Adding or removing bosons from each state is therefore analogous to exciting or de-exciting a quantum of energy in a harmonic oscillator.

Applying an annihilation operator a_k followed by its corresponding creation operator $a_k^\dagger$ returns the number N_k of particles in the kth single-particle eigenstate:

$$a_k^\dagger a_k | \ldots, N_k, \ldots \rangle = N_k | \ldots, N_k, \ldots \rangle.$$

The combination of operators $a_k^\dagger a_k$ is known as the number operator for the kth eigenstate.

The Hamiltonian operator of the quantum field (which, through the Schrödinger equation, determines its dynamics) can be written in terms of creation and annihilation operators. For instance, for a field of free (non-interacting) bosons, the total energy of the field is found by summing the energies of the bosons in each energy eigenstate. If the kth single-particle energy eigenstate has energy E_k and there are N_k bosons in this state, then the total energy of these bosons is $E_k N_k$. The energy in the *entire* field is then a sum over k :

$$E_{\text{tot}} = \sum_k E_k N_k$$

This can be turned into the Hamiltonian operator of the field by replacing N_k with the corresponding number operator, $a_k^\dagger a_k$. This yields

$$H = \sum_k E_k \, a_k^\dagger a_k.$$

Fermions

It turns out that a different definition of creation and annihilation must be used for describing fermions. According to the Pauli exclusion principle, fermions cannot share quantum states, so their occupation numbers Ni can only take on the value 0 or 1. The fermionic annihilation operators c and creation operators $c^\dagger$ are defined by their actions on a Fock state thus

$$c_j | N_1, N_2, \ldots, N_j = 0, \ldots \rangle = 0$$

$$c_j | N_1, N_2, \ldots, N_j = 1, \ldots \rangle = (-1)^{(N_1 + \cdots + N_{j-1})} | N_1, N_2, \ldots, N_j = 0, \ldots \rangle$$

$$c_j^\dagger | N_1, N_2, \ldots, N_j = 0, \ldots \rangle = (-1)^{(N_1 + \cdots + N_{j-1})} | N_1, N_2, \ldots, N_j = 1, \ldots \rangle$$

$c_j^\dagger |N_1, N_2, \ldots, N_j = 1, \ldots \rangle = 0.$

These obey an anticommutation relation:

$$\{c_i, c_j\} = 0 \quad , \quad \left\{c_i^\dagger, c_j^\dagger\right\} = 0 \quad , \quad \left\{c_i, c_j^\dagger\right\} = \delta_{ij}.$$

One may notice from this that applying a fermionic creation operator twice gives zero, so it is impossible for the particles to share single-particle states, in accordance with the exclusion principle.

Field operators

We have previously mentioned that there can be more than one way of indexing the degrees of freedom in a quantum field. Second quantization indexes the field by enumerating the single-particle quantum states. However, as we have discussed, it is more natural to think about a "field", such as the electromagnetic field, as a set of degrees of freedom indexed by position.

To this end, we can define *field operators* that create or destroy a particle at a particular point in space. In particle physics, these operators turn out to be more convenient to work with, because they make it easier to formulate theories that satisfy the demands of relativity.

Single-particle states are usually enumerated in terms of their momenta (as in the particle in a box problem.) We can construct field operators by applying the Fourier transform to the creation and annihilation operators for these states. For example, the bosonic field annihilation operator $\phi(\mathbf{r})$ is

$$\phi(\mathbf{r}) \overset{\text{def}}{=} \sum_j e^{i\mathbf{k}_j \cdot \mathbf{r}} a_j.$$

The bosonic field operators obey the commutation relation

$$[\phi(\mathbf{r}), \phi(\mathbf{r}')] = 0 \quad , \quad [\phi^\dagger(\mathbf{r}), \phi^\dagger(\mathbf{r}')] = 0 \quad , \quad [\phi(\mathbf{r}), \phi^\dagger(\mathbf{r}')] = \delta^3(\mathbf{r} - \mathbf{r}')$$

where $\delta(x)$ stands for the Dirac delta function. As before, the fermionic relations are the same, with the commutators replaced by anticommutators.

The field operator is not the same thing as a single-particle wavefunction. The former is an operator acting on the Fock space, and the latter is a quantum-mechanical amplitude for finding a particle in some position. However, they are closely related, and are indeed commonly denoted with the same symbol. If we have a Hamiltonian with a space representation, say

$$H = -\frac{\hbar^2}{2m} \sum_i \nabla_i^2 + \sum_{i<j} U(|\mathbf{r}_i - \mathbf{r}_j|)$$

where the indices i and j run over all particles, then the field theory Hamiltonian (in the non-relativistic limit and for negligible self-interactions) is

$$H = -\frac{\hbar^2}{2m} \int d^3r \, \phi^\dagger(\mathbf{r}) \nabla^2 \phi(\mathbf{r}) + \frac{1}{2} \int d^3r \int d^3r' \, \phi^\dagger(\mathbf{r}) \phi^\dagger(\mathbf{r}') U(|\mathbf{r} - \mathbf{r}'|) \phi(\mathbf{r}') \phi(\mathbf{r}).$$

This looks remarkably like an expression for the expectation value of the energy, with ϕ playing the role of the wavefunction. This relationship between the field operators and wavefunctions makes it very easy to formulate field theories starting from space-projected Hamiltonians.

15.4.4   Dynamics

Once the Hamiltonian operator is obtained as part of the canonical quantization process, the time dependence of the state is described with the Schrödinger equation, just as with other quantum theories. Alternatively, the Heisenberg picture can be used where the time dependence is in the operators rather than in the states.

15.4.5   Implications

Unification of fields and particles

The "second quantization" procedure that we have outlined in the previous section takes a set of single-particle quantum states as a starting point. Sometimes, it is impossible to define such single-particle states, and one must proceed directly to quantum field theory. For example, a quantum theory of the electromagnetic field *must* be a quantum field theory, because it is impossible (for various reasons) to define a wavefunction for a single photon.[10] In such situations, the quantum field theory can be constructed by examining the mechanical properties of the classical field and guessing the corresponding quantum theory. For free (non-interacting) quantum fields, the quantum field theories obtained in this way have the same properties as those obtained using second quantization, such as well-defined creation and annihilation operators obeying commutation or anticommutation relations.

Quantum field theory thus provides a unified framework for describing "field-like" objects (such as the electromagnetic field, whose excitations are photons) and "particle-like" objects (such as electrons, which are treated as excitations of an underlying electron field), so long as one can treat interactions as "perturbations" of free fields. There are still unsolved problems relating to the more general case of interacting fields that may or may not be adequately described by perturbation theory. For more on this topic, see Haag's theorem.

Physical meaning of particle indistinguishability

The second quantization procedure relies crucially on the particles being identical. We would not have been able to construct a quantum field theory from a distinguishable many-particle system, because there would have been no way of separating and indexing the degrees of freedom.

Many physicists prefer to take the converse interpretation, which is that *quantum field theory explains what identical particles are*. In ordinary quantum mechanics, there is not much theoretical motivation for using symmetric (bosonic) or antisymmetric (fermionic) states, and the need for such states is simply regarded as an empirical fact. From the point of view of quantum field theory, particles are identical if and only if they are excitations of the same underlying quantum field. Thus, the question "why are all electrons identical?" arises from mistakenly regarding individual electrons as fundamental objects, when in fact it is only the electron field that is fundamental.

Particle conservation and non-conservation

During second quantization, we started with a Hamiltonian and state space describing a fixed number of particles (N), and ended with a Hamiltonian and state space for an arbitrary number of particles. Of course, in many common situations N is an important and perfectly well-defined quantity, e.g. if we are describing a gas of atoms sealed in a box. From the point of view of quantum field theory, such situations are described by quantum states that are eigenstates of the number operator $\hat{N}$, which measures the total number of particles present. As with any quantum mechanical observable, $\hat{N}$ is conserved if it commutes with the Hamiltonian. In that case, the quantum state is trapped in the N-particle subspace of the total Fock space, and the situation could equally well be described by ordinary N-particle quantum mechanics. (Strictly speaking, this is only true in the noninteracting case or in the low energy density limit of renormalized quantum field theories)

For example, we can see that the free-boson Hamiltonian described above conserves particle number. Whenever the Hamiltonian operates on a state, each particle destroyed by an annihilation operator a_k is immediately put back by the creation operator $a_k^\dagger$.

On the other hand, it is possible, and indeed common, to encounter quantum states that are *not* eigenstates of $\hat{N}$, which do not have well-defined particle numbers. Such states are difficult or impossible to handle using ordinary quantum mechanics, but they can be easily described in quantum field theory as quantum superpositions of states having different values of N. For example, suppose we have a bosonic field whose particles can be created or destroyed by interactions with a fermionic field. The Hamiltonian of the combined system would be given by the Hamiltonians of the free boson and free fermion fields, plus a "potential energy" term such as

$$H_I = \sum_{k,q} V_q (a_q + a^{\dagger}_{-q}) c^{\dagger}_{k+q} c_k,$$

where $a^{\dagger}_k$ and a_k denotes the bosonic creation and annihilation operators, $c^{\dagger}_k$ and c_k denotes the fermionic creation and annihilation operators, and V_q is a parameter that describes the strength of the interaction. This "interaction term" describes processes in which a fermion in state k either absorbs or emits a boson, thereby being kicked into a different eigenstate $k + q$. (In fact, this type of Hamiltonian is used to describe interaction between conduction electrons and phonons in metals. The interaction between electrons and photons is treated in a similar way, but is a little more complicated because the role of spin must be taken into account.) One thing to notice here is that even if we start out with a fixed number of bosons, we will typically end up with a superposition of states with different numbers of bosons at later times. The number of fermions, however, is conserved in this case.

In condensed matter physics, states with ill-defined particle numbers are particularly important for describing the various superfluids. Many of the defining characteristics of a superfluid arise from the notion that its quantum state is a superposition of states with different particle numbers. In addition, the concept of a coherent state (used to model the laser and the BCS ground state) refers to a state with an ill-defined particle number but a well-defined phase.

15.4.6   Axiomatic approaches

The preceding description of quantum field theory follows the spirit in which most physicists approach the subject. However, it is not mathematically rigorous. Over the past several decades, there have been many attempts to put quantum field theory on a firm mathematical footing by formulating a set of axioms for it. These attempts fall into two broad classes.

The first class of axioms, first proposed during the 1950s, include the Wightman, Osterwalder–Schrader, and Haag–Kastler systems. They attempted to formalize the physicists' notion of an "operator-valued field" within the context of functional analysis, and enjoyed limited success. It was possible to prove that any quantum field theory satisfying these axioms satisfied certain general theorems, such as the spin-statistics theorem and the CPT theorem. Unfortunately, it proved extraordinarily difficult to show that any realistic field theory, including the Standard Model, satisfied these axioms. Most of the theories that could be treated with these analytic axioms were physically trivial, being restricted to low-dimensions and lacking interesting dynamics. The construction of theories satisfying one of these sets of axioms falls in the field of constructive quantum field theory. Important work was done in this area in the 1970s by Segal, Glimm, Jaffe and others.

During the 1980s, a second set of axioms based on geometric ideas was proposed. This line of investigation, which restricts its attention to a particular class of quantum field theories known as topological quantum field theories, is associated most closely with Michael Atiyah and Graeme Segal, and was notably expanded upon by Edward Witten, Richard Borcherds, and Maxim Kontsevich. However, most of the physically relevant quantum field theories, such as the Standard Model, are not topological quantum field theories; the quantum field theory of the fractional quantum Hall effect is a notable exception. The main impact of axiomatic topological quantum field theory has been on mathematics, with important applications in representation theory, algebraic topology, and differential geometry.

Finding the proper axioms for quantum field theory is still an open and difficult problem in mathematics. One of the Millennium Prize Problems—proving the existence of a mass gap in Yang–Mills theory—is linked to this issue.

15.5   Associated phenomena

In the previous part of the article, we described the most general features of quantum field theories. Some of the quantum field theories studied in various fields of theoretical physics involve additional special ideas, such as renormalizability, gauge symmetry, and supersymmetry. These are described in the following sections.

15.5.1   Renormalization

Main article: Renormalization

Early in the history of quantum field theory, it was found that many seemingly innocuous calculations, such as the perturbative shift in the energy of an electron due to the presence of the electromagnetic field, give infinite results. The reason is that the perturbation theory for the shift in an energy involves a sum over all other energy levels, and there are infinitely many levels at short distances that each give a finite contribution which results in a divergent series.

Many of these problems are related to failures in classical electrodynamics that were identified but unsolved in the 19th century, and they basically stem from the fact that many of the supposedly "intrinsic" properties of an electron are tied to the electromagnetic field that it carries around with it. The energy carried by a single electron—its self energy—is not simply the bare value, but also includes the energy contained in its electromagnetic field, its attendant cloud of photons. The energy in a field of a spherical source diverges in both classical and quantum mechanics, but as discovered by Weisskopf with help from Furry, in quantum mechanics the divergence is much milder, going only as the logarithm of the radius of the sphere.

The solution to the problem, presciently suggested by Stueckelberg, independently by Bethe after the crucial experiment by Lamb, implemented at one loop by Schwinger, and systematically extended to all loops by Feynman and Dyson, with converging work by Tomonaga in isolated postwar Japan, comes from recognizing that all the infinities in the interactions of photons and electrons can be isolated into redefining a finite number of quantities in the equations by replacing them with the observed values: specifically the electron's mass and charge: this is called renormalization. The technique of renormalization recognizes that the problem is essentially purely mathematical, that extremely short distances are at fault. In order to define a theory on a continuum, first place a cutoff on the fields, by postulating that quanta cannot have energies above some extremely high value. This has the effect of replacing continuous space by a structure where very short wavelengths do not exist, as on a lattice. Lattices break rotational symmetry, and one of the crucial contributions made by Feynman, Pauli and Villars, and modernized by 't Hooft and Veltman, is a symmetry-preserving cutoff for perturbation theory (this process is called regularization). There is no known symmetrical cutoff outside of perturbation theory, so for rigorous or numerical work people often use an actual lattice.

On a lattice, every quantity is finite but depends on the spacing. When taking the limit of zero spacing, we make sure that the physically observable quantities like the observed electron mass stay fixed, which means that the constants in the Lagrangian defining the theory depend on the spacing. Hopefully, by allowing the constants to vary with the lattice spacing, all the results at long distances become insensitive to the lattice, defining a continuum limit.

The renormalization procedure only works for a certain class of quantum field theories, called **renormalizable quantum field theories**. A theory is **perturbatively renormalizable** when the constants in the Lagrangian only diverge at worst as logarithms of the lattice spacing for very short spacings. The continuum limit is then well defined in perturbation theory, and even if it is not fully well defined non-perturbatively, the problems only show up at distance scales that are exponentially small in the inverse coupling for weak couplings. The Standard Model of particle physics is perturbatively renormalizable, and so are its component theories (quantum electrodynamics/electroweak theory and quantum chromodynamics). Of the three components, quantum electrodynamics is believed to not have a continuum limit, while the asymptotically free $SU(2)$ and $SU(3)$ weak hypercharge and strong color interactions are nonperturbatively well defined.

The renormalization group describes how renormalizable theories emerge as the long distance low-energy effective field theory for any given high-energy theory. Because of this, renormalizable theories are insensitive to the precise nature of the underlying high-energy short-distance phenomena. This is a blessing because it allows physicists to formulate low energy theories without knowing the details of high energy phenomenon. It is also a curse, because once a renormalizable theory like the standard model is found to work, it gives very few clues to higher energy processes. The only way high energy processes can be seen in the standard model is when they allow otherwise forbidden events, or if they predict quantitative relations between the coupling constants.

15.5.2 Haag's theorem

See also: Haag's theorem

From a mathematically rigorous perspective, there exists no interaction picture in a Lorentz-covariant quantum field theory. This implies that the perturbative approach of Feynman diagrams in QFT is not strictly justified, despite producing vastly precise predictions validated by experiment. This is called Haag's theorem, but most particle physicists relying on QFT largely shrug it off.

15.5.3 Gauge freedom

A gauge theory is a theory that admits a symmetry with a local parameter. For example, in every quantum theory the global phase of the wave function is arbitrary and does not represent something physical. Consequently, the theory is invariant under a global change of phases (adding a constant to the phase of all wave functions, everywhere); this is a global symmetry. In quantum electrodynamics, the theory is also invariant under a *local* change of phase, that is – one may shift the phase of all wave functions so that the shift may be different at every point in space-time. This is a *local* symmetry. However, in order for a well-defined derivative operator to exist, one must introduce a new field, the gauge field, which also transforms in order for the local change of variables (the phase in our example) not to affect the derivative. In quantum electrodynamics this gauge field is the electromagnetic field. The change of local gauge of variables is termed gauge transformation. It is worth noting that by Noether's theorem, for every such symmetry there exists an associated conserved current. The aforementioned symmetry of the wavefunction under global phase changes implies the conservation of electric charge.

In quantum field theory the excitations of fields represent particles. The particle associated with excitations of the gauge field is the gauge boson, which is the photon in the case of quantum electrodynamics.

The degrees of freedom in quantum field theory are local fluctuations of the fields. The existence of a gauge symmetry reduces the number of degrees of freedom, simply because some fluctuations of the fields can be transformed to zero by gauge transformations, so they are equivalent to having no fluctuations at all, and they therefore have no physical meaning. Such fluctuations are usually called "non-physical degrees of freedom" or *gauge artifacts*; usually some of them have a negative norm, making them inadequate for a consistent theory. Therefore, if a classical field theory has a gauge symmetry, then its quantized version (i.e. the corresponding quantum field theory) will have this symmetry as well. In other words, a gauge symmetry cannot have a quantum anomaly. If a gauge symmetry is anomalous (i.e. not kept in the quantum theory) then the theory is non-consistent: for example, in quantum electrodynamics, had there been a gauge anomaly, this would require the appearance of photons with longitudinal polarization and polarization in the time direction, the latter having a negative norm, rendering the theory inconsistent; another possibility would be for these photons to appear only in intermediate processes but not in the final products of any interaction, making the theory non-unitary and again inconsistent (see optical theorem).

In general, the gauge transformations of a theory consist of several different transformations, which may not be commutative. These transformations are together described by a mathematical object known as a gauge group. Infinitesimal gauge transformations are the gauge group generators. Therefore, the number of gauge bosons is the group dimension (i.e. number of generators forming a basis).

All the fundamental interactions in nature are described by gauge theories. These are:

- Quantum chromodynamics, whose gauge group is $SU(3)$. The gauge bosons are eight gluons.

- The electroweak theory, whose gauge group is $U(1) \times SU(2)$, (a direct product of $U(1)$ and $SU(2)$).

- Gravity, whose classical theory is general relativity, admits the equivalence principle, which is a form of gauge symmetry. However, it is explicitly non-renormalizable.

15.5.4   Multivalued gauge transformations

The gauge transformations which leave the theory invariant involve, by definition, only single-valued gauge functions $\Lambda(x_i)$ which satisfy the Schwarz integrability criterion

$$\partial_{x_i x_j}\Lambda = \partial_{x_j x_i}\Lambda.$$

An interesting extension of gauge transformations arises if the gauge functions $\Lambda(x_i)$ are allowed to be multivalued functions which violate the integrability criterion. These are capable of changing the physical field strengths and are therefore not proper symmetry transformations. Nevertheless, the transformed field equations describe correctly the physical laws in the presence of the newly generated field strengths. See the textbook by H. Kleinert cited below for the applications to phenomena in physics.

15.5.5   Supersymmetry

Main article: Supersymmetry

Supersymmetry assumes that every fundamental fermion has a superpartner that is a boson and vice versa. It was introduced in order to solve the so-called Hierarchy Problem, that is, to explain why particles not protected by any symmetry (like the Higgs boson) do not receive radiative corrections to its mass driving it to the larger scales (GUT, Planck...). It was soon realized that supersymmetry has other interesting properties: its gauged version is an extension of general relativity (Supergravity), and it is a key ingredient for the consistency of string theory.

The way supersymmetry protects the hierarchies is the following: since for every particle there is a superpartner with the same mass, any loop in a radiative correction is cancelled by the loop corresponding to its superpartner, rendering the theory UV finite.

Since no superpartners have yet been observed, if supersymmetry exists it must be broken (through a so-called soft term, which breaks supersymmetry without ruining its helpful features). The simplest models of this breaking require that the energy of the superpartners not be too high; in these cases, supersymmetry is expected to be observed by experiments at the Large Hadron Collider. The Higgs particle has been detected at the LHC, and no such superparticles have been discovered.

15.6   See also

- Abraham–Lorentz force

- Basic concepts of quantum mechanics

- Common integrals in quantum field theory

- Einstein–Maxwell–Dirac equations

- Form factor (quantum field theory)

- Green–Kubo relations

- Green's function (many-body theory)

- Invariance mechanics

- List of quantum field theories

- Quantization of a field

- Quantum electrodynamics

- Quantum field theory in curved spacetime

- Quantum flavordynamics

- Quantum hydrodynamics

- Quantum triviality

- Relation between Schrödinger's equation and the path integral formulation of quantum mechanics

- Relationship between string theory and quantum field theory

- Schwinger–Dyson equation

- Static forces and virtual-particle exchange

- Symmetry in quantum mechanics

- Theoretical and experimental justification for the Schrödinger equation

- Ward–Takahashi identity

- Wheeler–Feynman absorber theory

- Wigner's classification

- Wigner's theorem

15.7   Notes

15.8   References

[1] "Beautiful Minds, Vol. 20: Ed Witten". la Repubblica. 2010. Retrieved 22 June 2012. See here.

[2] J. J. Thorn et al. (2004). Observing the quantum behavior of light in an undergraduate laboratory. . J. J. Thorn, M. S. Neel, V. W. Donato, G. S. Bergreen, R. E. Davies, and M. Beck. American Association of Physics Teachers, 2004.DOI: 10.1119/1.1737397.

[3] David Tong, *Lectures on Quantum Field Theory*, chapter 1.

[4] Srednicki, Mark. *Quantum Field Theory* (1st ed.). p. 19.

[5] Srednicki, Mark. *Quantum Field Theory* (1st ed.). pp. 25–6.

[6] Zee, Anthony. *Quantum Field Theory in a Nutshell* (2nd ed.). p. 61.

[7] David Tong, *Lectures on Quantum Field Theory*, Introduction.

[8] Zee, Anthony. *Quantum Field Theory in a Nutshell* (2nd ed.). p. 3.

[9] Abraham Pais, *Inward Bound: Of Matter and Forces in the Physical World* ISBN 0-19-851997-4. Pais recounts how his astonishment at the rapidity with which Feynman could calculate using his method. Feynman's method is now part of the standard methods for physicists.

[10] Newton, T.D.; Wigner, E.P. (1949). "Localized states for elementary systems". *Reviews of Modern Physics* **21** (3): 400–406. Bibcode:1949RvMP...21..400N. doi:10.1103/RevModPhys.21.400.

15.9 Further reading

General readers

- Feynman, R.P. (2001) [1964]. *The Character of Physical Law*. MIT Press. ISBN 0-262-56003-8.

- Feynman, R.P. (2006) [1985]. *QED: The Strange Theory of Light and Matter*. Princeton University Press. ISBN 0-691-12575-9.

- Gribbin, J. (1998). *Q is for Quantum: Particle Physics from A to Z*. Weidenfeld & Nicolson. ISBN 0-297-81752-3.

- Schumm, Bruce A. (2004) *Deep Down Things*. Johns Hopkins Univ. Press. Chpt. 4.

Introductory texts

- McMahon, D. (2008). *Quantum Field Theory*. McGraw-Hill. ISBN 978-0-07-154382-8.

- Bogoliubov, N.; Shirkov, D. (1982). *Quantum Fields*. Benjamin-Cummings. ISBN 0-8053-0983-7.

- Frampton, P.H. (2000). *Gauge Field Theories. Frontiers in Physics (2nd ed.)*. Wiley.

- Greiner, W; Müller, B. (2000). *Gauge Theory of Weak Interactions*. Springer. ISBN 3-540-67672-4.

- Itzykson, C.; Zuber, J.-B. (1980). *Quantum Field Theory*. McGraw-Hill. ISBN 0-07-032071-3.

- Kane, G.L. (1987). *Modern Elementary Particle Physics*. Perseus Books. ISBN 0-201-11749-5.

- Kleinert, H.; Schulte-Frohlinde, Verena (2001). *Critical Properties of φ^4-Theories*. World Scientific. ISBN 981-02-4658-7.

- Kleinert, H. (2008). *Multivalued Fields in Condensed Matter, Electrodynamics, and Gravitation* (PDF). World Scientific. ISBN 978-981-279-170-2.

- Loudon, R (1983). *The Quantum Theory of Light*. Oxford University Press. ISBN 0-19-851155-8.

- Mandl, F.; Shaw, G. (1993). *Quantum Field Theory*. John Wiley & Sons. ISBN 978-0-471-94186-6.

- Peskin, M.; Schroeder, D. (1995). *An Introduction to Quantum Field Theory*. Westview Press. ISBN 0-201-50397-2.

- Ryder, L.H. (1985). *Quantum Field Theory*. Cambridge University Press. ISBN 0-521-33859-X.

- Schwartz, M.D. (2014). *Quantum Field Theory and the Standard Model*. Cambridge University Press. ISBN 978-1107034730.

- Srednicki, Mark (2007) *Quantum Field Theory*. Cambridge Univ. Press.

- Ynduráin, F.J. (1996). *Relativistic Quantum Mechanics and Introduction to Field Theory* (1st ed.). Springer. ISBN 978-3-540-60453-2.

- Zee, A. (2003). *Quantum Field Theory in a Nutshell*. Princeton University Press. ISBN 0-691-01019-6.

Advanced texts

- Brown, Lowell S. (1994). *Quantum Field Theory*. Cambridge University Press. ISBN 978-0-521-46946-3.

- Bogoliubov, N.; Logunov, A.A.; Oksak, A.I.; Todorov, I.T. (1990). *General Principles of Quantum Field Theory*. Kluwer Academic Publishers. ISBN 978-0-7923-0540-8.

- Weinberg, S. (1995). *The Quantum Theory of Fields* **1–3**. Cambridge University Press.

Articles:

- Gerard 't Hooft (2007) "The Conceptual Basis of Quantum Field Theory" in Butterfield, J., and John Earman, eds., *Philosophy of Physics, Part A*. Elsevier: 661–730.

- Frank Wilczek (1999) "Quantum field theory", *Reviews of Modern Physics* 71: S83–S95. Also doi=10.1103/Rev. Mod. Phys. 71.

15.10 External links

- Hazewinkel, Michiel, ed. (2001), "Quantum field theory", *Encyclopedia of Mathematics*, Springer, ISBN 978-1-55608-010-4

- Stanford Encyclopedia of Philosophy: "Quantum Field Theory", by Meinard Kuhlmann.

- Siegel, Warren, 2005. *Fields*. A free text, also available from arXiv:hep-th/9912205.

- Quantum Field Theory by P. J. Mulders

Chapter 16

Position and momentum space

In physics and geometry, there are two intertwined vector spaces.

Position space (also **real space** or **coordinate space**) is the set of all position vectors **r** of an object in space (usually 3D). The position vector defines a point in space. If the position vector varies with time it will trace out a path or surface, such as the trajectory of a particle.

Momentum space is the set of all momentum vectors **p** of an object in space (again usually 3D). The momentum vector corresponds to the motion of the object. The idea transcends all of physics, classical and quantum mechanics. However, in quantum mechanics, the De Broglie relation $\mathbf{p} = \hbar\mathbf{k}$ states that momentum and wavevectors for a free particle are proportional to each other. The set of all wavevectors **k** forms **k-space**.[1] and when it is unambiguous, the terms "momentum" (symbol **p**, also a vector) and "wavevector" are used interchangeably. The De Broglie relation is not true in a crystal.

The duality between position and momentum is an example of *Pontryagin duality*.

The position vector **r** has dimensions of length, the momentum vector has units of [mass][length][time]$^{-1}$, and the **k**-vector has dimensions of reciprocal length, so **k** is the frequency analogue of **r**, just as angular frequency ω is the inverse quantity and frequency analogue of time t. Physical phenomena can be described using either the positions of particles, or their momenta, both formulations equivalently provide the same information about the system in consideration. Usually **r** is more intuitive and simpler than **k**, though the converse is also true, such as in solid-state physics.

16.1 Position and momentum spaces in classical mechanics

16.1.1 Lagrangian mechanics

Most often in Lagrangian mechanics, the Lagrangian $L(\mathbf{q}, d\mathbf{q}/dt, t)$ is in configuration space, where $\mathbf{q} = (q_1, q_2,...., qn)$ is an n-tuple of the generalized coordinates. The Euler–Lagrange equations of motion are

$$\frac{d}{dt}\frac{\partial L}{\partial \dot{q}_i} = \frac{\partial L}{\partial q_i}, \quad \dot{q}_i \equiv \frac{dq_i}{dt}.$$

(One overdot indicates one time derivative). Introducing the definition of canonical momentum for each generalized coordinate

$$p_i = \frac{\partial L}{\partial \dot{q}_i},$$

the Euler–Lagrange equations take the form

$$\dot{p}_i = \frac{\partial L}{\partial q_i} \, .$$

The Lagrangian can be expressed in momentum space also,[2] $L'(\mathbf{p}, d\mathbf{p}/dt, t)$, where $\mathbf{p} = (p_1, p_2,..., pn)$ is an n-tuple of the generalized momenta. A Legendre transformation is performed to change the variables in the total differential of the generalized coordinate space Lagrangian;

$$dL = \sum_{i=1}^{n} \left(\frac{\partial L}{\partial q_i} dq_i + \frac{\partial L}{\partial \dot{q}_i} d\dot{q}_i \right) + \frac{\partial L}{\partial t} dt = \sum_{i=1}^{n} (\dot{p}_i dq_i + p_i d\dot{q}_i) + \frac{\partial L}{\partial t} dt \, ,$$

where the definition of generalized momentum and Euler–Lagrange equations have replaced the partial derivatives of L. The product rule for differentials[nb 1] allows the exchange of differentials in the generalized coordinates and velocities for the differentials in generalized momenta and their time derivatives,

$$\dot{p}_i dq_i = d(q_i \dot{p}_i) - q_i d\dot{p}_i$$

$$p_i d\dot{q}_i = d(\dot{q}_i p_i) - \dot{q}_i dp_i$$

which after substitution simplifies and rearranges to

$$d \left[L - \sum_{i=1}^{n} (q_i \dot{p}_i + \dot{q}_i p_i) \right] = - \sum_{i=1}^{n} (\dot{q}_i dp_i + q_i d\dot{p}_i) + \frac{\partial L}{\partial t} dt \, .$$

Now, the total differential of the momentum space Lagrangian L' is

$$dL' = \sum_{i=1}^{n} \left(\frac{\partial L'}{\partial p_i} dp_i + \frac{\partial L'}{\partial \dot{p}_i} d\dot{p}_i \right) + \frac{\partial L'}{\partial t} dt$$

so by comparison of differentials of the Lagrangians, the momenta, and their time derivatives, the momentum space Lagrangian L' and the generalized coordinates derived from L' are respectively

$$L' = L - \sum_{i=1}^{n} (q_i \dot{p}_i + \dot{q}_i p_i) \, , \quad -\dot{q}_i = \frac{\partial L'}{\partial p_i} \, , \quad -q_i = \frac{\partial L'}{\partial \dot{p}_i} \, .$$

Combining the last two equations gives the momentum space Euler–Lagrange equations

$$\frac{d}{dt} \frac{\partial L'}{\partial \dot{p}_i} = \frac{\partial L'}{\partial p_i} \, .$$

The advantage of the Legendre transformation is that the relation between the new and old functions and their variables are obtained in the process. Both the coordinate and momentum forms of the equation are equivalent and contain the same information about the dynamics of the system. This form may be more useful when momentum or angular momentum enters the Lagrangian.

16.1.2   Hamiltonian mechanics

In Hamiltonian mechanics, unlike Lagrangian mechanics which uses either the all the coordinates *or* the momenta, the Hamiltonian equations of motion place coordinates and momenta on equal footing. For a system with Hamiltonian $H(\mathbf{q}, \mathbf{p}, t)$, the equations are

$$\dot{q}_i = \frac{\partial H}{\partial p_i}, \quad \dot{p}_i = -\frac{\partial H}{\partial q_i}.$$

16.2   Position and momentum spaces in quantum mechanics

Further information: Momentum operator

In quantum mechanics, a particle is described by a quantum state. This quantum state can be represented as a superposition (i.e. a linear combination as a weighted sum) of basis states. In principle one is free to choose the set of basis states, as long as they span the space. If one chooses the eigenfunctions of the position operator as a set of basis functions, one speaks of a state as a wave function $\psi(\mathbf{r})$ in position space (our ordinary notion of space in terms of length). The familiar Schrödinger equation in terms of the position $\mathbf{r}$ is an example of quantum mechanics in the position representation.[3]

By choosing the eigenfunctions of a different operator as a set of basis functions, one can arrive at a number of different representations of the same state. If one picks the eigenfunctions of the momentum operator as a set of basis functions, the resulting wave function $\phi(\mathbf{k})$ is said to be the wave function in momentum space.[3]

16.3   Relation between space and reciprocal space

The momentum representation of a wave function is very closely related to the Fourier transform and the concept of frequency domain. Since a quantum mechanical particle has a frequency proportional to the momentum (de Broglie's equation given above), describing the particle as a sum of its momentum components is equivalent to describing it as a sum of frequency components (i.e. a Fourier transform).[4] This becomes clear when we ask ourselves how we can transform from one representation to another.

16.3.1   Functions and operators in position space

Suppose we have a three-dimensional wave function in position space $\psi(\mathbf{r})$, then we can write this functions as a weighted sum of orthogonal basis functions $\psi j(\mathbf{r})$:

$$\psi(\mathbf{r}) = \sum_j \phi_j \psi_j(\mathbf{r})$$

or, in the continuous case, as an integral

$$\psi(\mathbf{r}) = \int_{\mathbf{k}-\text{space}} \phi(\mathbf{k}) \psi_{\mathbf{k}}(\mathbf{r}) d^3\mathbf{k}$$

It is clear that if we specify the set of functions $\psi j(\mathbf{r})$, say as the set of eigenfunctions of the momentum operator, the function $\phi(\mathbf{k})$ holds all the information necessary to reconstruct $\psi(\mathbf{r})$ and is therefore an alternative description for the state ψ.

In quantum mechanics, the momentum operator is given by

$$\hat{\mathbf{p}} = -i\hbar \frac{\partial}{\partial \mathbf{r}}$$

(see matrix calculus for the denominator notation) with appropriate domain. The eigenfunctions are

$$\psi_{\mathbf{k}}(\mathbf{r}) = \frac{1}{(\sqrt{2\pi})^3} e^{i\mathbf{k}\cdot\mathbf{r}}$$

and eigenvalues $\hbar\mathbf{k}$. So

$$\psi(\mathbf{r}) = \frac{1}{(\sqrt{2\pi})^3} \int_{\mathbf{k}-\text{space}} \phi(\mathbf{k}) e^{i\mathbf{k}\cdot\mathbf{r}} d^3\mathbf{k}$$

and we see that the momentum representation is related to the position representation by a Fourier transform.[5]

16.3.2   Functions and operators in momentum space

Conversely, a three-dimensional wave function in momentum space $\phi(\mathbf{k})$ as a weighted sum of orthogonal basis functions $\phi_j(\mathbf{k})$:

$$\phi(\mathbf{k}) = \sum_j \psi_j \phi_j(\mathbf{k})$$

or as an integral:

$$\phi(\mathbf{k}) = \int_{\mathbf{r}-\text{space}} \psi(\mathbf{r}) \phi_{\mathbf{r}}(\mathbf{k}) d^3\mathbf{r}$$

the position operator is given by

$$\hat{\mathbf{r}} = i\hbar \frac{\partial}{\partial \mathbf{p}} = i \frac{\partial}{\partial \mathbf{k}}$$

with eigenfunctions

$$\phi_{\mathbf{r}}(\mathbf{k}) = \frac{1}{(\sqrt{2\pi})^3} e^{-i\mathbf{k}\cdot\mathbf{r}}$$

and eigenvalues $\mathbf{r}$. So a similar decomposition of $\phi(\mathbf{k})$ can be made in terms of the eigenfunctions of this operator, which turns out to be the inverse Fourier transform:[5]

$$\phi(\mathbf{k}) = \frac{1}{(\sqrt{2\pi})^3} \int_{\mathbf{r}-\text{space}} \psi(\mathbf{r}) e^{-i\mathbf{k}\cdot\mathbf{r}} d^3\mathbf{r}$$

16.4   Unitary equivalence between position and momentum operator

The $\mathbf{r}$ and $\mathbf{p}$ operators are unitarily equivalent, with the unitary operator being given explicitly by the Fourier transform. Thus they have the same spectrum. In physical language, $\mathbf{p}$ acting on momentum space wave functions is the same as $\mathbf{r}$ acting on position space wave functions (under the image of the Fourier transform).

16.5 Reciprocal space and crystals

Main article: Reciprocal lattice

For an electron (or other particle) in a crystal, its value of **k** relates almost always to its crystal momentum, not its normal momentum. Therefore **k** and **p** are not simply proportional but play different roles. See k·p perturbation theory for an example. Crystal momentum is like a wave envelope that describes how the wave varies from one unit cell to the next, but does *not* give any information about how the wave varies within each unit cell.

When **k** relates to crystal momentum instead of true momentum, the concept of **k**-space is still meaningful and extremely useful, but it differs in several ways from the non-crystal **k**-space discussed above. For example, in a crystal's **k**-space, there is an infinite set of points called the reciprocal lattice which are "equivalent" to **k** = 0 (this is analogous to aliasing). Likewise, the "first Brillouin zone" is a finite volume of **k**-space, such that every possible **k** is "equivalent" to exactly one point in this region.

For more details see reciprocal lattice.

16.6 See also

- Phase space
- Reciprocal space
- Configuration space

16.7 Footnotes

[1] For two functions u and v, the differential of the product is $d(uv) = u\,dv + v\,du$.

16.8 References

[1] Eisberg, R.; Resnick, R. (1985). *Quantum Physics of Atoms, Molecules, Solids, Nuclei, and Particles* (2nd ed.). John Wiley & Sons. ISBN 978-0-471-873730.

[2] Hand & Finch, p.190, Analytical Mechanics, 2008

[3] Peleg, Y.; Pnini, R.; Zaarur, E.; Hecht, E. (2010). *Quantum Mechanics (Schaum's Outline Series)* (2nd ed.). McGraw Hill. ISBN 978-0-071-623582.

[4] Abers, E. (2004). *Quantum Mechanics*. Addison Wesley, Prentice Hall Inc. ISBN 978-0-131-461000.

[5] R. Penrose (2007). *The Road to Reality*. Vintage books. ISBN 0-679-77631-1.

Chapter 17

Pilot wave

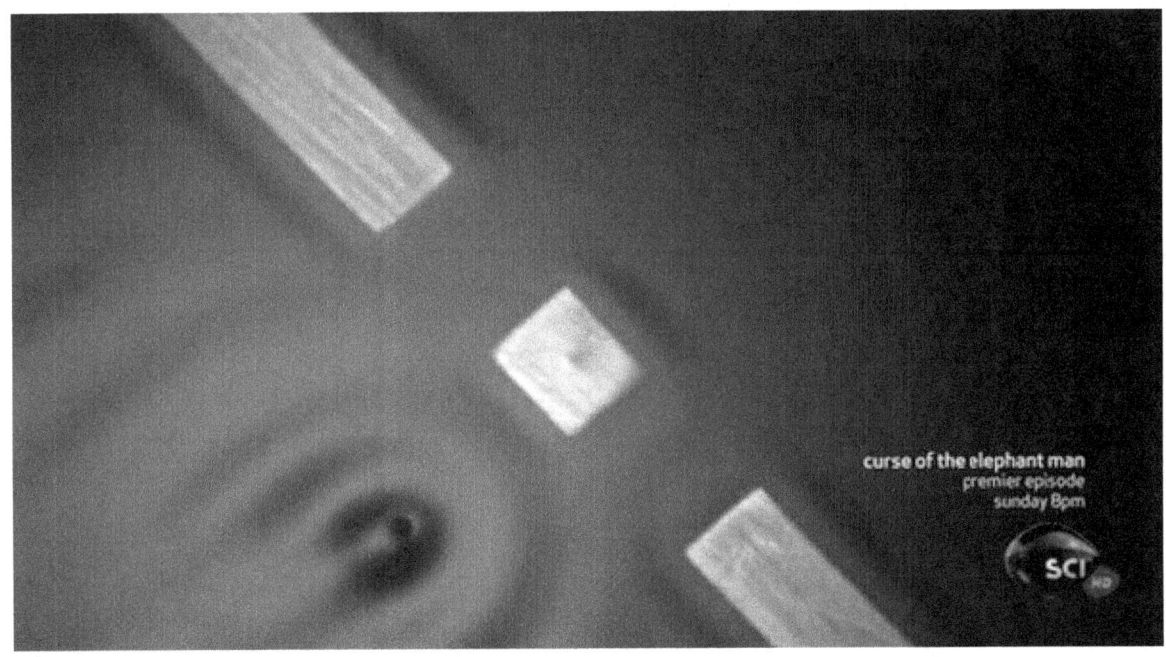

Couder experiments,[1][2] *"materializing" the* pilot wave *model.*

In theoretical physics, the **pilot wave theory** was the first known example of a hidden variable theory, presented by Louis de Broglie in 1927. Its more modern version, the de Broglie–Bohm theory, remains a non-mainstream attempt to interpret quantum mechanics as a deterministic theory, avoiding troublesome notions such as wave-particle duality, instantaneous wave function collapse and the paradox of Schrödinger's cat.

The De Broglie–Bohm pilot wave theory is one of several interpretations of quantum mechanics. It uses the same mathematics as other interpretations of quantum mechanics; consequently, it is also supported by the current experimental evidence to the same extent as the other interpretations.

In the 2000s, Couder and colleagues performed experiments[1][2] on hydrodynamic analogs of single-particle quantum systems.[3]

17.1 History

In his 1926 paper,[4] Max Born suggested that the wave function of Schrödinger's wave equation represents the probability density of finding a particle.

From this idea, de Broglie developed the pilot wave theory, and worked out a function for the guiding wave.[5] Initially, de Broglie proposed a *double solution* approach, in which the quantum object consists of a physical wave (*u*-wave) in real space which has a spherical singular region that gives rise to particle-like behaviour; in this initial form of his theory he did not have to postulate the existence of a quantum particle.[6] He later formulated it as a theory in which a particle is accompanied by a pilot wave. He presented the pilot wave theory at the 1927 Solvay Conference.[7] However, Wolfgang Pauli raised an objection to it at the conference, saying that it did not deal properly with the case of inelastic scattering. De Broglie was not able to find a response to this objection, and he and Born abandoned the pilot-wave approach. Unlike David Bohm years later, de Broglie did not complete his theory to encompass the many-particle case.[6] The many-particle case shows mathematically, that the energy dissipation in inelastic scattering could actually be distributed to the surrounding field structure by an as of yet unknown mechanism of the theory of hidden variables. One physicist has shown a dynamic Lie group geometrical object that could account for the unknown mechanism.[8]

Later, in 1932, John von Neumann published a paper claiming to prove that all hidden variable theories were impossible.[9] (A result found to be flawed by Grete Hermann three years later, though this went unnoticed by the physics community for over fifty years). However, in 1952, David Bohm, dissatisfied with the prevailing orthodoxy, rediscovered de Broglie's pilot wave theory. Bohm developed pilot wave theory into what is now called the de Broglie–Bohm theory.[10][11]

The de Broglie–Bohm theory itself might have gone unnoticed by most physicists, if it had not been championed by John Bell, who also countered the objections to it. In 1987, John Bell[12] rediscovered Grete Hermann's work, and thus showed the physics community that Pauli's and von Neumann's objections really only showed that the pilot wave theory did not have locality.

Yves Couder and co-workers in 2010 discovered a macroscopic pilot wave system in the form of *walking droplets*. This system exhibits behaviour of a pilot wave, heretofore considered to be reserved to microscopic phenomena.[1]

17.2 The pilot wave theory

17.2.1 Principles

The pilot wave theory is a hidden variable theory. Consequently:

- the theory has realism (meaning that its concepts exist independently of the observer);

- the theory has determinism.

The positions and momenta of the particles are considered to be the hidden variables. However, the observer not only doesn't know the precise value of these variables, but more importantly, cannot know them precisely because any measurement disturbs them.

A collection of particles has an associated matter wave, which evolves according to the Schrödinger equation. Each particle follows a deterministic trajectory, which is guided by the wave function; collectively, the density of the particles conforms to the magnitude of the wave function. The wave function is not influenced by the particle and can exist also as an empty wave function.[13]

The theory brings to light nonlocality that is implicit in the non-relativistic formulation of quantum mechanics and uses it to satisfy Bell's theorem. Interestingly, these nonlocal effects are compatible with the no-communication theorem, which prevents us from using them for faster-than-light communication.

17.2.2   Consequences

The pilot wave theory shows that it is possible to have a realistic and deterministic hidden variable theory, which reproduces the experimental results of ordinary quantum mechanics. The price which has to be paid for this is manifest nonlocality.

17.3   Mathematical foundations

To derive the de Broglie–Bohm pilot-wave for an electron, the quantum Lagrangian

$$L(t) = \frac{1}{2}mv^2 - (V + Q),$$

where Q is the potential associated with the quantum force (the particle being pushed by the wave function), is integrated along precisely one path (the one the electron actually follows). This leads to the following formula for the Bohm propagator:

$$K^Q(X_1, t_1; X_0, t_0) = \frac{1}{J(t)^{\frac{1}{2}}} \exp\left[\frac{i}{\hbar}\int_{t_0}^{t_1} L(t)\, dt\right].$$

This propagator allows to track the electron precisely over time under the influence of the quantum potential Q.

17.3.1   Derivation of the Schrödinger equation

Pilot Wave theory is based on Hamilton–Jacobi dynamics[14] rather than Lagrangian or Hamiltonian dynamics. Using the Hamilton–Jacobi equation

$$H\left(\mathbf{q}, \frac{\partial S}{\partial \mathbf{q}}, t\right) + \frac{\partial S}{\partial t}(\mathbf{q}, t) = 0$$

it is possible to derive the Schrödinger equation:

Consider a classical particle — the position of which is not known with certainty. We must deal with it statistically, so only the probability density $\rho(x,t)$ is known. Probability must be conserved, i.e. $\int \rho\, d^3x = 1$ for each t. Therefore it must satisfy the continuity equation

$$\partial\rho/\partial t = -\nabla \cdot (\rho v) \quad (1)$$

where $v(x,t)$ is the velocity of the particle.

In the Hamilton–Jacobi formulation of classical mechanics, velocity is given by $v(x,t) = \frac{\nabla S(x,t)}{m}$ where $S(x,t)$ is a solution of the Hamilton-Jacobi equation

$$-\frac{\partial S}{\partial t} = \frac{(\nabla S)^2}{2m} + V \quad (2)$$

We can combine (1) and (2) into a single complex equation by introducing the complex function $\psi = \sqrt{\rho}e^{\frac{iS}{\hbar}}$, then the two equations are equivalent to

$$i\hbar\frac{\partial\psi}{\partial t} = \left(-\frac{\hbar^2}{2m}\nabla^2 + V - Q\right)\psi \quad \text{with } Q = -\frac{\hbar^2}{2m}\frac{\nabla^2\sqrt{\rho}}{\sqrt{\rho}}$$

This is the time dependent Schrödinger equation with an extra potential, the quantum potential Q, which is the potential of the quantum force, which is proportional (in approximation) to the curvature of the amplitude of the wave function.

17.3.2 Mathematical formulation for a single particle

The matter wave of de Broglie is described by the time-dependent Schrödinger equation:

$$i\hbar\frac{\partial\psi}{\partial t} = \left(-\frac{\hbar^2}{2m}\nabla^2 + V\right)\psi$$

The complex wave function can be represented as:

$$\psi = \sqrt{\rho}\,\exp\left(\frac{iS}{\hbar}\right)$$

By plugging this into the Schrödinger equation, one can derive two new equations for the real variables. The first is the continuity equation for the probability density ρ : [10]

$$\partial\rho/\partial t + \nabla \cdot (\rho v) = 0 \,,$$

where the velocity field is defined by the guidance equation

$$\vec{v}(\vec{r},t) = \frac{\nabla S(\vec{r},t)}{m} \,.$$

According to pilot wave theory, the point particle and the matter wave are both real and distinct physical entities. (Unlike standard quantum mechanics, where particles and waves are considered to be the same entities, connected by wave–particle duality.) The pilot wave guides the motion of the point particles as described by the guidance equation.

Ordinary quantum mechanics and pilot wave theory are based on the same partial differential equation. The main difference is that in ordinary quantum mechanics, the Schrödinger equation is connected to reality by the Born postulate, which states that the probability density of the particle's position is given by $\rho = |\psi|^2$. Pilot wave theory considers the guidance equation to be the fundamental law, and sees the Born rule as a derived concept.

The second equation is a modified Hamilton–Jacobi equation for the action S :

$$-\frac{\partial S}{\partial t} = \frac{(\nabla S)^2}{2m} + V + Q \,,$$

where Q is the quantum potential defined by

$$Q = -\frac{\hbar^2}{2m}\frac{\nabla^2\sqrt{\rho}}{\sqrt{\rho}}$$

By neglecting Q, our equation is reduced to the Hamilton–Jacobi equation of a classical point particle. (Strictly speaking, this is only a semiclassical limit , because the superposition principle still holds and one needs a decoherence mechanism to get rid of it. Interaction with the environment can provide this mechanism.) So, the quantum potential is responsible for all the mysterious effects of quantum mechanics.

One can also combine the modified Hamilton–Jacobi equation with the guidance equation to derive a quasi-Newtonian equation of motion

$$m\frac{d}{dt}\vec{v} = -\nabla(V + Q) \,,$$

where the hydrodynamic time derivative is defined as

$$\frac{d}{dt} = \frac{\partial}{\partial t} + \vec{v}\cdot\nabla \,.$$

17.3.3   Mathematical formulation for multiple particles

The Schrödinger equation for the many-body wave function $\psi(\vec{r}_1, \vec{r}_2, \cdots, t)$ is given by

$$i\hbar \frac{\partial \psi}{\partial t} = \left(-\frac{\hbar^2}{2} \sum_{i=1}^{N} \frac{\nabla_i^2}{m_i} + V(\mathbf{r}_1, \mathbf{r}_2, \cdots \mathbf{r}_N) \right) \psi$$

The complex wave function can be represented as:

$$\psi = \sqrt{\rho}\, \exp\left(\frac{iS}{\hbar} \right)$$

The pilot wave guides the motion of the particles. The guidance equation for the jth particle is:

$$\vec{v}_j = \frac{\nabla_j S}{m} \, .$$

The velocity of the jth particle explicitly depends on the positions of the other particles. This means that the theory is nonlocal.

17.3.4   Empty wave function

Lucien Hardy[15] and John Stewart Bell[13] have emphasized that in the de Broglie–Bohm picture of quantum mechanics there can exist **empty waves**, represented by wave functions propagating in space and time but not carrying energy or momentum,[16] and not associated with a particle. The same concept was called *ghost waves* (or "Gespensterfelder", *ghost fields*) by Albert Einstein.[16]

The empty wave function notion has been discussed controversially.[17][18][19] In contrast, the many-worlds interpretation of quantum mechanics does not call for empty wave functions.[13]

17.4   References

[1]   Couder, Y.; Boudaoud, A.; Protière, S.; Moukhtar, J.; Fort, E. (2010). "Walking droplets: a form of wave–particle duality at macroscopic level?" (PDF). *Europhysics News* **41** (1): 14–18. Bibcode:2010ENews..41...14C. doi:10.1051/epn/2010101.

[2]   "Yves Couder experiments explains Wave/Particle Duality via silicon droplets". *How Does The Universe Work?*. *Through the Wormhole*. 13 July 2011.

[3]   "Pilot-Wave Hydrodynamics", John W.M. Bush, 2014

[4]   Born, M. (1926). "Quantenmechanik der Stoßvorgänge". *Zeitschrift für Physik* **38** (11–12): 803–827. Bibcode:1926ZPhy...38.. doi:10.1007/BF01397184.

[5]   de Broglie, L. (1927). "La mécanique ondulatoire et la structure atomique de la matière et du rayonnement". *Journal de Physique et le Radium* **8** (5): 225–241. Bibcode:1927JPhRa...8..225D. doi:10.1051/jphysrad:0192700805022500.

[6]   Dewdney, C.; Horton, G.; Lam, M. M.; Malik, Z.; Schmidt, M. (1992). "Wave-particle dualism and the interpretation of quantum mechanics". *Foundations of Physics* **22** (10): 1217–1265. Bibcode:1992FoPh...22.1217D. doi:10.1007/BF01889712.

[7]   Institut International de Physique Solvay (1928). *Electrons et Photons: Rapports et Discussions du Cinquième Conseil de Physique tenu à Bruxelles du 24 au 29 Octobre 1927*. Gauthier-Villars.

[8]   eldeeb, sherif. "The Simplest Ever Unified Field Theory". *www.academia.edu*.

[9] von Neumann, J. (1932). *Mathematische Grundlagen der Quantenmechanik*. Springer.

[10] Bohm, D. (1952). "A suggested Interpretation of the Quantum Theory in Terms of Hidden Variables, I". *Physical Review* **85** (2): 166–179. Bibcode:1952PhRv...85..166B. doi:10.1103/PhysRev.85.166.

[11] Bohm, D. (1952). "A suggested Interpretation of the Quantum Theory in Terms of Hidden Variables, II". *Physical Review* **85** (2): 180–193. Bibcode:1952PhRv...85..180B. doi:10.1103/PhysRev.85.180.

[12] Bell, J. S. (1987). *Speakable and Unspeakable in Quantum Mechanics*. Cambridge University Press. ISBN 978-0521334952.

[13] Bell, J. S. (1992). "Six possible worlds of quantum mechanics". *Foundations o f Physics* **22** (10): 1201–1215. Bibcode: doi:10.1007/BF01889711.

[14] Towler, M. (10 February 2009). "De Broglie-Bohm pilot-wave theory and the foundations of quantum mechanics". University of Cambridge. Retrieved 2014-07-03.

[15] Hardy, L. (1992). "On the existence of empty waves in quantum theory". *Physics Letters A* **167** (1): 11–16. Bibcode: doi:10.1016/0375-9601(92)90618-V.

[16] Selleri, F.; Van der Merwe, A. (1990). *Quantum paradoxes and physical reality*. Kluwer Academic Publishers. pp. 85–86. ISBN 0-7923-0253-2.

[17] Zukowski, M. (1993). ""On the existence of empty waves in quantum theory": a comment". *Physics Letters A* **175** (3–4): 257–258. Bibcode:1993PhLA..175..257Z. doi:10.1016/0375-9601(93)90837-P.

[18] Zeh, H. D. (1999). "Why Bohm's Quantum Theory?". *Foundations of Physics Letters* **12**: 197–200. arXiv:quant-ph/9812059. Bibcode:1999FoPhL..12..197Z. doi:10.1023/A:1021669308832.

[19] Vaidman, L. (2005). "The Reality in Bohmian Quantum Mechanics or Can You Kill with an Empty Wave Bullet?" *35* (2): 299–312. arXiv:quant-ph/0312227. Bibcode:2005FoPh...35..299V. doi:10.1007/s10701-004-1945-2.

17.5   External links

- "Pilot-wave hydrodynamics" Bush, J.W.M, 2014, Ann. Rev. Fluid Mech., 49, 269-292.

- "Quantum mechanics writ large", Bush, J.W.M, 2010.

- "Pilot waves, Bohmian metaphysics, and the foundations of quantum mechanics", lecture course on pilot wave theory by Mike Towler, Cambridge University (2009).

- "Hydrodynamic quantum analogues" Research on hydrodynamic quantum analogues and hydrodynamic pilot-wave theory, by John Bush (MIT) and coworkers.

- More complete HTML encyclopedic page about the subject.

- Hydrodynamic quantum analogues

Chapter 18

Hidden variable theory

This article is about a class of mechanics theories. For hidden variables in economics, see latent variable. For other uses, see Hidden variables (disambiguation).

Historically, in physics, **hidden variable theories** were espoused by some physicists who argued that the state of a physical system, as formulated by quantum mechanics, does not give a complete description for the system; i.e., that quantum mechanics is ultimately incomplete, and that a complete theory would provide descriptive categories to account for all observable behavior and thus avoid any indeterminism. The existence of indeterminacy for some measurements is a characteristic of prevalent interpretations of quantum mechanics; moreover, bounds for indeterminacy can be expressed in a quantitative form by the Heisenberg uncertainty principle.

Albert Einstein, the most famous proponent of hidden variables, objected to the fundamentally probabilistic nature of quantum mechanics,[1] and famously declared "I am convinced God does not play dice".[2] Einstein, Podolsky, and Rosen argued that "elements of reality" (hidden variables) must be added to quantum mechanics to explain entanglement without action at a distance.[3][4] Later, Bell's theorem would suggest that local hidden variables of certain types are impossible, or that they evolve non-locally. A famous non-local theory is De Broglie–Bohm theory.

18.1 Motivation

Under the standard Copenhagen interpretation, quantum mechanics is non-deterministic, meaning that it generally does not predict the outcome of any measurement with certainty. Instead, it indicates what the probabilities of the outcomes are, with the indeterminism of observable quantities constrained by the uncertainty principle. The question arises whether there might be some deeper reality hidden beneath quantum mechanics, to be described by a more fundamental theory that can always predict the outcome of each measurement with certainty: if the exact properties of every subatomic particle were known the entire system could be modeled exactly using deterministic physics similar to classical physics.

In other words, it is conceivable that the standard interpretation of quantum mechanics is an incomplete description of nature. The designation of variables as underlying "hidden" variables depends on the level of physical description (so, for example, "if a gas is described in terms of temperature, pressure, and volume, then the velocities of the individual atoms in the gas would be hidden variables".[5]). Physicists supporting De Broglie–Bohm theory maintain that underlying the observed probabilistic nature of the universe is a deterministic objective foundation/property—the hidden variable. Others, however, believe that there is no deeper deterministic reality in quantum mechanics—experiments have shown a vast class of hidden variable theories to be incompatible with observations.

Lack of a kind of realism (the latter understood here as asserting independent existence and evolution of physical quantities, such as position or momentum, without the process of measurement) is crucial in the Copenhagen interpretation. Realistic interpretations (which were already incorporated, to an extent, into the physics of Feynman[6]), on the other hand, assume that particles have certain trajectories. Under such view, these trajectories will almost always be continuous, which follows both from the finiteness of the perceived speed of light ("leaps" should rather be precluded) and, more importantly, from the principle of least action, as deduced in quantum physics by Dirac. But continuous movement, in accordance with the mathematical definition, implies deterministic movement for a range of time arguments;[7] and thus realism is, under modern physics, one more reason for seeking (at least certain limited) determinism and thus a hidden variable theory (especially that such theory exists: see De Broglie–Bohm interpretation).

Although determinism was initially a major motivation for physicists looking for hidden variable theories, nondeterministic theories trying to explain what the supposed reality underlying the quantum mechanics formalism looks like are also considered hidden variable theories; for example Edward Nelson's stochastic mechanics.

18.2 "God does not play dice"

In June 1926, Max Born published a paper, "Zur Quantenmechanik der Stoßvorgänge" ("Quantum Mechanics of Collision Phenomena") in the scientific journal *Zeitschrift für Physik*, in which he was the first to clearly enunciate the probabilistic interpretation of the quantum wavefunction, which had been introduced by Erwin Schrödinger earlier in the year. Born concluded the paper as follows:

> Here the whole problem of determinism comes up. From the standpoint of our quantum mechanics there is no quantity which in any individual case causally fixes the consequence of the collision; but also experimentally we have so far no reason to believe that there are some inner properties of the atom which conditions a definite outcome for the collision. Ought we to hope later to discover such properties ... and determine them in individual cases? Or ought we to believe that the agreement of theory and experiment—as to the impossibility of prescribing conditions for a causal evolution—is a pre-established harmony founded on the nonexistence of such conditions? I myself am inclined to give up determinism in the world of atoms. But that is a philosophical question for which physical arguments alone are not decisive.

Born's interpretation of the wavefunction was criticized by Schrödinger, who had previously attempted to interpret it in real physical terms, but Albert Einstein's response became one of the earliest and most famous assertions that quantum mechanics is incomplete:

> Quantum mechanics is very worthy of regard. But an inner voice tells me that this is not yet the right track. The theory yields much, but it hardly brings us closer to the Old One's secrets. I, in any case, am convinced that *He* does not play dice.[8][9]

18.3 Early attempts at hidden variable theories

Shortly after making his famous "God does not play dice" comment, Einstein attempted to formulate a deterministic counterproposal to quantum mechanics, presenting a paper at a meeting of the Academy of Sciences in Berlin, on 5 May 1927, titled "Bestimmt Schrödinger's Wellenmechanik die Bewegung eines Systems vollständig oder nur im Sinne der Statistik?" ("Does Schrödinger's wave mechanics determine the motion of a system completely or only in the statistical sense?").[10] However, as the paper was being prepared for publication in the academy's journal, Einstein decided to withdraw it, possibly because he discovered that, contrary to his intention, it implied non-separability of entangled systems, which he regarded as absurd.[11]

At the Fifth Solvay Congress, held in Belgium in October 1927 and attended by all the major theoretical physicists of the era, Louis de Broglie presented his own version of a deterministic hidden-variable theory, apparently unaware of Einstein's aborted attempt earlier in the year. In his theory, every particle had an associated, hidden "pilot wave" which served to guide its trajectory through space. The theory was subject to criticism at the Congress, particularly by Wolfgang Pauli, which de Broglie did not adequately answer. De Broglie abandoned the theory shortly thereafter.

18.4   Declaration of completeness of quantum mechanics

Also at the Fifth Solvay Congress, Max Born and Werner Heisenberg made a presentation summarizing the recent tremendous theoretical development of the subject. At the conclusion of the presentation, they declared:

> [W]hile we consider ... a quantum mechanical treatment of the electromagnetic field ... as not yet finished, we consider quantum mechanics to be a closed theory, whose fundamental physical and mathematical assumptions are no longer susceptible of any modification.... On the question of the 'validity of the law of causality' we have this opinion: as long as one takes into account only experiments that lie in the domain of our currently acquired physical and quantum mechanical experience, the assumption of indeterminism in principle, here taken as fundamental, agrees with experience.[12]

18.5   Bohr–Einstein debates

Main article: Bohr–Einstein debates

Although there is no record of Einstein responding to Born and Heisenberg during the technical sessions of the Fifth Solvay Congress, he did challenge the completeness of quantum mechanics during informal discussions over meals, presenting a thought experiment intended to demonstrate that quantum mechanics could not be entirely correct. He did likewise during the Sixth Solvay Congress held in 1930. Both times, Niels Bohr is generally considered to have successfully defended quantum mechanics by discovering errors in Einstein's arguments.

18.6   EPR paradox

Main article: EPR paradox

The debates between Bohr and Einstein essentially concluded in 1935, when Einstein finally expressed what is widely considered his best argument against the completeness of quantum mechanics. Einstein, Podolsky, and Rosen had proposed their definition of a "complete" description as one which uniquely determines the values of all its measurable properties. Einstein later summarized their argument as follows:

> Consider a mechanical system consisting of two partial systems A and B which interact with each other only during a limited time. Let the ψ function [i.e., wavefunction] before their interaction be given. Then the Schrödinger equation will furnish the ψ function after the interaction has taken place. Let us now determine the physical state of the partial system A as completely as possible by measurements. Then quantum mechanics allows us to determine the ψ function of the partial system B from the measurements made, and from the ψ function of the total system. This determination, however, gives a result which depends upon which of the physical quantities (observables) of A have been measured (for instance, coordinates or momenta). Since there can be only one physical state of B after the interaction which cannot reasonably be considered to depend on the particular measurement we perform on the system A separated from B it may be concluded that the ψ function is not unambiguously coordinated to the physical state. This coordination of several ψ functions to the same physical state of system B shows again that the ψ function cannot be interpreted as a (complete) description of a physical state of a single system.[13]

Bohr answered Einstein's challenge as follows:

> [The argument of] Einstein, Podolsky and Rosen contains an ambiguity as regards the meaning of the expression "without in any way disturbing a system." ... [E]ven at this stage [i.e., the measurement of, for example, a particle that is part of an entangled pair], there is essentially the question of an influence on the

very conditions which define the possible types of predictions regarding the future behavior of the system. Since these conditions constitute an inherent element of the description of any phenomenon to which the term "physical reality" can be properly attached, we see that the argumentation of the mentioned authors does not justify their conclusion that quantum-mechanical description is essentially incomplete."[14]

Bohr is here choosing to define a "physical reality" as limited to a phenomenon that is immediately observable by an arbitrarily chosen and explicitly specified technique, using his own special definition of the term 'phenomenon'. He wrote in 1948:

> As a more appropriate way of expression, one may strongly advocate limitation of the use of the word *phenomenon* to refer exclusively to observations obtained under specified circumstances, including an account of the whole experiment."[15][16]

This was, of course, in conflict with the definition used by the EPR paper, as follows:

> *If, without in any way disturbing a system, we can predict with certainty (i.e., with probability equal to unity) the value of a physical quantity, then there exists an element of physical reality corresponding to this physical quantity.* [Italics in original][3]

18.7 Bell's theorem

Main article: Bell's theorem

In 1964, John Bell showed through his famous theorem that if local hidden variables exist, certain experiments could be performed involving quantum entanglement where the result would satisfy a Bell inequality. If, on the other hand, statistical correlations resulting from quantum entanglement could not be explained by local hidden variables, the Bell inequality would be violated. Another no-go theorem concerning hidden variable theories is the Kochen–Specker theorem.

Physicists such as Alain Aspect and Paul Kwiat have performed experiments that have found violations of these inequalities up to 242 standard deviations[17] (excellent scientific certainty). This rules out local hidden variable theories, but does not rule out non-local ones. Theoretically, there could be experimental problems that affect the validity of the experimental findings.

Gerard 't Hooft has disputed the validity of Bell's theorem on the basis of the superdeterminism loophole and proposed some ideas to construct local deterministic models.[18]

18.8 Bohm's hidden variable theory

Main article: de Broglie–Bohm theory

Assuming the validity of Bell's theorem, any deterministic hidden-variable theory which is consistent with quantum mechanics would have to be non-local, maintaining the existence of instantaneous or faster-than-light relations (correlations) between physically separated entities. The currently best-known hidden-variable theory, the "causal" interpretation of the physicist and philosopher David Bohm, originally published in 1952, is a *non-local hidden variable theory*. Bohm unknowingly rediscovered (and extended) the idea that Louis de Broglie had proposed in 1927 (and abandoned) – hence this theory is commonly called "de Broglie-Bohm theory". Bohm posited *both* the quantum particle, e.g. an electron, and a hidden 'guiding wave' that governs its motion. Thus, in this theory electrons are quite clearly particles—when a double-slit experiment is performed, its trajectory goes through one slit rather than the other. Also, the slit passed through is not random but is governed by the (hidden) guiding wave, resulting in the wave pattern that is observed.

Such a view does not contradict the idea of local events that is used in both classical atomism and relativity theory as Bohm's theory (and quantum mechanics) are still locally causal (that is, information travel is still restricted to the speed of light) but allow nonlocal correlations. It points to a view of a more holistic, mutually interpenetrating and interacting world. Indeed, Bohm himself stressed the holistic aspect of quantum theory in his later years, when he became interested in the ideas of Jiddu Krishnamurti.

In Bohm's interpretation, the (nonlocal) quantum potential constitutes an implicate (hidden) order which organizes a particle, and which may itself be the result of yet a further implicate order: a *superimplicate order* which organizes a field.[19] Nowadays Bohm's theory is considered to be one of many interpretations of quantum mechanics which give a realist interpretation, and not merely a positivistic one, to quantum-mechanical calculations. Some consider it the simplest theory to explain quantum phenomena.[20] Nevertheless, it *is* a hidden variable theory, and necessarily so.[21] The major reference for Bohm's theory today is his book with Basil Hiley, published posthumously.[22]

A possible weakness of Bohm's theory is that some (including Einstein, Pauli, and Heisenberg) feel that it looks contrived.[23] (Indeed, Bohm thought this of his original formulation of the theory.[24]) It was deliberately designed to give predictions that are in all details identical to conventional quantum mechanics.[24] Bohm's original aim was not to make a serious counterproposal but simply to demonstrate that hidden-variable theories are indeed possible.[24] (It thus provided a supposed counterexample to the famous proof by John von Neumann that was generally believed to demonstrate that no deterministic theory reproducing the statistical predictions of quantum mechanics is possible.) Bohm said he considered his theory to be unacceptable as a physical theory due to the guiding wave's existence in an abstract multi-dimensional configuration space, rather than three-dimensional space.[24] His hope was that the theory would lead to new insights and experiments that would lead ultimately to an acceptable one;[24] his aim was not to set out a deterministic, mechanical viewpoint, but rather to show that it was possible to attribute properties to an underlying reality, in contrast to the conventional approach to quantum mechanics.[25]

18.9   Recent developments

In August 2011, Roger Colbeck and Renato Renner published a proof that any extension of quantum mechanical theory, whether using hidden variables or otherwise, cannot provide a more accurate prediction of outcomes, assuming that observers can freely choose the measurement settings.[26] Colbeck and Renner write: "In the present work, we have ... excluded the possibility that any extension of quantum theory (not necessarily in the form of local hidden variables) can help predict the outcomes of any measurement on any quantum state. In this sense, we show the following: under the assumption that measurement settings can be chosen freely, quantum theory really is complete".

In January 2013, GianCarlo Ghirardi and Raffaele Romano described a model which, "under a different free choice assumption [...] violates [the statement by Colbeck and Renner] for almost all states of a bipartite two-level system, in a possibly experimentally testable way".[27]

18.10   See also

- Local hidden variable theory

- Bell's theorem

- Bell test experiments

- Quantum mechanics

- Bohm interpretation

- Spekkens Toy Model

18.11   References

[1] *The Born-Einstein letters: correspondence between Albert Einstein and Max and Hedwig Born from 1916–1955, with commentaries by Max Born.* Macmillan. 1971. p. 158., (Private letter from Einstein to Max Born, 3 March 1947: "I admit, of course,

that there is a considerable amount of validity in the statistical approach which you were the first to recognize clearly as necessary given the framework of the existing formalism. I cannot seriously believe in it because the theory cannot be reconciled with the idea that physics should represent a reality in time and space, free from spooky actions at a distance.... I am quite convinced that someone will eventually come up with a theory whose objects, connected by laws, are not probabilities but considered facts, as used to be taken for granted until quite recently".)

[2] private letter to Max Born, 4 December 1926, Albert Einstein Archives reel 8, item 180

[3] Einstein, A.; Podolsky, B.; Rosen, N. (1935). "Can Quantum-Mechanical Description of Physical Reality Be Considered Complete?". *Physical Review* 47 (10): 777–780. Bibcode:1935PhRv...47..777E. doi:10.1103/PhysRev.47.777.

[4] "The debate whether Quantum Mechanics is a complete theory and probabilities have a non-epistemic character (i.e. nature is intrinsically probabilistic) or whether it is a statistical approximation of a deterministic theory and probabilities are due to our ignorance of some parameters (i.e. they are epistemic) dates to the beginning of the theory itself". See: arXiv:quant-ph/0701071v1 12 Jan 2007

[5] Senechal M, Cronin J (2001). "Social influences on quantum mechanics?-I". *The Mathematical Intelligencer* 23 (4): 15–17. doi:10.1007/BF03024596.

[6] Individual diagrams are often split into several parts, which may occur beyond observation; only the diagram as a whole describes an observed event.

[7] For every subset of points within a range, a value for every argument from the subset will be determined by the points in the neighbourhood. Thus, as a whole, the evolution in time can be described (for a specific time interval) as a function, e.g. a linear one or an arc. See Continuous function#Definition in terms of limits of functions

[8] *The Born–Einstein letters: correspondence between Albert Einstein and Max and Hedwig Born from 1916–1955, with commentaries by Max Born*. Macmillan. 1971. p. 91.

[9] Cache of the Einstein section of the American Museum of Natural History

[10] Albert Einstein Archives reel 2, item 100

[11] Baggott, Jim (2011). *The Quantum Story: A History in 40 Moments*. New York: Oxford University Press. pp. 116–117.

[12] Max Born and Werner Heisenberg, "Quantum mechanics", proceedings of the Fifth Solvay Congress.

[13] Einstein A (1936). "Physics and Reality". *Journal of the Franklin Institute* 221.

[14] Bohr N (1935). "Can Quantum-Mechanical Description of Physical Reality be Considered Complete?". *Physical Review* 48: 700. Bibcode:1935PhRv...48..696B. doi:10.1103/physrev.48.696.

[15] Bohr, N. (1948). 'On the notions of causality and complementarity', *Dialectica*, 2: 312–319, p. 317.

[16] Rosenfeld, L. (). 'Niels Bohr's contribution to epistemology', pp. 522–535 in *Selected Papers of Léon Rosenfeld*, Cohen, R.S., Stachel, J.J. (editors), D. Riedel, Dordrecht, ISBN 978-90-277-0652-2, p. 531: "Moreover, the complete definition of the phenomenon must essentially contain the indication of some permanent mark left upon a recording device which is part of the apparatus; only by thus envisaging the phenomenon as a closed event, terminated by a permanent record, can we do justice to the typical wholeness of the quantal processes."

[17] Kwiat, P. G., *et al.* (1999) "Ultrabright source of polarization-entangled photons", *Physical Review A* 60, R773–R776

[18] G 't Hooft, *The Free-Will Postulate in Quantum Mechanics ; Entangled quantum states in a local deterministic theory*

[19] David Pratt: "David Bohm and the Implicate Order". Appeared in *Sunrise magazine*, February/March 1993, Theosophical University Press

[20] Michael K.-H. Kiessling: "Misleading Signposts Along the de Broglie–Bohm Road to Quantum Mechanics", *Foundations of Physics*, volume 40, number 4, 2010, pp. 418–429 (abstract)

[21] "While the testable predictions of Bohmian mechanics are isomorphic to standard Copenhagen quantum mechanics, its underlying hidden variables have to be, in principle, unobservable. If one could observe them, one would be able to take advantage of that and signal faster than light, which – according to the special theory of relativity – leads to physical temporal paradoxes." J. Kofler and A. Zeilinger, "Quantum Information and Randomness", *European Review* (2010), Vol. 18, No. 4, 469–480.

[22] D. Bohm and B. J. Hiley, *The Undivided Universe*, Routledge, 1993, ISBN 0-415-06588-7.

[23] Wayne C. Myrvold (2003). "On some early objections to Bohm's theory" (PDF). *International Studies in the Philosophy of Science* **17** (1): 8–24. doi:10.1080/02698590305233.

[24] David Bohm (1957). *Causality and Chance in Modern Physics*. Routledge & Kegan Paul and D. Van Nostrand. p. 110. ISBN 0-8122-1002-6.

[25] B. J. Hiley: *Some remarks on the evolution of Bohm's proposals for an alternative to quantum mechanics*, 30 January 2010

[26] Roger Colbeck; Renato Renner (2011). "No extension of quantum theory can have improved predictive power". *Nature Communications* **2** (8). arXiv:1005.5173. Bibcode:2011NatCo...2E.411C. doi:10.1038/ncomms1416.

[27] Giancarlo Ghirardi; Raff aele Romano (2013). "Onthological models predictivel y inequivalent to quantum theory". arXiv: 1301. Bibcode:2013PhRvL.110q0404G. doi:10.1103/PhysRevLett.110.170404.

18.12   Bibliography

- **Albert Einstein, Boris Podolsky, and Nathan Rosen**, "Can Quantum-Mechanical Description of Physical Reality Be Considered Complete?" *Physical Review* **47**, 777–780 (1935).

- **John Stewart Bell**, "On the Einstein–Podolsky–Rosen paradox", *Physics* **1**, (1964) 195–200. Reprinted in *Speakable and Unspeakable in Quantum Mechanics*, Cambridge University Press, 2004.

- **D. Bohm and B. J. Hiley**, *The Undivided Universe*, Routledge, 1993

- **Wolfgang Pauli**, letter to M. Fierz dated 10 August 1954, reprinted and translated in K. V. Laurikainen, *Beyond the Atom: The Philosophical Thought of Wolfgang Pauli*, Springer-Verlag, Berlin, 1988, p. 226.

- **Werner Heisenberg**, *Physics and Beyond: Encounters and Conversations*, translated by A. J. Pomerans, Harper & Row, New York, 1971, pp. 63–64.

- **Claude Cohen-Tannoudji, Bernard Diu and Franck Laloë**, *Mecanique quantique* (see also *Quantum Mechanics* translated from the French by Susan Hemley, Nicole Ostrowsky, and Dan Ostrowsky; John Wiley & Sons 1982) Hermann, Paris, France. 1977.

- **P. S. Hanle**, *Indeterminacy before Heisenberg: The Case of Franz Exner and Erwin Schrödinger*, *Historical Studies in the Physical Sciences* **10**, 225 (1979).

- **Asher Peres and Wojciech Zurek**, "Is quantum theory universally valid?" *American Journal of Physics* **50**, 807 (1982).

- **Wojciech Zurek** *Physical Review* D **26** 1862. 1982.

- **Max Jammer**, "The EPR Problem in Its Historical Development", in *Symposium on the Foundations of Modern Physics: 50 years of the Einstein–Podolsky–Rosen Gedankenexperiment*, edited by P. Lahti and P. Mittelstaedt (World Scientific, Singapore, 1985), pp. 129–149.

- **Arthur Fine**, *The Shaky Game: Einstein Realism and the Quantum Theory*, University of Chicago Press, Chicago, 1986.

- **Thomas Kuhn**. *Black-Body Theory and the Quantum Discontinuity*, 1894–1912 Chicago University Press. 1987.

- **Asher Peres**, *Quantum Theory: Concepts and Methods*, Kluwer, Dordrecht, 1993.

- **Carlton M. Caves and Christopher A. Fuchs**, "Quantum Information: How Much Information in a State Vector?", in *The Dilemma of Einstein, Podolsky and Rosen – 60 Years Later*, edited by A. Mann and M. Revzen, *Ann. Israel Physical Society* **12**, 226–257 (1996).

- **Carlo Rovelli**. "Relational quantum mechanics" *International Journal of Theoretical Physics* **35** 1637–1678. 1996.

- Roland Omnès, *Understanding Quantum Mechanics*, Princeton University Press, 1999.

- Roman Jackiw and Daniel Kleppner, "One Hundred Years of Quantum Physics", *Science*, Vol. 289 Issue 5481, p. 893, August 2000.

- Orly Alter and Yoshihisa Yamamoto (2001). *Quantum Measurement of a Single System* (PDF). Wiley-Interscience. 136 pp. doi:10.1002/9783527617128. ISBN 9780471283089. Slides.

- Erich Joos, et al., *Decoherence and the Appearance of a Classical World in Quantum Theory*, 2nd ed., Berlin, Springer, 2003.

- Wojciech Zurek (2003). "Decoherence and the transition from quantum to classical — Revisited", arXiv:quant-ph/0306072 (An updated version of *Physics Today*, 44:36–44 (1991) article)

- Wojciech Zurek, "Decoherence, einselection, and the quantum origins of the classical" in *Reviews of Modern Physics*, vol.75, (715).

- Asher Peres and Daniel Terno, "Quantum Information and Relativity Theory", *Reviews of Modern Physics* 76 (2004) 93.

- Roger Penrose, *The Road to Reality: A Complete Guide to the Laws of the Universe*, Alfred Knopf 2004.

- Maximilian Schlosshauer, "Decoherence, the Measurement Problem, and Interpretations of Quantum Mechanics", in *Reviews of Modern Physics*, vol.76, pages 1267–1305, 2005.

- Federico Laudisa and Carlo Rovelli. "Relational Quantum Mechanics" *The Stanford Encyclopedia of Philosophy* (Fall 2005 Edition).

- Marco Genovese, "Research on hidden variable theories: a review of recent progresses", in *Physics Reports*, vol.413, 2005.

18.13   External links

- The David Bohm Society

Chapter 19

Quantum potential

The **quantum potential** or **quantum potentiality** is a central concept of the de Broglie–Bohm formulation of quantum mechanics, introduced by David Bohm in 1952.

Initially presented under the name *quantum-mechanical potential*, subsequently *quantum potential*, it was later elaborated upon by Bohm and Basil Hiley in its interpretation as an **information potential** which acts on a quantum particle. It is also referred to as *quantum potential energy*, *Bohm potential*, *quantum Bohm potential* or *Bohm quantum potential*.

In the framework of the de Broglie–Bohm theory, the quantum potential is a term within the Schrödinger equation which acts to guide the movement of quantum particles. The quantum potential approach introduced by Bohm[1][2] provides a formally more complete exposition of the idea presented by Louis de Broglie: de Broglie had postulated in 1926 that the wave function represents a pilot wave which guides a quantum particle, but had subsequently abandoned his approach due to objections raised by Wolfgang Pauli. The seminal articles of Bohm in 1952 introduced the quantum potential and included answers to the objections which had been raised against the pilot wave theory.

The Bohm quantum potential is closely linked with the results of other approaches, in particular relating to work by Erwin Madelung of 1927 and to work by Carl Friedrich von Weizsäcker of 1935.

Building on the interpretation of the quantum theory introduced by Bohm in 1952, David Bohm and Basil Hiley in 1975 presented how the concept of a *quantum potential* leads to the notion of an "unbroken wholeness of the entire universe", proposing that the fundamental new quality introduced by quantum physics is nonlocality.[3]

19.1 Quantum potential as part of the Schrödinger equation

The Schrödinger equation

$$ i\hbar \frac{\partial \psi}{\partial t} = \left(-\frac{\hbar^2}{2m} \nabla^2 + V \right) \psi $$

is re-written using the polar form for the wave function $\psi = R \exp iS/\hbar$ with real-valued functions R and S, where R is the amplitude (absolute value) of the wave function ψ and $S/\hbar$ its phase. This yields two equations: from the imaginary and real part of the Schrödinger equation follow the continuity equation and the quantum Hamilton-Jacobi equation, respectively.[1][4]

19.1.1 Continuity equation

The imaginary part of the Schrödinger equation in polar form yields:

$$\frac{\partial R}{\partial t} = -\frac{1}{2m}\left[R\nabla^2 S + 2\nabla R \cdot \nabla S\right],$$

which, provided $\rho = R^2$, can be interpreted as the continuity equation $\partial \rho / \partial t + \nabla \cdot (\rho v) = 0$ for the probability density ρ.

19.1.2 Quantum Hamilton-Jacobi equation

The real part of the Schrödinger equation in polar form yields a modified Hamilton–Jacobi equation

$$\frac{\partial S}{\partial t} = -\left[\frac{|\nabla S|^2}{2m} + V + Q\right],$$

also referred to as *quantum Hamilton–Jacobi equation*.[5] It differs from the classical Hamilton–Jacobi equation only by the term:

This term Q, called *quantum potential*, thus depends on the curvature of the amplitude of the wave function.[6] (See also: Pilot wave#Mathematical formulation for a single particle.)

In the limit $\hbar \to 0$, the function S is a solution of the (classical) Hamilton-Jacobi equation;[1] therefore, the function S is also called the Hamilton-Jacobi function or action extended to quantum physics.

19.2 Properties

Hiley emphasised several aspects[7] that regard the quantum potential of a quantum particle:

- it is derived mathematically from the real part of the Schrödinger equation under polar decomposition of the wave function,[8] is not derived from a Hamiltonian[9] or other external source, and could be said to be involved in a self-organising process involving a basic underlying field;

- it does not change if R is multiplied by a constant, as this term is also present in the denominator, so that Q is independent of the magnitude of ψ and thus of field intensity; therefore, the quantum potential fulfils a precondition for nonlocality: it need not fall off as distance increases;

- it carries information about the whole experimental arrangement in which the particle finds itself.

In 1979, Hiley and his co-workers Philippidis and Dewdney presented a full calculation on the explanation of the two-slit experiment in terms of Bohmian trajectories that arise for each particle moving under the influence of the quantum potential, resulting in the well-known interference patterns.[10]

Also the shift of the interference pattern which occurs in presence of a magnetic field in the Aharonov–Bohm effect could be explained as arising from the quantum potential.[11]

19.2.1 Relation to the measurement process

The collapse of the wave function of the Copenhagen interpretation of quantum theory is explained in the quantum potential approach by the demonstration that, after a measurement, "all the packets of the multi-dimensional wave function that do not correspond to the actual result of measurement have no effect on the particle" from then on.[12] Bohm and Hiley pointed out that

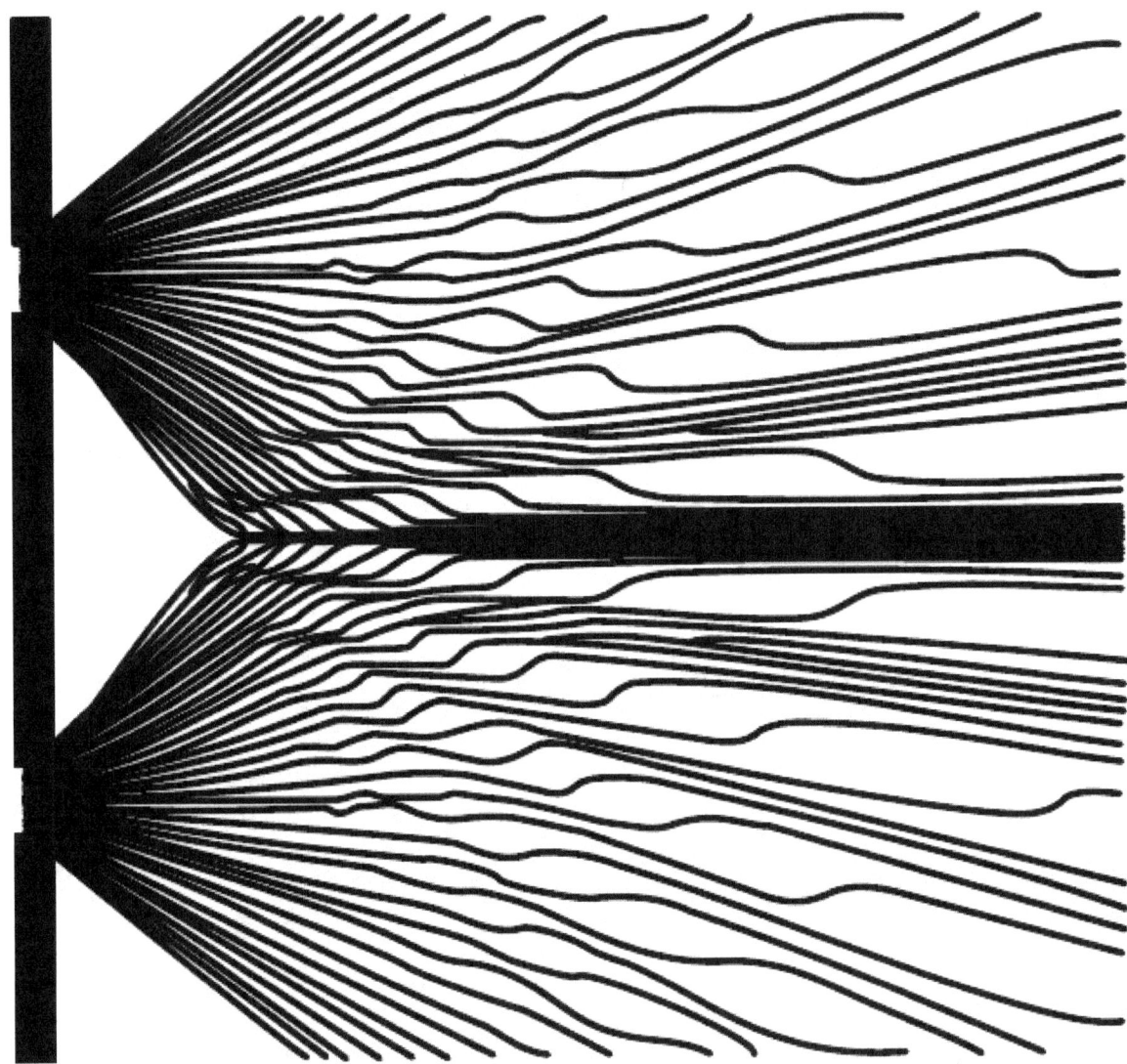

Bohm trajectories under the influence of the quantum potential, at the example of an electron going through the two-slit experiment.

'the quantum potential can develop unstable bifurcation points, which separate classes of particle trajectories according to the "channels" into which they eventually enter and within which they stay. This explains how measurement is possible without "collapse" of the wave function, and how all sorts of quantum processes, such as transitions between states, fusion of two states into one and fission of one system into two, are able to take place without the need for a human observer.'[13]

Measurement then "involves a participatory transformation in which both the system under observation and the observing apparatus undergo a mutual participation so that the trajectories behave in a correlated manner, becoming correlated and separated into different, non-overlapping sets (which we call 'channels')".[14]

19.2.2   Quantum potential of an n-particle system

The Schrödinger wave function of a many-particle quantum system cannot be represented in ordinary three-dimensional space. Rather, it is represented in configuration space, with three dimensions per particle. A single point in configuration space thus represents the configuration of the entire n-particle system as a whole.

A two-particle wave function $\psi(\mathbf{r}_1, \mathbf{r}_2, t)$ of identical particles of mass m has the quantum potential[15]

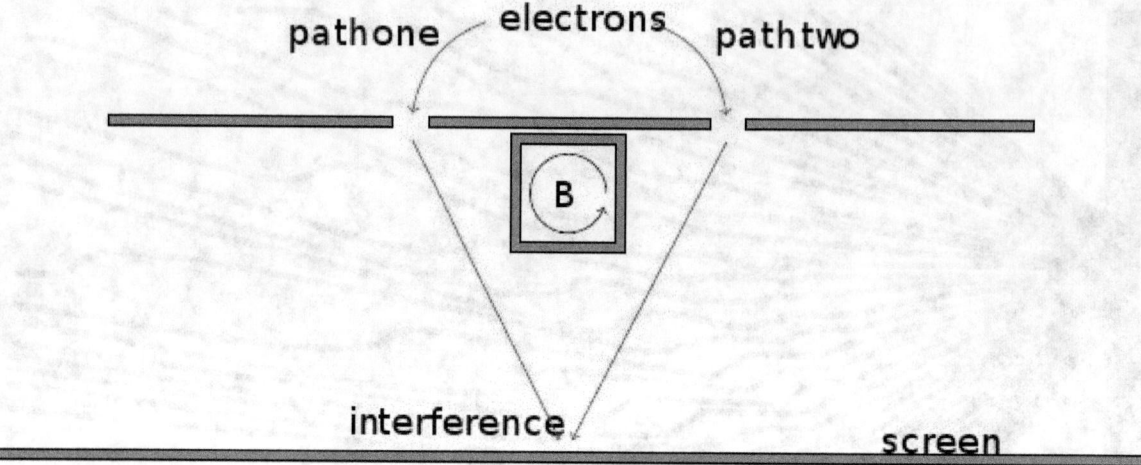

*Schematic of double-slit experiment in which Aharonov–Bohm effect can be observed: electrons pass through two slits, interfering at an observation screen, and the interference pattern undergoes a shift when a magnetic field **B** is turned on in the cylindrical solenoid.*

$$Q(\mathbf{r}_1, \mathbf{r}_2, t) = -\frac{\hbar^2}{2m} \frac{(\nabla_1^2 + \nabla_2^2) R(\mathbf{r}_1, \mathbf{r}_2, t)}{R(\mathbf{r}_1, \mathbf{r}_2, t)}$$

where ∇_1^2 and ∇_2^2 refer to particle 1 and particle 2 respectively. This expression generalizes in straightforward manner to n particles:

$$Q(\mathbf{r}_1, ..., \mathbf{r}_n, t) = -\frac{\hbar^2}{2} \sum_{i=1}^{n} \frac{\nabla_i^2}{m_i}$$

In case the wave function of two or more particles is separable, then the system's total quantum potential becomes the sum of the quantum potentials of the two particles. Exact separability is extremely unphysical given that interactions between the system and its environment destroy the factorization; however, a wave function that is a superposition of several wave functions of approximately disjoint support will factorize approximately.[16]

Derivation for a separable quantum system

That the wave function is separable means that ψ factorizes in the form $\psi(\mathbf{r}_1, \mathbf{r}_2, t) = \psi_A(\mathbf{r}_1, t) \psi_B(\mathbf{r}_2, t)$. Then it follows that also R factorizes, and the system's total quantum potential becomes the sum of the quantum potentials of the two particles.[17]

$$Q(\mathbf{r}_1, \mathbf{r}_2, t) = -\frac{\hbar^2}{2m} \left(\frac{\nabla_1^2 R_A(\mathbf{r}_1, t)}{R_A(\mathbf{r}_1, t)} + \frac{\nabla_2^2 R_B(\mathbf{r}_2, t)}{R_B(\mathbf{r}_2, t)} \right) = Q_A(\mathbf{r}_1, t) + Q_B(\mathbf{r}_2, t)$$

In case the wave function is separable, that is, if ψ factorizes in the form $\psi(\mathbf{r}_1, \mathbf{r}_2, t) = \psi_A(\mathbf{r}_1, t) \psi_B(\mathbf{r}_2, t)$, the two one-particle systems behave independently. More generally, the quantum potential of an n-particle system with separable wave function is the sum of n quantum potentials, separating the system into n independent one-particle systems.[18]

19.3   Formulation in terms of probability density

19.3.1   Quantum potential in terms of the probability density function

Bohm, as well as other physicists after him including Antony Valentini, have sought to provide evidence that the Born rule linking R to the probability density function

$$\rho = R^2$$

can be understood, in a pilot wave formulation, as not representing a basic law, but rather a *theorem* (called quantum equilibrium hypothesis) which applies when a *quantum equilibrium* is reached during the course of the time development under the Schrödinger equation. With Born's rule, and straightforward application of the chain and product rules

$$\nabla^2 \sqrt{\rho} = \nabla\nabla\rho^{1/2} = \nabla(\frac{1}{2}\rho^{-1/2}\nabla\rho) = \frac{1}{2}\nabla(\rho^{-1/2}\nabla\rho) = \frac{1}{2}\left[(\nabla\rho^{-1/2})\nabla\rho + \rho^{-1/2}\nabla^2\rho\right]$$

the quantum potential, expressed in terms of the probability density function, becomes:[19]

$$Q = -\frac{\hbar^2}{2m}\frac{\nabla^2\sqrt{\rho}}{\sqrt{\rho}} = -\frac{\hbar^2}{4m}\left[\frac{\nabla^2\rho}{\rho} - \frac{1}{2}\frac{(\nabla\rho)^2}{\rho^2}\right]$$

19.3.2   Quantum force

The quantum force $F_Q = -\nabla Q$, expressed in terms of the probability distribution, amounts to:[20]

$$F_Q = \frac{\hbar^2}{4m}\left[\frac{\nabla(\nabla^2\rho)}{\rho} - \frac{\nabla(\nabla\rho \cdot \nabla\rho)}{2\rho^2} - \left(\frac{\nabla^2\rho}{\rho} - \frac{\nabla\rho \cdot \nabla\rho}{\rho^2}\right)\frac{\nabla\rho}{\rho}\right]$$

19.3.3   Formulation in configuration space and in momentum space, as the result of projections

M. R. Brown and B. Hiley showed that, as alternative to its formulation terms of configuration space (x-space), the quantum potential can also be formulated in terms of momentum space (p-space).[21][22]

In line with David Bohm's approach, Basil Hiley and mathematician Maurice de Gosson showed that the quantum potential can be seen as a consequence of a projection of an underlying structure, more specifically of a non-commutative algebraic structure, onto a subspace such as ordinary space (x-space). In algebraic terms, the quantum potential can be seen as arising from the relation between implicate and explicate orders: if a non-commutative algebra is employed to describe the non-commutative structure of the quantum formalism, it turns out that it is impossible to define an underlying space, but that rather "shadow spaces" (homomorphic spaces) can be constructed and that in so doing the quantum potential appears.[22][23][24][25][26] The quantum potential approach can be seen as a way to construct the shadow spaces.[24] The quantum potential thus results as a distortion due to the projection of the underlying space into x-space, in similar manner as a Mercator projection inevitably results in a distortion in a geographical map.[27][28] There exists complete symmetry between the x-representation, and the quantum potential as it appears in configuration space can be seen as arising from the dispersion of the momentum p-representation.[29]

The approach has been applied to extended phase space, also in terms of a Duffin–Kemmer–Petia u algebra approach.

19.4   Relation to other quantities and theories

19.4.1   Relation to the Fisher information

It can be shown[33] that the mean value of the quantum potential $Q = -\hbar^2 \, \nabla^2 \sqrt{\rho}/(2m\sqrt{\rho})$ is proportional to the probability density's Fisher information about the observable $\hat{x}$

$$\mathcal{I} = \int \rho \, (\nabla \ln \rho)^2 d^3x = -\int \rho \, \nabla^2 (\ln \rho) \, d^3x$$

Using this definition for the Fisher Information, we can write[34]

$$<Q> = \int \psi^* Q \, \psi \, d^3x = \int \rho \, Q \, d^3x = \frac{\hbar^2}{8m}\mathcal{I}$$

19.4.2   Relation to the Madelung pressure tensor

In the Madelung equations presented by Erwin Madelung in 1927, the non-local quantum pressure tensor has the same mathematical form as the quantum potential. The underlying theory is different in that the Bohm approach describes particle trajectories whereas the equations of Madelung quantum hydrodynamics are the Euler equations of a fluid that describe its averaged statistical characteristics.[35]

19.4.3   Relation to the von Weizsäcker correction

In 1935,[36] Carl Friedrich von Weizsäcker proposed the addition of an inhomogeneity term (sometimes referred to as a *von Weizsäcker correction*) to the kinetic energy of the Thomas–Fermi (TF) theory of atoms.[37]

The von Weizsäcker correction term is:[38]

$$E_W[\rho] = \int dr \rho \hbar^2 (\nabla \ln \rho)^2/8m = (\hbar^2/8m) \int dr (\nabla \rho)^2/\rho = \int dr \rho \, Q$$

The correction term has also been derived as the first order correction to the TF kinetic energy in a semi-classical correction to the Hartree–Fock theory.[39]

It has been pointed out[38] that the von Weizsäcker correction term at low density takes on the same form as the quantum potential.

19.4.4   Quantum potential as energy of internal motion associated with spin

Givanni Salesi, Erasmo Recami and co-workers showed in 1998 that, in agreement with the König's theorem, the quantum potential can be identified with the kinetic energy of the internal motion ("zitterbewegung") associated with the spin of a spin-½ particle observed in a center-of-mass frame. More specifically, they showed that the internal *zitterbewegung* velocity for a spinning, non-relativistic particle of constant spin with no precession, and in absence of an external field, has the squared value:[40]

$$\mathbf{V}^2 = \frac{(\nabla \rho \wedge \mathbf{s})^2}{(m\rho)^2} = \frac{(\nabla \rho)^2 \mathbf{s}^2 - (\nabla \rho \cdot \mathbf{s})}{(m\rho)^2}$$

from which the second term is shown to be of negligible size; then with $|\mathbf{s}| = \hbar/2$ it follows that

$$|\mathbf{V}| = \frac{\hbar}{2} \frac{\nabla \rho}{m\rho}$$

Salesi gave further details on this work in 2009.[41]

In 1999, Salvatore Esposito generalized their result from spin-½ particles to particles of arbitrary spin, confirming the interpretation of the quantum potential as a kinetic energy for an internal motion. Esposito showed that (using the notation $\hbar = 1$) the quantum potential can be written as:[42]

$$Q = -\frac{1}{2}m\mathbf{v}_S^2 - \frac{1}{2}\nabla \cdot \mathbf{v}_S$$

and that the causal interpretation of quantum mechanics can be reformulated in terms of a particle velocity

$$\mathbf{v} = \mathbf{v}_B + \mathbf{v}_S \times \mathbf{s}$$

where the "drift velocity" is

$$\mathbf{v}_B = \frac{\nabla S}{m}$$

and the "relative velocity" is $\mathbf{v}_S \times \mathbf{s}$, with

$$\mathbf{v}_S = \frac{\nabla R^2}{2mR^2}$$

and $\mathbf{s}$ representing the spin direction of the particle. In this formulation, according to Esposito, quantum mechanics must necessarily be interpreted in probabilistic terms, for the reason that a system's initial motion condition cannot be exactly determined.[42] Esposito explained that "the quantum effects present in the Schrödinger equation are due to the presence of a peculiar spatial direction associated with the particle that, assuming the isotropy of space, can be identified with the spin of the particle itself".[43] Esposito generalized it from matter particles to gauge particles, in particular photons, for which he showed that, if modelled as $\psi = (\mathbf{E} - i\mathbf{B})/\sqrt{2}$, with probability function $\psi^* \cdot \psi = (\mathbf{E}^2 + \mathbf{B}^2)/2$, they can be understood in a quantum potential approach.[44]

James R. Bogan, in 2002, published the derivation of a reciprocal transformation from the Hamilton-Jacobi equation of classical mechanics to the time-dependent Schrödinger equation of quantum mechanics which arises from a gauge transformation representing spin, under the simple requirement of conservation of probability. This spin-dependent transformation is a function of the quantum potential.[45]

19.4.5   EP quantum mechanics with quantum potential as Schwarzian derivative

In a different approach, the EP quantum mechanics formulate on the basis of an Equivalence Principle (EP), a quantum potential is written as:[46][47]

$$Q(q) = \frac{\hbar^2}{4m}\{S; q\}$$

where $\{ ; \}$ is the Schwarzian derivative, that is, $\{S; q\} = (S'''/S') - (3/2)(S''/S')^2$. However, even in cases where this may equal

$$Q(q) = -\frac{\hbar^2}{2m}\frac{\Delta R}{R}$$

it is stressed by E. Faraggi and M. Matone that this does not correspond with the usual quantum potential, as in their approach $R \exp(iS/\hbar)$ is a solution to the Schrödinger equation but does *not* correspond to the wave function.[46] This has been investigated further by E.R. Floyd for the classical limit $\hbar \to 0$,[48] as well as by Robert Carroll.[49]

19.5 Re-interpretation in terms of Clifford algebras

B. Hiley and R. E. Callaghan re-interpret the role of the Bohm model and its notion of quantum potential in the framework of Clifford algebra, taking account of recent advances that include the work of David Hestenes on spacetime algebra. They show how, within a nested hierarchy of Clifford algebras $C\ell_{i,j}$, for each Clifford algebra an element of a minimal left ideal $\Phi_L(\mathbf{r},t)$ and an element of a right ideal representing its Clifford conjugation $\Phi_R(\mathbf{r},t) = \tilde{\Phi}_L(\mathbf{r},t)$ can be constructed, and from it the *Clifford density element* (CDE) $\rho_c(\mathbf{r},t) = \Phi_L(\mathbf{r},t)\tilde{\Phi}_L(\mathbf{r},t)$, an element of the Clifford algebra which is isomorphic to the standard density matrix but independent of any specific representation.[50] On this basis, bilinear invariants can be formed which represent properties of the system. Hiley and Callaghan distinguish bilinear invariants of a first kind, of which each stands for the expectation value of an element B of the algebra which can be formed as $\text{Tr}B\rho_c$, and bilinear invariants of a second kind which are constructed with derivatives and represent momentum and energy. Using these terms, they reconstruct the results of quantum mechanics without depending on a particular representation in terms of a wave function nor requiring reference to an external Hilbert space. Consistent with earlier results, the quantum potential of a non-relativistic particle with spin (Pauli particle) is shown to have an additional spin-dependent term, and the momentum of a relativistic particle with spin (Dirac particle) is shown to consist in a linear motion and a rotational part.[51] The two dynamical equations governing the time evolution are re-interpreted as conservation equations. One of them stands for the conservation of energy; the other stands for the conservation of probability and of spin.[52] The quantum potential plays the role of an internal energy[53] which ensures the conservation of total energy.[52]

19.6 Relativistic and field-theoretic extensions

19.6.1 Quantum potential and relativity

Bohm and Hiley demonstrated that the non-locality of quantum theory can be understood as limit case of a purely local theory, provided the transmission of *active information* is allowed to be greater than the speed of light, and that this limit case yields approximations to both quantum theory and relativity.[54]

The quantum potential approach was extended by Hiley and co-workers to quantum field theory in Minkowski spacetime[55][56][57][58] and to curved spacetime.[59]

Carlo Castro and Jorge Mahecha derived the Schrödinger equation from the Hamilton-Jacobi equation in conjunction with the continuity equation, and showed that the properties of the relativistic Bohm quantum potential in terms of the ensemble density can be described by the Weyl properties of space. In Riemann flat space, the Bohm potential is shown to equal the Weyl curvature. According to Castro and Mahecha, in the relativistic case, the quantum potential (using the d'Alembert operator $\Box$ and in the notation $\hbar = 1$) takes the form

$$Q = -\frac{1}{2m}\frac{\Box\sqrt{\rho}}{\sqrt{\rho}}$$

and the quantum force exerted by the relativistic quantum potential is shown to depend on the Weyl gauge potential and its derivatives. Furthermore, the relationship among Bohm's potential and the Weyl curvature in flat spacetime corresponds to a similar relationship among Fisher Information and Weyl geometry after introduction of a complex momentum.[60]

Diego L. Rapoport, on the other hand, associates the relativistic quantum potential with the metric scalar curvature (Riemann curvature).[61]

In relation to the Klein–Gordon equation for a particle with mass and charge, Peter R. Holland spoke in his book of 1993 of a 'quantum potential-like term' that is proportional $\Box R/R$. He emphasized however that to give the Klein–Gordon theory a single-particle interpretation in terms of trajectories, as can be done for nonrelativistic Schrödinger quantum mechanics, would lead to unacceptable inconsistencies. For instance, wave functions $\psi(\mathbf{x},t)$ that are solutions to the Klein–Gordon or the Dirac equation cannot be interpreted as the probability amplitude for a particle to *be found in* a given volume d^3x at time t in accordance with the usual axioms of quantum mechanics, and similarly in the causal interpretation it cannot be interpreted as the probability for the particle to *be in* that volume at that time. Holland pointed out that, while efforts have been made to determine a Hermitian position operator that would allow an interpretation of configuration space quantum field theory, in particular using the Newton–Wigner localization approach, but that no connection with possibilities for an empirical determination of position in terms of a relativistic measurement theory or for a trajectory interpretation has so far been established. Yet according to Holland this does not mean that the trajectory concept is to be discarded from considerations of relativistic quantum mechanics.[62]

Hrvoje Nikolić derived $Q = -(1/2m)\,\square R/R$ as expression for the quantum potential, and he proposed a Lorentz-covariant formulation of the Bohmian interpretation of many-particle wave functions.[63] He also developed a generalized relativistic-invariant probabilistic interpretation of quantum theory,[64][65][66] in which $|\psi|^2$ is no longer a probability density in space but a probability density in space-time.[67]

See also: De Broglie–Bohm theory § Relativity

19.6.2 Quantum potential in quantum field theory

Starting from the space representation of the field coordinate, a causal interpretation of the Schrödinger picture of relativistic quantum theory has been constructed starting from the space representation of the field coordinate. The Schrödinger picture for a neutral, spin 0, massless field $\Psi\left[\psi(\mathbf{x},t)\right] = R\left[\psi(\mathbf{x},t)\right] e^{S[\psi(\mathbf{x},t)]}$, with $R\left[\psi(\mathbf{x},t)\right]$, $S\left[\psi(\mathbf{x},t)\right]$ real-valued functionals, can be shown[68] to lead to

$$Q\left[\psi(\mathbf{x},t)\right] = -(1/2R) \int d^3 x \, \delta^2 R / \delta\psi^2$$

This has been called the **superquantum potential** by Bohm and his co-workers.[69]

Basil Hiley showed that the energy-momentum-relations in the Bohm model can be obtained directly from the energy-momentum tensor of quantum field theory and that the quantum potential is an energy term that is required for local energy-momentum conservation.[70] He has also hinted that for particle with energies equal to or higher than the pair creation threshold, Bohm's model constitutes a many-particle theory that describes also pair creation and annihilation processes.[71]

19.7 Interpretation and naming of the quantum potential

In his article of 1952, providing an alternative interpretation of quantum mechanics, Bohm already spoke of a "quantum-mechanical" potential.[72]

Bohm and Basil Hiley also called the quantum potential an *information potential*, given that it influences the form of processes and is itself shaped by the environment.[9] Bohm indicated "The ship or aeroplane (with its automatic Pilot) is a *self-active* system, i.e. it has its own energy. But the form of its activity is determined by the *information content* concerning its environment that is carried by the radar waves. This is independent of the intensity of the waves. We can similarly regard the quantum potential as containing *active information*. It is potentially active everywhere, but actually active only where and when there is a particle." (italics in original).[73]

Hiley refers to the quantum potential as internal energy[24] and as "a new quality of energy only playing a role in quantum processes".[74] He explains that the quantum potential is a further energy term aside the well-known kinetic energy and the (classical) potential energy and that it is a nonlocal energy term that arises necessarily in view of the requirement of energy conservation; he added that much of the physics community's resistance against the notion of the quantum potential may have been due to scientists' expectations that energy should be local.[75]

Hiley has emphasized that the quantum potential, for Bohm, was "a key element in gaining insights into what could underlie the quantum formalism. Bohm was convinced by his deeper analysis of this aspect of the approach that the theory could not be mechanical. Rather, it is organic in the sense of Whitehead. Namely, that it was the whole that determined the properties of the individual particles and their relationship, not the other way round."[76] (*See also:* Bohm and Hiley's work on quantum potential and active information)

Peter R. Holland, in his comprehensive textbook, also refers to it as *quantum potential energy*.[77] The quantum potential is also referred to in association with Bohm's name as *Bohm potential*, *quantum Bohm potential* or *Bohm quantum potential*.

19.8 Applications

The quantum potential approach can be used to model quantum effects without requiring the Schrödinger equation to be explicitly solved, and it can be integrated in simulations, such as Monte Carlo simulations using the hydrodynamic and drift diffusion equations.[78] This is done in form of a "hydrodynamic" calculation of trajectories: starting from the density at each "fluid element", the acceleration of each "fluid element" is computed from the gradient of V and Q , and the resulting divergence of the velocity field determines the change to the density.[79]

The approach using Bohmian trajectories and the quantum potential is used for calculating properties of quantum systems which cannot be solved exactly, which are often approximated using semi-classical approaches. Whereas in mean field approaches the potential for the classical motion results from an average over wave functions, this approach does not require the computation of an integral over wave functions.[80]

The expression for the quantum force has been used, together with Bayesian statistical analysis and Expectation-maximisation methods, for computing ensembles of trajectories that arise under the influence of classical and quantum forces.[20]

19.9 Further reading

Fundamental articles

- Bohm, David (1952). "A Suggested Interpretation of the Quantum Theory in Terms of "Hidden Variables" I". *Physical Review* 85: 166–179. Bibcode:1952PhRv...85..166B. doi:10.1103/PhysRev.85.166. (full text)

- Bohm, David (1952). "A Suggested Interpretation of the Quantum Theory in Terms of "Hidden Variables", II". *Physical Review* 85: 180–193. Bibcode:1952PhRv...85..180B. doi:10.1103/PhysRev.85.180. (full text)

- D. Bohm, B. J. Hiley, P. N. Kaloyerou: *An ontological basis for the quantum theory*, Physics Reports (Review section of Physics Letters), volume 144, number 6, pp. 321–375, 1987 (full text), therein: D. Bohm, B. J. Hiley: *I. Non-relativistic particle systems*, pp. 321–348, and D. Bohm, B. J. Hiley, P. N. Kaloyerou: *II. A causal interpretation of quantum fields*, pp. 349–375

Recent articles

- *Spontaneous creation of the universe from nothing*, arXiv:1404.1207v1, 4 April 2014

- Maurice de Gosson, Basil Hiley: *Short Time Quantum Propagator and Bohmian Trajectories*, arXiv:1304.4771v1 (submitted 17 April 2013)

- Robert Carroll: *Fluctuations, gravity, and the quantum potential*, 13 January 2005, asXiv:gr-qc/0501045v1

Overview

- Davide Fiscaletti: *About the Different Approaches to Bohm's Quantum Potential in Non-Relativistic Quantum Mechanics*, Quantum Matter, Volume 3, Number 3, June 2014, pp. 177–199(23), DOI 10.1166/qm.2014.1113.

- Ignazio Licata, Davide Fiscaletti (with a foreword by B.J. Hiley): *Quantum potential: Physics, Geometry and Algebra*, AMC, Springer, 2013, ISBN 978-3-319-00332-0 (print) / ISBN 978-3-319-00333-7 (online)

- Peter R. Holland: *The Quantum Theory of Motion: An Account of the De Broglie-Bohm Causal Interpretation of Quantum Mechanics*, Cambridge University Press, Cambridge (first published June 25, 1993), ISBN 0-521-35404-8 hardback, ISBN 0-521-48543-6 paperback, transferred to digital printing 2004

- David Bohm, Basil Hiley: *The Undivided Universe: An Ontological Interpretation of Quantum Theory*, Routledge, 1993, ISBN 0-415-06588-7

- David Bohm, F. David Peat: *Science, Order and Creativity*, 1987, Routledge, 2nd ed. 2000 (transferred to digital printing 2008, Routledge), ISBN 0-415-17182-2

19.10   References

[1] Bohm, David (1952). "A Suggested Interpretation of the Quantum Theory in Terms of "Hidden Variables" I". *Physical Review* **85**: 166–179. Bibcode:1952PhRv...85..166B. doi:10.1103/PhysRev.85.166. (full text)

[2] Bohm, David (1952). "A Suggested Interpretation of the Quantum Theory in Terms of "Hidden Variables", II". *Physical Review* **85**: 180–193. Bibcode:1952PhRv...85..180B. doi:10.1103/PhysRev.85.180. (full text)

[3] D. Bohm, B. J. Hiley: *On the intuitive understanding of nonlocality as implied by quantum theory*, Foundations of Physics, Volume 5, Number 1, pp. 93-109, 1975, doi:10.1007/BF01100319 (abstract)

[4] David Bohm, Basil Hiley: *The Undivided Universe: An Ontological Interpretation of Quantum Theory*, Routledge, 1993, ISBN 0-415-06588-7, therein Chapter 3.1. *The main points of the causal interpretation*, p. 22–23

[5] David Bohm, Basil Hiley: *The Undivided Universe: An Ontological Interpretation of Quantum Theory*, Routledge, 1993, ISBN 0-415-06588-7, also as cited in: B. J. Hiley and R. E. Callaghan: *Clifford Algebras and the Dirac-Bohm Quantum Hamilton-Jacobi Equation*, Foundations of Physics, January 2012, Volume 42, Issue 1, pp 192-208 (published online 20 May 2011), doi:10.1007/s10701-011-9558-z (abstract, 2010 preprint by B. Hiley)

[6] See for ex. Robert E. Wyatt, Eric R. Bittner: *Quantum wave packet dynamics with trajectories: Implementation with adaptive Lagrangian grids of the amplitude of the wave function*, Journal of Chamical Physics, vol. 113, no. 20, 22 November 2000, p. 8898

[7] B. J. Hiley: *Active Information and Teleportation*, p. 7; appeared in: Epistemological and Experimental Perspectives on Quantum Physics, D. Greenberger et al. (eds.), pages 113-126, Kluwer, Netherlands, 1999

[8] B.J. Hiley: *From the Heisenberg picture to Bohm: A New Perspective on Active Information and it Relation to Shannon Information*, pp. 2 and 5. Published in: A. Khrennikov (ed.): *Proc. Conf. Quantum Theory: reconsideration of foundations*, pp. 141–162, Vaxjö University Press, Sweden, 2002

[9] B. J. Hiley: *Information, quantum theory and the brain*. In: Gordon G. Globus (ed.), Karl H. Pribram (ed.), Giuseppe Vitiello (ed.): Brain and being: at the boundary between science, philosophy, language and arts, Advances in Consciousness Research, John Benjamins B.V., 2004, ISBN 90-272-5194-0, pp. 197-214, p. 207

[10] C. Philippidis, C. Dewdney, B. J. Hiley: *Quantum interference and the quantum potential*, Il nuovo cimento B, vol. 52, no. 1, 1979, pp.15-28, doi:10.1007/BF02743566

[11] C. Philippidis, D. Bohm, R. D. Kaye: *The Aharonov-Bohm effect and the quantum potential*, Il nuovo cimento B, vol. 71, no. 1, pp. 75-88, 1982, doi:10.1007/BF02721695

[12] Basil J. Hiley: *The role of the quantum potential*. In: G. Tarozzi, Alwyn Van der Merwe: *Open questions in quantum physics: invited papers on the foundations of microphysics*, Springer, 1985, pages 237 ff., therein page 239

[13] D. Bohm, B. J. Hiley, P. N. Kaloyerou: *An ontological basis for the quantum theory*, Physics Reports (Review section of Physics Letters), volume 144, number 6, pp. 323–348, 1987 (abstract)

[14] B. J. Hiley: *The conceptual structure of the Bohm interpretation of quantum mechanics*, In: K.V. Laurikainen, C. Montonen, K. Sunnarborg (eds.): Symposium on the Foundations of Modern Physics 1994 – 70 years of Matter Waves, Editions Frontières, pp. 99–118, ISBN 2-86332-169-2, p. 106

[15] B. J. Hiley: *Active Information and Teleportation*, p. 10; appeared in: Epistemological and Experimental Perspectives on Quantum Physics, D. Greenberger et al. (eds.), pages 113-126, Kluwer, Netherlands, 1999

[16] See for instance Detlef Dürr et al: *Quantum equilibrium and the origin of absolute uncertainty*, arXiv:quant-ph/0308039v1 6 August 2003, p. 23 ff.

[17] David Bohm, Basil Hiley: *The Undivided Universe: An Ontological Interpretation of Quantum Theory*, Routledge, 1993, ISBN 0-415-06588-7, transferred to digital printing 2005, therein Chapter 4.1. *The ontological interpretation of the many-body system*, p. 59

[18] D. Bohm, B. J. Hiley, P. N. Kaloyerou: *An ontological basis for the quantum theory*, Physics Reports (Review section of Physics Letters), volume 144, number 6, pp. 323–348, 1987 (p. 351, eq. (12)<--page=31 p. 351 is not(!) a typo-->)

[19] See for example the *Introduction* section of: Fernando Ogiba: *Phenomenological derivation of the Schrödinger equation*, Progress in Physics (indicated date: October 2011, but retrieved online earlier: July 31, 2011)

[20] Jeremy B. Maddox, Eric R. Bittner: *Estimating Bohm's quantum force using Bayesian statistics*, Journal of Chemical Physics, October 2003, vol. 119, no. 13, p. 6465–6474, therein p. 6472, eq.(38)

[21] M. R. Brown: *The quantum potential: the breakdown of classical symplectic symmetry and the energy of localisation and dispersion*, arXiv.org (submitted on 6 Mar 1997, version of 5 Feb 2002, retrieved 24 July 2011) (abstract)

[22] M. R. Brown, B. J. Hiley: *Schrodinger revisited: an algebraic approach*, arXiv.org (submitted 4 May 2000, version of 19 July 2004, retrieved June 3, 2011) (abstract)

[23] Maurice A. de Gosson: *"The Principles of Newtonian and Quantum Mechanics – The Need for Planck's Constant, h"*, Imperial College Press, World Scientific Publishing, 2001, ISBN 1-86094-274-1

[24] B. J. Hiley: *Non-commutative quantum geometry: A reappraisal of the Bohm approach to quantum theory*, in: A. Elitzur et al. (eds.): *Quo vadis quantum mechanics*, Springer, 2005, ISBN 3-540-22188-3, p. 299–324

[25] B.J. Hiley: *Non-Commutative Quantum Geometry: A Reappraisal of the Bohm Approach to Quantum Theory*. In: Avshalom C. Elitzur, Shahar Dolev, Nancy Kolenda (eds.): *Quo Vadis Quantum Mechanics? The Frontiers Collection*, 2005, pp. 299-324, doi:10.1007/3-540-26669-0_16 (abstract, preprint)

[26] B.J. Hiley: *Phase space description of quantum mechanics and non-commutative geometry: Wigner–Moyal and Bohm in a wider context*, In: Theo M. Nieuwenhuizen et al (eds.): *Beyond the quantum*, World Scientific Publishing, 2007, ISBN 978-981-277-117-9, pp. 203–211, therein p. 204

[27] Basil J. Hiley: *Towards a Dynamics of Moments: The Role of Algebraic Deformation and Inequivalent Vacuum States*, published in: Correlations ed. K. G. Bowden, Proc. ANPA 23, 104-134, 2001 (PDF)

[28] B. J. Hiley, R. E. Callaghan: *The Clifford Algebra approach to Quantum Mechanics A: The Schroedinger and Pauli Particles*, arXiv.org (submitted on 17 Nov 2010 - abstract)

[29] B. Hiley: *Phase space description of quantum mechanics and non-commutative geometry: Wigner-Moyal and Bohm in a wider context*, in: Th. M. Nieuwenhuizen et al. (eds.): *Beyond the Quantum*, World Scientific, 2007, ISBN 978-981-277-117-9, p. 203–211, therein: p. 207 ff.

[30] S. Nasiri: *Quantum potential and symmetries in extended phase space*, SIGMA 2 (2006), 062, quant-ph/0511125

[31] Marco Cezar B. Fernandes, J. David M. Vianna: *On the Generalized Phase Space Approach to Duffin–Kemmer–Petiau Particles*, Brazilian Journal of Physics, vol. 28, no. 4, December 1998, doi:10.1590/S0103-97331998000400024

[32] M.C.B. Fernandes, J.D.M. Vianna: *On the Duffin-Kemmer-Petiau algebra and the generalized phase space*, Foundations of Physics, vol. 29, no. 2, 1999 (abstract)

[33] M. Reginatto, Phys. Rev. A 58, 1775 (1998), cited after: Roumen Tsekov: *Towards nonlinear quantum Fokker-Planck equations*, Int. J. Theor. Phys. 48 (2009) 1431-1435 (arXiv 0808.0326, p. 4)

[34] Robert Carroll: *On the Emergence Theme of Physics*, World Scientific, 2010, ISBN 981-4291-79-X, Chapter 1 *Some quantum background*, p. 1

[35] Tsekov, R. (2012) Bohmian Mechanics versus Madelung Quantum Hydrodynamics doi:10.13140/RG.2.1.3663.8245

[36] C.F. von Weizsäcker: *Zur Theorie der Kernmassen*, Zeitschrift für Physik, Volume 96, pp. 431-458, 1935

[37] See also section "Introduction" of: Rafael Benguria, Haim Brezis, Elliott H. Lieb: *The Thomas–Fermi–von Weizsäcker theory of atoms and molecules*, Commun. Math. Phys., Volume 79, pp. 167–180 (1981) doi:10.1007/BF01942059

[38] See also Roumen Tsekov: *Dissipative time dependent density functional theory*, Int. J. Theor. Phys., Vol. 48, pp. 2660–2664 (2009) [arXiv 0903.3644]

[39] Kompaneets, A.S., Pavlovskii, E.S.: Sov. Phys. JETP, Volume 4, pp. 328–336 (1957). Cited in section "Introduction" of: Rafael Benguria, Haim Brezis, Elliott H. Lieb: *The Thomas–Fermi–von Weizsäcker theory of atoms and molecules*, Commun. Math. Phys., Volume 79, pp. 167–180 (1981) doi:10.1007/BF01942059

[40] G. Salesi, E. Recami, H. E. Hernández F., Luis C. Kretly: *Hydrodynamics of spinning particles*, submitted 15 February 1998, arXiv.org, arXiv:hep-th/9802106v1

[41] G. Salesi: *Spin and Madelung fluid*, submitted 23 June 2009, arXiv:quant-ph/0906.4147v1

[42] Salvatore Esposito: *On the role of spin in quantum mechanics*, submitted 5 February 1999, arXiv:quant-ph/9902019v1

[43] p. 7

[44] S. Esposito: *Photon wave mechanics: A de Broglie–Bohm approach*, p. 8 ff.

[45] James R. Bogan: *Spin: The classical to quantum connection*, arXiv.org, submitted 19 December 2002, arXiv:quant-ph/0212110

[46] Alon E. Faraggi, M. Matone: *The Equivalence Postulate of Quantum Mechanics*, International Journal of Modern Physics A, vol. 15, no. 13, pp. 1869–2017. arXiv hep-th/9809127 of 6 August 1999

[47] Robert Carroll: *Aspects of quantum groups and integrable systems*, Proceedings of Institute of Mathematics of NAS of Ukraine, vo. 50, part 1, 2004, pp. 356–367, p. 357

[48] Edward R. Floyd: *Classical limit of the trajectory representation of quantum mechanics, loss of information and residual indeterminacy*, arXiv:quant-ph/9907092v3

[49] R. Carroll: *Some remarks on time, uncertainty, and spin*, arXiv:quant-ph/9903081v1

[50] B. Hiley, R. E. Callaghan: *The Clifford algebra approach to quantum mechanics A: The Schrödinger and Pauli particles*, 14 March 2010, p. 6

[51] B. Hiley, R. E. Callaghan: *The Clifford algebra approach to quantum mechanics A: The Schrödinger and Pauli particles*, 14 March 2010, p. 1-29

[52] B. Hiley: *Clifford algebras and the Dirac–Bohm Hamilton–Jacobi equation*, 2 March 2010, p. 22

[53] B. J. Hiley: *Non-commutative geometry, the Bohm interpretation and the mind–matter relationship*, p. 14

[54] D. Bohm, B. J. Hiley: *Non-locality and locality in the stochastic interpretation of quantum mechanics*, Physics Reports, Volume 172, Issue 3, January 1989, Pages 93-122, doi:10.1016/0370-1573(89)90160-9 (abstract)

[55] P.N. Kaloyerou, *Investigation of the Quantum Potential in the Relativistic Domain*, PhD. Thesis, Birkbeck College, London (1985)

[56] P.N. Kaloyerou, Phys. Rep. 244, 288 (1994).

[57] P.N. Kaloyerou, in "Bohmian Mechanics and Quantum Theory: An Appraisal", eds. J.T. Cushing, A. Fine and S. Goldstein, Kluwer, Dordrecht, 155 (1996).

[58] D. Bohm, B. J. Hiley, P. N. Kaloyerou: *An ontological basis for the quantum theory*, Physics Reports (Review section of Physics Letters), volume 144, number 6, pp. 323–348, 1987 (PDF)

[59] B. J. Hiley, A. H. Aziz Muft: *The ontological interpretation of quantum field theory applied in a cosmological context*. In: Miguel Ferrero, Alwyn Van der Merwe (eds.): *Fundamental problems in quantum physics*, Fundamental theories of physics, Kluwer Academic Publishers, 1995, ISBN 0-7923-3670-4, pages 141-156

[60] Carlo Castro, Jorge Mahecha: *On nonlinear quantum mechanics, Brownian motion, Weyl geometry and Fisher information*, submitted February 2005, In: F. Smarandache and V. Christianto (Eds.): *Quantization in Astrophysics, Brownian Motion, and Supersymmetry*, pp.73–87, MathTiger, 2007, Chennai, Tamil Nadu, ISBN 81-902190-9-X, page 82, eq.(37) ff.

[61] Rapoport, Diego L. (2007). "Torsion fields, Cartan-Weyl space-time, and state-space quantum geometries, Brownian motion, and their topological dimension". In Smarandache, F.; Christianto, V. *Quantization in Astrophysics, Brownian Motion, and Supersymmetry*. Chennai, Tamil Nadu: MathTiger. pp. 276–328. ISBN 81-902190-9-X. CiteSeerX: 10.1.1.75.6580.

[62] Peter R. Holland: *The quantum theory of motion*, Cambridge University Press, 1993 (re-printed 2000, transferred to digital printing 2004), ISBN 0-521-48543-6, p. 498 ff.

[63] Hrvoje Nikolić: *Relativistic Quantum Mechanics and the Bohmian Interpretation*, Foundations of Physics Letters, vol. 18, no. 6, November 2005, pp. 549-561, doi:10.1007/s10702-005-1128-1

[64] Hrvoje Nikolić: *Time in relativistic and nonrelativistic quantum mechanics*, arXiv:0811/0811.1905v2 (submitted 12 November 2008 (v1), revised 12 Jan 2009)

[65] Nikolic, H. 2010 "QFT as pilot-wave theory of particle creation and destruction", Int. J. Mod. Phys. A 25, 1477 (2010)

[66] Hrvoje Nikolić: *Making nonlocal reality compatible with relativity*, arXiv:1002.3226v2 [quant-ph] (submitted on 17 Feb 2010, version of 31 May 2010)

[67] Hrvoje Nikolić: *Bohmian mechanics in relativistic quantum mechanics, quantum field theory and string theory*, 2007 J. Phys.: Conf. Ser. 67 012035

[68] Peter R. Holland: *The quantum theory of motion*, Cambridge University Press, 1993 (re-printed 2000, transferred to digital printing 2004), ISBN 0-521-48543-6, p. 520 ff.

[69] Basil Hiley: *The conceptual structure of the Bohm interpretation of quantum mechanics*, Kalervo Vihtori Laurikainen et al (ed.): Symposium on the Foundations of Modern Physics 1994: 70 years of matter waves, *Editions Frontières, ISBN 2-86332-169-2, p. 99–117, p. 144*

[70] B. J. Hiley: *The Bohm approach re-assessed* (2010 preprint), p. 6

[71] B. J. Hiley. "Bohmian Non-commutative Dynamics: History and New Developments". Pre-print arXiv:1303.6057 (submitted 25 March 2013)

[72] Bohm, David (1952). "A Suggested Interpretation of the Quantum Theory in Terms of "Hidden Variables" I". *Physical Review* 85: 166–179. Bibcode:1952PhRv...85..166B. doi:10.1103/PhysRev.85.166. p. 170

[73] David Bohm: *Meaning And Information*, In: P. Pylkkänen (ed.): *The Search for Meaning: The New Spirit in Science and Philosophy*, Crucible, The Aquarian Press, 1989, ISBN 978-1-85274-061-0

[74] B.J. Hiley: *Non-commutative quantum geometry: A reappraisal of the Bohm approach to quantum theory*. In: Avshalom C. Elitzur, Shahar Dolev, Nancy Kolenda (es.): *Quo vadis quantum mechanics?* Springer, 2005, ISBN 3-540-22188-3, pp. 299 ff., therein p. 310

[75] Basil Hiley & Taher Gozel, episode 5, YouTube (downloaded 8 September 2013)

[76] B. J. Hiley: *Some remarks on the evolution of Bohm's proposals for an alternative to quantum mechanics*, 30 January 2010

[77] Peter R. Holland: *The quantum theory of motion*, Cambridge University Press, 1993 (re-printed 2000, transferred to digital printing 2004), ISBN 0-521-48543-6, p. 72

[78] G. Iannaccone, G. Curatola, G. Fiori: *Effective Bohm Quantum Potential for device simulators based on drift-diffusion and energy transport*, Simulation of Semiconductor Processes and Devices, 2004, vol. 2004, pp. 275–278

[79] Eric R. Bittner: *Quantum tunneling dynamics using hydrodynamic trajectories*, arXiv:quant-ph/0001119v2, 18 February 2000, p. 3.

[80] E. Gindensberger, C. Meier, J.A. Beswick: *Mixing quantum and classical dynamics using Bohmian trajectories*, Journal of Chemical Physics, vol. 113, no. 21, 1 December 2000, pp. 9369–9372

Chapter 20

Interference (wave propagation)

For interference in radio communications, see Interference (communication).
"Interference pattern" redirects here. For Moiré patterns, see Moiré pattern.
In physics, **interference** is a phenomenon in which two waves superpose to form a resultant wave of greater or lower

The iridescence of soap bubbles is due to thin-film interference.

amplitude. Interference usually refers to the interaction of waves that are correlated or coherent with each other, either because they come from the same source or because they have the same or nearly the same frequency. Interference effects can be observed with all types of waves, for example, light, radio, acoustic, surface water waves or matter waves.

20.1 Mechanism

Interference of left traveling (green) and right traveling (blue) waves in one dimension, resulting in final (red) wave

The principle of superposition of waves states that when two or more propagating waves of same type are incident on the same point, the total displacement at that point is equal to the pointwise sum of the displacements of the individual waves. If a crest of a wave meets a crest of another wave of the same frequency at the same point, then the magnitude of the displacement is the sum of the individual magnitudes – this is constructive interference. If a crest of one wave meets a trough of another wave then the magnitude of the displacements is equal to the difference in the individual magnitudes – this is known as destructive interference.

Constructive interference occurs when the phase difference between the waves is a multiple of 2π, whereas destructive interference occurs when the difference is an **odd multiple** of π. If the difference between the phases is intermediate between these two extremes, then the magnitude of the displacement of the summed waves lies between the minimum and maximum values.

Consider, for example, what happens when two identical stones are dropped into a still pool of water at different locations. Each stone generates a circular wave propagating outwards from the point where the stone was dropped. When the two waves overlap, the net displacement at a particular point is the sum of the displacements of the individual waves. At some points, these will be in phase, and will produce a maximum displacement. In other places, the waves will be in anti-phase, and there will be no net displacement at these points. Thus, parts of the surface will be stationary—these are seen in the figure above and to the right as stationary blue-green lines radiating from the center.

20.1.1 Between two plane waves

A simple form of interference pattern is obtained if two plane waves of the same frequency intersect at an angle. Interference is essentially an energy redistribution process. The energy which is lost at the destructive interference is regained at the constructive interference. One wave is travelling horizontally, and the other is travelling downwards at an angle θ to the first wave. Assuming that the two waves are in phase at the point **B**, then the relative phase changes along the x-axis. The phase difference at the point **A** is given by

$$\Delta\varphi = \frac{2\pi d}{\lambda} = \frac{2\pi x \sin\theta}{\lambda}$$

It can be seen that the two waves are in phase when

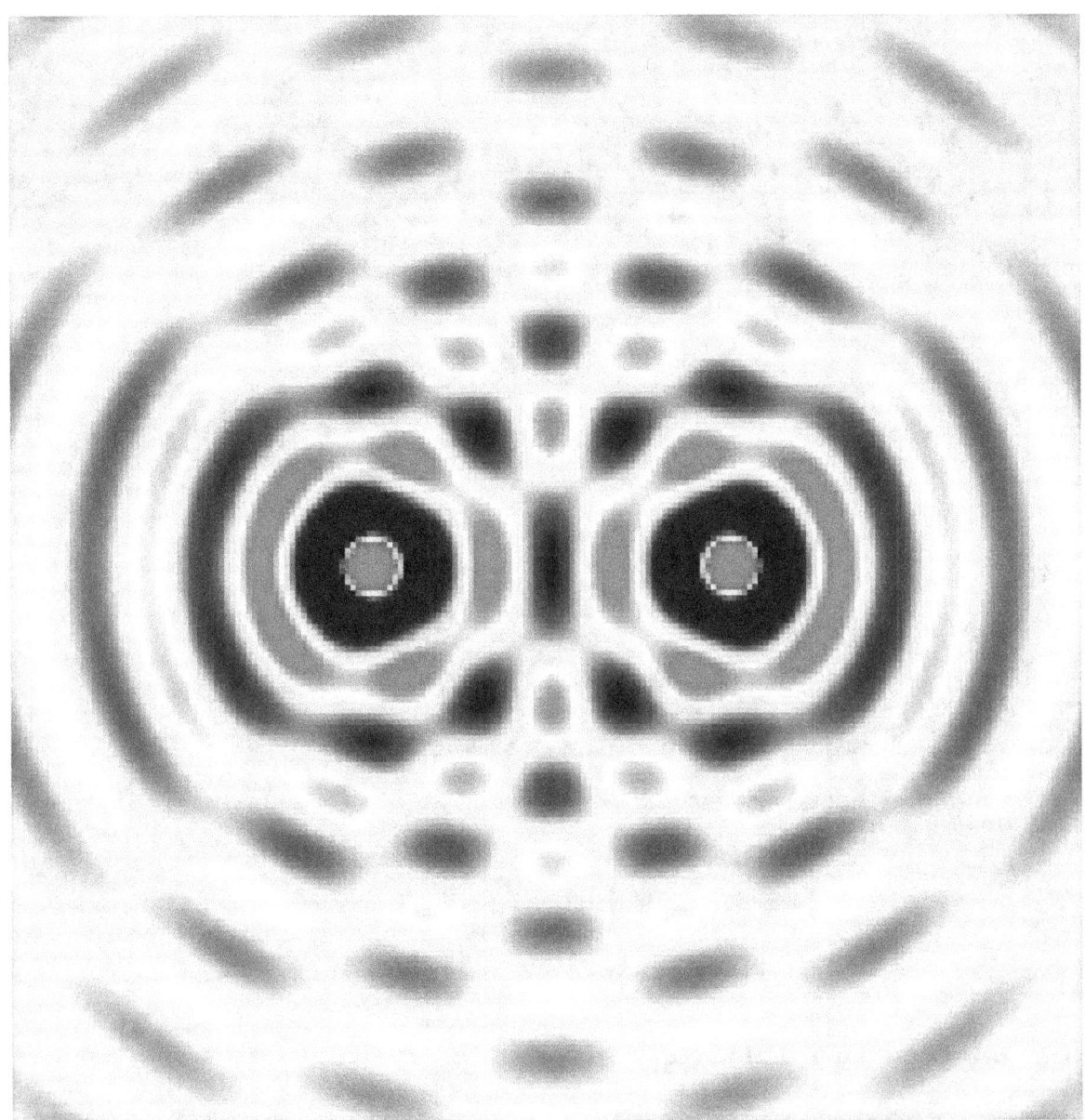

Interference of waves from two point sources.

$$\frac{x \sin \theta}{\lambda} = 0, \pm 1, \pm 2, ...$$

and are half a cycle out of phase when

$$\frac{x \sin \theta}{\lambda} = \pm \frac{1}{2}, \pm \frac{3}{2}, ...$$

Constructive interference occurs when the waves are in phase, and destructive interference when they are half a cycle out of phase. Thus, an interference fringe pattern is produced, where the separation of the maxima is

$$d_f = \frac{\lambda}{\sin \theta}$$

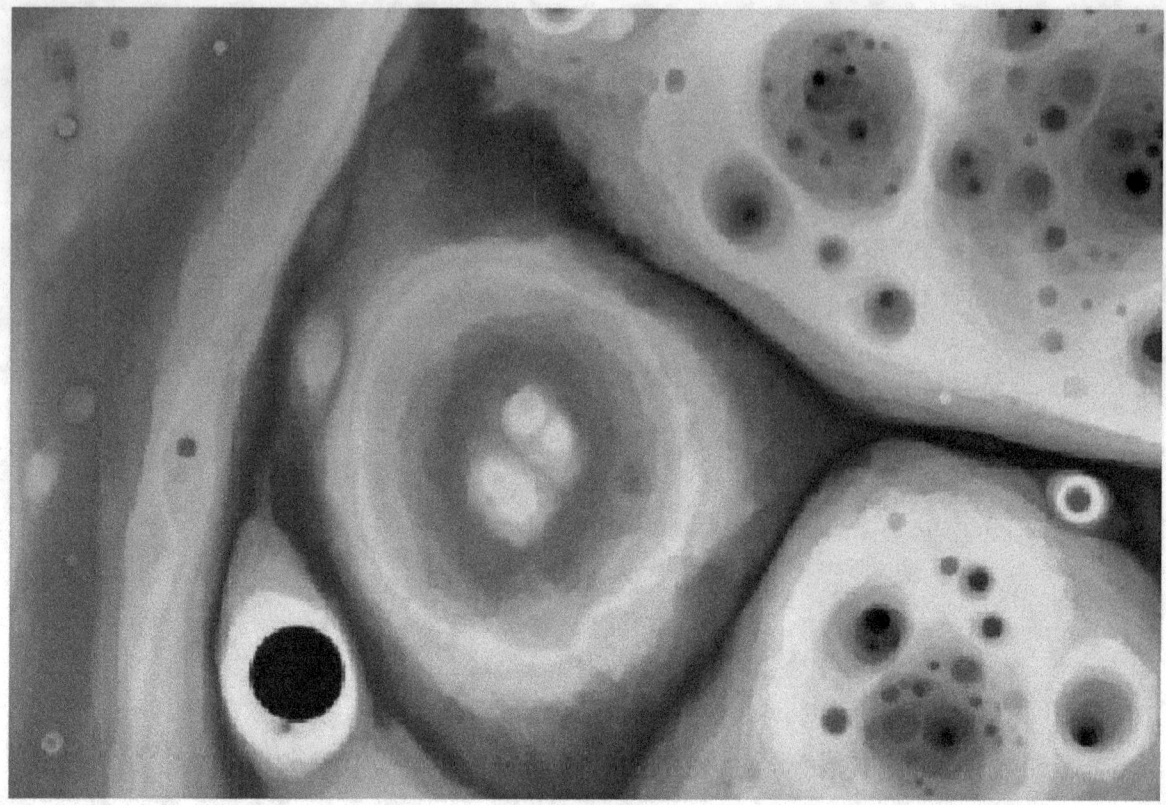

Magnified-image of coloured interference-pattern in soap-film. The black areas ("holes") are areas where the film is very thin and there is a nearly total destructive interference.

and df is known as the fringe spacing. The fringe spacing increases with increase in wavelength, and with decreasing angle θ.

The fringes are observed wherever the two waves overlap and the fringe spacing is uniform throughout.

20.1.2 Between two spherical waves

A point source produces a spherical wave. If the light from two point sources overlaps, the interference pattern maps out the way in which the phase difference between the two waves varies in space. This depends on the wavelength and on the separation of the point sources. The figure to the right shows interference between two spherical waves. The wavelength increases from top to bottom, and the distance between the sources increases from left to right.

When the plane of observation is far enough away, the fringe pattern will be a series of almost straight lines, since the waves will then be almost planar.

20.1.3 Multiple beams

Interference occurs when several waves are added together provided that the phase differences between them remain constant over the observation time.

It is sometimes desirable for several waves of the same frequency and amplitude to sum to zero (that is, interfere destructively, cancel). This is the principle behind, for example, 3-phase power and the diffraction grating. In both of these cases, the result is achieved by uniform spacing of the phases.

It is easy to see that a set of waves will cancel if they have the same amplitude and their phases are spaced equally in angle. Using phasors, each wave can be represented as $Ae^{i\varphi_n}$ for N waves from $n = 0$ to $n = N - 1$, where

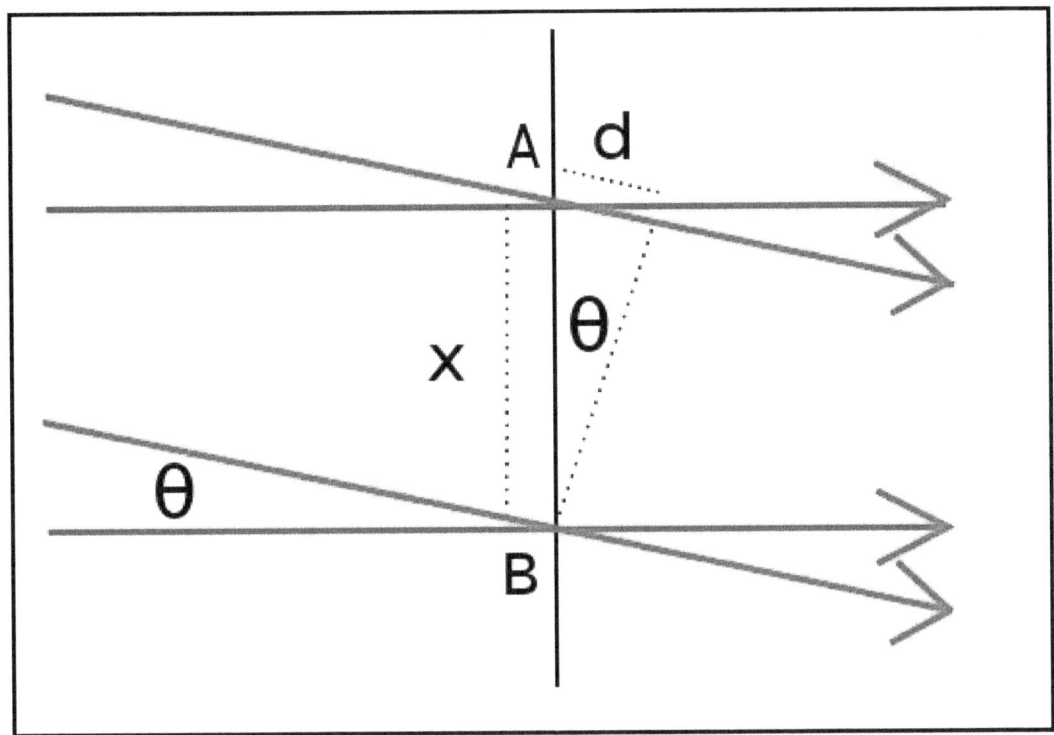

Geometrical arrangement for two plane wave interference

$$\varphi_n - \varphi_{n-1} = \frac{2\pi}{N}$$

To show that

$$\sum_{n=0}^{N-1} A e^{i\varphi_n} = 0$$

one merely assumes the converse, then multiplies both sides by $e^{i\frac{2\pi}{N}}$

The Fabry–Pérot interferometer uses interference between multiple reflections.

A diffraction grating can be considered to be a multiple-beam interferometer, since the peaks which it produces are generated by interference between the light transmitted by each of the elements in the grating; see interference vs. diffraction for further discussion.

20.2   Optical interference

Because the frequency of light waves ($\sim 10^{14}$ Hz) is too high to be detected by currently available detectors, it is possible to observe only the intensity of an optical interference pattern. The intensity of the light at a given point is proportional to the square of the average amplitude of the wave. This can be expressed mathematically as follows. The displacement of the two waves at a point $\mathbf{r}$ is:

Interference fringes in overlapping plane waves

$$U_1(\mathbf{r}, t) = A_1(\mathbf{r})e^{i[\varphi_1(\mathbf{r}) - \omega t]}$$

$$U_2(\mathbf{r}, t) = A_2(\mathbf{r})e^{i[\varphi_2(\mathbf{r}) - \omega t]}$$

where A represents the magnitude of the displacement, φ represents the phase and ω represents the angular frequency. The displacement of the summed waves is

$$U(\mathbf{r}, t) = A_1(\mathbf{r})e^{i[\varphi_1(\mathbf{r}) - \omega t]} + A_2(\mathbf{r})e^{i[\varphi_2(\mathbf{r}) - \omega t]}$$

The intensity of the light at $\mathbf{r}$ is given by

$$I(\mathbf{r}) = \int U(\mathbf{r}, t)U^*(\mathbf{r}, t)dt \propto A_1^2(\mathbf{r}) + A_2^2(\mathbf{r}) + 2A_1(\mathbf{r})A_2(\mathbf{r})\cos\left[\varphi_1(\mathbf{r}) - \varphi_2(\mathbf{r})\right]$$

This can be expressed in terms of the intensities of the individual waves as

Optical interference between two point sources for different wavelengths and source separations

$$I(\mathbf{r}) = I_1(\mathbf{r}) + I_2(\mathbf{r}) + 2\sqrt{I_1(\mathbf{r})I_2(\mathbf{r})}\cos\left[\varphi_1(\mathbf{r}) - \varphi_2(\mathbf{r})\right]$$

Thus, the interference pattern maps out the difference in phase between the two waves, with maxima occurring when the phase difference is a multiple of 2π. If the two beams are of equal intensity, the maxima are four times as bright as the individual beams, and the minima have zero intensity.

The two waves must have the same polarization to give rise to interference fringes since it is not possible for waves of different polarizations to cancel one another out or add together. Instead, when waves of different polarization are added together, they give rise to a wave of a different polarization state.

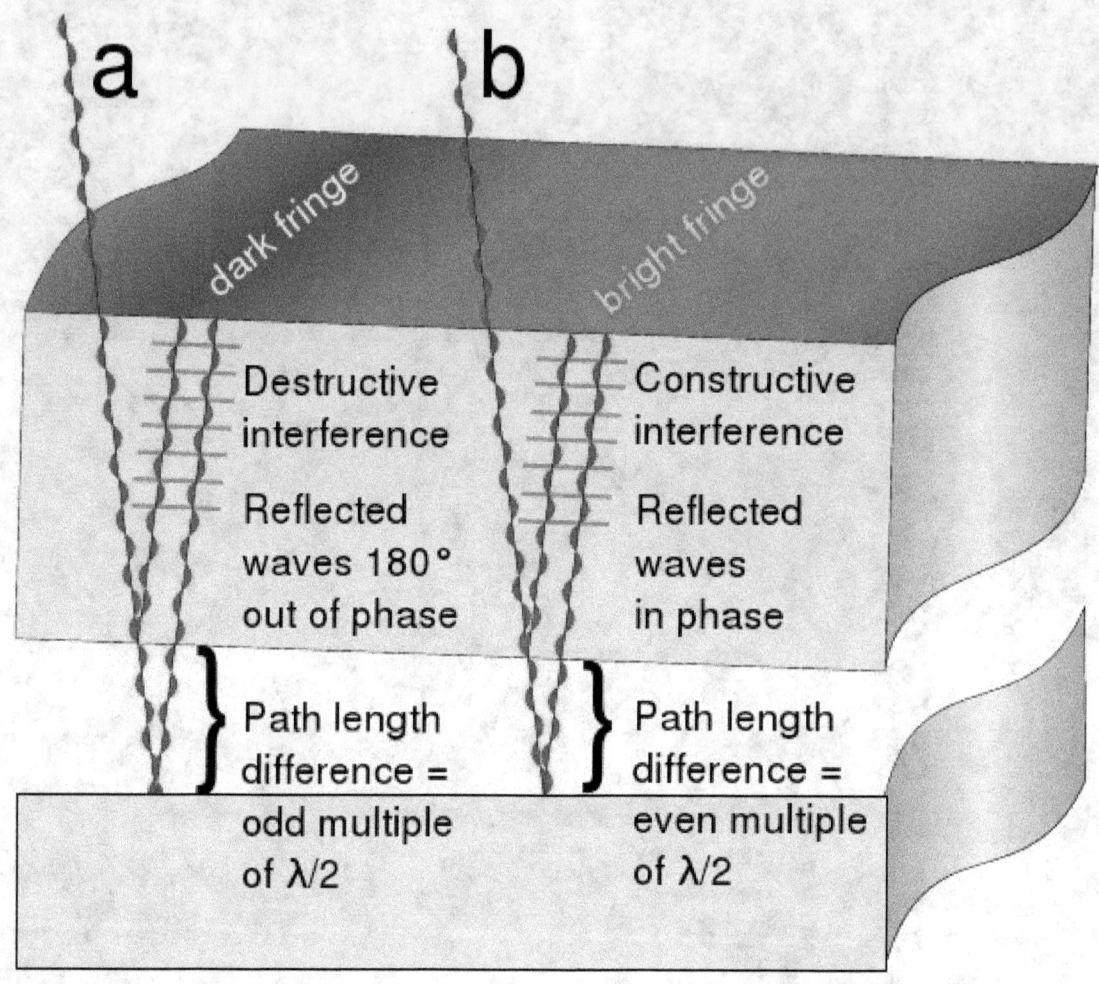

Creation of interference fringes by an optical flat on a reflective surface. Light rays from a monochromatic source pass through the glass and reflect off both the bottom surface of the flat and the supporting surface. The tiny gap between the surfaces mean the two reflected rays have different path lengths and interfere when they combine. At locations (b) where the path difference is an even multiple of λ/2, the waves reinforce. At locations (a) where the path difference is an odd multiple of λ/2 the waves cancel. Since the gap between the surfaces varies slightly in width at different points, a series of alternating bright and dark bands are seen.

20.2.1 Light source requirements

The discussion above assumes that the waves which interfere with one another are monochromatic, i.e. have a single frequency—this requires that they are infinite in time. This is not, however, either practical or necessary. Two identical waves of finite duration whose frequency is fixed over that period will give rise to an interference pattern while they overlap. Two identical waves which consist of a narrow spectrum of frequency waves of finite duration, will give a series of fringe patterns of slightly differing spacings, and provided the spread of spacings is significantly less than the average fringe spacing, a fringe pattern will again be observed during the time when the two waves overlap.

Conventional light sources emit waves of differing frequencies and at different times from different points in the source. If the light is split into two waves and then re-combined, each individual light wave may generate an interference pattern with its other half, but the individual fringe patterns generated will have different phases and spacings, and normally no overall fringe pattern will be observable. However, single-element light sources, such as sodium- or mercury-vapor lamps have emission lines with quite narrow frequency spectra. When these are spatially and colour filtered, and then split into two waves, they can be superimposed to generate interference fringes.[1] All interferometry prior to the invention of the laser was done using such sources and had a wide range of successful applications.

A laser beam generally approximates much more closely to a monochromatic source, and it is much more straightforward to generate interference fringes using a laser. The ease with which interference fringes can be observed with a laser beam can sometimes cause problems in that stray reflections may give spurious interference fringes which can result in errors.

Normally, a single laser beam is used in interferometry, though interference has been observed using two independent lasers whose frequencies were sufficiently matched to satisfy the phase requirements.[2]

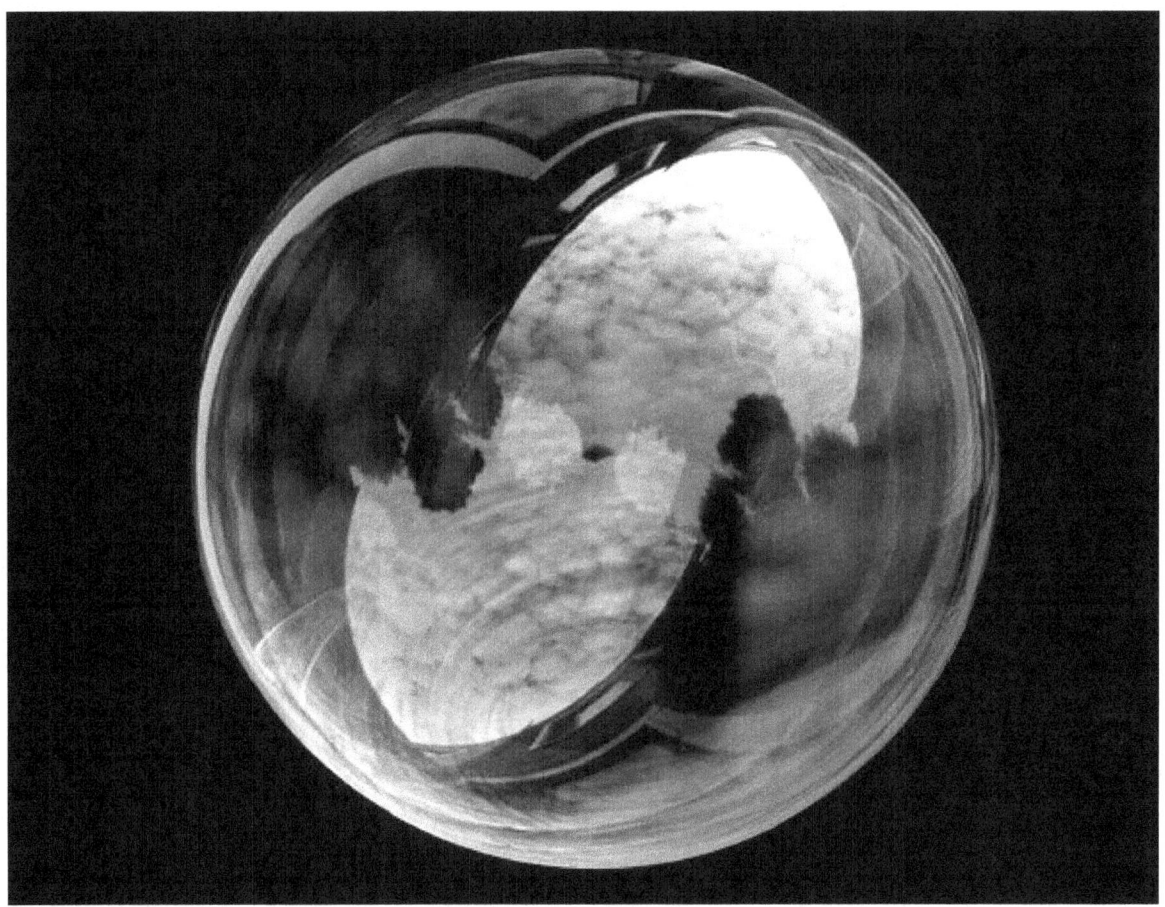

White light interference in a soap bubble

It is also possible to observe interference fringes using white light. A white light fringe pattern can be considered to be made up of a 'spectrum' of fringe patterns each of slightly different spacing. If all the fringe patterns are in phase in the centre, then the fringes will increase in size as the wavelength decreases and the summed intensity will show three to four fringes of varying colour. Young describes this very elegantly in his discussion of two slit interference. Some fine examples of white light fringes can be seen here. Since white light fringes are obtained only when the two waves have travelled equal distances from the light source, they can be very useful in interferometry, as they allow the zero path difference fringe to be identified.[3]

20.2.2   Optical arrangements

To generate interference fringes, light from the source has to be divided into two waves which have then to be re-combined. Traditionally, interferometers have been classified as either amplitude-division or wavefront-division systems.

In an amplitude-division system, a beam splitter is used to divide the light into two beams travelling in different directions, which are then superimposed to produce the interference pattern. The Michelson interferometer and the Mach-Zehnder interferometer are examples of amplitude-division systems.

In wavefront-division systems, the wave is divided in space—examples are Young's double slit interferometer and Lloyd's mirror.

Interference can also be seen in everyday phenomena such as iridescence and structural coloration. For example, the colours seen in a soap bubble arise from interference of light reflecting off the front and back surfaces of the thin soap film. Depending on the thickness of the film, different colours interfere constructively and destructively.

20.3 Applications

Main article: Interferometry

20.3.1 Optical interferometry

Main article: Optical interferometry

Interferometry has played an important role in the advancement of physics, and also has a wide range of applications in physical and engineering measurement.

Thomas Young's double slit interferometer in 1803 demonstrated interference fringes when two small holes were illuminated by light from another small hole which was illuminated by sunlight. Young was able to estimate the wavelength of different colours in the spectrum from the spacing of the fringes. The experiment played a major role in the general acceptance of the wave theory of light.[3] In quantum mechanics, this experiment is considered to demonstrate the inseparability of the wave and particle natures of light and other quantum particles (wave–particle duality). Richard Feynman was fond of saying that all of quantum mechanics can be gleaned from carefully thinking through the implications of this single experiment.[4]

The results of the Michelson–Morley experiment, are generally considered to be the first strong evidence against the theory of a luminiferous aether and in favor of special relativity.

Interferometry has been used in defining and calibrating length standards. When the metre was defined as the distance between two marks on a platinum-iridium bar, Michelson and Benoît used interferometry to measure the wavelength of the red cadmium line in the new standard, and also showed that it could be used as a length standard. Sixty years later, in 1960, the metre in the new SI system was defined to be equal to 1,650,763.73 wavelengths of the orange-red emission line in the electromagnetic spectrum of the krypton-86 atom in a vacuum. This definition was replaced in 1983 by defining the metre as the distance travelled by light in vacuum during a specific time interval. Interferometry is still fundamental in establishing the calibration chain in length measurement.

Interferometry is used in the calibration of slip gauges (called gauge blocks in the US) and in coordinate-measuring machines. It is also used in the testing of optical components.[5]

20.3.2 Radio interferometry

Main article: Astronomical interferometer

In 1946, a technique called astronomical interferometry was developed. Astronomical radio interferometers usually consist either of arrays of parabolic dishes or two-dimensional arrays of omni-directional antennas. All of the telescopes in the array are widely separated and are usually connected together using coaxial cable, waveguide, optical fiber, or other type of transmission line. Interferometry increases the total signal collected, but its primary purpose is to vastly increase the resolution through a process called Aperture synthesis. This technique works by superposing (interfering) the signal waves from the different telescopes on the principle that waves that coincide with the same phase will add to each other while two waves that have opposite phases will cancel each other out. This creates a combined telescope that is equivalent

The Very Large Array, an interferometric array formed from many smaller telescopes, like many larger radio telescopes.

in resolution (though not in sensitivity) to a single antenna whose diameter is equal to the spacing of the antennas furthest apart in the array.

20.3.3   Acoustic interferometry

An acoustic interferometer is an instrument for measuring the physical characteristics of sound wave in a gas or liquid. It may be used to measure velocity, wavelength, absorption, or impedance. A vibrating crystal creates the ultrasonic waves that are radiated into the medium. The waves strike a reflector placed parallel to the crystal. The waves are then reflected back to the source and measured.

20.4   Quantum interference

See also: Quantum version of double-slit experiment

If a system is in state ψ, its wavefunction is described in Dirac or bra–ket notation as:

$$|\psi\rangle = \sum_i |i\rangle \psi_i$$

where the $|i\rangle$ s specify the different quantum "alternatives" available (technically, they form an eigenvector basis) and the ψ_i are the probability amplitude coefficients, which are complex numbers.

The probability of observing the system making a transition or quantum leap from state ψ to a new state φ is the square of the modulus of the scalar or inner product of the two states:

$$\text{prob}(\psi \Rightarrow \varphi) = |\langle \psi | \varphi \rangle|^2 = \left| \sum_i \psi_i^* \varphi_i \right|^2$$

$$= \sum_{ij} \psi_i^* \psi_j \varphi_j^* \varphi_i = \sum_i |\psi_i|^2 |\varphi_i|^2 + \sum_{ij;i \neq j} \psi_i^* \psi_j \varphi_j^* \varphi_i$$

where $\psi_i = \langle i | \psi \rangle$ (as defined above) and similarly $\varphi_i = \langle i | \varphi \rangle$ are the coefficients of the final state of the system. * is the complex conjugate so that $\psi_i^* = \langle \psi | i \rangle$, etc.

Now let's consider the situation classically and imagine that the system transited from $|\psi\rangle$ to $|\varphi\rangle$ via an intermediate state $|i\rangle$. Then we would *classically* expect the probability of the two-step transition to be the sum of all the possible intermediate steps. So we would have

$$\text{prob}(\psi \Rightarrow \varphi) = \sum_i \text{prob}(\psi \Rightarrow i \Rightarrow \varphi)$$

$$= \sum_i |\langle \psi | i \rangle|^2 |\langle i | \varphi \rangle|^2 = \sum_i |\psi_i|^2 |\varphi_i|^2$$

The classical and quantum derivations for the transition probability differ by the presence, in the quantum case, of the extra terms $\sum_{ij;i \neq j} \psi_i^* \psi_j \varphi_j^* \varphi_i$; these extra quantum terms represent *interference* between the different $i \neq j$ intermediate "alternatives". These are consequently known as the quantum interference terms, or cross terms. This is a purely quantum effect and is a consequence of the non-additivity of the probabilities of quantum alternatives.

The interference terms vanish, via the mechanism of quantum decoherence, if the intermediate state $|i\rangle$ is measured or coupled with the environment.[6][7]

20.5 See also

- Active noise control

- Beat (acoustics)

- Coherence (physics)

- Diffraction

- Double-slit experiment

- Young's Double Slit Interferometer

- Haidinger fringes

- Hong–Ou–Mandel effect

- Interference lithography

- Interferometer

- List of types of interferometers

- Lloyd's Mirror

- Moiré pattern

- Newton's rings

- Thin-film interference

- Optical feedback

- Retroreflector

- Upfade

- Multipath interference

- Inter-flow interference

- Intra-flow interference

- Bio-Layer Interferometry

- N-slit interferometric equation

20.6   References

[1] WH Steel, *Interferometry*, 1986, Cambridge University Press, Cambridge

[2] R. L. Pfleegor and L. Mandel, 1967, "Interference of independent photon beams", *Phys. Rev.*, Volume 159, Issue 5. pp. 1084–1088.

[3] Max Born and Emil Wolf, 1999, *Principles of Optics*, Cambridge University Press, Cambridge.

[4] Greene, Brian (1999). *The Elegant Universe: Superstrings, Hidden Dimensions, and the Quest for the Ultimate Theory*. New York: W.W. Norton. pp. 97–109. ISBN 0-393-04688-5.

[5] RS Longhurst, *Geometrical and Physical Optics*, 1968, Longmans, London.

[6] Wojciech H. Zurek, "Decoherence and the transition from quantum to classical", *Physics Today*, 44, pp 36–44 (1991)

[7] Wojciech H. Zurek, "Decoherence, einselection, and the quantum origins of the classical", *Reviews of Modern Physics* 2003, 75, 715.

20.7   External links

- Easy JavaScript Simulation Model of One Dimensional Wave Interference

- Expressions of position and fringe spacing

- Java demonstration of interference

- Java simulation of interference of water waves 1

- Java simulation of interference of water waves 2

- Flash animations demonstrating interference

- Lissajous Curves: Interactive simulation of graphical representations of musical intervals, beats, interference, vibrating strings

- Animations demonstrating optical interference by QED

Chapter 21

Unified field theory

"Unified theory" redirects here. For the band, see Unified Theory (band).

In physics, a **unified field theory** (**UFT**), occasionally referred to as a **uniform field theory**,[1] is a type of field theory that allows all that is usually thought of as fundamental forces and elementary particles to be written in terms of a single field.

There is no accepted unified field theory, and thus it remains an open line of research. The term was coined by Einstein, who attempted to unify the general theory of relativity with electromagnetism. The "theory of everything" and Grand Unified Theory are closely related to unified field theory, but differ by not requiring the basis of nature to be fields, and often by attempting to explain physical constants of nature.

This article describes unified field theory as it is currently understood in connection with quantum theory. Earlier attempts based on classical physics are described in the article on classical unified field theories.

There may be no *a priori* reason why the correct description of nature has to be a unified field theory. However, this goal has led to a great deal of progress in modern theoretical physics and continues to motivate research.

21.1   Introduction

According to the current understanding of physics, forces are not transmitted directly between interacting objects, but instead are described by intermediary entities called fields. All four of the known fundamental forces are mediated by fields, which in the Standard Model of particle physics result from exchange of gauge bosons. Specifically the four fundamental interactions to be unified are:

- Strong interaction: the interaction responsible for holding quarks together to form hadrons, and holding neutrons and also protons together to form atomic nuclei. The exchange particle that mediates this force is the gluon.

- Electromagnetic interaction: the familiar interaction that acts on electrically charged particles. The photon is the exchange particle for this force.

- Weak interaction: a short-range interaction responsible for some forms of radioactivity, that acts on electrons, neutrinos, and quarks. It is mediated by the W and Z bosons.

- Gravitational interaction: a long-range attractive interaction that acts on *all* particles. The postulated exchange particle has been named the graviton.

Modern unified field theory attempts to bring these four interactions together into a single framework.

21.2   History

The first successful classical unified field theory was developed by James Clerk Maxwell. In 1820 Hans Christian Ørsted discovered that electric currents exerted forces on magnets, while in 1831, Michael Faraday made the observation that time-varying magnetic fields could induce electric currents. Until then, electricity and magnetism had been thought of as unrelated phenomena. In 1864, Maxwell published his famous paper on a dynamical theory of the electromagnetic field. This was the first example of a theory that was able to encompass previously separate field theories (namely electricity and magnetism) to provide a unifying theory of electromagnetism. By 1905, Albert Einstein had used the constancy of the speed of light in Maxwell's theory to unify our notions of space and time into an entity we now call spacetime and in 1915 he expanded this theory of special relativity to a description of gravity, General Relativity, using a field to describe the curving geometry of four-dimensional spacetime.

In the years following the creation of the general theory, a large number of physicists and mathematicians enthusiastically participated in the attempt to unify the then-known fundamental interactions.[2] In view of later developments in this domain, of particular interest are the theories of Hermann Weyl of 1919, who introduced the concept of an (electromagnetic) gauge field in a classical field theory[3] and, two years later, that of Theodor Kaluza, who extended General Relativity to five dimensions.[4] Continuing in this latter direction, Oscar Klein proposed in 1926 that the fourth spatial dimension be curled up into a small, unobserved circle. In Kaluza–Klein theory, the gravitational curvature of the extra spatial direction behaves as an additional force similar to electromagnetism. These and other models of electromagnetism and gravity were pursued by Albert Einstein in his attempts at a classical unified field theory. By 1930 Einstein had already considered the Einstein–Maxwell–Dirac System [Dongen]. This system is (heuristically) the super-classical [Varadarajan] limit of (the not mathematically well-defined) Quantum Electrodynamics. One can extend this system to include the weak and strong nuclear forces to get the Einstein–Yang–Mills–Dirac System.

21.3   Modern progress

In 1963 American physicist Sheldon Glashow proposed that the weak nuclear force and electricity and magnetism could arise from a partially unified electroweak theory. In 1967, Pakistani Abdus Salam and American Steven Weinberg independently revised Glashow's theory by having the masses for the W particle and Z particle arise through spontaneous symmetry breaking with the Higgs mechanism. This unified theory modeled the electroweak interaction as a force mediated by four particles: the photon for the electromagnetic aspect, and a neutral Z particle and two charged W particles for weak aspect. As a result of the spontaneous symmetry breaking, the weak force becomes short-range and the Z and W bosons acquire masses of 80.4 and 91.2 GeV/c^2, respectively. Their theory was first given experimental support by the discovery of weak neutral currents in 1973. In 1983, the Z and W bosons were first produced at CERN by Carlo Rubbia's team. For their insights, Glashow, Salam, and Weinberg were awarded the Nobel Prize in Physics in 1979. Carlo Rubbia and Simon van der Meer received the Prize in 1984.

After Gerardus 't Hooft showed the Glashow–Weinberg–Salam electroweak interactions to be mathematically consistent, the electroweak theory became a template for further attempts at unifying forces. In 1974, Sheldon Glashow and Howard Georgi proposed unifying the strong and electroweak interactions into Georgi–Glashow model, the first Grand Unified Theory, which would have observable effects for energies much above 100 GeV.

Since then there have been several proposals for Grand Unified Theories, e.g. the Pati–Salam model, although none is currently universally accepted. A major problem for experimental tests of such theories is the energy scale involved, which is well beyond the reach of current accelerators. Grand Unified Theories make predictions for the relative strengths of the strong, weak, and electromagnetic forces, and in 1991 LEP determined that supersymmetric theories have the correct ratio of couplings for a Georgi–Glashow Grand Unified Theory.

Many Grand Unified Theories (but not Pati–Salam) predict that the proton can decay, and if this were to be seen, details of the decay products could give hints at more aspects of the Grand Unified Theory. It is at present unknown if the proton can decay, although experiments have determined a lower bound of 10^{35} years for its lifetime.

21.4 Current status

Gravity has yet to be successfully included in a theory of everything.

Simply trying to combine the graviton with the strong and electroweak interactions runs into fundamental difficulties since the resulting theory is not renormalizable. Theoretical physicists have not yet formulated a widely accepted, consistent theory that combines general relativity and quantum mechanics. The incompatibility of the two theories remains an outstanding problem in the field of physics.

Some theoretical physicists currently believe that a quantum theory of general relativity may require frameworks other than field theory itself, such as string theory or loop quantum gravity. Some models in string theory that are promising by way of realizing our familiar standard model are the perturbative heterotic string models, 11-dimensional M-theory, Singular geometries (e.g. orbifold and orientifold), D-branes and other branes, flux compactification and warped geometry, and non-perturbative type IIB superstring solutions (F-theory).[5]

21.5 Notes

[1] *See, e.g., Beyond Art: A Third Culture* page 199. *Compare* Uniform field theory.

[2] See Catherine Goldstein & Jim Ritter (2003) "The varieties of unity: sounding unified theories 1920-1930" in A. Ashtekar, et al. (eds.), *Revisiting the Foundations of Relativistic Physics*, Dordrecht, Kluwer, p. 93-149; Vladimir Vizgin (1994), *Unified Field Theories in the First Third of the 20th Century*, Basel, Birkhäuser; Hubert Goenner On the History of Unified Field Theories.

[3] Erhard Scholtz (ed) (2001), *Hermann Weyl's* Raum - Zeit- Materie *and a General Introduction to His Scientific Work*, Basel, Birkhäuser.

[4] Daniela Wuensch (2003), "The fifth dimension: Theodor Kaluza's ground-breaking idea", *Annalen der Physik*, vol. 12, p. 519–542.

[5] http://arxiv.org/abs/0812.1372

21.6 References

- Shushi Tomer, *A Possible Connection between Quantum and General Relativity Theories*, SSRN: http://papers.ssrn.com/sol3/papers.cfm?abstract_id=2538728 (December 15, 2014)

- Jeroen van van Dongen *Einstein's Unification*, Cambridge University Press (July 26, 2010)

- Varadarajan, V.S. *Supersymmetry for Mathematicians: An Introduction (Courant Lecture Notes)*, American Mathematical Society (July 2004)

21.7 External links

- On the History of Unified Field Theories, by Hubert F. M. Goenner

Chapter 22

Many-worlds interpretation

The quantum-mechanical "Schrödinger's cat" paradox according to the many-worlds interpretation. In this interpretation, every event is a branch point; the cat is both alive and dead, even before the box is opened, but the "alive" and "dead" cats are in different branches of the universe, both of which are equally real, but which do not interact with each other.[1]

The **many-worlds interpretation** is an interpretation of quantum mechanics that asserts the objective reality of the universal wavefunction and denies the actuality of wavefunction collapse. Many-worlds implies that all possible alternate histories and futures are real, each representing an actual "world" (or "universe"). In lay terms, the hypothesis states there is a very large—perhaps infinite[2]—number of universes, and everything that could possibly have happened in our past, but did not, has occurred in the past of some other universe or universes. The theory is also referred to as **MWI**, the **relative state formulation**, the **Everett interpretation**, the **theory of the universal wavefunction, many-universes interpretation**, or just **many-worlds**.

The original relative state formulation is due to Hugh Everett in 1957.[3][4] Later, this formulation was popularized and renamed *many-worlds* by Bryce Seligman DeWitt in the 1960s and 1970s.[1][5][6][7] The decoherence approaches to interpreting quantum theory have been further explored and developed,[8][9][10] becoming quite popular. MWI is one of many multiverse hypotheses in physics and philosophy. It is currently considered a mainstream interpretation along with the other decoherence interpretations, collapse theories (including the historical Copenhagen interpretation),[11] and hidden variable theories such as the Bohmian mechanics.

Before many-worlds, reality had always been viewed as a single unfolding history. Many-worlds, however, views reality as a many-branched tree, wherein every possible quantum outcome is realised.[12] Many-worlds reconciles the observation of non-deterministic events, such as random radioactive decay, with the fully deterministic equations of quantum physics.

In many-worlds, the subjective appearance of wavefunction collapse is explained by the mechanism of quantum decoherence, and this is supposed to resolve all of the correlation paradoxes of quantum theory, such as the EPR paradox[13][14] and Schrödinger's cat,[1] since every possible outcome of every event defines or exists in its own "history" or "world".

22.1 Outline

Although several versions of many-worlds have been proposed since Hugh Everett's original work,[4] they all contain one key idea: the equations of physics that model the time evolution of systems *without* embedded observers are sufficient for modelling systems which *do* contain observers; in particular there is no observation-triggered wave function collapse which the Copenhagen interpretation proposes. Provided the theory is linear with respect to the wavefunction, the exact form of the quantum dynamics modelled, be it the non-relativistic Schrödinger equation, relativistic quantum field theory or some form of quantum gravity or string theory, does not alter the validity of MWI since MWI is a metatheory applicable to all linear quantum theories, and there is no experimental evidence for any non-linearity of the wavefunction in physics.[15][16] MWI's main conclusion is that the universe (or multiverse in this context) is composed of a quantum superposition of very many, possibly even non-denumerably infinitely[2] many, increasingly divergent, non-communicating parallel universes or quantum worlds.[7]

The idea of MWI originated in Everett's Princeton Ph.D. thesis "The Theory of the Universal Wavefunction",[7] developed under his thesis advisor John Archibald Wheeler, a shorter summary of which was published in 1957 entitled "Relative State Formulation of Quantum Mechanics" (Wheeler contributed the title "relative state";[17] Everett originally called his approach the "Correlation Interpretation", where "correlation" refers to quantum entanglement). The phrase "many-worlds" is due to Bryce DeWitt,[7] who was responsible for the wider popularisation of Everett's theory, which had been largely ignored for the first decade after publication. DeWitt's phrase "many-worlds" has become so much more popular than Everett's "Universal Wavefunction" or Everett–Wheeler's "Relative State Formulation" that many forget that this is only a difference of terminology; the content of both of Everett's papers and DeWitt's popular article is the same.

The many-worlds interpretation shares many similarities with later, other "post-Everett" interpretations of quantum mechanics which also use decoherence to explain the process of measurement or wavefunction collapse. MWI treats the other histories or worlds as real since it regards the universal wavefunction as the "basic physical entity"[18] or "the fundamental entity, obeying at all times a deterministic wave equation".[19] The other decoherent interpretations, such as consistent histories, the Existential Interpretation etc., either regard the extra quantum worlds as metaphorical in some sense, or are agnostic about their reality; it is sometimes hard to distinguish between the different varieties. MWI is distinguished by two qualities: it assumes realism,[18][19] which it assigns to the wavefunction, and it has the minimal formal structure possible, rejecting any hidden variables, quantum potential, any form of a collapse postulate (i.e., Copenhagenism) or mental postulates (such as the many-minds interpretation makes).

Decoherent interpretations of many-worlds using einselection to explain how a small number of classical pointer states can emerge from the enormous Hilbert space of superpositions have been proposed by Wojciech H. Zurek. "Under scrutiny of the environment, only pointer states remain unchanged. Other states decohere into mixtures of stable pointer states that can persist, and, in this sense, exist: They are einselected."[20] These ideas complement MWI and bring the interpretation in line with our perception of reality.

Many-worlds is often referred to as a theory, rather than just an interpretation, by those who propose that many-worlds can make testable predictions (such as David Deutsch) or is falsifiable (such as Everett) or by those who propose that all the other, non-MW interpretations, are inconsistent, illogical or unscientific in their handling of measurements; Hugh Everett argued that his formulation was a metatheory, since it made statements about other interpretations of quantum theory; that it was the "only completely coherent approach to explaining both the contents of quantum mechanics and the appearance of the world."[21] Deutsch is dismissive that many-worlds is an "interpretation", saying that calling it an interpretation "is like talking about dinosaurs as an 'interpretation' of fossil records."[22]

22.2   Interpreting wavefunction collapse

As with the other interpretations of quantum mechanics, the many-worlds interpretation is motivated by behavior that can be illustrated by the double-slit experiment. When particles of light (or anything else) are passed through the double slit, a calculation assuming wave-like behavior of light can be used to identify where the particles are likely to be observed. Yet when the particles are observed in this experiment, they appear as particles (i.e., at definite places) and not as non-localized waves.

Some versions of the Copenhagen interpretation of quantum mechanics proposed a process of "collapse" in which an indeterminate quantum system would probabilistically collapse down onto, or select, just one determinate outcome to "explain" this phenomenon of observation. Wavefunction collapse was widely regarded as artificial and *ad hoc*, so an alternative interpretation in which the behavior of measurement could be understood from more fundamental physical principles was considered desirable.

Everett's Ph.D. work provided such an alternative interpretation. Everett stated that for a composite system – for example a subject (the "observer" or measuring apparatus) observing an object (the "observed" system, such as a particle) – the statement that either the observer or the observed has a well-defined state is meaningless; in modern parlance, the observer and the observed have become entangled; we can only specify the state of one *relative* to the other, i.e., the state of the observer and the observed are correlated *after* the observation is made. This led Everett to derive from the unitary, deterministic dynamics alone (i.e., without assuming wavefunction collapse) the notion of a *relativity of states*.

Everett noticed that the unitary, deterministic dynamics alone decreed that after an observation is made each element of the quantum superposition of the combined subject–object wavefunction contains two "relative states": a "collapsed" object state and an associated observer who has observed the same collapsed outcome; what the observer sees and the state of the object have become correlated by the act of measurement or observation. The subsequent evolution of each pair of relative subject–object states proceeds with complete indifference as to the presence or absence of the other elements, *as if* wavefunction collapse has occurred, which has the consequence that later observations are always consistent with the earlier observations. Thus the *appearance* of the object's wavefunction's collapse has emerged from the unitary, deterministic theory itself. (This answered Einstein's early criticism of quantum theory, that the theory should define what is observed, not for the observables to define the theory).[23] Since the wavefunction merely appears to have collapsed then, Everett reasoned, there was no need to actually assume that it had collapsed. And so, invoking Occam's razor, he removed the postulate of wavefunction collapse from the theory.

22.3   Probability

A consequence of removing wavefunction collapse from the quantum formalism is that the Born rule requires derivation, since many-worlds derives its interpretation from the formalism. Attempts have been made, by many-world advocates and others, over the years to *derive* the Born rule, rather than just conventionally *assume* it, so as to reproduce all the required statistical behaviour associated with quantum mechanics. There is no consensus on whether this has been successful.[24][25][26]

22.3.1   Everett, Gleason and Hartle

Everett (1957) briefly derived the Born rule by showing that the Born rule was the only possible rule, and that its derivation was as justified as the procedure for defining probability in classical mechanics. Everett stopped doing research in theoretical physics shortly after obtaining his Ph.D., but his work on probability has been extended by a number of people. Andrew Gleason (1957) and James Hartle (1965) independently reproduced Everett's work, known as Gleason's theorem[27][28] which was later extended.[29][30]

22.3.2   DeWitt and Graham

Bryce DeWitt and his doctoral student R. Neill Graham later provided alternative (and longer) derivations to Everett's derivation of the Born rule. They demonstrated that the norm of the worlds where the usual statistical rules of quantum theory broke down vanished, in the limit where the number of measurements went to infinity.

22.3.3   Deutsch *et al.*

An information-theoretic derivation of the Born rule from Everettarian assumptions, was produced by David Deutsch (1999)[31] and refined by Wallace (2002–2009)[32][33][34][35] and Saunders (2004).[36][37] Deutsch's derivation is a two-stage proof: first he shows that the number of orthonormal Everett-worlds after a branching is proportional to the conventional probability density. Then he uses game theory to show that these are all equally likely to be observed. The last step in particular has been criticised for circularity.[38][39] Some other reviews have been positive, although the status of these arguments remains highly controversial; some theoretical physicists have taken them as supporting the case for parallel universes.[40][41] In the *New Scientist* article, reviewing their presentation at a September 2007 conference,[42][43] Andy Albrecht, a physicist at the University of California at Davis, is quoted as saying "This work will go down as one of the most important developments in the history of science."[40]

Wojciech H. Zurek (2005)[44] has produced a derivation of the Born rule, where decoherence has replaced Deutsch's informatic assumptions.[45] Lutz Polley (2000) has produced Born rule derivations where the informatic assumptions are replaced by symmetry arguments.[46][47]

The Born rule and the collapse of the wave function have been obtained in the framework of the relative-state formulation of quantum mechanics by Armando V.D.B. Assis. He has proved that the Born rule and the collapse of the wave function follow from a game-theoretical strategy, namely the Nash equilibrium within a von Neumann zero-sum game between nature and observer.[48]

22.4   Brief overview

In Everett's formulation, a measuring apparatus **M** and an object system **S** form a composite system, each of which prior to measurement exists in well-defined (but time-dependent) states. Measurement is regarded as causing **M** and **S** to interact. After **S** interacts with **M**, it is no longer possible to describe either system by an independent state. According to Everett, the only meaningful descriptions of each system are relative states: for example the relative state of **S** given the state of **M** or the relative state of **M** given the state of **S**. In DeWitt's formulation, the state of **S** after a sequence of measurements is given by a quantum superposition of states, each one corresponding to an alternative measurement history of **S**.

For example, consider the smallest possible truly quantum system **S**, as shown in the illustration. This describes for instance, the spin-state of an electron. Considering a specific axis (say the z-axis) the north pole represents spin "up" and the south pole, spin "down". The superposition states of the system are described by (the surface of) a sphere called the Bloch sphere. To perform a measurement on **S**, it is made to interact with another similar system **M**. After the interaction, the combined system is described by a state that ranges over a six-dimensional space (the reason for the number six is explained in the article on the Bloch sphere). This six-dimensional object can also be regarded as a quantum superposition of two "alternative histories" of the original system **S**, one in which "up" was observed and the other in which "down" was observed. Each subsequent binary measurement (that is interaction with a system **M**) causes a similar split in the history tree. Thus after three measurements, the system can be regarded as a quantum superposition of $8 = 2 \times 2 \times 2$ copies of the original system **S**.

The accepted terminology is somewhat misleading because it is incorrect to regard the universe as splitting at certain times; at any given instant there is one state in one universe.

22.5 Relative state

In his 1957 doctoral dissertation, Everett proposed that rather than modeling an isolated quantum system subject to external observation, one could mathematically model an object as well as its observers as purely physical systems within the mathematical framework developed by Paul Dirac, von Neumann and others, discarding altogether the *ad hoc* mechanism of wave function collapse. Since Everett's original work, there have appeared a number of similar formalisms in the literature. One such idea is discussed in the next section.

The relative state formulation makes two assumptions. The first is that the wavefunction is not simply a description of the object's state, but that it actually is entirely equivalent to the object, a claim it has in common with some other interpretations. The second is that observation or measurement has no special laws or mechanics, unlike in the Copenhagen interpretation which considers the wavefunction collapse as a special kind of event which occurs as a result of observation. Instead, measurement in the relative state formulation is the consequence of a configuration change in the memory of an observer described by the same basic wave physics as the object being modeled.

The many-worlds interpretation is DeWitt's popularisation of Everett's work, who had referred to the combined observer–object system as being split by an observation, each split corresponding to the different or multiple possible outcomes of an observation. These splits generate a possible tree as shown in the graphic below. Subsequently DeWitt introduced the term "world" to describe a complete measurement history of an observer, which corresponds roughly to a single branch of that tree. Note that "splitting" in this sense, is hardly new or even quantum mechanical. The idea of a space of complete alternative histories had already been used in the theory of probability since the mid-1930s for instance to model Brownian motion.

Under the many-worlds interpretation, the Schrödinger equation, or relativistic analog, holds all the time everywhere. An observation or measurement of an object by an observer is modeled by applying the wave equation to the entire system comprising the observer *and* the object. One consequence is that every observation can be thought of as causing the combined observer–object's wavefunction to change into a quantum superposition of two or more non-interacting branches, or split into many "worlds". Since many observation-like events have happened, and are constantly happening, there are an enormous and growing number of simultaneously existing states.

If a system is composed of two or more subsystems, the system's state will be a superposition of products of the subsystems' states. Once the subsystems interact, their states are no longer independent. Each product of subsystem states in the overall superposition evolves over time independently of other products. The subsystems states have become correlated or entangled and it is no longer possible to consider them independent of one another. In Everett's terminology each subsystem state was now *correlated* with its *relative state*, since each subsystem must now be considered relative to the other subsystems with which it has interacted.

22.6 Properties of the theory

- MWI removes the observer-dependent role in the quantum measurement process by replacing wavefunction collapse with quantum decoherence. Since the role of the observer lies at the heart of most if not all "quantum paradoxes," this automatically resolves a number of problems; see for example Schrödinger's cat thought experiment, the EPR paradox, von Neumann's "boundary problem" and even wave-particle duality. Quantum cosmology also becomes intelligible, since there is no need anymore for an observer outside of the universe.

- MWI is a realist, deterministic, local theory, akin to classical physics (including the theory of relativity), at the expense of losing counterfactual definiteness. MWI achieves this by removing wavefunction collapse, which is indeterministic and non-local, from the deterministic and local equations of quantum theory.[49]

- MWI (or other, broader multiverse considerations) provides a context for the anthropic principle which may provide an explanation for the fine-tuned universe.[50][51]

- MWI, being a decoherent formulation, is axiomatically more streamlined than the Copenhagen and other collapse interpretations; and thus favoured under certain interpretations of Occam's razor.[52] Of course there are other decoherent interpretations that also possess this advantage with respect to the collapse interpretations.

22.7 Comparative properties and possible experimental tests

One of the salient properties of the many-worlds interpretation is that it does not require an exceptional method of wave function collapse to explain it. "It seems that there is no experiment distinguishing the MWI from other no-collapse theories such as Bohmian mechanics or other variants of MWI... In most no-collapse interpretations, the evolution of the quantum state of the Universe is the same. Still, one might imagine that there is an experiment distinguishing the MWI from another no-collapse interpretation based on the difference in the correspondence between the formalism and the experience (the results of experiments)."[53]

However, in 1985, David Deutsch published three related thought experiments which could test the theory vs the Copenhagen interpretation.[54] The experiments require macroscopic quantum state preparation and quantum erasure by a hypothetical quantum computer which is currently outside experimental possibility. Since then Lockwood (1989), Vaidman and others have made similar proposals.[53] These proposals also require an advanced technology which is able to place a macroscopic object in a coherent superposition, another task for which it is uncertain whether it will ever be possible. Many other controversial ideas have been put forward though, such as a recent claim that cosmological observations could test the theory,[55] and another claim by Rainer Plaga (1997), published in *Foundations of Physics*, that communication might be possible between worlds.[56]

22.7.1 Copenhagen interpretation

In the Copenhagen interpretation, the mathematics of quantum mechanics allows one to predict probabilities for the occurrence of various events. When an event occurs, it becomes part of the definite reality, and alternative possibilities do not. There is no necessity to say anything definite about what is not observed.

22.7.2 The universe decaying to a new vacuum state

Any event that changes the number of observers in the universe may have experimental consequences.[57] Quantum tunnelling to a new vacuum state would reduce the number of observers to zero (i.e., kill all life). Some cosmologists argue that the universe is in a false vacuum state and that consequently the universe should have already experienced quantum tunnelling to a true vacuum state. This has not happened and is cited as evidence in favor of many-worlds. In some worlds, quantum tunnelling to a true vacuum state has happened but most other worlds escape this tunneling and remain viable. This can be thought of as a variation on quantum suicide.

22.7.3 Many-minds

Main article: Many-minds interpretation

The *many-minds* interpretation is a multi-world interpretation that defines the splitting of reality on the level of the observers' minds. In this, it differs from Everett's many-worlds interpretation, in which there is no special role for the observer's mind.[56]

22.8 Common objections

- The many-worlds interpretation is very vague about the ways to determine when splitting happens, and nowadays usually the criterion is that the two branches have decohered. However, present day understanding of decoherence does not allow a completely precise, self-contained way to say when the two branches have decohered/"do not interact", and hence many-worlds interpretation remains arbitrary. This objection is saying that it is not clear what is precisely meant by branching, and point to the lack of self-contained criteria specifying branching.

    MWI response: the decoherence or "splitting" or "branching" is complete when the measurement is complete. In Dirac notation a measurement is complete when:

$$\langle O_i | O_j \rangle = \delta_{ij} \text{ [58]}$$

where O_i represents the observer having detected the object system in the ith state. Before the measurement has started the observer states are identical; after the measurement is complete the observer states are orthonormal.[4][7] Thus a measurement defines the branching process: the branching is as well- or ill-defined as the measurement is; the branching is as complete as the measurement is complete – which is to say that the delta function above represents an idealised measurement. Although true "for all practical purposes" in reality the measurement, and hence the branching, is never fully complete, since delta functions are unphysical.[59]

Since the role of the observer and measurement per se plays no special role in MWI (measurements are handled as all other interactions are) there is no need for a precise definition of what an observer or a measurement is — just as in Newtonian physics no precise definition of either an observer or a measurement was required or expected. In all circumstances the universal wavefunction is still available to give a complete description of reality.

Also, it is a common misconception to think that branches are completely separate. In Everett's formulation, they may in principle quantum interfere (i.e., "merge" instead of "splitting") with each other in the future,[60] although this requires all "memory" of the earlier branching event to be lost, so no observer ever sees two branches of reality.[61][62]

- MWI states that there is no special role nor need for precise definition of measurement in MWI, yet Everett uses the word "measurement" repeatedly throughout its exposition.

  MWI response: "measurements" are treated as a subclass of interactions, which induce subject–object correlations in the combined wavefunction. There is nothing special about measurements (such as the ability to trigger a wave function collapse), that cannot be dealt with by the usual unitary time development process.[3] This is why there is no precise definition of measurement in Everett's formulation, although some other formulations emphasize that measurements must be effectively irreversible or create classical information.

- The splitting of worlds forward in time, but not backwards in time (i.e., merging worlds), is time asymmetric and incompatible with the time symmetric nature of Schrödinger's equation, or CPT invariance in general.[63]

  MWI response: The splitting is time asymmetric; this observed temporal asymmetry is due to the boundary conditions imposed by the Big Bang[64]

- There is circularity in Everett's measurement theory. Under the assumptions made by Everett, there are no 'good observations' as defined by him, and since his analysis of the observational process depends on the latter, it is void of any meaning. The concept of a 'good observation' is the projection postulate in disguise and Everett's analysis simply derives this postulate by having assumed it, without any discussion.[65]

  MWI response: Everett's treatment of observations / measurements covers *both* idealised good measurements and the more general bad or approximate cases.[66] Thus it is legitimate to analyse probability in terms of measurement; no circularity is present.

- Talk of probability in Everett presumes the existence of a preferred basis to identify measurement outcomes for the probabilities to range over. But the existence of a preferred basis can only be established by the process of decoherence, which is itself probabilistic[38] or arbitrary.[67]

  MWI response: Everett analysed branching using what we now call the "measurement basis". It is fundamental theorem of quantum theory that nothing measurable or empirical is changed by adopting a different basis. Everett was therefore free to choose whatever basis he liked. The measurement basis was simply the simplest basis in which to analyse the measurement process.[68][69]

- We cannot be sure that the universe is a quantum multiverse until we have a theory of everything and, in particular, a successful theory of quantum gravity.[70] If the final theory of everything is non-linear with respect to wavefunctions then many-worlds would be invalid.[1][4][5][6][7]

  > MWI response: All accepted quantum theories of fundamental physics are linear with respect to the wavefunction. While quantum gravity or string theory may be non-linear in this respect there is no evidence to indicate this at the moment.[15][16]

- Conservation of energy is grossly violated if at every instant near-infinite amounts of new matter are generated to create the new universes.

  > MWI response: There are two responses to this objection. First, the law of conservation of energy says that energy is conserved *within each universe*. Hence, even if "new matter" were being generated to create new universes, this would not violate conservation of energy. Second, conservation of energy is not violated since the energy of each branch has to be weighted by its probability, according to the standard formula for the conservation of energy in quantum theory. This results in the total energy of the multiverse being conserved.[71]

- Occam's Razor rules against a plethora of unobservable universes – Occam would prefer just one universe; i.e., any non-MWI.

  > MWI response: Occam's razor actually is a constraint on the complexity of physical theory, not on the number of universes. MWI is a simpler theory since it has fewer postulates.[52] Occams's razor is often cited by MWI adherents as an advantage of MWI.

- Unphysical universes: If a state is a superposition of two states Ψ_A and Ψ_B, i.e., $\Psi = (a\Psi_A + b\Psi_B)$, i.e., weighted by coefficients a and b, then if $b \ll a$, what principle allows a universe with vanishingly small probability b to be instantiated on an equal footing with the much more probable one with probability a? This seems to throw away the information in the probability amplitudes.

  > MWI response: The magnitude of the coefficients provides the weighting that makes the branches or universes "unequal", as Everett and others have shown, leading the emergence of the conventional probabilistic rules.[1][4][5][6][7][72]

- Violation of the principle of locality, which contradicts special relativity: MWI splitting is instant and total: this may conflict with relativity, since an alien in the Andromeda galaxy can't know I collapse an electron over here before she collapses hers there: the relativity of simultaneity says we can't say which electron collapsed first – so which one splits off another universe first? This leads to a hopeless muddle with everyone splitting differently. Note: EPR is not a get-out here, as the alien's and my electrons need never have been part of the same quantum, i.e., entangled.

  > MWI response: the splitting can be regarded as causal, local and relativistic, spreading at, or below, the speed of light (e.g., we are not split by Schrödinger's cat until we look in the box).[73] For spacelike separated splitting you can't say which occurred first — but this is true of all spacelike separated events, simultaneity is not defined for them. Splitting is no exception; many-worlds is a local theory.[49]

22.9 Reception

There is a wide range of claims that are considered "many-worlds" interpretations. It was often claimed by those who do not believe in MWI[74] that Everett himself was not entirely clear[75] as to what he believed; however, MWI adherents (such as DeWitt, Tegmark, Deutsch and others) believe they fully understand Everett's meaning as implying the literal existence of the other worlds. Additionally, recent biographical sources make it clear that Everett believed in the literal reality of the other quantum worlds.[22] Everett's son reported that Hugh Everett "never wavered in his belief over his many-worlds theory".[76] Also Everett was reported to believe "his many-worlds theory guaranteed him immortality".[77]

One of MWI's strongest advocates is David Deutsch.[78] According to Deutsch, the single photon interference pattern observed in the double slit experiment can be explained by interference of photons in multiple universes. Viewed in this way, the single photon interference experiment is indistinguishable from the multiple photon interference experiment. In a more practical vein, in one of the earliest papers on quantum computing,[79] he suggested that parallelism that results from the validity of MWI could lead to "*a method by which certain probabilistic tasks can be performed faster by a universal quantum computer than by any classical restriction of it*". Deutsch has also proposed that when reversible computers become conscious that MWI will be testable (at least against "naive" Copenhagenism) via the reversible observation of spin.[61]

Asher Peres was an outspoken critic of MWI; for example, a section in his 1993 textbook had the title *Everett's interpretation and other bizarre theories*. In fact, Peres not only questioned whether MWI is really an "interpretation", but rather, if *any* interpretations of quantum mechanics are needed at all. Indeed, an interpretation can be regarded as a purely formal transformation, which adds nothing to the rules of the quantum mechanics. Peres seems to suggest that positing the existence of an infinite number of non-communicating parallel universes is highly suspect per those who interpret it as a violation of Occam's razor, i.e., that it does not minimize the number of hypothesized entities. However, it is understood that the number of elementary particles are not a gross violation of Occam's Razor, one counts the types, not the tokens. Max Tegmark remarks that the alternative to many-worlds is "many words", an allusion to the complexity of von Neumann's collapse postulate. On the other hand, the same derogatory qualification "many words" is often applied to MWI by its critics who see it as a word game which obfuscates rather than clarifies by confounding the von Neumann branching of possible worlds with the Schrödinger parallelism of many worlds in superposition.

MWI is considered by some to be unfalsifiable and hence unscientific because the multiple parallel universes are non-communicating, in the sense that no information can be passed between them. Others[61] claim MWI is directly testable. Everett regarded MWI as falsifiable since any test that falsifies conventional quantum theory would also falsify MWI.[21]

According to Martin Gardner, the "other" worlds of MWI have two different interpretations: real or unreal; he claims that Stephen Hawking and Steve Weinberg both favour the unreal interpretation.[80] Gardner also claims that the nonreal interpretation is favoured by the majority of physicists, whereas the "realist" view is only supported by MWI experts such as Deutsch and Bryce DeWitt. Hawking has said that "according to Feynman's idea", all the other histories are as "equally real" as our own,[81] and Martin Gardner reports Hawking saying that MWI is "trivially true".[82] In a 1983 interview, Hawking also said he regarded the MWI as "self-evidently correct" but was dismissive towards questions about the interpretation of quantum mechanics, saying, "When I hear of Schrödinger's cat, I reach for my gun." In the same interview, he also said, "But, look: All that one does, really, is to calculate conditional probabilities—in other words, the probability of A happening, given B. I think that that's all the many worlds interpretation is. Some people overlay it with a lot of mysticism about the wave function splitting into different parts. But all that you're calculating is conditional probabilities."[83] Elsewhere Hawking contrasted his attitude towards the "reality" of physical theories with that of his colleague Roger Penrose, saying, "He's a Platonist and I'm a positivist. He's worried that Schrödinger's cat is in a quantum state, where it is half alive and half dead. He feels that can't correspond to reality. But that doesn't bother me. I don't demand that a theory correspond to reality because I don't know what it is. Reality is not a quality you can test with litmus paper. All I'm concerned with is that the theory should predict the results of measurements. Quantum theory does this very successfully."[84] For his own part, Penrose agrees with Hawking that QM applied to the universe implies MW, although he considers the current lack of a successful theory of quantum gravity negates the claimed universality of conventional QM.[70]

22.9.1   Polls

Advocates of MWI often cite a poll of 72 "leading cosmologists and other quantum field theorists"[85] conducted by the American political scientist David Raub in 1995 showing 58% agreement with "Yes, I think MWI is true".[86]

The poll is controversial: for example, Victor J. Stenger remarks that Murray Gell-Mann's published work explicitly rejects the existence of simultaneous parallel universes. Collaborating with James Hartle, Gell-Mann is working toward the development a more "palatable" *post-Everett quantum mechanics*. Stenger thinks it's fair to say that most physicists dismiss the many-world interpretation as too extreme, while noting it "has merit in finding a place for the observer inside the system being analyzed and doing away with the troublesome notion of wave function collapse".[87]

Max Tegmark also reports the result of a "highly unscientific" poll taken at a 1997 quantum mechanics workshop.[88] According to Tegmark, "The many worlds interpretation (MWI) scored second, comfortably ahead of the consistent histories and Bohm interpretations." Such polls have been taken at other conferences, for example, in response to Sean Carroll's observation, "As crazy as it sounds, most working physicists buy into the many-worlds theory"[89] Michael Nielsen counters: "at a quantum computing conference at Cambridge in 1998, a many-worlder surveyed the audience of approximately 200 people... Many-worlds did just fine, garnering support on a level comparable to, but somewhat below, Copenhagen and decoherence." However, Nielsen notes that it seemed most attendees found it to be a waste of time: Asher Peres "got a huge and sustained round of applause... when he got up at the end of the polling and asked 'And who here believes the laws of physics are decided by a democratic vote?'"[90]

A 2005 poll of fewer than 40 students and researchers taken after a course on the Interpretation of Quantum Mechanics at the Institute for Quantum Computing University of Waterloo found "Many Worlds (and decoherence)" to be the least favored.[91]

A 2011 poll of 33 participants at an Austrian conference found 6 endorsed MWI, 8 "Information-based/information-theoretical", and 14 Copenhagen;[92] the authors remark that the results are similar to Tegmark's 1998 poll.

22.10   Speculative implications

Speculative physics deals with questions which are also discussed in science fiction.

22.10.1   Quantum suicide thought experiment

Main article: Quantum suicide and immortality

Quantum suicide, as a thought experiment, was published independently by Hans Moravec in 1987[93][94] and Bruno Marchal in 1988[95][96] and was independently developed further by Max Tegmark in 1998.[97] It attempts to distinguish between the Copenhagen interpretation of quantum mechanics and the Everett many-worlds interpretation by means of a variation of the Schrödinger's cat thought experiment, from the cat's point of view. **Quantum immortality** refers to the subjective experience of surviving quantum suicide regardless of the odds.[98]

22.10.2   Weak coupling

Another speculation is that the separate worlds remain weakly coupled (e.g., by gravity) permitting "communication between parallel universes". A possible test of this using quantum-optical equipment is described in a 1997 *Foundations of Physics* article by Rainer Plaga.[56] It involves an isolated ion in an ion trap, a quantum measurement that would yield two parallel worlds (their difference just being in the detection of a single photon), and the excitation of the ion from only one of these worlds. If the excited ion can be detected from the other parallel universe, then this would constitute direct evidence in support of the many-worlds interpretation and would automatically exclude the orthodox, "logical", and "many-histories" interpretations. The reason the ion is isolated is to make it not participate immediately in the decoherence which insulates the parallel world branches, therefore allowing it to act as a gateway between the two worlds, and if the measure apparatus could perform the measurements quickly enough before the gateway ion is decoupled then the test would succeed (with electronic computers the necessary time window between the two worlds would be in a time scale of milliseconds or nanoseconds, and if the measurements are taken by humans then a few seconds would still be enough). R. Plaga shows that macroscopic decoherence timescales are a possibility. The proposed test is based on technical equipment described in a 1993 *Physical Review* article by Itano et al.[99] and R. Plaga says that this level of technology is enough to realize the proposed inter-world communication experiment. The necessary technology for precision measurements of single ions already exists since the 1970s, and the ion recommended for excitation is $^{199}Hg^{+}$. The excitation methodology is described by Itano et al. and the time needed for it is given by the Rabi flopping formula[100]

Such a test as described by R. Plaga would mean that energy transfer is possible between parallel worlds. This does not violate the fundamental principles of physics because these require energy conservation only for the whole universe and not for the single parallel branches.[56] Neither the excitation of the single ion (which is a degree of freedom of the proposed system) leads to decoherence, something which is proven by Welcher Weg detectors which can excite atoms without momentum transfer (which causes the loss of coherence).[101]

The proposed test would allow for low-bandwidth inter-world communication, the limiting factors of bandwidth and time being dependent on the technology of the equipment. Because of the time needed to determine the state of the partially decohered isolated excited ion based on Itano et al.'s methodology, the ion would decohere by the time its state is determined during the experiment, so Plaga's proposal would pass just enough information between the two worlds to confirm their parallel existence and nothing more. The author contemplates that with increased bandwidth, one could even transfer television imagery across the parallel worlds.[56] For example, Itano et al.'s methodology could be improved (by lowering the time needed for state determination of the excited ion) if a more efficient process were found for the detection of fluorescence radiation using 194 nm photons.[56]

A 1991 article by J.Polchinski also supports the view that inter-world communication is a theoretical possibility.[102] Other authors in a 1994 preprint article also contemplated similar ideas.[103]

The reason inter-world communication seems like a possibility is because decoherence which separates the parallel worlds is never fully complete,[104][105] therefore weak influences from one parallel world to another can still pass between them,[104][106] and these should be measurable with advanced technology. Deutsch proposed such an experiment in a 1985 *International Journal of Theoretical Physics* article,[107] but the technology it requires involves human-level artificial intelligence.[56]

22.10.3 Similarity to modal realism

The many-worlds interpretation has some similarity to modal realism in philosophy, which is the view that the possible worlds used to interpret modal claims exist and are of a kind with the actual world. Unlike the possible worlds of philosophy, however, in quantum mechanics counterfactual alternatives can influence the results of experiments, as in the Elitzur–Vaidman bomb-testing problem or the Quantum Zeno effect. Also, while the worlds of the many-worlds interpretation all share the same physical laws, modal realism postulates a world for every way things could conceivably have been.

22.10.4 Time travel

The many-worlds interpretation could be one possible way to resolve the paradoxes[78] that one would expect to arise *if* time travel turns out to be permitted by physics (permitting closed timelike curves and thus violating causality). Entering the past would itself be a quantum event causing branching, and therefore the timeline accessed by the time traveller simply would be another timeline of many. In that sense, it would make the Novikov self-consistency principle unnecessary.

22.11 Many-worlds in literature and science fiction

Main article: Parallel universe (fiction)
See also: Alternate History

The many-worlds interpretation (and the somewhat related concept of possible worlds) has been associated to numerous themes in literature, art and science fiction.

Some of these stories or films violate fundamental principles of causality and relativity, and are extremely misleading since the information-theoretic structure of the path space of multiple universes (that is information flow between different paths) is very likely extraordinarily complex. Also see Michael Clive Price's FAQ referenced in the external links section below where these issues (and other similar ones) are dealt with more decisively.

Another kind of popular illustration of many-worlds splittings, which does not involve information flow between paths, or information flow backwards in time considers alternate outcomes of historical events. According to the many-worlds interpretation, all of the historical speculations entertained within the alternate history genre are realized in parallel universes.[1]

The many-worlds interpretation of reality was anticipated with remarkable fidelity in Olaf Stapledon's 1937 science fiction novel Star Maker, in a paragraph describing one of the many universes created by the Star Maker god of the title. "In one inconceivably complex cosmos, whenever a creature was faced with several possible courses of action, it took them all, thereby creating many distinct temporal dimensions and distinct histories of the cosmos. Since in every evolutionary sequence of the cosmos there were very many creatures, and each was constantly faced with many possible courses, and the combinations of all their courses were innumerable, an infinity of distinct universes exfoliated from every moment of every temporal sequence in this cosmos."

22.12   See also

- Consistent histories

- EPR paradox

- *Fabric of Reality*

- Garden of Forking Paths

- Interpretations of quantum mechanics

- Many-minds interpretation

- Multiverse

- Multiple histories

- *The Beginning of Infinity*

- Quantum immortality – a thought experiment.

- Wave function collapse

22.13   Notes

[1] Bryce Seligman DeWitt, Quantum Mechanics and Reality: Could the solution to the dilemma of indeterminism be a universe in which all possible outcomes of an experiment actually occur?, *Physics Today*, 23(9) pp 30–40 (September 1970) "every quantum transition taking place on every star, in every galaxy, in every remote corner of the universe is splitting our local world on earth into myriads of copies of itself." See also *Physics Today*, letters followup, 24(4), (April 1971), pp 38–44

[2] Osnaghi, Stefano; Freitas, Fabio; Olival Freire, Jr (2009). "The Origin of the Everettian Heresy" (PDF). *Studies in History and Philosophy of Modern Physics* 40: 97–123. doi:10.1016/j.shpsb.2008.10.002.

[3] Hugh Everett Theory of the Universal Wavefunction, Thesis, Princeton University, (1956, 1973), pp 1–140

[4] Everett, Hugh (1957). "Relative State Formulation of Quantum Mechanics". *Reviews of Modern Physics* 29: 454–462. Bibcode:1957RvMP...29..454E. doi:10.1103/RevModPhys.29.454.

[5] Cecile M. DeWitt, John A. Wheeler eds, The Everett–Wheeler Interpretation of Quantum Mechanics, *Battelle Rencontres: 1967 Lectures in Mathematics and Physics* (1968)

[6] Bryce Seligman DeWitt, The Many-Universes Interpretation of Quantum Mechanics, *Proceedings of the International School of Physics "Enrico Fermi" Course IL: Foundations of Quantum Mechanics*, Academic Press (1972)

[7] Bryce Seligman DeWitt, R. Neill Graham, eds, *The Many-Worlds Interpretation of Quantum Mechanics*, Princeton Series in Physics, Princeton University Press (1973), ISBN 0-691-08131-X Contains Everett's thesis: The Theory of the Universal Wavefunction, pp 3–140.

[8] H. Dieter Zeh, On the Interpretation of Measurement in Quantum Theory, *Foundation of Physics*, vol. 1, pp. 69–76, (1970).

[9] Wojciech Hubert Zurek, Decoherence and the transition from quantum to classical, *Physics Today*, vol. 44, issue 10, pp. 36–44, (1991).

[10] Wojciech Hubert Zurek, Decoherence, einselection, and the quantum origins of the classical, *Reviews of Modern Physics*, 75, pp 715–775, (2003)

[11] The Many Worlds Interpretation of Quantum Mechanics

[12] David Deutsch argues that a great deal of fiction is close to a fact somewhere in the so called multiverse, *Beginning of Infinity*, p. 294

[13] Bryce Seligman DeWitt, R. Neill Graham, eds, *The Many-Worlds Interpretation of Quantum Mechanics*, Princeton Series in Physics, Princeton University Press (1973), ISBN 0-691-08131-X Contains Everett's thesis: The Theory of the Universal Wavefunction, where the claim to resolves all paradoxes is made on pg 118, 149.

[14] Hugh Everett, Relative State Formulation of Quantum Mechanics, *Reviews of Modern Physics* vol 29, (July 1957) pp 454–462. The claim to resolve EPR is made on page 462

[15] Steven Weinberg, *Dreams of a Final Theory: The Search for the Fundamental Laws of Nature* (1993), ISBN 0-09-922391-0, pg 68–69

[16] Steven Weinberg *Testing Quantum Mechanics*, Annals of Physics Vol 194 #2 (1989), pg 336–386

[17] John Archibald Wheeler, *Geons, Black Holes & Quantum Foam*, ISBN 0-393-31991-1. pp 268–270

[18] Everett 1957, section 3, 2nd paragraph, 1st sentence

[19] Everett [1956]1973, "Theory of the Universal Wavefunction", chapter 6 (e)

[20] Zurek, Wojciech (March 2009). "Quantum Darwinism". *Nature Physics* 5 (3): 181–188. arXiv:0903.5082. Bibcode:2009 doi:10.1038/nphys1202.

[21] Everett

[22] Peter Byrne, The Many Worlds of Hugh Everett III: Multiple Universes, Mutual Assured Destruction, and the Meltdown of a Nuclear Family, ISBN 978-0-19-955227-6

[23] "Whether you can observe a thing or not depends on the theory which you use. It is the theory which decides what can be observed." Albert Einstein to Werner Heisenberg, objecting to placing observables at the heart of the new quantum mechanics, during Heisenberg's 1926 lecture at Berlin; related by Heisenberg in 1968, quoted by Abdus Salam, *Unification of Fundamental Forces*, Cambridge University Press (1990) ISBN 0-521-37140-6, pp 98–101

[24] N.P. Landsman, "The conclusion seems to be that no generally accepted derivation of the Born rule has been given to date, but this does not imply that such a derivation is impossible in principle.", in *Compendium of Quantum Physics* (eds.) F.Weinert, K. Hentschel, D.Greenberger and B. Falkenburg (Springer, 2008), ISBN 3-540-70622-4

[25] Adrian Kent (May 5, 2009), *One world versus many: the inadequacy of Everettian accounts of evolution, probability, and scientific confirmation*

[26] Kent, Adrian (1990). "Against Many-Worlds Interpretations". *Int.J.Mod.Phys* A5: 1745–1762. arXiv:gr-qc/9703089. doi:10.1142/S0217751X90000805.

[27] Gleason, A. M. (1957). "Measures on the closed subspaces of a Hilbert space". *Journal of Mathematics and Mechanics* 6: 885–893. doi:10.1512/iumj.1957.6.56050. MR 0096113.

[28] James Hartle, Quantum Mechanics of Individual Systems, *American Journal of Physics*, 1968, vol 36 (#8), pp. 704–712

[29] E. Farhi, J. Goldstone & S. Gutmann. *How probability arises in quantum mechanics.*, Ann. Phys. (N.Y.) 192, 368–382 (1989).

[30] Pitowsky, I. (2005). "Quantum mechanics as a theory of probability". *Eprint arXiv:quant-ph/0510095*: 10095. arXiv:quant-ph/0510095. Bibcode:2005quant.ph.10095P.

[31] Deutsch, D. (1999). Quantum Theory of Probability and Decisions. *Proceedings of the Royal Society of London* A455, 3129–3137. .

[32] David Wallace: Quantum Probability and Decision Theory, Revisited

[33] David Wallace. Everettian Rationality: defending Deutsch's approach to probability in the Everett interpretation. Stud. Hist. Phil. Mod. Phys. 34 (2003), 415–438.

[34] David Wallace (2003), Quantum Probability from Subjective Likelihood: improving on Deutsch's proof of the probability rule

[35] David Wallace, 2009,A formal proof of the Born rule from decision-theoretic assumptions

[36] Simon Saunders: Derivation of the Born rule from operational assumptions. Proc. Roy. Soc. Lond. A460, 1771–1788 (2004).

[37] Simon Saunders, 2004: What is Probability?

[38] David J Baker, Measurement Outcomes and Probability in Everettian Quantum Mechanics, *Studies In History and Philosophy of Science Part B: Studies In History and Philosophy of Modern Physics*, Volume 38, Issue 1, March 2007, Pages 153–169

[39] H. Barnum, C. M. Caves, J. Finkelstein, C. A. Fuchs, R. Schack: Quantum Probability from Decision Theory? *Proc. Roy. Soc. Lond.* A456, 1175–1182 (2000).

[40] Merali, Zeeya (2007-09-21). "Parallel universes make quantum sense". *New Scientist* (2622). Retrieved 2013-11-22. (Summary only).

[41] Breitbart.com, Parallel universes exist – study, Sept 23 2007

[42] Perimeter Institute, Seminar overview, Probability in the Everett interpretation: state of play, David Wallace – Oxford University, 21 Sept 2007

[43] Perimeter Institute, Many worlds at 50 conference, September 21–24, 2007

[44] Wojciech H. Zurek: Probabilities from entanglement, Born's rule from envariance, *Phys. Rev.* A71, 052105 (2005).

[45] Schlosshauer, M.; Fine, A. (2005). "On Zurek's derivation of the Born rule". *Found. Phys.* 35: 197–213. arXiv:quant-ph/0312058. Bibcode:2005FoPh...35..197S. doi:10.1007/s10701-004-1941-6.

[46] Lutz Polley, Position eigenstates and the statistical axiom of quantum mechanics, contribution to conference *Foundations of Probability and Physics*, Vaxjo, Nov 27 – Dec 1, 2000

[47] Lutz Polley, Quantum-mechanical probability from the symmetries of two-state systems

[48] Armando V.D.B. Assis (2011). "Assis, Armando V.D.B. On the nature of $a_k^* a_k$ and the emergence of the Born rule. Annalen der Physik, 2011.". *Annalen der Physik (Berlin)* 523: 883–897. arXiv:1009.1532. Bibcode:2011AnP...523..883A. doi:10.1002/andp.201100062.

[49] Mark A. Rubin, Locality in the Everett Interpretation of Heisenberg-Picture Quantum Mechanics, *Foundations of Physics Letters*, 14, (2001) , pp. 301–322, arXiv:quant-ph/0103079

[50] Paul C.W. Davies, *Other Worlds*, chapters 8 & 9 *The Anthropic Principle & Is the Universe an accident?*, (1980) ISBN 0-460-04400-1

[51] Paul C.W. Davies, *The Accidental Universe*, (1982) ISBN 0-521-28692-1

[52] Everett FAQ "Does many-worlds violate Ockham's Razor?"

[53] Vaidman, Lev. "Many-Worlds Interpretation of Quantum Mechanics". The Stanford Encyclopedia of Philosophy.

[54] Deutsch, D., (1986) 'Three experimental implications of the Everett interpretation', in R. Penrose and C.J. Isham (eds.), Quantum Concepts of Space and Time, Oxford: The Clarendon Press, pp. 204–214.

[55] Page, D., (2000) 'Can Quantum Cosmology Give Observational Consequences of Many-Worlds Quantum Theory?'

[56] Plaga, R. (1997). "On a possibility to find experimental evidence for the many-worlds interpretation of quantum mechanics". *Foundations of Physics* 27: 559–577. arXiv:quant-ph/9510007. Bibcode:1997FoPh...27..559P. doi:10.1007/BF02550677.

[57] Page, Don N. (2000). "Can Quantum Cosmology Give Observational Consequences of Many-Worlds Quantum Theory?". arXiv:gr-qc/0001001. doi:10.1063/1.1301589.

[58] Bryce Seligman DeWitt, Quantum Mechanics and Reality: Could the solution to the dilemma of indeterminism be a universe in which all possible outcomes of an experiment actually occur?, *Physics Today*, 23(9) pp 30–40 (September 1970); see equation 10

[59] Penrose, R. *The Road to Reality*, §21.11

[60] Tegmark, Max The Interpretation of Quantum Mechanics: Many Worlds or Many Words?, 1998. To quote: "What Everett does NOT postulate: *"At certain magic instances, the world undergoes some sort of metaphysical 'split' into two branches that subsequently never interact."* This is not only a misrepresentation of the MWI, but also inconsistent with the Everett postulate, since the subsequent time evolution could in principle make the two terms...interfere. According to the MWI, there is, was and always will be only one wavefunction, and only decoherence calculations, not postulates, can tell us when it is a good approximation to treat two terms as non-interacting."

[61] Paul C.W. Davies, J.R. Brown, *The Ghost in the Atom* (1986) ISBN 0-521-31316-3, pp. 34–38: "The Many-Universes Interpretation", pp 83–105 for David Deutsch's test of MWI and reversible quantum memories

[62] Christoph Simon, 2009, *Conscious observers clarify many worlds*

[63] Joseph Gerver, The past as backward movies of the future, *Physics Today*, letters followup, 24(4), (April 1971), pp 46–7

[64] Bryce Seligman DeWitt, *Physics Today*, letters followup, 24(4), (April 1971), pp 43

[65] Arnold Neumaier's comments on the Everett FAQ, 1999 & 2003

[66] Everett [1956] 1973, *"Theory of the Universal Wavefunction"*, chapter V, section 4 "Approximate Measurements", pp. 100–103 (e)

[67] Stapp, Henry (2002). "The basis problem in many-world theories" (PDF). *Canadian Journal of Physics* 80: 1043–1052. arXiv:quant-ph/0110148. Bibcode:2002CaJPh..80.1043S. doi:10.1139/p02-068.

[68] Brown, Harvey R; Wallace, David (2005). "Solving the measurement problem: de Broglie–Bohm loses out to Everett" (PDF). *Foundations of Physics* 35: 517–540. arXiv:quant-ph/0403094. Bibcode:2005FoPh...35..517B. doi:10.1007/s10701-004-2009-3.

[69] Mark A Rubin (2005), There Is No Basis Ambiguity in Everett Quantum Mechanics, *Foundations of Physics Letters*, Volume 17, Number 4 / August, 2004, pp 323–341

[70] Penrose, Roger (August 1991). "Roger Penrose Looks Beyond the Classic-Quantum Dichotomy". Sciencewatch. Retrieved 2007-10-21.

[71] Everett FAQ "Does many-worlds violate conservation of energy?"

[72] Everett FAQ "How do probabilities emerge within many-worlds?"

[73] Everett FAQ "When does Schrodinger's cat split?"

[74] Jeffrey A. Barrett, *The Quantum Mechanics of Minds and Worlds*, Oxford University Press, 1999. According to Barrett (loc. cit. Chapter 6) "There are many many-worlds interpretations."

[75] Barrett, Jeffrey A. (2010). Zalta, Edward N., ed. "Everett's Relative-State Formulation of Quantum Mechanics" (Fall 2010 ed.). The Stanford Encyclopedia of Philosophy. Again, according to Barrett "It is... unclear precisely how this was supposed to work."

[76] Aldhous, Peter (2007-11-24). "Parallel lives can never touch". *New Scientist* (2631). Retrieved 2007-11-21.

[77] Eugene Shikhovtsev's Biography of Everett, in particular see *"Keith Lynch remembers 1979–1980"*

[78] David Deutsch, *The Fabric of Reality: The Science of Parallel Universes And Its Implications*, Penguin Books (1998), ISBN 0-14-027541-X

[79] Deutsch, David (1985). "Quantum theory, the Church–Turing principle and the universal quantum computer". *Proceedings of the Royal Society of London A* **400**: 97–117. Bibcode:1985RSPSA.400...97D. doi:10.1098/rspa.1985.0070.

[80] A response to Bryce DeWitt, Martin Gardner, May 2002

[81] Award winning 1995 Channel 4 documentary "Reality on the rocks: Beyond our Ken" where, in response to Ken Campbell's question "all these trillions of Universes of the Multiverse, are they as real as this one seems to be to me?" Hawking states, "Yes.... According to Feynman's idea, every possible history (of Ken) is equally real."

[82] Gardner, Martin (2003). *Are universes thicker than blackberries?*. W.W. Norton. p. 10. ISBN 978-0-393-05742-3.

[83] Ferris, Timothy (1997). *The Whole Shebang*. Simon & Schuster. pp. 345. ISBN 978-0-684-81020-1.

[84] Hawking, Stephen; Roger Penrose (1996). *The Nature of Space and Time*. Princeton University Press. pp. 121. ISBN 978-0-691-03791-2.

[85] Elvridge., Jim (2008-01-02). *The Universe – Solved!*. pp. 35–36. ISBN 978-1-4243-3626-5. OCLC 247614399. 58% believed that the Many Worlds Interpretation (MWI) was true, including Stephen Hawking and Nobel Laureates Murray Gell-Mann and Richard Feynman

[86] Bruce., Alexandra. "How does reality work?". *Beyond the bleep : the definitive unauthorized guide to What the bleep do we know!?*. p. 33. ISBN 978-1-932857-22-1. [the poll was] published in the French periodical *Sciences et Avenir* in January 1998

[87] Stenger, V.J. (1995). *The Unconscious Quantum: Metaphysics in Modern Physics and Cosmology*. Prometheus Books. p. 176. ISBN 978-1-57392-022-3. LCCN lc95032599. Gell-Mann and collaborator James Hartle, along with a score of others, have been working to develop a more palatable interpretation of quantum mechanics that is free of the problems that plague all the interpretations we have considered so far. This new interpretation is called, in its various incarnations, **post-Everett quantum mechanics**, alternate histories, consistent histories, or decoherent histories. I will not be overly concerned with the detailed differences between these characterizations and will use the terms more or less interchangeably.

[88] Max Tegmark on many-worlds (contains MWI poll)

[89] Caroll, Sean (1 April 2004). "Preposterous Universe". Archived from the original on 8 September 2004.

[90] Nielsen, Michael (3 April 2004). "Michael Nielsen: The Interpretation of Quantum Mechanics". Archived from the original on 20 May 2004.

[91] Interpretation of Quantum Mechanics class survey

[92] "A Snapshot of Foundational Attitudes Toward Quantum Mechanics", Schlosshauer et al 2013

[93] "The Many Minds Approach". 25 October 2010. Retrieved 7 December 2010. This idea was first proposed by Austrian mathematician Hans Moravec in 1987...

[94] Moravec, Hans (1988). "The Doomsday Device". *Mind Children: The Future of Robot and Human Intelligence*. Harvard: Harvard University Press. p. 188. ISBN 978-0-674-57618-6. (If MWI is true, apocalyptic particle accelerators won't function as advertised).

[95] Marchal, Bruno (1988). "Informatique théorique et philosophie de l'esprit" [Theoretical Computer Science and Philosophy of Mind]. *Acte du 3ème colloque international Cognition et Connaissance [Proceedings of the 3rd International Conference Cognition and Knowledge]* (Toulouse): 193–227.

[96] Marchal, Bruno (1991). De Glas, M.; Gabbay, D., eds. "Mechanism and personal identity" (PDF). *Proceedings of WOCFAI 91* (Paris. Angkor.): 335–345.

[97] Tegmark, Max The Interpretation of Quantum Mechanics: Many Worlds or Many Words?, 1998

[98] Tegmark, Max (November 1998). "Quantum immortality". Retrieved 25 October 2010.

[99] W.M.Itano et al., Phys.Rev. A47,3354 (1993).

[100] M.SargentIII,M.O.Scully and W.E.Lamb, Laser physics (Addison-Wesley, Reading, 1974), p.27.

[101] M.O.Scully and H.Walther, Phys.Rev. A39,5229 (1989).

[102] J.Polchinski, Phys.Rev.Lett. 66,397 (1991).

[103] M.Gell-Mann and J.B.Hartle, Equivalent Sets of Histories and Multiple Quasiclassical Domains, preprint University of California at Santa Barbara UCSBTH-94-09 (1994).

[104] H.D.Zeh, Found.Phys. 3,109 (1973).

[105] H.D.Zeh, Phys.Lett.A 172,189 (1993).

[106] A.Albrecht, Phys.Rev. D48,3768 (1993).

[107] D.Deutsch, Int.J.theor.Phys. 24,1 (1985).

22.14   Further reading

- Jeffrey A. Barrett, *The Quantum Mechanics of Minds and Worlds*, Oxford University Press, Oxford, 1999.

- Peter Byrne, *The Many Worlds of Hugh Everett III: Multiple Universes, Mutual Assured Destruction, and the Meltdown of a Nuclear Family*, Oxford University Press, 2010.

- Jeffrey A. Barrett and Peter Byrne, eds., "The Everett Interpretation of Quantum Mechanics: Collected Works 1955–1980 with Commentary", Princeton University Press, 2012.

- Julian Brown, *Minds, Machines, and the Multiverse*, Simon & Schuster, 2000, ISBN 0-684-81481-1

- Paul C.W. Davies, *Other Worlds*, (1980) ISBN 0-460-04400-1

- James P. Hogan, *The Proteus Operation* (science fiction involving the many-worlds interpretation, time travel and World War 2 history), Baen, Reissue edition (August 1, 1996) ISBN 0-671-87757-7

- Adrian Kent, One world versus many: the inadequacy of Everettian accounts of evolution, probability, and scientific confirmation

- Andrei Linde and Vitaly Vanchurin, How Many Universes are in the Multiverse?

- Osnaghi, Stefano; Freitas, Fabio; Olival Freire, Jr (2009). "The Origin of the Everettian Heresy" (PDF). *Studies in History and Philosophy of Modern Physics* **40**: 97–123. doi:10.1016/j.shpsb.2008.10.002. A study of the painful three-way relationship between Hugh Everett, John A Wheeler and Niels Bohr and how this affected the early development of the many-worlds theory.

- Asher Peres, *Quantum Theory: Concepts and Methods*, Kluwer, Dordrecht, 1993.

- Mark A. Rubin, Locality in the Everett Interpretation of Heisenberg-Picture Quantum Mechanics, *Foundations of Physics Letters*, 14, (2001), pp. 301–322, arXiv:quant-ph/0103079

- David Wallace, Harvey R. Brown, Solving the measurement problem: de Broglie–Bohm loses out to Everett, *Foundations of Physics*, arXiv:quant-ph/0403094

- David Wallace, Worlds in the Everett Interpretation, *Studies in the History and Philosophy of Modern Physics*, 33, (2002), pp. 637–661, arXiv:quant-ph/0103092

- John A. Wheeler and Wojciech Hubert Zurek (eds), *Quantum Theory and Measurement*, Princeton University Press, (1983), ISBN 0-691-08316-9

- Sean M. Carroll, Charles T. Sebens, *Many Worlds, the Born Rule, and Self-Locating Uncertainty*, arXiv:1405.7907

22.15 External links

- Everett's Relative-State Formulation of Quantum Mechanics – Jeffrey A. Barrett's article on Everett's formulation of quantum mechanics in the Stanford Encyclopedia of Philosophy.

- Many-Worlds Interpretation of Quantum Mechanics – Lev Vaidman's article on the many-worlds interpretation of quantum mechanics in the Stanford Encyclopedia of Philosophy.

- Hugh Everett III Manuscript Archive (UC Irvine) – Jeffrey A. Barrett, Peter Byrne, and James O. Weatherall (eds.).

- Michael C Price's Everett FAQ – a clear FAQ-style presentation of the theory.

- The Many-Worlds Interpretation of Quantum Mechanics – a description for the lay reader with links.

- Against Many-Worlds Interpretations by Adrian Kent

- Many-Worlds is a "lost cause" according to R. F. Streater

- The many worlds of quantum mechanics John Sankey

- Max Tegmark's web page

- Henry Stapp's critique of MWI, focusing on the basis problem Canadian J. Phys. 80,1043–1052 (2002).

- Everett hit count on arxiv.org

- Many Worlds 50th anniversary conference at Oxford

- "Many Worlds at 50" conference at Perimeter Institute

- Scientific American report on the Many Worlds 50th anniversary conference at Oxford

- Highfield, Roger (September 21, 2007). "Parallel universe proof boosts time travel hopes". The Daily Telegraph. Archived from the original on 2007-10-20. Retrieved 2007-10-26..

- HowStuffWorks article

- Physicists Calculate Number of Parallel Universes Physorg.com October 16, 2009.

- TED-Education video – How many universes are there?.

Hugh Everett III (1930–1982) was the first physicist who proposed the many-worlds interpretation (MWI) of quantum physics, which he termed his "relative state" formulation.

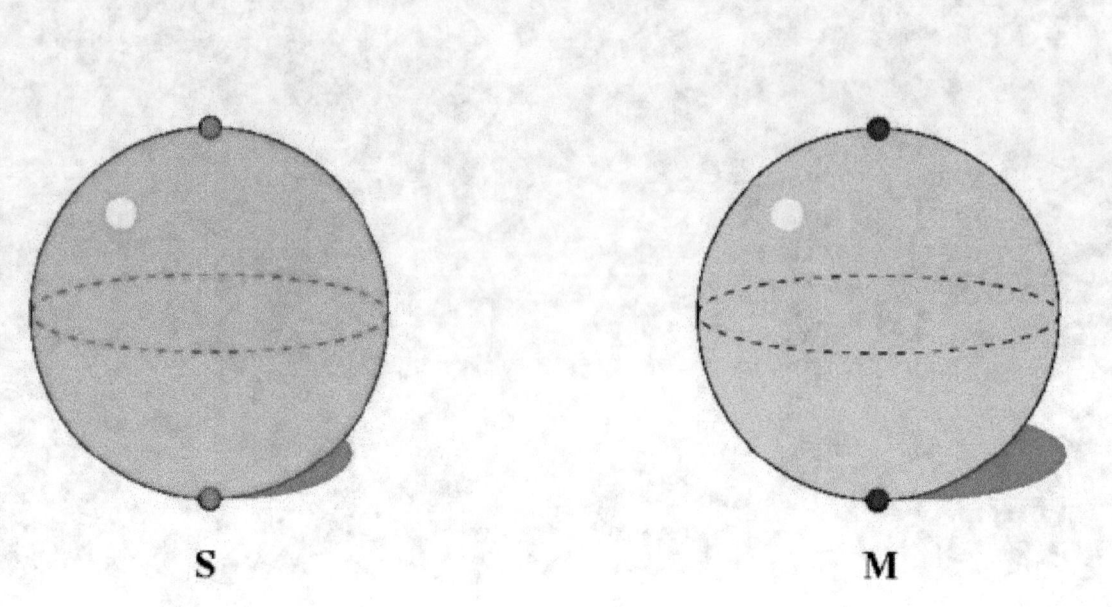

Schematic representation of pair of "smallest possible" quantum mechanical systems prior to interaction: Measured system S and measurement apparatus M. Systems such as S are referred to as 1-qubit systems.

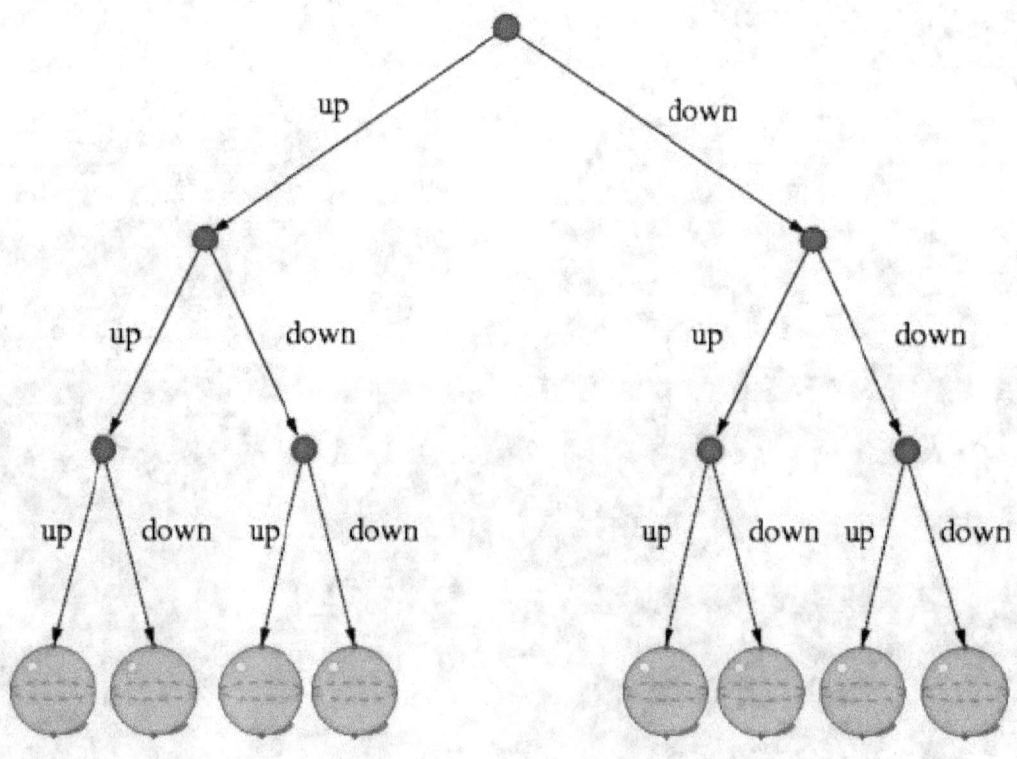

Schematic illustration of splitting as a result of a repeated measurement.

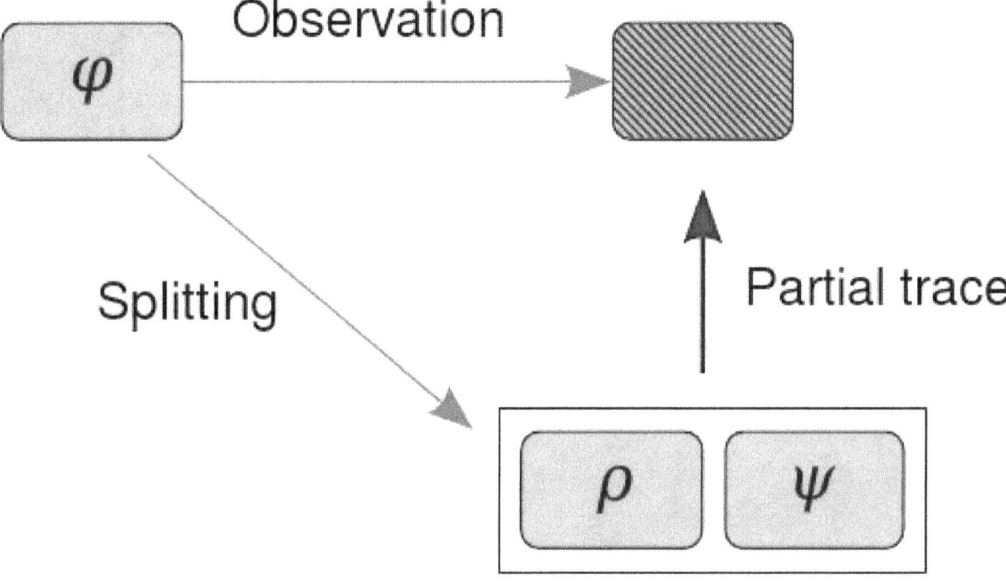

Partial trace as relative state. Light blue rectangle on upper left denotes system in pure state. Trellis shaded rectangle in upper right denotes a (possibly) mixed state. Mixed state from observation is partial trace of a linear superposition of states as shown in lower right-hand corner.

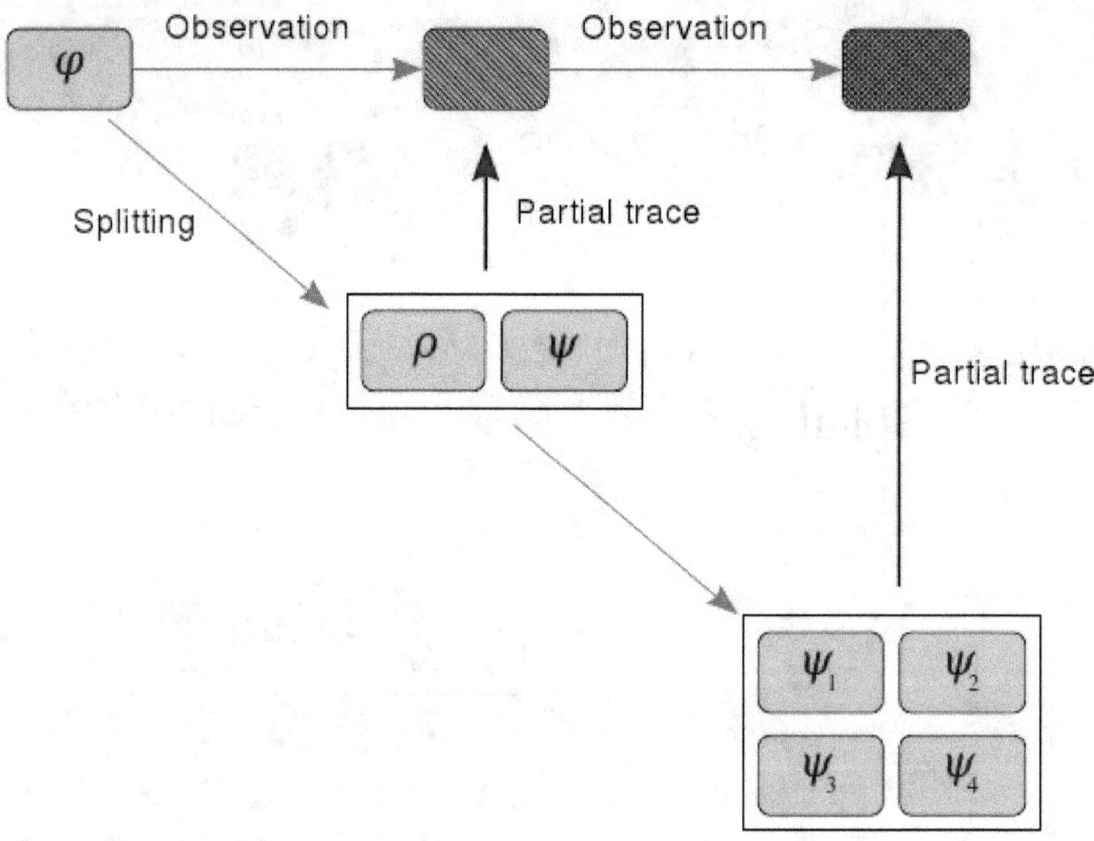

Successive measurements with successive splittings

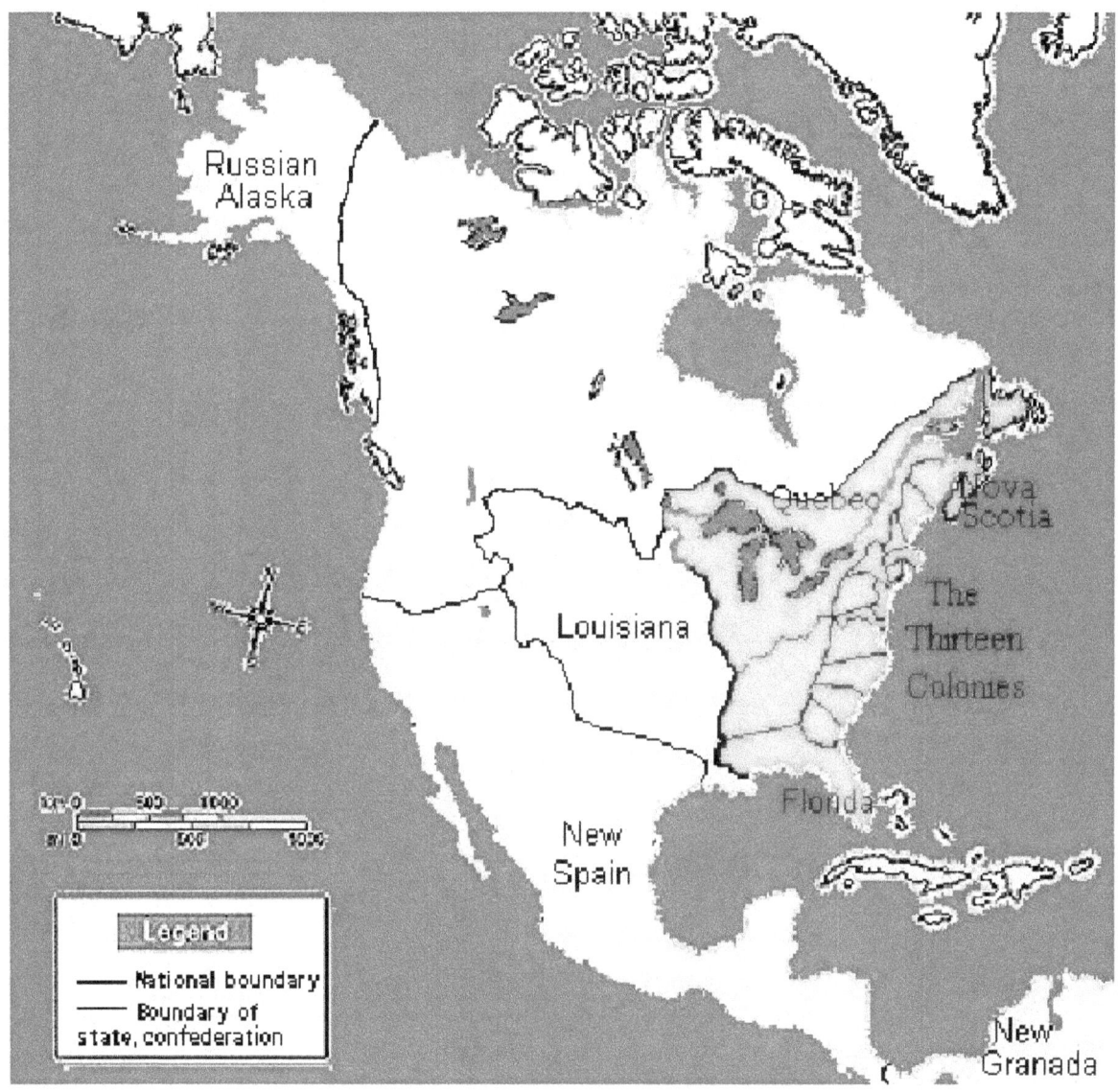

A map from Robert Sobel's novel For Want of a Nail, *an artistic illustration of how small events – in this example the branching or point of divergence from our timeline's history is in October 1777 – can profoundly alter the course of history. According to the many-worlds interpretation every event, even microscopic, is a branch point; all possible alternative histories actually exist.*[1]

Chapter 23

Dirac delta function

"Delta function" redirects here. For other uses, see Delta function (disambiguation).

In mathematics, the **Dirac delta function**, or **δ function**, is a generalized function, or distribution, on the real number

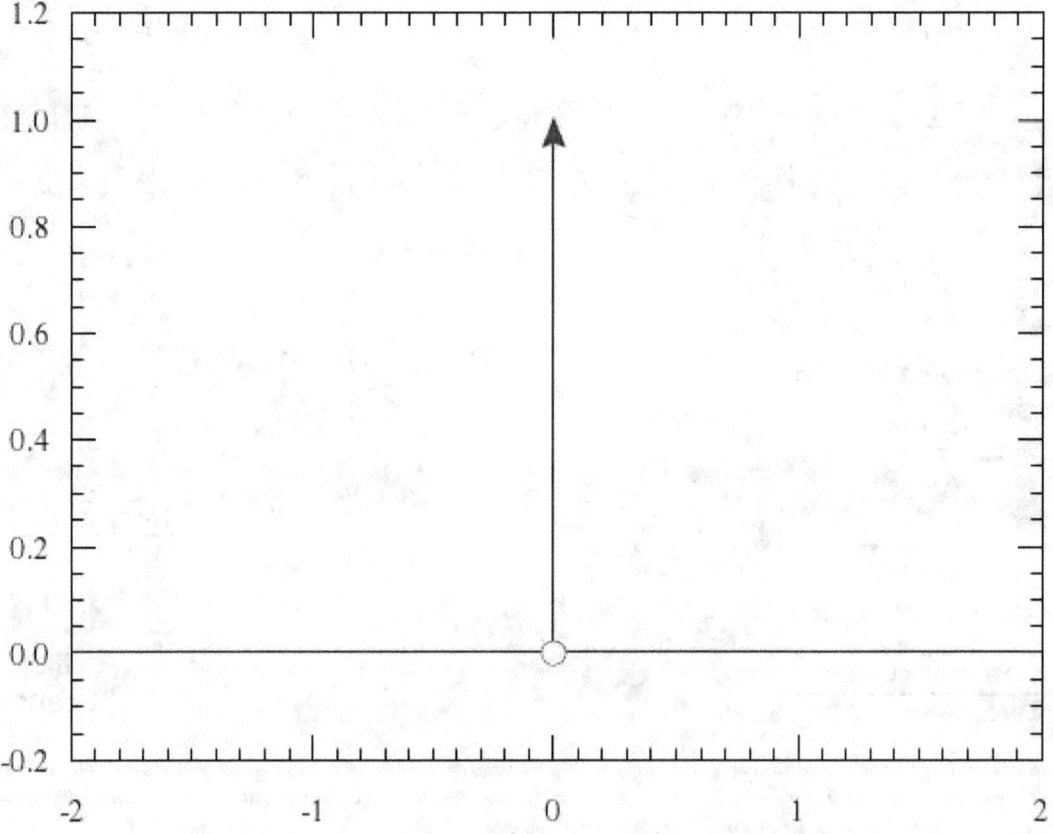

Schematic representation of the Dirac delta function by a line surmounted by an arrow. The height of the arrow is usually used to specify the value of any multiplicative constant, which will give the area under the function. The other convention is to write the area next to the arrowhead.

line that is zero everywhere except at zero, with an integral of one over the entire real line.[1][2][3] The delta function is sometimes thought of as an infinitely high, infinitely thin spike at the origin, with total area one under the spike, and physically represents the density of an idealized point mass or point charge.[4] It was introduced by theoretical physicist

Paul Dirac. In the context of signal processing it is often referred to as the **unit impulse symbol** (or function).[5] Its discrete analog is the Kronecker delta function, which is usually defined on a discrete domain and takes values 0 and 1.

From a purely mathematical viewpoint, the Dirac delta is not strictly a function, because any extended-real function that is equal to zero everywhere but a single point must have total integral zero.[6] The delta function only makes sense as a mathematical object when it appears inside an integral. While from this perspective the Dirac delta can usually be manipulated as though it were a function, formally it must be defined as a distribution that is also a measure. In many applications, the Dirac delta is regarded as a kind of limit (a weak limit) of a sequence of functions having a tall spike at the origin. The approximating functions of the sequence are thus "approximate" or "nascent" delta functions.

23.1 Overview

The graph of the delta function is usually thought of as following the whole x-axis and the positive y-axis. Despite its name, the delta function is not truly a function, at least not a usual one with range in real numbers. For example, the objects $f(x) = \delta(x)$ and $g(x) = 0$ are equal everywhere except at $x = 0$ yet have integrals that are different. According to Lebesgue integration theory, if f and g are functions such that $f = g$ almost everywhere, then f is integrable if and only if g is integrable and the integrals of f and g are identical. Rigorous treatment of the Dirac delta requires measure theory or the theory of distributions.

The Dirac delta is used to model a tall narrow spike function (an *impulse*), and other similar abstractions such as a point charge, point mass or electron point. For example, to calculate the dynamics of a baseball being hit by a bat, one can approximate the force of the bat hitting the baseball by a delta function. In doing so, one not only simplifies the equations, but one also is able to calculate the motion of the baseball by only considering the total impulse of the bat against the ball rather than requiring knowledge of the details of how the bat transferred energy to the ball.

In applied mathematics, the delta function is often manipulated as a kind of limit (a weak limit) of a sequence of functions, each member of which has a tall spike at the origin: for example, a sequence of Gaussian distributions centered at the origin with variance tending to zero.

23.2 History

Joseph Fourier presented what is now called the Fourier integral theorem in his treatise *Théorie analytique de la chaleur* in the form:[7]

$$f(x) = \frac{1}{2\pi} \int_{-\infty}^{\infty} d\alpha f(\alpha) \int_{-\infty}^{\infty} dp \ \cos(px - p\alpha) \,,$$

which is tantamount to the introduction of the δ-function in the form:[8]

$$\delta(x - \alpha) = \frac{1}{2\pi} \int_{-\infty}^{\infty} dp \ \cos(px - p\alpha) \,.$$

Later, Augustin Cauchy expressed the theorem using exponentials:[9][10]

$$f(x) = \frac{1}{2\pi} \int_{-\infty}^{\infty} e^{ipx} \left(\int_{-\infty}^{\infty} e^{-ip\alpha} f(\alpha) \, d\alpha \right) \, dp.$$

Cauchy pointed out that in some circumstances the *order* of integration in this result was significant.[11][12]

As justified using the theory of distributions, the Cauchy equation can be rearranged to resemble Fourier's original formulation and expose the δ-function as:

$$f(x) = \frac{1}{2\pi} \int_{-\infty}^{\infty} e^{ipx} \left(\int_{-\infty}^{\infty} e^{-ip\alpha} f(\alpha)\, d\alpha \right) dp$$

$$= \frac{1}{2\pi} \int_{-\infty}^{\infty} \left(\int_{-\infty}^{\infty} e^{ipx} e^{-ip\alpha}\, dp \right) f(\alpha)\, d\alpha = \int_{-\infty}^{\infty} \delta(x - \alpha) f(\alpha)\, d\alpha,$$

where the δ-function is expressed as:

$$\delta(x - \alpha) = \frac{1}{2\pi} \int_{-\infty}^{\infty} e^{ip(x-\alpha)}\, dp \, .$$

A rigorous interpretation of the exponential form and the various limitations upon the function f necessary for its application extended over several centuries. The problems with a classical interpretation are explained as follows:[13]

> The greatest drawback of the classical Fourier transformation is a rather narrow class of functions (originals) for which it can be effectively computed. Namely, it is necessary that these functions decrease sufficiently rapidly to zero (in the neighborhood of infinity) in order to ensure the existence of the Fourier integral. For example, the Fourier transform of such simple functions as polynomials does not exist in the classical sense. The extension of the classical Fourier transformation to distributions considerably enlarged the class of functions that could be transformed and this removed many obstacles.

Further developments included generalization of the Fourier integral, "beginning with Plancherel's pathbreaking L^2-theory (1910), continuing with Wiener's and Bochner's works (around 1930) and culminating with the amalgamation into L. Schwartz's theory of distributions (1945) ...",[14] and leading to the formal development of the Dirac delta function.

An infinitesimal formula for an infinitely tall, unit impulse delta function (infinitesimal version of Cauchy distribution) explicitly appears in an 1827 text of Augustin Louis Cauchy.[15] Siméon Denis Poisson considered the issue in connection with the study of wave propagation as did Gustav Kirchhoff somewhat later. Kirchhoff and Hermann von Helmholtz also introduced the unit impulse as a limit of Gaussians, which also corresponded to Lord Kelvin's notion of a point heat source. At the end of the 19th century, Oliver Heaviside used formal Fourier series to manipulate the unit impulse.[16] The Dirac delta function as such was introduced as a "convenient notation" by Paul Dirac in his influential 1930 book *The Principles of Quantum Mechanics*.[17] He called it the "delta function" since he used it as a continuous analogue of the discrete Kronecker delta.

23.3 Definitions

The Dirac delta can be loosely thought of as a function on the real line which is zero everywhere except at the origin, where it is infinite,

$$\delta(x) = \begin{cases} +\infty, & x = 0 \\ 0, & x \neq 0 \end{cases}$$

and which is also constrained to satisfy the identity

$$\int_{-\infty}^{\infty} \delta(x)\, dx = 1. \text{ [18]}$$

This is merely a heuristic characterization. The Dirac delta is not a function in the traditional sense as no function defined on the real numbers has these properties.[17] The Dirac delta function can be rigorously defined either as a distribution or as a measure.

23.3.1 As a measure

One way to rigorously define the delta function is as a measure, which accepts as an argument a subset A of the real line $\mathbf{R}$, and returns $\delta(A) = 1$ if $0 \in A$, and $\delta(A) = 0$ otherwise.[19] If the delta function is conceptualized as modeling an idealized point mass at 0, then $\delta(A)$ represents the mass contained in the set A. One may then define the integral against δ as the integral of a function against this mass distribution. Formally, the Lebesgue integral provides the necessary analytic device. The Lebesgue integral with respect to the measure δ satisfies

$$\int_{-\infty}^{\infty} f(x)\, \delta\{dx\} = f(0)$$

for all continuous compactly supported functions f. The measure δ is not absolutely continuous with respect to the Lebesgue measure — in fact, it is a singular measure. Consequently, the delta measure has no Radon–Nikodym derivative — no true function for which the property

$$\int_{-\infty}^{\infty} f(x)\delta(x)\, dx = f(0)$$

holds.[20] As a result, the latter notation is a convenient abuse of notation, and not a standard (Riemann or Lebesgue) integral.

As a probability measure on $\mathbf{R}$, the delta measure is characterized by its cumulative distribution function, which is the unit step function[21]

$$H(x) = \begin{cases} 1 & \text{if } x \geq 0 \\ 0 & \text{if } x < 0. \end{cases}$$

This means that $H(x)$ is the integral of the cumulative indicator function $\mathbf{1}_{(-\infty,\, x]}$ with respect to the measure δ; to wit,

$$H(x) = \int_{\mathbf{R}} \mathbf{1}_{(-\infty,x]}(t)\, \delta\{dt\} = \delta(-\infty, x].$$

Thus in particular the integral of the delta function against a continuous function can be properly understood as a Stieltjes integral:[22]

$$\int_{-\infty}^{\infty} f(x)\delta\{dx\} = \int_{-\infty}^{\infty} f(x)\, dH(x).$$

All higher moments of δ are zero. In particular, characteristic function and moment generating function are both equal to one.

23.3.2 As a distribution

In the theory of distributions a generalized function is thought of not as a function itself, but only in relation to how it affects other functions when it is "integrated" against them. In keeping with this philosophy, to define the delta function properly, it is enough to say what the "integral" of the delta function against a sufficiently "good" test function is. If the delta function is already understood as a measure, then the Lebesgue integral of a test function against that measure supplies the necessary integral.

A typical space of test functions consists of all smooth functions on $\mathbf{R}$ with compact support. As a distribution, the Dirac delta is a linear functional on the space of test functions and is defined by[23] for every test function φ.

For δ to be properly a distribution, it must be "continuous" in a suitable sense. In general, for a linear functional S on the space of test functions to define a distribution, it is necessary and sufficient that, for every positive integer N there is an integer MN and a constant CN such that for every test function φ, one has the inequality[24]

$$|S[\varphi]| \leq C_N \sum_{k=0}^{M_N} \sup_{x \in [-N,N]} |\varphi^{(k)}(x)|.$$

With the δ distribution, one has such an inequality (with $CN = 1$) with $MN = 0$ for all N. Thus δ is a distribution of order zero. It is, furthermore, a distribution with compact support (the support being $\{0\}$).

The delta distribution can also be defined in a number of equivalent ways. For instance, it is the distributional derivative of the Heaviside step function. This means that, for every test function φ, one has

$$\delta[\varphi] = -\int_{-\infty}^{\infty} \varphi'(x)H(x)\,dx.$$

Intuitively, if integration by parts were permitted, then the latter integral should simplify to

$$\int_{-\infty}^{\infty} \varphi(x)H'(x)\,dx = \int_{-\infty}^{\infty} \varphi(x)\delta(x)\,dx,$$

and indeed, a form of integration by parts is permitted for the Stieltjes integral, and in that case one does have

$$-\int_{-\infty}^{\infty} \varphi'(x)H(x)\,dx = \int_{-\infty}^{\infty} \varphi(x)\,dH(x).$$

In the context of measure theory, the Dirac measure gives rise to a distribution by integration. Conversely, equation (1) defines a Daniell integral on the space of all compactly supported continuous functions φ which, by the Riesz representation theorem, can be represented as the Lebesgue integral of φ with respect to some Radon measure.

23.3.3   Generalizations

The delta function can be defined in n-dimensional Euclidean space $\mathbf{R}^n$ as the measure such that

$$\int_{\mathbf{R}^n} f(\mathbf{x})\delta\{d\mathbf{x}\} = f(\mathbf{0})$$

for every compactly supported continuous function f. As a measure, the n-dimensional delta function is the product measure of the 1-dimensional delta functions in each variable separately. Thus, formally, with $\mathbf{x} = (x_1, x_2, ..., xn)$, one has[5]

The delta function can also be defined in the sense of distributions exactly as above in the one-dimensional case.[25] However, despite widespread use in engineering contexts, (2) should be manipulated with care, since the product of distributions can only be defined under quite narrow circumstances.[26]

The notion of a **Dirac measure** makes sense on any set.[19] Thus if X is a set, $x_0 \in X$ is a marked point, and Σ is any sigma algebra of subsets of X, then the measure defined on sets $A \in \Sigma$ by

$$\delta_{x_0}(A) = \begin{cases} 1 & \text{if } x_0 \in A \\ 0 & \text{if } x_0 \notin A \end{cases}$$

is the delta measure or unit mass concentrated at x_0.

Another common generalization of the delta function is to a differentiable manifold where most of its properties as a distribution can also be exploited because of the differentiable structure. The delta function on a manifold M centered at the point $x_0 \in M$ is defined as the following distribution:

for all compactly supported smooth real-valued functions φ on M.[27] A common special case of this construction is when M is an open set in the Euclidean space $\mathbf{R}^n$.

On a locally compact Hausdorff space X, the Dirac delta measure concentrated at a point x is the Radon measure associated with the Daniell integral (3) on compactly supported continuous functions φ. At this level of generality, calculus as such is no longer possible, however a variety of techniques from abstract analysis are available. For instance, the mapping $x_0 \mapsto \delta_{x_0}$ is a continuous embedding of X into the space of finite Radon measures on X, equipped with its vague topology. Moreover, the convex hull of the image of X under this embedding is dense in the space of probability measures on X.[28]

23.4   Properties

23.4.1   Scaling and symmetry

The delta function satisfies the following scaling property for a non-zero scalar α:[29]

$$\int_{-\infty}^{\infty} \delta(\alpha x)\, dx = \int_{-\infty}^{\infty} \delta(u)\, \frac{du}{|\alpha|} = \frac{1}{|\alpha|}$$

and so

In particular, the delta function is an even distribution, in the sense that

$$\delta(-x) = \delta(x)$$

which is homogeneous of degree -1.

23.4.2   Algebraic properties

The distributional product of δ with x is equal to zero:

$$x\delta(x) = 0.$$

Conversely, if $xf(x) = xg(x)$, where f and g are distributions, then

$$f(x) = g(x) + c\delta(x)$$

for some constant c.[30]

23.4.3 Translation

The integral of the time-delayed Dirac delta is given by:

$$\int_{-\infty}^{\infty} f(t)\delta(t - T)\, dt = f(T).$$

This is sometimes referred to as the *sifting property*[31] or the *sampling property*. The delta function is said to "sift out" the value at $t = T$.

It follows that the effect of convolving a function $f(t)$ with the time-delayed Dirac delta is to time-delay $f(t)$ by the same amount:

This holds under the precise condition that f be a tempered distribution (see the discussion of the Fourier transform below). As a special case, for instance, we have the identity (understood in the distribution sense)

$$\int_{-\infty}^{\infty} \delta(\xi - x)\delta(x - \eta)\, dx = \delta(\xi - \eta).$$

23.4.4 Composition with a function

More generally, the delta distribution may be composed with a smooth function $g(x)$ in such a way that the familiar change of variables formula holds, that

$$\int_{\mathbf{R}} \delta\big(g(x)\big) f\big(g(x)\big) |g'(x)|\, dx = \int_{g(\mathbf{R})} \delta(u) f(u)\, du$$

provided that g is a continuously differentiable function with g' nowhere zero.[32] That is, there is a unique way to assign meaning to the distribution $\delta \circ g$ so that this identity holds for all compactly supported test functions f. Therefore, the domain must be broken up to exclude the $g' = 0$ point. This distribution satisfies $\delta(g(x)) = 0$ if g is nowhere zero, and otherwise if g has a real root at x_0, then

$$\delta(g(x)) = \frac{\delta(x - x_0)}{|g'(x_0)|}.$$

It is natural therefore to *define* the composition $\delta(g(x))$ for continuously differentiable functions g by

$$\delta(g(x)) = \sum_i \frac{\delta(x - x_i)}{|g'(x_i)|}$$

where the sum extends over all roots of $g(x)$, which are assumed to be simple.[32] Thus, for example

$$\delta\left(x^2 - \alpha^2\right) = \frac{1}{2|\alpha|}\Big[\delta\left(x + \alpha\right) + \delta\left(x - \alpha\right)\Big].$$

In the integral form the generalized scaling property may be written as

$$\int_{-\infty}^{\infty} f(x)\,\delta(g(x))\,dx = \sum_i \frac{f(x_i)}{|g'(x_i)|}.$$

23.4.5 Properties in n dimensions

The delta distribution in an n-dimensional space satisfies the following scaling property instead:

$$\delta(\alpha\mathbf{x}) = |\alpha|^{-n}\delta(\mathbf{x})$$

so that δ is a homogeneous distribution of degree $-n$. Under any reflection or rotation ρ, the delta function is invariant:

$$\delta(\rho\mathbf{x}) = \delta(\mathbf{x}).$$

As in the one-variable case, it is possible to define the composition of δ with a bi-Lipschitz function[33] $g\colon \mathbf{R}^n \to \mathbf{R}^n$ uniquely so that the identity

$$\int_{\mathbf{R}^n} \delta(g(\mathbf{x}))\, f(g(\mathbf{x}))\,|\det g'(\mathbf{x})|\,d\mathbf{x} = \int_{g(\mathbf{R}^n)} \delta(\mathbf{u}) f(\mathbf{u})\,d\mathbf{u}$$

for all compactly supported functions f.

Using the coarea formula from geometric measure theory, one can also define the composition of the delta function with a submersion from one Euclidean space to another one of different dimension; the result is a type of current. In the special case of a continuously differentiable function $g\colon \mathbf{R}^n \to \mathbf{R}$ such that the gradient of g is nowhere zero, the following identity holds[34]

$$\int_{\mathbf{R}^n} f(\mathbf{x})\,\delta(g(\mathbf{x}))\,d\mathbf{x} = \int_{g^{-1}(0)} \frac{f(\mathbf{x})}{|\nabla g|}\,d\sigma(\mathbf{x})$$

where the integral on the right is over $g^{-1}(0)$, the $(n-1)$-dimensional surface defined by $g(\mathbf{x}) = 0$ with respect to the Minkowski content measure. This is known as a simple layer integral.

More generally, if S is a smooth hypersurface of $\mathbf{R}^n$, then we can associate to S the distribution that integrates any compactly supported smooth function g over S:

$$\delta_S[g] = \int_S g(\mathbf{s})\,d\sigma(\mathbf{s})$$

where σ is the hypersurface measure associated to S. This generalization is associated with the potential theory of simple layer potentials on S. If D is a domain in $\mathbf{R}^n$ with smooth boundary S, then δS is equal to the normal derivative of the indicator function of D in the distribution sense:

$$-\int_{\mathbf{R}^n} g(\mathbf{x})\,\frac{\partial 1_D(\mathbf{x})}{\partial n}\,d\mathbf{x} = \int_S g(\mathbf{s})\,d\sigma(\mathbf{s}),$$

where n is the outward normal.[35][36] For a proof, see e.g. the article on the surface delta function.

23.5 Fourier transform

The delta function is a tempered distribution, and therefore it has a well-defined Fourier transform. Formally, one finds[37]

$$\hat{\delta}(\xi) = \int_{-\infty}^{\infty} e^{-2\pi i x \xi} \delta(x)\, dx = 1.$$

Properly speaking, the Fourier transform of a distribution is defined by imposing self-adjointness of the Fourier transform under the duality pairing $\langle \cdot, \cdot \rangle$ of tempered distributions with Schwartz functions. Thus $\hat{\delta}$ is defined as the unique tempered distribution satisfying

$$\langle \hat{\delta}, \varphi \rangle = \langle \delta, \hat{\varphi} \rangle$$

for all Schwartz functions φ. And indeed it follows from this that $\hat{\delta} = 1$.

As a result of this identity, the convolution of the delta function with any other tempered distribution S is simply S:

$$S * \delta = S.$$

That is to say that δ is an identity element for the convolution on tempered distributions, and in fact the space of compactly supported distributions under convolution is an associative algebra with identity the delta function. This property is fundamental in signal processing, as convolution with a tempered distribution is a linear time-invariant system, and applying the linear time-invariant system measures its impulse response. The impulse response can be computed to any desired degree of accuracy by choosing a suitable approximation for δ, and once it is known, it characterizes the system completely. See *LTI system theory:Impulse response and convolution*.

The inverse Fourier transform of the tempered distribution $f(\xi) = 1$ is the delta function. Formally, this is expressed

$$\int_{-\infty}^{\infty} 1 \cdot e^{2\pi i x \xi}\, d\xi = \delta(x)$$

and more rigorously, it follows since

$$\langle 1, f^{\vee} \rangle = f(0) = \langle \delta, f \rangle$$

for all Schwartz functions f.

In these terms, the delta function provides a suggestive statement of the orthogonality property of the Fourier kernel on **R**. Formally, one has

$$\int_{-\infty}^{\infty} e^{i2\pi \xi_1 t} \left[e^{i2\pi \xi_2 t} \right]^* dt = \int_{-\infty}^{\infty} e^{-i2\pi(\xi_2 - \xi_1)t}\, dt = \delta(\xi_2 - \xi_1).$$

This is, of course, shorthand for the assertion that the Fourier transform of the tempered distribution

$$f(t) = e^{i2\pi \xi_1 t}$$

is

$$\hat{f}(\xi_2) = \delta(\xi_1 - \xi_2)$$

which again follows by imposing self-adjointness of the Fourier transform.

By analytic continuation of the Fourier transform, the Laplace transform of the delta function is found to be[38]

$$\int_0^\infty \delta(t-a)e^{-st}\,dt = e^{-sa}.$$

23.6   Distributional derivatives

The distributional derivative of the Dirac delta distribution is the distribution δ' defined on compactly supported smooth test functions φ by[39]

$$\delta'[\varphi] = -\delta[\varphi'] = -\varphi'(0).$$

The first equality here is a kind of integration by parts, for if δ were a true function then

$$\int_{-\infty}^\infty \delta'(x)\varphi(x)\,dx = -\int_{-\infty}^\infty \delta(x)\varphi'(x)\,dx.$$

The k-th derivative of δ is defined similarly as the distribution given on test functions by

$$\delta^{(k)}[\varphi] = (-1)^k \varphi^{(k)}(0).$$

In particular, δ is an infinitely differentiable distribution.

The first derivative of the delta function is the distributional limit of the difference quotients:[40]

$$\delta'(x) = \lim_{h\to 0} \frac{\delta(x+h) - \delta(x)}{h}.$$

More properly, one has

$$\delta' = \lim_{h\to 0} \frac{1}{h}(\tau_h \delta - \delta)$$

where τh is the translation operator, defined on functions by $\tau h \varphi(x) = \varphi(x+h)$, and on a distribution S by

$$(\tau_h S)[\varphi] = S[\tau_{-h}\varphi].$$

In the theory of electromagnetism, the first derivative of the delta function represents a point magnetic dipole situated at the origin. Accordingly, it is referred to as a dipole or the doublet function.[41]

The derivative of the delta function satisfies a number of basic properties, including:

- $\frac{d}{dx}\delta(-x) = \frac{d}{dx}\delta(x)$
- $\delta'(-x) = -\delta'(x)$
- $x\delta'(x) = -\delta(x)$. [42]

Furthermore, the convolution of δ' with a compactly supported smooth function f is

$$\delta' * f = \delta * f' = f',$$

which follows from the properties of the distributional derivative of a convolution.

23.6.1 Higher dimensions

More generally, on an open set U in the n-dimensional Euclidean space $\mathbf{R}^n$, the Dirac delta distribution centered at a point $a \in U$ is defined by[43]

$$\delta_a[\varphi] = \varphi(a)$$

for all $\varphi \in S(U)$, the space of all smooth compactly supported functions on U. If $\alpha = (\alpha_1, ..., \alpha n)$ is any multi-index and ∂^α denotes the associated mixed partial derivative operator, then the αth derivative $\partial^\alpha \delta a$ of δa is given by[43]

$$\langle \partial^\alpha \delta_a, \varphi \rangle = (-1)^{|\alpha|} \langle \delta_a, \partial^\alpha \varphi \rangle = (-1)^{|\alpha|} \partial^\alpha \varphi(x) \Big|_{x=a} \quad \text{for all } \varphi \in S(U).$$

That is, the αth derivative of δa is the distribution whose value on any test function φ is the αth derivative of φ at a (with the appropriate positive or negative sign).

The first partial derivatives of the delta function are thought of as double layers along the coordinate planes. More generally, the normal derivative of a simple layer supported on a surface is a double layer supported on that surface, and represents a laminar magnetic monopole. Higher derivatives of the delta function are known in physics as multipoles.

Higher derivatives enter into mathematics naturally as the building blocks for the complete structure of distributions with point support. If S is any distribution on U supported on the set $\{a\}$ consisting of a single point, then there is an integer m and coefficients $c\alpha$ such that[44]

$$S = \sum_{|\alpha| \leq m} c_\alpha \partial^\alpha \delta_a.$$

23.7 Representations of the delta function

The delta function can be viewed as the limit of a sequence of functions

$$\delta(x) = \lim_{\varepsilon \to 0^+} \eta_\varepsilon(x),$$

where $\eta\varepsilon(x)$ is sometimes called a **nascent delta function**. This limit is meant in a weak sense: either that

for all continuous functions f having compact support, or that this limit holds for all smooth functions f with compact support. The difference between these two slightly different modes of weak convergence is often subtle: the former is convergence in the vague topology of measures, and the latter is convergence in the sense of distributions.

23.7.1 Approximations to the identity

Typically a nascent delta function $\eta\varepsilon$ can be constructed in the following manner. Let η be an absolutely integrable function on $\mathbf{R}$ of total integral 1, and define

$$\eta_\varepsilon(x) = \varepsilon^{-1} \eta\left(\frac{x}{\varepsilon}\right).$$

In n dimensions, one uses instead the scaling

$$\eta_\varepsilon(x) = \varepsilon^{-n} \eta\left(\frac{x}{\varepsilon}\right).$$

Then a simple change of variables shows that $\eta\varepsilon$ also has integral 1.[45] One shows easily that (5) holds for all continuous compactly supported functions f, and so $\eta\varepsilon$ converges weakly to δ in the sense of measures.

The $\eta\varepsilon$ constructed in this way are known as an **approximation to the identity**.[46] This terminology is because the space $L^1(\mathbf{R})$ of absolutely integrable functions is closed under the operation of convolution of functions: $f * g \in L^1(\mathbf{R})$ whenever f and g are in $L^1(\mathbf{R})$. However, there is no identity in $L^1(\mathbf{R})$ for the convolution product: no element h such that $f * h = f$ for all f. Nevertheless, the sequence $\eta\varepsilon$ does approximate such an identity in the sense that

$$f * \eta_\varepsilon \to f \quad \text{as } \varepsilon \to 0.$$

This limit holds in the sense of mean convergence (convergence in L^1). Further conditions on the $\eta\varepsilon$, for instance that it be a mollifier associated to a compactly supported function,[47] are needed to ensure pointwise convergence almost everywhere.

If the initial $\eta = \eta_1$ is itself smooth and compactly supported then the sequence is called a mollifier. The standard mollifier is obtained by choosing η to be a suitably normalized bump function, for instance

$$\eta(x) = \begin{cases} e^{-\frac{1}{1-|x|^2}} & \text{if } |x| < 1 \\ 0 & \text{if } |x| \geq 1. \end{cases}$$

In some situations such as numerical analysis, a piecewise linear approximation to the identity is desirable. This can be obtained by taking η_1 to be a hat function. With this choice of η_1, one has

$$\eta_\varepsilon(x) = \varepsilon^{-1} \max\left(1 - \left|\frac{x}{\varepsilon}\right|, 0\right)$$

which are all continuous and compactly supported, although not smooth and so not a mollifier.

23.7.2   Probabilistic considerations

In the context of probability theory, it is natural to impose the additional condition that the initial η_1 in an approximation to the identity should be positive, as such a function then represents a probability distribution. Convolution with a probability distribution is sometimes favorable because it does not result in overshoot or undershoot, as the output is a convex combination of the input values, and thus falls between the maximum and minimum of the input function. Taking η_1 to be any probability distribution at all, and letting $\eta\varepsilon(x) = \eta_1(x/\varepsilon)/\varepsilon$ as above will give rise to an approximation to the identity. In general this converges more rapidly to a delta function if, in addition, η has mean 0 and has small higher moments. For instance, if η_1 is the uniform distribution on $[-1/2, 1/2]$, also known as the rectangular function, then:[48]

$$\eta_\varepsilon(x) = \frac{1}{\varepsilon} \text{rect}\left(\frac{x}{\varepsilon}\right) = \begin{cases} \frac{1}{\varepsilon}, & -\frac{\varepsilon}{2} < x < \frac{\varepsilon}{2} \\ 0, & \text{otherwise.} \end{cases}$$

Another example is with the Wigner semicircle distribution

$$\eta_\varepsilon(x) = \begin{cases} \frac{2}{\pi\varepsilon^2}\sqrt{\varepsilon^2 - x^2}, & -\varepsilon < x < \varepsilon \\ 0, & \text{otherwise} \end{cases}$$

This is continuous and compactly supported, but not a mollifier because it is not smooth.

23.7.3   Semigroups

Nascent delta functions often arise as convolution semigroups. This amounts to the further constraint that the convolution of $\eta\varepsilon$ with $\eta\delta$ must satisfy

$$\eta_\varepsilon * \eta_\delta = \eta_{\varepsilon+\delta}$$

for all $\varepsilon, \delta > 0$. Convolution semigroups in L^1 that form a nascent delta function are always an approximation to the identity in the above sense, however the semigroup condition is quite a strong restriction.

In practice, semigroups approximating the delta function arise as fundamental solutions or Green's functions to physically motivated elliptic or parabolic partial differential equations. In the context of applied mathematics, semigroups arise as the output of a linear time-invariant system. Abstractly, if A is a linear operator acting on functions of x, then a convolution semigroup arises by solving the initial value problem

$$\begin{cases} \frac{\partial}{\partial t}\eta(t,x) = A\eta(t,x), & t > 0 \\ \lim_{t \to 0^+} \eta(t,x) = \delta(x) \end{cases}$$

in which the limit is as usual understood in the weak sense. Setting $\eta\varepsilon(x) = \eta(\varepsilon, x)$ gives the associated nascent delta function.

Some examples of physically important convolution semigroups arising from such a fundamental solution include the following.

The heat kernel

The heat kernel, defined by

$$\eta_\varepsilon(x) = \frac{1}{\sqrt{2\pi\varepsilon}}e^{-\frac{x^2}{2\varepsilon}}$$

represents the temperature in an infinite wire at time $t > 0$, if a unit of heat energy is stored at the origin of the wire at time $t = 0$. This semigroup evolves according to the one-dimensional heat equation:

$$\frac{\partial u}{\partial t} = \frac{1}{2}\frac{\partial^2 u}{\partial x^2}.$$

In probability theory, $\eta\varepsilon(x)$ is a normal distribution of variance ε and mean 0. It represents the probability density at time $t = \varepsilon$ of the position of a particle starting at the origin following a standard Brownian motion. In this context, the semigroup condition is then an expression of the Markov property of Brownian motion.

In higher-dimensional Euclidean space $\mathbf{R}^n$, the heat kernel is

$$\eta_\varepsilon = \frac{1}{(2\pi\varepsilon)^{n/2}}e^{-\frac{x\cdot x}{2\varepsilon}},$$

and has the same physical interpretation, *mutatis mutandis*. It also represents a nascent delta function in the sense that $\eta\varepsilon \to \delta$ in the distribution sense as $\varepsilon \to 0$.

The Poisson kernel

The Poisson kernel

$$\eta_\varepsilon(x) = \frac{1}{\pi}\frac{\varepsilon}{\varepsilon^2 + x^2} = \int_{-\infty}^{\infty} e^{2\pi i \xi x - |\varepsilon \xi|}\, d\xi$$

is the fundamental solution of the Laplace equation in the upper half-plane.[49] It represents the electrostatic potential in a semi-infinite plate whose potential along the edge is held at fixed at the delta function. The Poisson kernel is also closely related to the Cauchy distribution. This semigroup evolves according to the equation

$$\frac{\partial u}{\partial t} = -\left(-\frac{\partial^2}{\partial x^2}\right)^{\frac{1}{2}} u(t, x)$$

where the operator is rigorously defined as the Fourier multiplier

$$\mathcal{F}\left[\left(-\frac{\partial^2}{\partial x^2}\right)^{\frac{1}{2}} f\right](\xi) = |2\pi\xi|\mathcal{F}f(\xi).$$

23.7.4 Oscillatory integrals

In areas of physics such as wave propagation and wave mechanics, the equations involved are hyperbolic and so may have more singular solutions. As a result, the nascent delta functions that arise as fundamental solutions of the associated Cauchy problems are generally oscillatory integrals. An example, which comes from a solution of the Euler–Tricomi equation of transonic gas dynamics,[50] is the rescaled Airy function

$$\varepsilon^{-\frac{1}{3}}\,\mathrm{Ai}\left(x\varepsilon^{-\frac{1}{3}}\right).$$

Although using the Fourier transform, it is easy to see that this generates a semigroup in some sense, it is not absolutely integrable and so cannot define a semigroup in the above strong sense. Many nascent delta functions constructed as oscillatory integrals only converge in the sense of distributions (an example is the Dirichlet kernel below), rather than in the sense of measures.

Another example is the Cauchy problem for the wave equation in $\mathbf{R}^{1+1}$.[51]

$$c^{-2}\frac{\partial^2 u}{\partial t^2} - \Delta u = 0$$

$$u = 0, \quad \frac{\partial u}{\partial t} = \delta \qquad \text{for } t = 0.$$

The solution u represents the displacement from equilibrium of an infinite elastic string, with an initial disturbance at the origin.

Other approximations to the identity of this kind include the sinc function (used widely in electronics and telecommunications)

$$\eta_\varepsilon(x) = \frac{1}{\pi x}\sin\left(\frac{x}{\varepsilon}\right) = \frac{1}{2\pi}\int_{-\frac{1}{\varepsilon}}^{\frac{1}{\varepsilon}} \cos(kx)\, dk$$

and the Bessel function

$$\eta_\varepsilon(x) = \frac{1}{\varepsilon}J_{\frac{1}{\varepsilon}}\left(\frac{x+1}{\varepsilon}\right).$$

23.7.5 Plane wave decomposition

One approach to the study of a linear partial differential equation

$$L[u] = f,$$

where L is a differential operator on $\mathbf{R}^n$, is to seek first a fundamental solution, which is a solution of the equation

$$L[u] = \delta.$$

When L is particularly simple, this problem can often be resolved using the Fourier transform directly (as in the case of the Poisson kernel and heat kernel already mentioned). For more complicated operators, it is sometimes easier first to consider an equation of the form

$$L[u] = h$$

where h is a plane wave function, meaning that it has the form

$$h = h(x \cdot \xi)$$

for some vector ξ. Such an equation can be resolved (if the coefficients of L are analytic functions) by the Cauchy–Kovalevskaya theorem or (if the coefficients of L are constant) by quadrature. So, if the delta function can be decomposed into plane waves, then one can in principle solve linear partial differential equations.

Such a decomposition of the delta function into plane waves was part of a general technique first introduced essentially by Johann Radon, and then developed in this form by Fritz John (1955).[52] Choose k so that $n + k$ is an even integer, and for a real number s, put

$$g(s) = \mathrm{Re}\left[\frac{-s^k \log(-is)}{k!(2\pi i)^n} \right] = \begin{cases} \frac{|s|^k}{4k!(2\pi i)^{n-1}} & n\ odd \\[2ex] -\frac{|s|^k \log|s|}{k!(2\pi i)^n} & n\ even. \end{cases}$$

Then δ is obtained by applying a power of the Laplacian to the integral with respect to the unit sphere measure $d\omega$ of $g(x \cdot \xi)$ for ξ in the unit sphere S^{n-1}:

$$\delta(x) = \Delta_x^{\frac{n+k}{2}} \int_{S^{n-1}} g(x \cdot \xi) \, d\omega_\xi.$$

The Laplacian here is interpreted as a weak derivative, so that this equation is taken to mean that, for any test function φ,

$$\varphi(x) = \int_{\mathbf{R}^n} \varphi(y) \, dy \, \Delta_x^{\frac{n+k}{2}} \int_{S^{n-1}} g((x - y) \cdot \xi) \, d\omega_\xi.$$

The result follows from the formula for the Newtonian potential (the fundamental solution of Poisson's equation). This is essentially a form of the inversion formula for the Radon transform, because it recovers the value of $\varphi(x)$ from its integrals over hyperplanes. For instance, if n is odd and $k = 1$, then the integral on the right hand side is

$$c_n \Delta_x^{\frac{n+1}{2}} \int\!\!\int_{S^{n-1}} \varphi(y)|(y - x) \cdot \xi| \, d\omega_\xi \, dy = c_n \Delta_x^{\frac{n+1}{2}} \int_{S^{n-1}} d\omega_\xi \int_{-\infty}^{\infty} |p| R\varphi(\xi, p + x \cdot \xi) \, dp$$

where $R\varphi(\xi, p)$ is the Radon transform of φ:

$$R\varphi(\xi, p) = \int_{x \cdot \xi = p} f(x) \, d^{n-1}x.$$

An alternative equivalent expression of the plane wave decomposition, from Gel'fand & Shilov (1966–1968, I, §3.10), is

$$\delta(x) = \frac{(n-1)!}{(2\pi i)^n} \int_{S^{n-1}} (x \cdot \xi)^{-n} \, d\omega_\xi$$

for *n* even, and

$$\delta(x) = \frac{1}{2(2\pi i)^{n-1}} \int_{S^{n-1}} \delta^{(n-1)}(x \cdot \xi) \, d\omega_\xi$$

for *n* odd.

23.7.6   Fourier kernels

See also: Convergence of Fourier series

In the study of Fourier series, a major question consists of determining whether and in what sense the Fourier series associated with a periodic function converges to the function. The *n*th partial sum of the Fourier series of a function f of period 2π is defined by convolution (on the interval $[-\pi,\pi]$) with the Dirichlet kernel:

$$D_N(x) = \sum_{n=-N}^{N} e^{inx} = \frac{\sin\left((N + \frac{1}{2})x\right)}{\sin(x/2)}.$$

Thus,

$$s_N(f)(x) = D_N * f(x) = \sum_{n=-N}^{N} a_n e^{inx}$$

where

$$a_n = \frac{1}{2\pi} \int_{-\pi}^{\pi} f(y)e^{-iny} \, dy.$$

A fundamental result of elementary Fourier series states that the Dirichlet kernel tends to the a multiple of the delta function as $N \to \infty$. This is interpreted in the distribution sense, that

$$s_N(f)(0) = \int_{\mathbf{R}} D_N(x)f(x) \, dx \to 2\pi f(0)$$

for every compactly supported *smooth* function f. Thus, formally one has

$$\delta(x) = \frac{1}{2\pi} \sum_{n=-\infty}^{\infty} e^{inx}$$

on the interval $[-\pi,\pi]$.

In spite of this, the result does not hold for all compactly supported *continuous* functions: that is DN does not converge weakly in the sense of measures. The lack of convergence of the Fourier series has led to the introduction of a variety of summability methods in order to produce convergence. The method of Cesàro summation leads to the Fejér kernel[53]

$$F_N(x) = \frac{1}{N} \sum_{n=0}^{N-1} D_n(x) = \frac{1}{N} \left(\frac{\sin \frac{Nx}{2}}{\sin \frac{x}{2}} \right)^2 .$$

The Fejér kernels tend to the delta function in a stronger sense that[54]

$$\int_{\mathbf{R}} F_N(x) f(x)\, dx \to 2\pi f(0)$$

for every compactly supported *continuous* function f. The implication is that the Fourier series of any continuous function is Cesàro summable to the value of the function at every point.

23.7.7 Hilbert space theory

The Dirac delta distribution is a densely defined unbounded linear functional on the Hilbert space L^2 of square integrable functions. Indeed, smooth compactly support functions are dense in L^2, and the action of the delta distribution on such functions is well-defined. In many applications, it is possible to identify subspaces of L^2 and to give a stronger topology on which the delta function defines a bounded linear functional.

Sobolev spaces

The Sobolev embedding theorem for Sobolev spaces on the real line $\mathbf{R}$ implies that any square-integrable function f such that

$$\|f\|_{H^1}^2 = \int_{-\infty}^{\infty} |\hat{f}(\xi)|^2 (1 + |\xi|^2)\, d\xi < \infty$$

is automatically continuous, and satisfies in particular

$$\delta[f] = |f(0)| < C\|f\|_{H^1}.$$

Thus δ is a bounded linear functional on the Sobolev space H^1. Equivalently δ is an element of the continuous dual space H^{-1} of H^1. More generally, in n dimensions, one has $\delta \in H^{-s}(\mathbf{R}^n)$ provided $s > n / 2$.

Spaces of holomorphic functions

In complex analysis, the delta function enters via Cauchy's integral formula which asserts that if D is a domain in the complex plane with smooth boundary, then

$$f(z) = \frac{1}{2\pi i} \oint_{\partial D} \frac{f(\zeta)\, d\zeta}{\zeta - z}, \quad z \in D$$

for all holomorphic functions f in D that are continuous on the closure of D. As a result, the delta function δz is represented on this class of holomorphic functions by the Cauchy integral:

$$\delta_z[f] = f(z) = \frac{1}{2\pi i} \oint_{\partial D} \frac{f(\zeta)\, d\zeta}{\zeta - z}.$$

More generally, let $H^2(\partial D)$ be the Hardy space consisting of the closure in $L^2(\partial D)$ of all holomorphic functions in D continuous up to the boundary of D. Then functions in $H^2(\partial D)$ uniquely extend to holomorphic functions in D, and the Cauchy integral formula continues to hold. In particular for $z \in D$, the delta function δz is a continuous linear functional on $H^2(\partial D)$. This is a special case of the situation in several complex variables in which, for smooth domains D, the Szegő kernel plays the role of the Cauchy integral.

Resolutions of the identity

Given a complete orthonormal basis set of functions $\{\varphi n\}$ in a separable Hilbert space, for example, the normalized eigenvectors of a compact self-adjoint operator, any vector f can be expressed as:

$$f = \sum_{n=1}^{\infty} \alpha_n \varphi_n.$$

The coefficients $\{\alpha_n\}$ are found as:

$$\alpha_n = \langle \varphi_n, f \rangle,$$

which may be represented by the notation:

$$\alpha_n = \varphi_n^\dagger f,$$

a form of the bra–ket notation of Dirac.[55] Adopting this notation, the expansion of f takes the dyadic form:[56]

$$f = \sum_{n=1}^{\infty} \varphi_n \left(\varphi_n^\dagger f \right).$$

Letting I denote the identity operator on the Hilbert space, the expression

$$I = \sum_{n=1}^{\infty} \varphi_n \varphi_n^\dagger,$$

is called a resolution of the identity. When the Hilbert space is the space $L^2(D)$ of square-integrable functions on a domain D, the quantity:

$$\varphi_n \varphi_n^\dagger,$$

is an integral operator, and the expression for f can be rewritten as:

$$f(x) = \sum_{n=1}^{\infty} \int_D \left(\varphi_n(x) \varphi_n^*(\xi) \right) f(\xi) \, d\xi.$$

The right-hand side converges to f in the L^2 sense. It need not hold in a pointwise sense, even when f is a continuous function. Nevertheless, it is common to abuse notation and write

$$f(x) = \int \delta(x - \xi) f(\xi) \, d\xi,$$

resulting in the representation of the delta function:[57]

$$\delta(x - \xi) = \sum_{n=1}^{\infty} \varphi_n(x) \varphi_n^*(\xi).$$

With a suitable rigged Hilbert space $(\Phi, L^2(D), \Phi^*)$ where $\Phi \subset L^2(D)$ contains all compactly supported smooth functions, this summation may converge in Φ^*, depending on the properties of the basis φn. In most cases of practical interest, the orthonormal basis comes from an integral or differential operator, in which case the series converges in the distribution sense.[58]

23.7.8 Infinitesimal delta functions

Cauchy used an infinitesimal α to write down a unit impulse, infinitely tall and narrow Dirac-type delta function $\delta\alpha$ satisfying $\int F(x)\delta_\alpha(x) = F(0)$ in a number of articles in 1827.[59] Cauchy defined an infinitesimal in Cours d'Analyse (1827) in terms of a sequence tending to zero. Namely, such a null sequence becomes an infinitesimal in Cauchy's and Lazare Carnot's terminology.

Non-standard analysis allows one to rigorously treat infinitesimals. The article by Yamashita (2007) contains a bibliography on modern Dirac delta functions in the context of an infinitesimal-enriched continuum provided by the hyperreals. Here the Dirac delta can be given by an actual function, having the property that for every real function F one has $\int F(x)\delta_\alpha(x) = F(0)$ as anticipated by Fourier and Cauchy.

23.8 Dirac comb

Main article: Dirac comb

A so-called uniform "pulse train" of Dirac delta measures, which is known as a Dirac comb, or as the Shah distribution, creates a sampling function, often used in digital signal processing (DSP) and discrete time signal analysis. The Dirac comb is given as the infinite sum, whose limit is understood in the distribution sense,

$$\Delta(x) = \sum_{n=-\infty}^{\infty} \delta(x - n),$$

which is a sequence of point masses at each of the integers.

Up to an overall normalizing constant, the Dirac comb is equal to its own Fourier transform. This is significant because if f is any Schwartz function, then the periodization of f is given by the convolution

$$(f * \Delta)(x) = \sum_{n=-\infty}^{\infty} f(x - n).$$

In particular,

$$(f * \Delta)^\wedge = \hat{f}\hat{\Delta} = \hat{f}\Delta$$

is precisely the Poisson summation formula.[60]

23.9   Sokhotski–Plemelj theorem

The Sokhotski–Plemelj theorem, important in quantum mechanics, relates the delta function to the distribution p.v.$1/x$, the Cauchy principal value of the function $1/x$, defined by

$$\left\langle \mathrm{p.\,v.}\,\frac{1}{x}, \varphi \right\rangle = \lim_{\varepsilon \to 0^+} \int_{|x|>\varepsilon} \frac{\varphi(x)}{x}\, dx.$$

Sokhotsky's formula states that[61]

$$\lim_{\varepsilon \to 0^+} \frac{1}{x \pm i\varepsilon} = \mathrm{p.\,v.}\,\frac{1}{x} \mp i\pi\delta(x),$$

Here the limit is understood in the distribution sense, that for all compactly supported smooth functions f,

$$\lim_{\varepsilon \to 0^+} \int_{-\infty}^{\infty} \frac{f(x)}{x \pm i\varepsilon}\, dx = \mp i\pi f(0) + \lim_{\varepsilon \to 0^+} \int_{|x|>\varepsilon} \frac{f(x)}{x}\, dx.$$

23.10   Relationship to the Kronecker delta

The Kronecker delta δ_{ij} is the quantity defined by

$$\delta_{ij} = \begin{cases} 1 & i = j \\ 0 & i \neq j \end{cases}$$

for all integers i, j. This function then satisfies the following analog of the sifting property: if $(a_i)_{i\in\mathbf{Z}}$ is any doubly infinite sequence, then

$$\sum_{i=-\infty}^{\infty} a_i \delta_{ik} = a_k.$$

Similarly, for any real or complex valued continuous function f on $\mathbf{R}$, the Dirac delta satisfies the sifting property

$$\int_{-\infty}^{\infty} f(x)\delta(x - x_0)\, dx = f(x_0).$$

This exhibits the Kronecker delta function as a discrete analog of the Dirac delta function.[62]

23.11 Applications

23.11.1 Probability theory

In probability theory and statistics, the Dirac delta function is often used to represent a discrete distribution, or a partially discrete, partially continuous distribution, using a probability density function (which is normally used to represent fully continuous distributions). For example, the probability density function $f(x)$ of a discrete distribution consisting of points $\mathbf{x} = \{x_1, ..., xn\}$, with corresponding probabilities $p_1, ..., pn$, can be written as

$$f(x) = \sum_{i=1}^{n} p_i \delta(x - x_i).$$

As another example, consider a distribution which 6/10 of the time returns a standard normal distribution, and 4/10 of the time returns exactly the value 3.5 (i.e. a partly continuous, partly discrete mixture distribution). The density function of this distribution can be written as

$$f(x) = 0.6 \, \frac{1}{\sqrt{2\pi}} e^{-\frac{x^2}{2}} + 0.4 \, \delta(x - 3.5).$$

The delta function is also used in a completely different way to represent the local time of a diffusion process (like Brownian motion). The local time of a stochastic process $B(t)$ is given by

$$\ell(x, t) = \int_0^t \delta(x - B(s)) \, ds$$

and represents the amount of time that the process spends at the point x in the range of the process. More precisely, in one dimension this integral can be written

$$\ell(x, t) = \lim_{\varepsilon \to 0^+} \frac{1}{2\varepsilon} \int_0^t \mathbf{1}_{[x-\varepsilon, x+\varepsilon]}(B(s)) \, ds$$

where $1[x_-\varepsilon, x_+\varepsilon]$ is the indicator function of the interval $[x-\varepsilon, x+\varepsilon]$.

23.11.2 Quantum mechanics

We give an example of how the delta function is expedient in quantum mechanics. The wave function of a particle gives the probability amplitude of finding a particle within a given region of space. Wave functions are assumed to be elements of the Hilbert space L^2 of square-integrable functions, and the total probability of finding a particle within a given interval is the integral of the magnitude of the wave function squared over the interval. A set $\{\varphi n\}$ of wave functions is orthonormal if they are normalized by

$$\langle \varphi_n | \varphi_m \rangle = \delta_{nm}$$

where δ here refers to the Kronecker delta. A set of orthonormal wave functions is complete in the space of square-integrable functions if any wave function ψ can be expressed as a combination of the φn:

$$\psi = \sum c_n \varphi_n,$$

with $c_n = \langle \varphi_n | \psi \rangle$. Complete orthonormal systems of wave functions appear naturally as the eigenfunctions of the Hamiltonian (of a bound system) in quantum mechanics that measures the energy levels, which are called the eigenvalues. The set of eigenvalues, in this case, is known as the spectrum of the Hamiltonian. In bra–ket notation, as above, this equality implies the resolution of the identity:

$$I = \sum |\varphi_n\rangle\langle\varphi_n|.$$

Here the eigenvalues are assumed to be discrete, but the set of eigenvalues of an observable may be continuous rather than discrete. An example is the position observable, $Q\psi(x) = x\psi(x)$. The spectrum of the position (in one dimension) is the entire real line, and is called a continuous spectrum. However, unlike the Hamiltonian, the position operator lacks proper eigenfunctions. The conventional way to overcome this shortcoming is to widen the class of available functions by allowing distributions as well: that is, to replace the Hilbert space of quantum mechanics by an appropriate rigged Hilbert space.[63] In this context, the position operator has a complete set of eigen-distributions, labeled by the points y of the real line, given by

$$\varphi_y(x) = \delta(x - y).$$

The eigenfunctions of position are denoted by $\varphi_y = |y\rangle$ in Dirac notation, and are known as position eigenstates.

Similar considerations apply to the eigenstates of the momentum operator, or indeed any other self-adjoint unbounded operator P on the Hilbert space, provided the spectrum of P is continuous and there are no degenerate eigenvalues. In that case, there is a set Ω of real numbers (the spectrum), and a collection φy of distributions indexed by the elements of Ω, such that

$$P\varphi_y = y\varphi_y.$$

That is, φy are the eigenvectors of P. If the eigenvectors are normalized so that

$$\langle \varphi_y, \varphi_{y'} \rangle = \delta(y - y')$$

in the distribution sense, then for any test function ψ,

$$\psi(x) = \int_\Omega c(y)\varphi_y(x)\,dy$$

where

$$c(y) = \langle \psi, \varphi_y \rangle.$$

That is, as in the discrete case, there is a resolution of the identity

$$I = \int_\Omega |\varphi_y\rangle\,\langle\varphi_y|\,dy$$

where the operator-valued integral is again understood in the weak sense. If the spectrum of P has both continuous and discrete parts, then the resolution of the identity involves a summation over the discrete spectrum *and* an integral over the continuous spectrum.

The delta function also has many more specialized applications in quantum mechanics, such as the delta potential models for a single and double potential well.

23.11.3 Structural mechanics

The delta function can be used in structural mechanics to describe transient loads or point loads acting on structures. The governing equation of a simple mass–spring system excited by a sudden force impulse I at time $t = 0$ can be written

$$m\frac{\mathrm{d}^2\xi}{\mathrm{d}t^2} + k\xi = I\delta(t),$$

where m is the mass, ξ the deflection and k the spring constant.

As another example, the equation governing the static deflection of a slender beam is, according to Euler–Bernoulli theory,

$$EI\frac{\mathrm{d}^4 w}{\mathrm{d}x^4} = q(x),$$

where EI is the bending stiffness of the beam, w the deflection, x the spatial coordinate and $q(x)$ the load distribution. If a beam is loaded by a point force F at $x = x_0$, the load distribution is written

$$q(x) = F\delta(x - x_0).$$

As integration of the delta function results in the Heaviside step function, it follows that the static deflection of a slender beam subject to multiple point loads is described by a set of piecewise polynomials.

Also a point moment acting on a beam can be described by delta functions. Consider two opposing point forces F at a distance d apart. They then produce a moment $M = Fd$ acting on the beam. Now, let the distance d approach the limit zero, while M is kept constant. The load distribution, assuming a clockwise moment acting at $x = 0$, is written

$$
\begin{aligned}
q(x) &= \lim_{d\to 0}\Big(F\delta(x) - F\delta(x - d)\Big) \\
&= \lim_{d\to 0}\left(\frac{M}{d}\delta(x) - \frac{M}{d}\delta(x - d)\right) \\
&= M\lim_{d\to 0}\frac{\delta(x) - \delta(x - d)}{d} \\
&= M\delta'(x).
\end{aligned}
$$

Point moments can thus be represented by the derivative of the delta function. Integration of the beam equation again results in piecewise polynomial deflection.

23.12   See also

- Atom (measure theory)

- Delta potential

- Dirac measure

- Fundamental solution

- Green's function

- Laplacian of the indicator

23.13   Notes

[1] Dirac 1958, §15 The δ function, p. 58

[2] Gel'fand & Shilov 1968, Volume I, §§1.1, 1.3

[3] Schwartz 1950, p. 3

[4] Arfken & Weber 2000, p. 84

[5] Bracewell 1986, Chapter 5

[6] Vladimirov 1971, §5.1

[7] JB Fourier (1822). *The Analytical Theory of Heat* (English translation by Alexander Freeman, 1878 ed.). The University Press. p. 408., cf p 449 and pp 546-551. The original French text can be found here.

[8] Hikosaburo Komatsu (2002). "Fourier's hyperfunctions and Heaviside's pseudodifferential operators". In Takahiro Kawai, Keiko Fujita, eds. *Microlocal Analysis and Complex Fourier Analysis*. World Scientific. p. 200. ISBN 981-238-161-9.

[9] Tyn Myint-U., Lokenath Debnath (2007). *Linear Partial Differential Equations for Scientists And Engineers* (4th ed.). Springer. p. 4. ISBN 0-8176-4393-1.

[10] Lokenath Debnath, Dambaru Bhatta (2007). *Integral Transforms And Their Applications* (2nd ed.). CRC Press. p. 2. ISBN 1-58488-575-0.

[11] Ivor Grattan-Guinness (2009). *Convolutions in French Mathematics, 1800–1840: From the Calculus and Mechanics to Mathematical Analysis and Mathematical Physics, Volume 2*. Birkhäuser. p. 653. ISBN 3-7643-2238-1.

[12] See, for example, *Des intégrales doubles qui se présentent sous une forme indéterminée*

[13] Dragiša Mitrović, Darko Žubrinić (1998). *Fundamentals of Applied Functional Analysis: Distributions, Sobolev Spaces*. CRC Press. p. 62. ISBN 0-582-24694-6.

[14] Manfred Kracht, Erwin Kreyszig (1989). "On singular integral operators and generalizations". In Themistocles M. Rassias, ed. *Topics in Mathematical Analysis: A Volume Dedicated to the Memory of A.L. Cauchy*. World Scientific. p. 553. ISBN 9971-5-0666-1.

[15] Laugwitz 1989, p. 230

[16] A more complete historical account can be found in van der Pol & Bremmer 1987, §V.4.

[17] Dirac 1958, §15

[18] Gel'fand & Shilov 1968, Volume I, §1.1, p. 1

[19] Rudin 1966, §1.20

[20] Hewitt & Stromberg 1963, §19.61

[21] Driggers 2003, p. 2321. See also Bracewell 1986, Chapter 5 for a different interpretation. Other conventions for the assigning the value of the Heaviside function at zero exist, and some of these are not consistent with what follows.

[22] Hewitt & Stromberg 1965, §9.19

[23] Strichartz 1994, §2.2

[24] Hörmander 1983, Theorem 2.1.5

[25] Hörmander 1983, §3.1

[26] Strichartz 1994, §2.3; Hörmander 1983, §8.2

[27] Dieudonné 1972, §17.3.3

[28] Federer 1969, §2.5.19

[29] Strichartz 1994, Problem 2.6.2

[30] Vladimirov 1971, Chapter 2, Example 3(d)

[31] Weisstein, Eric W., "Sifting Property", *MathWorld*.

[32] Gel'fand & Shilov 1966–1968, Vol. 1, §II.2.5

[33] Further refinement is possible, namely to submersions, although these require a more involved change of variables formula.

[34] Hörmander 1983, §6.1

[35] Lange 2012, pp.29–30

[36] Gelfand Shilov, p. 212

[37] In some conventions for the Fourier transform.

[38] Bracewell 1986

[39] Gel'fand & Shilov 1966, p. 26

[40] Gel'fand & Shilov 1966, §2.1

[41] Weisstein, Eric W., "Doublet Function", *MathWorld*.

[42] The property follows by applying a test function and integration by parts.

[43] Hörmander 1983, p. 56

[44] Hörmander 1983, p. 56; Rudin 1991, Theorem 6.25

[45] Stein Weiss, Theorem 1.18

[46] Rudin 1991, §II.6.31

[47] More generally, one only needs $\eta = \eta_1$ to have an integrable radially symmetric decreasing rearrangement.

[48] Saichev & Woyczyński 1997, §1.1 The "delta function" as viewed by a physicist and an engineer, p. 3

[49] Stein & Weiss 1971, §I.1

[50] Vallée & Soares 2004, §7.2

[51] Hörmander 1983, §7.8

[52] See also Courant & Hilbert 1962, §14.

[53] Lang 1997, p. 312

[54] In the terminology of Lang (1997), the Fejér kernel is a Dirac sequence, whereas the Dirichlet kernel is not.

[55] The development of this section in bra–ket notation is found in (Levin 2002, Coordinate-space wave functions and completeness, pp.=109*ff*)

[56] Davis & Thomson 2000, Perfect operators, p.344

[57] Davis & Thomson 2000, Equation 8.9.11, p. 344

[58] de la Madrid, Bohm & Gadella 2002

[59] See Laugwitz (1989).

[60] Córdoba 1988; Hörmander 1983, §7.2

[61] Vladimirov 1971, §5.7

[62] Hartmann 1997, pp. 154–155

[63] Isham 1995, §6.2

23.14   References

- Aratyn, Henrik; Rasinariu, Constantin (2006), *A short course in mathematical methods with Maple*, World Scientific, ISBN 981-256-461-6.

- Arfken, G. B.; Weber, H. J. (2000), *Mathematical Methods for Physicists* (5th ed.), Boston, Massachusetts: Academic Press, ISBN 978-0-12-059825-0.

- Bracewell, R. (1986), *The Fourier Transform and Its Applications* (2nd ed.), McGraw-Hill.

- Córdoba, A., "La formule sommatoire de Poisson", *C.R. Acad. Sci. Paris, Series I* **306**: 373–376.

- Courant, Richard; Hilbert, David (1962), *Methods of Mathematical Physics, Volume II*, Wiley-Interscience.

- Davis, Howard Ted; Thomson, Kendall T (2000), *Linear algebra and linear operators in engineering with applications in Mathematica*, Academic Press, ISBN 0-12-206349-X

- Dieudonné, Jean (1976), *Treatise on analysis. Vol. II*, New York: Academic Press [Harcourt Brace Jovanovich Publishers], ISBN 978-0-12-215502-4, MR 0530406.

- Dieudonné, Jean (1972), *Treatise on analysis. Vol. III*, Boston, Massachusetts: Academic Press, MR 0350769

- Dirac, Paul (1958), *The Principles of Quantum Mechanics* (4th ed.), Oxford at the Clarendon Press, ISBN 978-0-19-852011-5.

- Driggers, Ronald G. (2003), *Encyclopedia of Optical Engineering*, CRC Press, ISBN 978-0-8247-0940-2.

- Federer, Herbert (1969), *Geometric measure theory*, Die Grundlehren der mathematischen Wissenschaften **153**, New York: Springer-Verlag, pp. xiv+676, ISBN 978-3-540-60656-7, MR 0257325.

- Gel'fand, I. M.; Shilov, G. E. (1966–1968), *Generalized functions* **1–5**, Academic Press.

- Hartman, William M. (1997), *Signals, sound, and sensation*, Springer, ISBN 978-1-56396-283-7.

- Hewitt, E; Stromberg, K (1963), *Real and abstract analysis*, Springer-Verlag.

- Hörmander, L. (1983), *The analysis of linear partial differential operators I*, Grundl. Math. Wissenschaft. **256**, Springer, ISBN 3-540-12104-8, MR 0717035.

- Isham, C. J. (1995), *Lectures on quantum theory: mathematical and structural foundations*, Imperial College Press, ISBN 978-81-7764-190-5.

- John, Fritz (1955), *Plane waves and spherical means applied to partial differential equations*, Interscience Publishers, New York-London, MR 0075429.

- Lang, Serge (1997), *Undergraduate analysis*, Undergraduate Texts in Mathematics (2nd ed.), Berlin, New York: Springer-Verlag, ISBN 978-0-387-94841-6, MR 1476913.

- Lange, Rutger-Jan (2012), "Potential theory, path integrals and the Laplacian of the indicator", *Journal of High Energy Physics* (Springer) **2012** (11): 29–30, arXiv:1302.0864, Bibcode:2012JHEP...11..032L, doi:10.1007/JHE

- Laugwitz, D. (1989), "Definite values of infinite sums: aspects of the foundations of infinitesimal analysis around 1820", *Arch. Hist. Exact Sci.* **39** (3): 195–245, doi:10.1007/BF00329867.

- Levin, Frank S. (2002), "Coordinate-space wave functions and completeness", *An introduction to quantum theory*, Cambridge University Press, pp. 109*ff*, ISBN 0-521-59841-9

- Li, Y. T.; Wong, R. (2008), "Integral and series representations of the Dirac delta function", *Commun. Pure Appl. Anal.* **7** (2): 229–247, doi:10.3934/cpaa.2008.7.229, MR 2373214.

- de la Madrid, R.; Bohm, A.; Gadella, M. (2002), "Rigged Hilbert Space Treatment of Continuous Spectrum", *Fortschr. Phys.* **50** (2): 185–216, arXiv:quant-ph/0109154, Bibcode:2002ForPh..50..185D, doi:10.1002/1521-3978(200203)50:2<185::AID-PROP185>3.0.CO;2-S.

- McMahon, D. (2005-11-22), "An Introduction to State Space", *Quantum Mechanics Demystified, A Self-Teaching Guide*, Demystified Series, New York: McGraw-Hill, p. 108, doi:10.1036/0071455469, ISBN 0-07-145546-9, retrieved 2008-03-17.

- van der Pol, Balth.; Bremmer, H. (1987), *Operational calculus* (3rd ed.), New York: Chelsea Publishing Co., ISBN 978-0-8284-0327-6, MR 904873.

- Rudin, W. (1991), *Functional Analysis* (2nd ed.), McGraw-Hill, ISBN 0-07-054236-8.

- Soares, Manuel; Vallée, Olivier (2004), *Airy functions and applications to physics*, London: Imperial College Press.

- Saichev, A I; Woyczyński, Wojbor Andrzej (1997), "Chapter1: Basic definitions and operations", *Distributions in the Physical and Engineering Sciences: Distributional and fractal calculus, integral transforms, and wavelets*, Birkhäuser, ISBN 0-8176-3924-1

- Schwartz, L. (1950), *Théorie des distributions* **1**, Hermann.

- Schwartz, L. (1951), *Théorie des distributions* **2**, Hermann.

- Stein, Elias; Weiss, Guido (1971), *Introduction to Fourier Analysis on Euclidean Spaces*, Princeton University Press, ISBN 0-691-08078-X.

- Strichartz, R. (1994), *A Guide to Distribution Theory and Fourier Transforms*, CRC Press, ISBN 0-8493 8273 4.

- Vladimirov, V. S. (1971), *Equations of mathematical physics*, Marcel Dekker, ISBN 0-8247-1713-9.

- Weisstein, Eric W., "Delta Function", *MathWorld*.

- Yamashita, H. (2006), "Pointwise analysis of scalar fields: A nonstandard approach", *Journal of Mathematical Physics* **47** (9): 092301, Bibcode:2006JMP....47i2301Y, doi:10.1063/1.2339017

- Yamashita, H. (2007), "Comment on "Pointwise analysis of scalar fields: A nonstandard approach" [J. Math. Phys. 47, 092301 (2006)]", *Journal of Mathematical Physics* **48** (8): 084101, Bibcode:2007JMP....48h4101Y, doi:10.1063/1.2771422

23.15 External links

- Hazewinkel, Michiel, ed. (2001), "Delta-function", *Encyclopedia of Mathematics*, Springer, ISBN 978-1-55608-010-4

- KhanAcademy.org video lesson

- The Dirac Delta function, a tutorial on the Dirac delta function.

- Video Lectures – Lecture 23, a lecture by Arthur Mattuck.

- Dirac Delta Function on PlanetMath

- The Dirac delta measure is a hyperfunction

- We show the existence of a unique solution and analyze a finite element approximation when the source term is a Dirac delta measure

- Non-Lebesgue measures on R. Lebesgue-Stieltjes measure, Dirac delta measure.

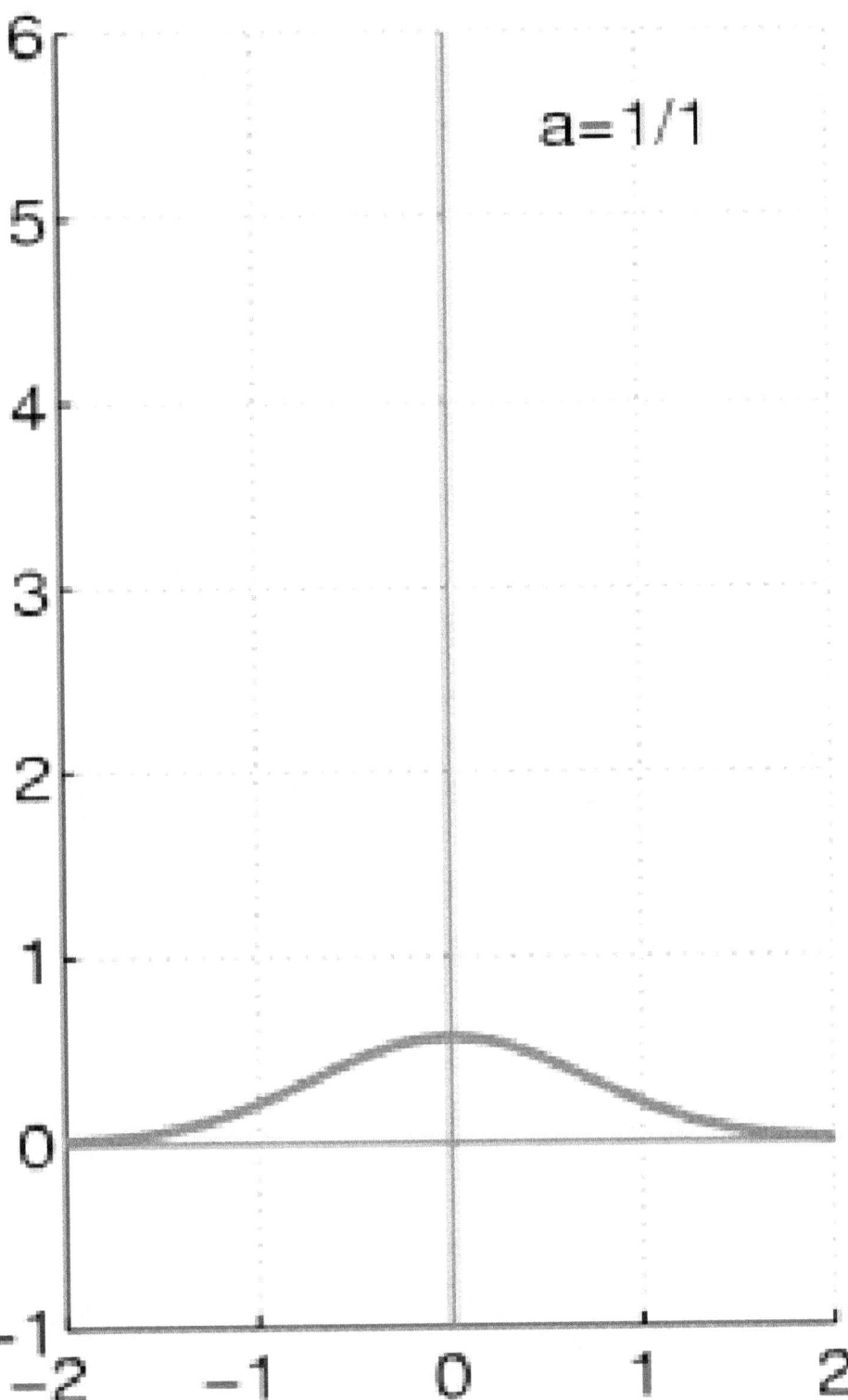

The Dirac delta function as the limit (in the sense of distributions) of the sequence of zero-centered normal distributions \delta_a(x) = \frac{1} {a \sqrt(\pi)} \mathrm{e}^{-x^2/a^2} as a \rightarrow 0.

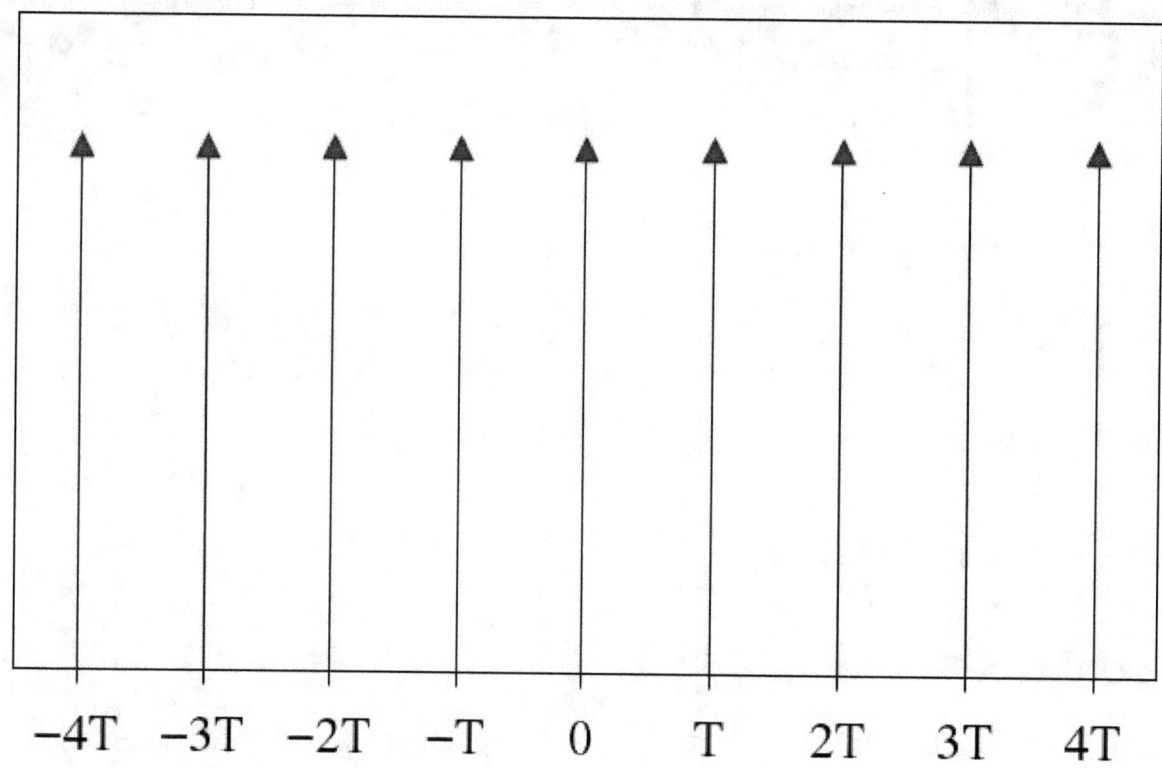

A Dirac comb is an infinite series of Dirac delta functions spaced at intervals of T

23.16 Text and image sources, contributors, and licenses

23.16.1 Text

- **Wave–particle duality** *Source:* https://en.wikipedia.org/wiki/Wave%E2%80%93particle_duality?oldid=697162231 *Contributors:* AxelBoldt,Tobias Hoevekamp, Dere k Ross, MarXidad, The Anome, Manning Bartlett, Wayne Hardman, Andre Engels, Jos h Grosse, Miguel~enwiki,ChangChienFu, DrBob, Heron, KF, Stevertigo, Michae l Hardy, Gabbe, Dgrant, William M. Connolley, Pizza Puzzle, Hike395, Charles Matthews, Timwi, Geoff, Reddi, El~enwiki, ErikStewart, Timc, Maximus Rex, Populus, BenRG, Francs2000, Phil Boswell, SJRubenstein, Robbot, Hankwang, Cdang, Owain, Fredrik, Chris 73, Bkalafut, Sverdrup, Roscoe x, Anthony, Giftlite, BenFrantzDale, Lethe, AJim, JorgeStolfi , Quamaretto, Eequor, Pcarbonn, Antandrus, Unquantum, Lumidek, Chadernook, Adashiel, Trevor MacInnis, Eep², Chris Howard,Ric h Farmbrough, Filthybutter, Hidaspal, Vsmith, Mjpieters, Bender235, Cyclopia, JustinWick, Pt, El C, Laurascudder, RoyBoy, Spoon!,Army1987, Enri c Naval, Atlant, Ricky81682, AzaToth, Fritzpoll, PAR, BRW, Jheald, Gene Nygaard, Afshar, Oleg Alexandrov, Velho,Woohookitty, Linas, Davidkazuhiro, Zealander, Pol098, Ruud Koot, Cbdorsett, Ch'marr, Mandarax, BD2412, Qwertyus, Thierry Dugnolle~enwiki,Drbogdan, Rjwilmsi, KYPark, Strait, Jmcc150, Cjpuffin, Arnero, Musical Linguist, Nihiltres, Alfred Centauri, Fresheneesz, Srle ffler, Kri,Snailwalker, CiaPan, Chobot, Siddhant, Vyrogliph, YurikBot, Wavelength, X42bn6, Wolfmankurd, Hede2000, Gaius Cornelius, Salsb, Schlafl y,Expensivehat , SCZenz, Jb849, Danlaycock, Chichui, Zwobot, Ospalh, Wknight94, Ott2, Rwxrwxrwx, Smkolins, Light current, Enormous-dude, StuRat, Oysteinp, Keithd, Mavaddat, Sbyrnes321, That Guy, From That Show!, Luk, KnightRider~enwiki, SmackBot, Ashenai, Dmc-caig, Joonhon, Delldot, Eskimbot, Gilliam, Hmains, Rrscott, Melburnian, Complexica, DHN-bot~enwiki, Colonies Chris, Fiziker, Duncombe,Lesnail, Kittybrewster, ThreeAnswers, King Vegita, Rich.lewis, Turms, DMacks, Bentreuherz~enwiki, SciBrad, Lambiam, ArglebargleIV,Petr Kopač, Feraudyh, Jpawloski, SMasters, Phanc y Physicist, Dicklyon, Spiel496, Hetar, Wizard191, Newone, IRevLinas, Pathosbot, Gar-rettcobb, JRSpriggs, Chetvorno, Hairyfairycarpetfluff, CRGreathouse, BeenAroundAWhile, Linus M., John courtneidge, LPerson, Myasuda,Act333, Yzphub, Krauss, Nic k Y., Bvcrist, Peterdjones, Michae l C Price, Viridae, Abract, Raoul NK, Letranova, Thijs!bot, Michae l D.Wolok, Mbell, Headbomb, JustAGal, GordonRoss, Thadius856, Nipisiquit, JAnDbot, MER-C, .anacondabot, VoABot II, SHCarter, Docduke,AllenDowney, Tserton, R'n'B, Andrej.westermann, Rie ff el, J.delanoy, AstroHurricane001, Maurice Carbonaro, Kevin aylward, TimLong2001,Azus~enwiki, Hakuf u Sonsaku, Tarotcards, Aqm2241, Fountains of Bryn Mawr, Laurenpass, SirHolo, Lseixas, TraceyR, Sheliak, Club house,Maniaphobic, DParlevliet, AlnoktaBOT, Macspaunday, Philip Trueman, Revilo314, TXiKiBoT, The Original Wildbear, Red Act, Tom239,Voorlandt, RedAndr, Robert1947, Kpedersen1, Rshob, YohanN7, SieBot, ShiftFn, Hertz1888, Caltas, Yuefairchild, Julianva, Anfi eldman,Androi d Mouse, Flyer22 Reborn, Oxymoron83, Janfri, Szalagloria, Hamiltondaniel, Velvetron, MenoBot, Martarius, ClueBot, Binksternet,Razimantv, Mild Bill Hiccup, Isirr, DanielDeibler, Djr32, TheUNOFFICIALvandalpolice, Tyler, NuclearWarfare, Iohannes Animosus, Zil-liput, MelonBot, Mchaddock, James Kanjo, Ost316, Wyatt915, Addbot, DOI bot, Captain-tucker, Moshamoot, WMdeMuynck, CarsracBot,AnnaFrance, Favonian, LinkFA-Bot, ATOE, Luckas-bot, Yobot, TaBOT-zerem, THEN WHO WA S PHONE?, Silca678, AnomieBOT, Pianonon troppo, EryZ, Andaza, Danno uk, Citation bot, KT-2500, Peterdx, J JMesserly, Vonharris, RibotBOT, Flaviusvulso, LazyMapleSunday,樣樣, Chjoaygame, Tank.hasmukh, Machine Elf 1735, Ysyoon, Citation bot 1, Vanzac11, Cnwilliams, Trappist the monk, Wdanbae, Vrena-tor, Fivedoughnut, RjwilmsiBot, Jonlegere, EmausBot, John of Reading, Dewritech, Tommy2010, Slawekb, AvicBot, 1howardsr1, H3llBot,Fizicist, Chrisman62, Maschen, Bulwersator, Chewings72, Sudozero, Will Bebac k Auto, ClueBot NG, Jostikas, Andybiddulph, Widr, Help-ful Pixi e Bot, Verberate, Jubobroff, Bibcode Bot, Bm gub2, BG19bot, Gu y vandegrift, Vokesk, PhnomPencil, MusikAnimal, F=q(E+v^B),Jivey81, Elemenat, BattyBot, Samanthaclark11, Mdann52, Thojuf, Dexbot, Hmainsbot1, Purplearc, CuriousMind01, Nishantarya98, Pde-calculus, Iztwoz, Perseus.3,14, Jsresearch, My name is not dave, Mdominguez611, Johnfranciscollins, Monkbot, PlaidPolarity, Master Pok,Rhynhardtk, Pasten2, Vespro Latuna, Knife-in-the-drawer, DireNeed and Anonymous: 362

- **Elementary particle** *Source:* https://en.wikipedia.org/wiki/Elementary_particle?oldid=695842630 *Contributors:* CYD, Mav, Bryan Derksen, XJaM, Heron, Stevertigo, Patrick, Fbjon, Looxix~enwiki, Александър, Julesd, Glenn, AugPi, Mxn, Timwi, Reddi, Tpbradbury, Furrykef, Bevo, Donarreiskoffer, Robbot, Craig Stuntz, Nurg, Papadopc, Wikibot, Jimduck, Anthony, Ancheta Wis, Giftlite, DavidCary, Mikez, Haselhurst, Monedula, Xerxes314, Alison, Guanaco, Greydream, Anythingyouwant, Bodnotbod, Kate, Brianjd, Mormegil, Urvabara, Rich Farmbrough, Guanabot, Qutezuce, Hidaspal, Dmr2, Goplat, RJHall, RoyBoy, Robotje, Neonumbers, לזוזל, Dirac1933, DV8 2XL, Azmaverick623, Blaxthos, Kay Dekker, Joriki, Simetrical, TomTheHand, Mpatel, Isnow, Ggonnell, Palica, Strait, Miserlou, Ligulem, Naraht, DannyWilde, Lmatt, Srleffler, Chobot, Cactus.man, Roboto de Ajvol, YurikBot, Hairy Dude, NTBot~enwiki, Ohwilleke, Albert Einsteins pipe, Stephenb, Chaos, Vibritannia, SCZenz, Edwardlalone, Larsobrien, Bota47, BraneJ, Dna-webmaster, Arthur Rubin, Oyvind, GrinBot~enwiki, Smack-Bot, Mrcoolbp, Bomac, GrGBL~enwiki, Chris the speller, MalafayaBot, George Rodney Maruri Game, Silly rabbit, Complexica, MovGP0, Fmalan, Scwlong, Amazins490, Cybercobra, EPM, Garry Denke, Drphilharmonic, Sadi Carnot, ArglebargleIV, Tktktk, NongBot~enwiki, WhiteHatLurker, Jonhall, Dekaels~enwiki, Jynus, Newone, Courcelles, Laplace's Demon, SchmittM, J Milburn, Fordmadoxfraud, Cydebot, Bvcrist, Kozuch, Thijs!bot, Lord Hawk, Headbomb, MichaelMaggs, Escarbot, Ssr, JAnDbot, Eurobas, Acroterion, VoABot II, Appraiser, BatteryIncluded, R'n'B, Sgreddin, MikeBaharmast, Lk69, Acalamari, DraakUSA, TomasBat, Joshua Issac, Kenneth M Burke, Ken g6, Idioma-bot, VolkovBot, SarahLawrence Scott, Nxavar, JhsBot, Abdullais4u, Lejarrag, Antixt, PGWG, SieBot, Timb66, Sonicology, PlanetStar, Bamkin, Dhatfield, Byrialbot, Svick, Perfectapproach, Thorncrag, Big55e, ClueBot, Jmorris84, Maxtitan, Alexbot, Dekisugi, Paradoxalterist, Saintlu-cifer2008, Cockshut12345, Rreagan007, RP459, Truthnlove, Addbot, Yakiv Gluck, Draco 2k, Mac Dreamstate, Funky Fantom, CarsracBot, HerculeBot, Legobot, Blah28948, Luckas-bot, Zhitelew, KamikazeBot, Kulmalukko, Orion11M87, AnomieBOT, Girl Scout cookie, Tem-platehater, Icalanise, Citation bot, Onesius, Vuerqex, Bci2, ArthurBot, Rightly, Xqbot, Phazvmk, Kirin13, FrescoBot, Pepper, Delphinus1997, Steve Quinn, Robo37, SuperJew, HRoestBot, Sthyne, Hellknowz, Yahia.barie, Skyerise, Tobi - Tobsen, FoxBot, Physics therapist, Think!97, Bj norge, RjwilmsiBot, Beyond My Ken, EmausBot, John of Reading, Mnkyman, GoingBatty, Mthorndill, ZéroBot, Bollyjeff, StringTheory11, Markinvancouver, Quantumor, Maschen, RolteVolte, Negovori, NTox, I hate whitespace, ClueBot NG, CocuBot, Widr, Micah.yannatos1, Helpful Pixie Bot, Guzman.c, Bibcode Bot, BG19bot, Spaceawesome, Rainbot, Leaverward, Let'sBuildTheFuture, Eduardofeld, Sha-256, Dr.RobertTweed, ZX95, Joeinwiki, Mark viking, Cephas Atheos, Yo butt, Snakeboy666, Psyruby42, Haminoon, Sardeth42, TaiSakuma, LadyCailin, Morph dtlr, Delbert7, Karam adel, Liance, Isambard Kingdom, Vegitø-λmericium, KasparBot, Are you freaking kidding me, Kurousagi and Anonymous: 189

- **Quantum** *Source:* https://en.wikipedia.org/wiki/Quantum?oldid=693853902 *Contributors:* The Anome, Stevertigo, Ahoerstemeier, Smack, RodC, Charles Matthews, Jitse Niesen, Topbanana, Robbot, Sverdrup, Academic Challenger, Pengo, Graeme Bartlett, Lethe, Alison, Bensac-count, Finn-Zoltan, Jaan513, Smartcowboy, Raylu, Tsemii, Andreas Kaufmann, Freakofnurture, EugeneZelenko, Masudr, Vsmith, Too Old, El

Venot, Giftlite, Dratman, Pne, Delta G, DemonThing, Unquantum, JimWae, Mschlindwein, Rich Farmbrough, Paul August, Nabla, Guidod, Kjkolb, Photonique, Eric Kvaalen, Keflavich, PAR, Jheald, Count Iblis, Lerdsuwa, Gene Nygaard, Dennis Bratland, Falcorian, Stemonitis, Simetrical, Linas, Mindmatrix, Rparson, Johan Lont, Li-sung, Eteq, Nanite, Drbogdan, Rjwilmsi, R.e.b., Dudegalea, The wub, Alejo2083, Srleffler, Chobot, DVdm, Unc.hbar, Bgwhite, Adoniscik, Roboto de Ajvol, YurikBot, Gaius Cornelius, Ojcit, R'son-W, Moe Epsilon, Light current, Enormousdude, 2over0, Reyk, Fram, Katieh5584, Bo Jacoby, SmackBot, Varunbhalerao, Bggoldie~enwiki, Anthony Liguori, Robin Whittle, Njerseyguy, Metacomet, Rizzardi, DaveFoster110@hotmail.com, Cybercobra, Daniel.o.jenkins, Hgilbert, Sadi Carnot, Andrei Stroe, Sashato-Bot, Jaganath, Kreuzfeld, IronGargoyle, Dicklyon, Vaughan Pratt, Mikiemike, Thermochap, Diegueins, Gtxfrance, Michael C Price, Dchristle, Mikewax, HappyInGeneral, Headbomb, Jojan, Second Quantization, Dr. Submillimeter, Naturalnumber, Magioladitis, Bongwarrior, Ethanminot, Necrofear, Mbweissman, HEL, C. Trifle, Martyjmch, Drphysics, Cmelsheimer, Inwind, VolkovBot, Pasquale.Carelli, Suprcel, Aborghgr, Maxim, Q Science, Andy Dingley, Dmcq, Mike4ty4, AlleborgoBot, Ilzuhg3n1u5, Damorbel, Yintan, Texliebmann, Jm smits, Pac72, Agur bar Jacé, StewartMH, ClueBot, EoGuy, R000t, Djr32, Francisco Albani, Muro Bot, Wnt, DumZiBoT, TimothyRias, Nathan Johnson, Salam32, MaizeAndBlue86, Phidus, Addbot, Jkastrup, Favonian, Lightbot, Zorrobot, Luckas-bot, Yobot, Gigaphd, AnomieBOT, מוקיינו, Muhali, Materialscientist, Citation bot, Xqbot, Tornado79, Spancek, Liberty821, Waleswatcher, Amaury, Chjoaygame, FrescoBot, Omniscientest, D'ohBot, Pythagoras0, Citation bot 1, Pinethicket, Holy Masamune, Bookerj, FoxBot, Trappist the monk, Wdanbae, PleaseStand, Joseph449008, Jiffles1, RjwilmsiBot, John of Reading, Kosta Dean, Dewritech, GoingBatty, Robertbuckles, Hhhippo, ZéroBot, GianniG46, HCPotter, Toomuchrockcankill, 28bot, ClueBot NG, Uli Zappe, Bernhlav, Helpful Pixie Bot, Bibcode Bot, Lowercase sigmabot, Skarmenadius, Happyboy2011, Sparkie82, BattyBot, Ariaveeg, Mdann52, Jimw338, ChrisGualtieri, AK456, Dexbot, Planetree, Webclient101, Enyokoyama, Ruby Murray, Haohaolihaohao, Monkbot, Ashwiniyengar, 33wantittrue, Boehm, Nertuop, Nlebed32, Chemistry1111 and Anonymous: 187

- **Photoelectric effect** *Source:* https://en.wikipedia.org/wiki/Photoelectric_effect?oldid=696373539 *Contributors:* AxelBoldt, Brion VIBBER, Bryan Derksen, Rjstott, Css, XJaM, Fredbauder, William Avery, DrBob, Heron, Ewen, Stevertigo, Michael Hardy, Looxix~enwiki, Ahoerstemeier, Mac, Stevenj, William M. Connolley, Theresa knott, Julesd, Glenn, AugPi, Tristanb, Pizza Puzzle, Hashar, Emperorbma, Reddi, Itai, Omegatron, Pstudier, Robbot, Fredrik, Npettiaux, Chris 73, Chancemill, Hadal, SpellBott, Enochlau, Giftlite, Ferkelparade, Bensaccount, Glenn Koenig, ConradPino, Csmiller, Karol Langner, Unquantum, Mikko Paananen, Zfr, Urhixidur, Joyous!, Deglr6328, M1ss1ontomars2k4, Brianjd, Twinxor, Bosteen, Pjacobi, Roo72, Pavel Vozenilek, Nabla, Mwanner, Gilgamesh he, Robamler, Sietse Snel, David kitson, Honeycake, Atlant, Axl, Pion, Vedantm, Wtshymanski, RainbowOfLight, Lerdsuwa, Forteblast, Dennis Bratland, Linas, StradivariusTV, Kzollman, Urod, Lensovet, Eilthireach, SeventyThree, Palica, Jan.bannister, Li-sung, Volland, Saperaud~enwiki, Rjwilmsi, Zbxgscqf, Vegaswikian, Nneonneo, InFairness, Sferrier, Ems57fcva, Sjlegg, Yamamoto Ichiro, Anurup, Arnero, Doc glasgow, RexNL, Srleffler, Chobot, DaGizza, YurikBot, Wavelength, Sceptre, Jimp, Wolfmankurd, Lucinos~enwiki, CambridgeBayWeather, NawlinWiki, Wiki alf, Grafen, Kdkeller, E2mb0t~enwiki, STufaro, Ospalh, Kkmurray, Ott2, Georgewilliamherbert, Light current, Lt-wiki-bot, Imaninjapirate, Ageekgal, Petri Krohn, GraemeL, CWenger, Cjfsyntropy, Junglecat, SmackBot, InverseHypercube, KnowledgeOfSelf, Joonhon, Melchoir, Nickst, Delldot, Gary2863, Yamaguchi先生, Gilliam, JSpudeman, Saros136, SMP, Jprg1966, SchfiftyThree, Droll, Complexica, KJie.Neo, CMacMillan, VirtualSteve, ThePromenader, Chlewbot, TheKMan, GeorgeMoney, PsychoCola, Waprap, Cybercobra, Nakon, James084, Sadi Carnot, Chymicus, The undertow, SashatoBot, Zahid Abdassabur, Kuru, John, Loodog, OutSales, JorisvS, Mgiganteus1, Cielomobile, Ben Moore, Dicklyon, Chady-Wady, Tawkerbot2, CmdrObot, Mattbr, Van helsing, Laplacian, Harriemkali, Tex, Kanags, A876, Gravitroid, Jedonnelley, Michael C Price, Tawkerbot4, Roberta F., Branclem, Thijs!bot, Poorleno, Mojo Hand, Headbomb, West Brom 4ever, Tapir Terrific, Brichcja, Hcobb, D.H, Escarbot, AntiVandalBot, KP Botany, Mrshaba, LibLord, Rico402, JAnDbot, Montparnasse, IJMacD, Gumby600, Magioladitis, Pedro, Bongwarrior, VoABot II, Delta107~enwiki, WODUP, Giggy, Dirac66, Cpl Syx, Khalid Mahmood, Gwern, Gjd001, MartinBot, Nono64, J.delanoy, Trusilver, Yabbadabbadoo, AstroHurricane001, Numbo3, Mostlymostly, Tdadamemd, McSly, AntiSpamBot, Luke FM, Uberdude85, Treisijs, Deor, Tourbillon, Meaningful Username, Coldplasma, Philip Trueman, TXiKiBoT, The Original Wildbear, Red Act, Calwiki, FDominec, Nxavar, Anna Lincoln, Martin451, Jackfork, Tarun06071987, UnitedStatesian, Zain Ebrahim111, Entropy1963, Andy Dingley, Enviroboy, Junkinbomb, Planet-man828, Neparis, SieBot, AquaDTRS, YonaBot, Krawi, Matthew Yeager, The way, the truth, and the light, JerrySteal, Dattebayo321, Happysailor, Proofhand, Janfri, CultureDrone, Tesi1700, Sfan00 IMG, ClueBot, Rumping, Binksternet, Yaleks, The Thing That Should Not Be, Mriya, Frdayeen, Niceguyedc, DragonBot, Djr32, Jamespitt, ChrisHodgesUK, DumZiBoT, XLinkBot, BodhisattvaBot, Azaz129, Avoided, Mifter, Jd027, SkyLined, Truthnlove, Tayste, Addbot, Yousou, ProjectTux, CarsracBot, Debresser, Norman21, Numbo3-bot, Tide rolls, Lightbot, QuadrivialMind, Gail, Legobot, Luckas-bot, Yobot, Tamtamar, THEN WHO WAS PHONE?, AmeliorationBot, KamikazeBot, Mhmolitor, Tempodivalse, AnomieBOT, Galoubet, EryZ, Materialscientist, Citation bot, Soldarat, ArthurBot, Xqbot, Capricorn42, Pandaninja91, Idegmcsa, Turk oğlan, GrouchoBot, Sujitmahj, SobaNoodleForYou, 救火, FrescoBot, Cwtiyar, Paine Ellsworth, VI, Amadeus666, Dr John Kalien, Endofskull, Steve Quinn, Left Coast Bernard, Dscraggs, Metastabil01, Louperibot, Citation bot 1, Pinethicket, I dream of horses, Clivebeale, Boobarkee, Jeans.a, Earthandmoon, Marie Poise, Mean as custard, RjwilmsiBot, Rajettan, EmausBot, Dewritech, Racerx11, Mitartep, Kiran Gopi, Old nic, Hhhippo, M1arvin, Harddk, Dizchnik, Wayne Slam, Looscan, RaptureBot, Donner60, Surajt88, Damirgraffiti, Orange Suede Sofa, ChuispastonBot, RockMagnetist, ClueBot NG, Elcubano91, Lmoding, A520, SusikMkr, Crlance, Prateek.sondhi, Fkhwang, SolidStateDD, NuclearEnergy, Helpful Pixie Bot, Mulhollant, Wbm1058, Bibcode Bot, Shesinastro, KlausWilhelm, Jarad619, Zedshort, Shawn Worthington Laser Plasma, TheGoodBadWorst, Kisokj, Abdulllahumar, ChrisGualtieri, JYBot, Sflintg, Dexbot, Erinepwright, Lugia2453, Mohpof, Kptech, NeapleBerlina, SJ Defender, Torqu3e, Diegonolovich, Sreynoldsbros2, SkateTier, HiYahhFriend, Kanawishi, Sofia Koutsouveli, Pentazoid, LemarqueSadler, TerryAlex, Laxman z, LauraIsabelleDB, Kush bansal, Saltimoore, Katgirl2000, KasparBot, Aaseeshdatla, Frvfrferv, Jeffwu64, Shaik sunain and Anonymous: 551

- **Matter wave** *Source:* https://en.wikipedia.org/wiki/Matter_wave?oldid=687006378 *Contributors:* Heron, Stevertigo, Tim Starling, Dcljr, Paul A, Drxenocide, Gentgeen, Cdang, Dratman, Edcolins, DemonThing, Zeimusu, Jossi, Anythingyouwant, DragonflySixtyseven, Brianjd, Pjacobi, Bender235, Nabla, Dataphile, Aranel, El C, Kwamikagami, Laurascudder, Robotje, Reinyday, Jag123, Blinken, Haham hanuka, Free Bear, Keenan Pepper, PAR, Gene Nygaard, Siafu, Jtauber, Linas, Rjwilmsi, Winkels, Pleiotrop3, FlaBot, Meeve, DVdm, Korg, YurikBot, Ugha, Nmondal, JabberWok, Artur Lion~enwiki, Mingshey~enwiki, Chichui, Enormousdude, Fram, Migdejong, Carlosguitar, Teply, KasugaHuang, SmackBot, Eskimbot, Gaff, Robin Whittle, Amatulic, Pieter Kuiper, Complexica, Tobywhcheng, Jbergquist, Daniel.Cardenas, Andrei Stroe, Attys, Calum MacÙisdean, JorisvS, Hemmingsen, Nijdam, Melody Concerto, BillFlis, Dl2000, UncleDouggie, Domitori, Rhetth, Zipz0p, Chetvorno, Linus M., Myasuda, A876, RZ heretic, Quibik, DumbBOT, Iliank, Mckinlayr, Adechau, Thijs!bot, Epbr123, Headbomb, Second Quantization, Erik Baas, Qwerty Binary, Steelpillow, JAnDbot, Deflective, Shayno, Igodard, Micahnewman, Dirac66, Warren Dew, STBot, Obscurans, Jqar, Pulsarphysics, TXiKiBoT, Mathwhiz 29, Hhkaviani, Aymatth2, Anandramanathan, SieBot, Graham Beards, Ljagerman, Yintan, Patamia, ClueBot, The Thing That Should Not Be, Niceguyedc, Djr32, Excirial, Alexbot, Thingg, Tjako, DumZi-

talate, Timwi, Stone, Jitse Niesen, Rednblu, Wik, Dtgm, Patrick0Moran, Tpbradbury, Nv8200pa, Phys, Bevo, Jecar, Fvw, Stormie, Sokane, Optim, Bcorr, Johnleemk, Jni, Rogper~enwiki, Robbot, Ke4roh, Midom, MrJones, Jaleho, Astronautics~enwiki, Fredrik, Chris 73, Moncrief, Goethean, Bkalafut, Lowellian, Centic, Gandalf61, StefanPernar, Academic Challenger, Rursus, Texture, Matty j, Moink, Hadal, Papadopc, Johnstone, Fuelbottle, Lupo, HaeB, Mcdutchie, Xanzzibar, Tobias Bergemann, David Gerard, Enochlau, Ancheta Wis, Decumanus, Giftlite, Donvinzk, DocWatson42, ScudLee, Awolf002, Barbara Shack, Harp, Fudoreaper, Lethe, Fastfission, Zigger, Monedula, Wwoods, Anville, Alison, Bensaccount, Tromer, Sukael, Andris, Jason Quinn, Gracefool, Solipsist, Nathan Hamblen, Foobar, SWAdair, Mckaysalisbury, AdamJacobMuller, Utcursch, CryptoDerk, Knutux, Yath, Amarvc, Pcarbonn, Stephan Leclercq, Antandrus, JoJan, Savant1984, Jossi, Karol Langner, CSTAR, Rdsmith4, APH, Anythingyouwant, Thincat, Aaron Einstein, Edsanville, Robin klein, Muijz, Zondor, Guybrush, Grunt, Lacrimosus, Chris Howard, L-H, Ta bu shi da yu, Freakofnurture, Sfngan, Venu62, Spiffy sperry, CALR, Ultratomio, KeyStroke, Noisy, Discospinster, Caroline Thompson, Rich Farmbrough, H0riz0n, FT2, Pj.de.bruin, Hidaspal, Pjacobi, Vsmith, Wk muriithi, Silence, Smyth, Phil179, Moogoo, WarEagleTH, Smear~enwiki, Paul August, Dmr2, Bender235, ESkog, Nabla, Dataphile, Dpotter, Floorsheim, El C, Lankiveil, Kross, Laurascudder, Edward Z. Yang, Shanes, Spearhead, RoyBoy, Femto, MPS, Bobo192, Army1987, John Vandenberg, AugustinMa, Geek84, GTubio, Clarkbhm, SpaceMonkey, Sjoerd visscher, I9Q79oL78KiL0QTFHgyc, Sriram sh, Matt McIrvin, Sasquatch, BM, Firewheel, MtB, Nsaa, Storm Rider, Alansohn, Gary, ChristopherWillis, Tek022, ZiggyZig, Keenan Pepper, La hapalo, Gpgarrettboast, Pippu d'Angelo, PAR, Batmanand, Hdeasy, Bart133, Snowolf, Wtmitchell, Tycho, Leoadec, Jon Cates, Mikeo, Dominic, Bsadowski1, BlastOButter42, GabrielF, DV8 2XL, Alai, Nick Mks, KTC, Dan100, Chughtai, Falcorian, Oleg Alexandrov, Ashujo, Ott, Feezo, OwenX, Woohookitty, Linas, Superstring, Tripodics, Shoyer, StradivariusTV, Kzollman, Kosher Fan, JeremyA, Tylerni7, Pchov, GeorgeOrr, Mpatel, Adhalanay, Firien, Wikiklrsc, GregorB, AndriyK, SeventyThree, Wayward, Prashanthns, DL5MDA, Palica, Pfalstad, Graham87, Magister Mathematicae, Chun-hian, FreplySpang, Baker APS, JIP, RxS, Search4Lancer, Canderson7, Sjö, Saperaud~enwiki, Rjwilmsi, Jake Wartenberg, Linuxbeak, Tangotango, Bruce1ee, Darguz Parsilvan, Mike Peel, Pasky, HappyCamper, Ligulem, The wub, Reinis, Hermione1980, Sango123, Oo64eva, St33lbird, Kevmitch, Titoxd, Das Nerd, Alejo2083, FlaBot, Moskvax, RobertG, Urbansky~enwiki, Arnero, Latka, Nihiltres, Pathoschild, Quuxplusone, Srleffler, Kri, Cpcheung, Phoenix2~enwiki, Chobot, DVdm, Gwernol, Niz, YurikBot, Wavelength, Paulraine, Arado, Loom91, Xihr, GLaDOS, Khatharr, Firas@user, Gaius Cornelius, Chaos, Rsrikanth05, Rodier, Wimt, Anomalocaris, Royalbroil, David R. Ingham, NawlinWiki, Grafen, NickBush24, RazorICE, Stephen e nelson, JocK, SCZenz, Randolf Richardson, Vb, E2mb0t~enwiki, Tony1, Syrthiss, SFC9394, DeadEyeArrow, Bota47, Werdna, Bmju, Wknight94, WAS 4.250, FF2010, Donbert, Light current, Enormousdude, 21655, Zzuuzz, TheKoG, Lt-wiki-bot, Nielad, Closedmouth, Ketsuekigata, E Wing, Brina700, Modify, Dspradau, Netrapt, Petri Krohn, Badgettrg, Peter, Willtron, Mebden, RG2, GrinBot~enwiki, Mejor Los Indios, Sbyrnes321, CIreland, Luk, Itub, Hvitlys, SmackBot, Paulc1001, Moeron, Rex the first, InverseHypercube, KnowledgeOfSelf, Royalguard11, K-UNIT, Lagalag, Pgk, Jagged 85, Clpo13, Chairman S., Pxfbird, Grey Shadow, Delldot, Petgraveyard, Weiguxp, David Woolley, Lithium412, Philmurray, Yamaguchi⬚⬚, Robbjedi, Gilliam, Slaniel, Betacommand, Skizzik, Dauto, Holy Ganga, JSpudeman, Modusoperandi, Amatulic, Stevenwagner, DetlevSchm, MK8, Jprg1966, MalafayaBot, Marks87, Silly rabbit, Complexica, Colonies Chris, Darth Panda, Sajendra, Warbirdadmiral, El Chupacabra, Zhinz, Can't sleep, clown will eat me, Physika~enwiki, Scott3, Scray, ApolloCreed, Ackbeet, Le fantome de l'opera, Onorem, Surfcuba, Voyajer, Addshore, Stiangk, Paul E T, Huon, Khoikhoi, Kingdon, DenisDiderot, Cybercobra, Nakon, Nick125, SnappingTurtle, Dreadstar, Richard001, Akriasas, Freemarket, Weregerbil, DeFoaBuSe, DMacks, Salamurai, LeoNomis, Sadi Carnot, Pilotguy, Byelf2007, Xezlec, DJIndica, Akubra, Rory096, Bcasterline, Harryboyles, JzG, Richard L. Peterson, Rodri316, AmiDaniel, UberCryxic, Wtwilson3, Zslevi, LWF, Gobonobo, Jaganath, JorisvS, Evan Robidoux, Mgiganteus1, Zarniwoot, Goodnightmush, Jordan M, Ex nihil, SirFozzie, Waggers, MarphyBlack, Caiaffa, Asyndeton, Dan Gluck, BranStark, Iridescent, JMK, Dreftymac, Joseph Solis in Australia, UncleDouggie, Rnb, Hikui87~enwiki, Cain47, Mbenzdabest, Nturton, Civil Engineer III, Cleric12121, Tawkerbot2, Chetvorno, Carborn1, Mustbe, SkyWalker, JForget, Frovingslosh, Ale jrb, Peace love and feminism, Wafulz, Sir Vicious, Asmackey, Dycedarg, Lavateraguy, Van helsing, The ed17, Bad2101, Jayunderscorezero, BeenAroundAWhile, JohnCD, Nunquam Dormio, Harriemkali, Swwright, Wquester, Melicans, Smallpond, Myasuda, Gregbard, Xana's Servant, Dragon's Blood, Cydebot, Wrwrwr, Beek man, Meznaric, Jack O'Lantern, Peterdjones, Meno25, Gogo Dodo, Islander, DangApricot, NijaMunki, Pascal.Tesson, Hughgr, Benvogel, Michael C Price, Christian75, DumbBOT, FastLizard4, Waxigloo, Amit Moscovich, FrancoGG, CieloEstrellado, Thijs!bot, Epbr123, Derval Sloan, Koeplinger, Mbell, N5iln, Headbomb, Marek69, Ujm, Second Quantization, Martin Hedegaard, Philippe, CharlotteWebb, Nick Number, MichaelMaggs, Sbandrews, Mentifisto, Austin Maxwell, Cyclonenim, AntiVandalBot, Luna Santin, Widefox, Tkirkman, Eveross, Lontax, Grafnita, Rakniz, Prolog, Gnixon, CStar, TimVickers, Dylan Lake, Casomerville, Danger, Farosdaughter, Tim Shuba, North Shoreman, Yellowdesk, Glennwells, Byrgenwulf, GaaraMsg, Figma, JAnDbot, Leuko, Husond, Superior IQ Genius, MER-C, CosineKitty, Matthew Fennell, Eurobas, IJMacD, Andonic, Dcooper, Hut 8.5, 100110100, Skewwhiffy, Four Dog Night, Acroterion, Magioladitis, Connormah, Mattb112885, Bongwarrior, VoABot II, AtticusX, JamesBWatson, SHCarter, FagChops, Bfiene, Rivertorch, Michele123, Zooloo, Jmartinsson, Thunderhead~enwiki, Couki, Catgut, Indon, ClovisPt, Dirac66, 28421u2232nfenfcenc, Joe hill, Schumi555, Adventurer, Cpl Syx, Robb37, Quantummotion, DerHexer, Chaujie328, Khalid Mahmood, Teardrop onthefire, Guitarspecs, Info D, Seba5618, Gjd001, CiA10386, MartinBot, Arjun01, Rettetast, Mike6271, Fpaiano~enwiki, CommonsDelinker, AlexiusHoratius, Andrej.westermann, Tgeairn, Dinkytown, J.delanoy, DrKay, Trusilver, Kaesle, Numbo3, NightFalcon90909, Uncle Dick, Maurice Carbonaro, Kevin aylward, 5Q5, StonedChipmunk, Foober, Acalamari, Metaldev, Bot-Schafter, Katalaveno, DarkFalls, McSly, Bustamonkey2003, Ignatzmice, Tarotcards, JayJasper, Gcad92, Detah, LucianLachance, Midnight Madness, NewEnglandYankee, Nin0rz4u 2nv, SJP, MKoltnow, KCinDC, Han Solar de Harmonics, Cmichael, Juliancolton, Cometstyles, MoForce, Chao129, Elenseel, Wfaze, Samlyn.josfyn, Martial75, GrahamHardy, CardinalDan, Sheliak, Spellcast, Signalhead, Pgb23, Zakuragi, MBlue2020, Pleasantville, LokiClock, Lear's Fool, Soliloquial, Philip Trueman, TXiKiBoT, Oshwah, Maximillion Pegasus, SanfordEsq, RyanB88, SCriBu, Nxavar, Sean D Martin, Sankalpdravid, ChooseAnother, Qxz, Someguy1221, Liko81, Bsharvy, XeniaKon, Clarince63, Seraphim, Saibod, Fizzackerly, Zolot, Raymondwinn, Handsome Pete, Geometry guy, Ilyushka88, Leavage, Krazywrath, V81, Sodicadl, RandomXYZb, Lerdthenerd, Andy Dingley, Enigmaman, Meters, Lindsaiv, Synthebot, Antixt, Falcon8765, Enviroboy, Spinningspark, H1nomaru senshi, The Devil's Advocate, Monty845, AlleborgoBot, Nagy, The Mad Genius, Logan, PGWG, DarthBotto, Vitalikk, Belsazar, Katzmik, EmxBot, Givegains, Kbrose, Mk2rhino, YohanN7, SieBot, Ivan Štambuk, Nibbleboob, WereSpielChequers, Dawn Bard, AdevarTruth, RJaguar3, Hekoshi, Yintan, 4RM0~enwiki, Ujjwol, Bentogoa, Jc-S0CO, JSpung, Arjen Dijksman, Oxymoron83, Antonio Lopez, Henry Delforn (old), Hello71, AnonGuy, Lightmouse, Radzewicz, Hobartimus, Jaquesthehunter, Michael Courtney, Macy, Hatster301, Swegei, Curlymeatball38, Quackbumper, Coldcreation, Zenbullets, StaticGull, Heptarchy of teh Anglo-Saxons, baby, Mygerardromance, Fishnet37222, Stentor7, Mouselb, Randy Kryn, Velvetron, ElectronicsEnthusiast, Darrellpenta, Soporaeternus, Martarius, ClueBot, NickCT, Mod.torrentrealm, Scottstensland, Yeahyeahkickball, The Thing That Should Not Be, EMC125, Zero over zero, Infrasonik, MichaelVernonDavis, Herakles01, Drmies, Cp111, Diafanakrina, Mackafi92, Mrsastrochicken, VandalCruncher, Agge1000, Otolemur crassicaudatus, Ridge Runner, Neverquick, Asdf1990, DragonBot, Djr32, Ondon, Excirial, HounsGut, Welsh-girl-Lowri, Quercus basaseachicensis,

Antipodean Contributor, Tedickey, R'n'B, CommonsDelinker, Andrej.westermann, HEL, MITBeaverRocks, Maurice Carbonaro, Clackmannanshireman, (jarbarf), M-le-mot-dit, Juliancolton, DH85868993, Sheliak, Cuzkatzimhut, LokiClock, Philip Trueman, TXiKiBoT, Anonymous Dissident, Spoisp, Xnquist, EmxBot, Psymun747, YohanN7, SieBot, Keilana, Hxhbot, Physics one, Hoof47, Scorpion451, MiNombreDeGuerra, Dravecky, ClueBot, EoGuy, Niceguyedc, Saraiva.if, MorrisRob, Muhandes, NuclearWarfare, PhySusie, Phenylphenol, SchreiberBike, DumZiBoT, Boleyn, Truthnlove, Ttimespan, Realworth, Addbot, Pyfan, Mac Dreamstate, Numbo3-bot, Tide rolls, Yobot, Turiacus, Kan8eDie, Yngvadottir, Gongshow, Tonyrex, AnomieBOT, Rubinbot, Citation bot, Xqbot, Haljolad, GrouchoBot, Omnipaedista, RibotBOT, Adrignola, 丸丸, Chjoaygame, FrescoBot, Zero Thrust, Steve Quinn, Machine Elf 1735, Robo37, I dream of horses, Allthingstoallpeople, Thomas1134, Tiberius Curtainsmith, Orenburg1, Augustus the Pony, Puzl bustr, Miracle Pen, Dandrestor, EmausBot, Beatnik8983, Tpudlik, Dewritech, Socob, Stanford96, Brazmyth, Quondum, Milad pourrahmani, Higgs Teilchen, Lukedmor, Maschen, CountMacula, RockMagnetist, Llightex, E. Fokker, One really angry guy, Rememberway, Gilderien, Hermajesty21, Snotbot, TeXnocrat, Helpful Pixie Bot, Bibcode Bot, Krishnaprasaths, BG19bot, F=q(E+v^B), Nvallejo, Davidcpearce, BattyBot, Hebert Peró, Eflatmajor7th, Besprnt, ChrisGualtieri, DonnieSwanson, Mogism, LTWoods, Rick from Richmond, AHusain314, YanikB, Samhg, Hossieni2013, Muhsenphysics, Ant.ton.t, JCMPC, Monkbot, Tigercompanion25, Yaymaths, Heinerj, Akaazhar, Isambard Kingdom, Ethansolly, Pengyulong7, Phonon112358, Usman chep and Anonymous: 161

- **Schrödinger equation** *Source:* https://en.wikipedia.org/wiki/Schr%C3%B6dinger_equation?oldid=697886287 *Contributors:* AxelBoldt, Marj Tiefert, CYD, Brion VIBBER, The Anome, Ap, Josh Grosse, Toby Bartels, Mbays, David spector, FlorianMarquardt, Camembert, Stevertigo, Spiff~enwiki, Bdesham, Nommonomanac, Michael Hardy, Tim Starling, Oliver Pereira, Kku, Liftarn, Gabbe, TakuyaMurata, JeremyR, Skysmith, Dgrant, Looxix~enwiki, Lupinoid, Nikai, Andres, RodC, Charles Matthews, Laussy, Patrick0Moran, Maximus Rex, Fibonacci, Joseaperez, BenRG, Frish, Donarreiskoffer, Robbot, Rahuljp, AceMyth, Diderot, Halibutt, Fuelbottle, Syntax~enwiki, Dmn, Rubber hound, Giftlite, Wolfkeeper, BenFrantzDale, Lethe, MathKnight, Dratman, Alison, Waltpohl, Ferren~enwiki, DemonThing, Geni, Antandrus, Beland, Mako098765, Karol Langner, CSTAR, Lvl, Gscshoyru, Edsanville, Mschlindwein, Klemen Kocjancic, Vivacissamamente, Atlastawake, Garrison, Discospinster, Vsmith, Gianluigi, Pavel Vozenilek, Bender235, Neko-chan, Billlion, El C, Edward Z. Yang, Shoujun, Bobo192, Army1987, John Vandenberg, Nk, Rje, JellyWorld, HasharBot~enwiki, Passw0rd, JohnyDog, Keenan Pepper, PAR, Burn, HenkvD, Count Iblis, Egg, H2g2bob, Cmprince, Kusma, Ianblair23, Gene Nygaard, Simetrical, Linas, Elansey, StradivariusTV, Jwanders, Mpatel, MFH, Zzyzx11, Pfalstad, Torquil~enwiki, Graham87, Canderson7, Rjwilmsi, JVz, MarSch, HappyCamper, Ligulem, Bubba73, Boccobrock, Gseryakov, John Baez, Dimator, Fresheneesz, SoccerHooligan, Srleffler, Kri, Sbrools, Mhking, Daveblack, YurikBot, Wavelength, Wolfmankurd, RussBot, Bhny, JabberWok, KSmrq, Blutfink, Witger, Chichui, Larsobrien, Hirak 99, Enormousdude, Lt-wiki-bot, Arthur Rubin, Modify, Flowersofnight, RG2, Sbyrnes321, That Guy, From That Show!, Marquez~enwiki, Quadpus, Itub, SmackBot, AngelovdS, KocjoBot~enwiki, Eskimbot, Timotheus Canens, Fueled~enwiki, Jjalexand, Bduke, Georgelulu, GregRM, Papa November, Complexica, Gutworth, Threepounds, Sbharris, Hongooi, Berland, Voyajer, Wine Guy, Drphilharmonic, Daniel.Cardenas, DJIndica, SashatoBot, Vanished user 9i39j3, JorisvS, Jacobdyer, Mets501, Dan Gluck, Alex Selby, Muhammet.alp, MOBle, Mfrosz, IvanLanin, UncleDouggie, JRSpriggs, Chetvorno, Dc3~enwiki, Myncknm, Peter1c, Vaughan Pratt, BeenAroundAWhile, Tiglet, ShelfSkewed, Myasuda, Cydebot, Xxanthippe, Farrellm, Michael C Price, Karl-H, Dragonflare82, Epbr123, Kablammo, Mojo Hand, Headbomb, Second Quantization, Voyaging, GPeterL, Spartaz, MER-C, CosineKitty, Colliand, Grimlock, Kevinmon, First Harmonic, CliffC, Reden~enwiki, Arjun01, Anaxial, BrainFeverMusic, ScorpO, Leyo, Mbweissman, Freeboson, SilentCricket, Maurice Carbonaro, Yonidebot, Kevin aylward, NYCRuss, Reneedownie, P.wormer, TomyDuby, Samtheboy, Policron, Jkeohane, Tygrrr, Y2H, Izno, Sheliak, VolkovBot, JohnBlackburne, Thurth, Kjt522, DoorsAjar, TXiKiBoT, The Original Wildbear, Floddinn, Zatch zeno, A4bot, Rei-bot, Anna Lincoln, Kresadlo, PDFbot, Lokster, Dirc, Andy Dingley, Gabrielsleitao, Synthebot, EmxBot, Ian Glenn, Cerman07, Coffee, WereSpielChequers, Wing gundam, The way, the truth, and the light, Likebox, Word4michael, Paolo.dL, Michael Courtney, Cschim, Emil Wiedemann, Jfromcanada, Curtdbz, Neo., WikiBotas, FlamingSilmaril, ClueBot, GorillaWarfare, The Thing That Should Not Be, EoGuy, Warbler271, Metiscus, Dainkaplan, Morseite, Arakunem, Blanchardb, Rotational, DragonBot, Excirial, Eeekster, Muhandes, Lartoven, Sun Creator, Tıç, PhySusie, Stypex, Thingg, Crowsnest, Mdeby, RMFan1, TimothyRias, Markgeza, Truthnlove, Addbot, Substar, DOI bot, ElMaison, Wickey-nl, R85, WMdeMuynck, Wernhervonbraun, Bspasov, LaaknorBot, Glane23, CUSENZA Mario, Bob K31416, Tayzhian, Tide rolls, Matěj Grabovský, Legobot, Yobot, Kan8eDie, Yngvadottir, MichalKotowski, Moensee, AnomieBOT, 1exec1, BsBsBs, Jim1138, WWsemir, Jo3sampl, Stann78s, Citation bot, Fatkyle369, ArthurBot, Xqbot, Psyoptix, Gap9551, VVVasconcelos, Astro Reeves, RibotBOT, Blueawr, Skogssvinet, Shadowjams, 丸丸, Undsoweiter, FrescoBot, 7yl4r, Ponpan, Vuldoraq, Steve Quinn, Machine Elf 1735, Laaa200, Citation bot 1, Khashishi, I dream of horses, Spidey104, Yuri Kornyushin, XmiyumiyuX, Serols, Gryllida, Trappist the monk, Puzl bustr, Wikeno, Egemont, Cardinality, Robster95gmailcom, VernoWhitney, EmausBot, John of Reading, WikitanvirBot, Beatnik8983, Nerissa-Marie, Fly by Night, Benlansdell, KHamsun, Lequi7, TuHan-Bot, Solomonfromfinland, Gradatmit, Quantumavik, Quondum, SporkBot, Jarodalien, Wmayner, Milad pourrahmani, Mattedia, Qmtead, Maschen, C5and19, Rahul kaushik, Chewings72, Scientific29, Notolder, CountMacula, ItsTheEquations, Kurolong, RockMagnetist, Zo3a, Sitar Physics, AUN4, ClueBot NG, Lord Roem, Muon, Xinbone, Helpful Pixie Bot, Bibcode Bot, JentheGeek, BG19bot, BeRo999, Archrith, Aetheos, Solomon7968, Thdl, F=q(E+v^B), Brad7777, Randomguess, Jnracv, Freesodas, Pratyush Sarkar, Mahia9w911, Qashqaiilove, LHcheM, Mikepssd, ChinchillaBread, Dexbot, SoledadKabocha, Falktan, Garuda0001, TwoTwoHello, Chetan666, Pawnsplat, Lae985, Chandler wiland, Tgoodman656, The Herald, Prubbens, Aubreybardo, W. P. Uzer, Manul, Abhayshastry, Parushio, Anrnusna, Jamisonsloan, H Eugene Kingsley, QuantumMatt101, Monkbot, CarlosCalatayud, Trfrm123, Jmaes25, Dompenn2010, TD712, Isambard Kingdom, Punctuated11235, KasparBot, Erjas3oa, Boehm, Turtlelover22, QFT38, Nleroy917, KyleQM and Anonymous: 486

- **Quantum field theory** *Source:* https://en.wikipedia.org/wiki/Quantum_field_theory?oldid=697015959 *Contributors:* AxelBoldt, CYD, Mav, The Anome, XJaM, Roadrunner, Stevertigo, Michael Hardy, Tim Starling, IZAK, TakuyaMurata, SebastianHelm, Looxix~enwiki, Ahoerstemeier, Cyp, Glenn, Rotem Dan, Stupidmoron, Charles Matthews, Timwi, Jitse Niesen, Kbk, Rudminjd, Wik, Phys, Bevo, BenRG, Northgrove, Robbot, Bkalafut, Gandalf61, Rursus, Fuelbottle, Tobias Bergemann, Ancheta Wis, Giftlite, Lethe, Dratman, Alison, St3vo, Mboverload, DefLog~enwiki, ConradPino, Amarvc, Pcarbonn, Karol Langner, APH, AmarChandra, D6, CALR, Urvabara, Discospinster, Guanabot, Igorivanov-enwiki, Masudr, Pjacobi, Vsmith, Nvj, MuDavid, Bender235, Pt, El C, Shanes, Sietse Snel, Physicistjedi, KarlHallowell, PWilkinson, Helix84, Thialfi, Varuna, Gcbirzan, Docboat, Count Iblis, Egg, Mpatel, Marudubshinki, Graham87, Opie, Vanderdecken, Rjwilmsi, MarSch, Earin, R.e.b., RE, Strobilomyces, Arnero, Itinerant1, Alfred Centauri, Srleffler, Chobot, UkPaolo, Wavelength, Bambaiah, Hairy Dude, RussBot, TimNelson, Archelon, CambridgeBayWeather, SCZenz, Odddmonster, E2mb0t~enwiki, Semperf, Tetracube, Garion96, Erik J, Robert L, Banus, RG2, SmackBot, Stephan Schneider, Tom Lougheed, Melchoir, KocjoBot~enwiki, Mcld, Dauto, Chris the speller, Complexica, Threepounds, RuudVisser, QFT, Jmnbatista, Cybercobra, Rebooted, Victor Eremita, DJIndica, Lambiam, Mgiganteus1, Zarniwoot,

Jim.belk, Stwalkerster, SirFozzie, Hu12, Dan Gluck, Iridescent, Joseph Solis in Australia, Albertod4, Van helsing, BeenAroundAWhile, Witten Is God, Cydebot, Jamie Lokier, Meno25, Michael C Price, The 80s chick, Mendicus~enwiki, AstroPig7, Msebast~enwiki, Mbell, Headbomb, Nick Number, Mentifisto, AntiVandalBot, Bt414, Bananan~enwiki, Martin Kostner, Moltrix, Kasimann, Kromatol, Puksik, Lerman, LLHolm, RogueNinja, Tlabshier, JEH, Nikolas Karalis, Storkk, JAnDbot, Igodard, Four Dog Night, N shaji, Bongwarrior, Andrea Allais, Soulbot, Etale, Maliz, Custos0, HEL, J.delanoy, Maurice Carbonaro, Acalamari, Jeepday, Policron, Blckavnger, Juliancolton, Skou, Telecomtom, GrahamHardy, Sheliak, Cuzkatzimhut, VolkovBot, Pleasantville, Bktennis2006, Marksr, HowardFrampton, Oshwah, The Original Wildbear, Dj thegreat, Markisgreen, TBond, Lejarrag, Moose-32, Raphtee, Sue Rangell, Neparis, Drschawrz, YohanN7, SieBot, TCO, Yintan, Likebox, Paolo.dL, Tugjob, Henry Delforn (old), Jecht (Final Fantasy X), OKBot, StewartMH, ClueBot, EoGuy, Wwheaton, The Wild West guy, Shvav~enwiki, Bob108, Brews ohare, Thingg, Count Truthstein, XLinkBot, PSimeon, SilvonenBot, Truthnlove, HexaChord, Addbot, ConCompS, Pinkgoanna, Leapold~enwiki, Dmhowarth26, Glane23, Hanish.polavarapu, Lightbot, Scientryst, R.ductor, Ettrig, Yndurain, Legobot, Luckas-bot, Yobot, Ht686rg90, Niout, Tamtamar, AnomieBOT, Ciphers, Palpher, IRP, Gjsreejith, Materialscientist, Citation bot, Bci2, ArthurBot, Northryde, LilHelpa, Caracolillo, Amareto2, MIRROR, Professor J Lawrence, Plasmon1248, Omnipaedista, RibotBOT, Spellage, JayJay, FrescoBot, Kenneth Dawson, D'ohBot, Knowandgive, N4tur4le, Hyqeom, Newt Winkler, Hickorybark, Lotje, Dinamik-bot, LilyKitty, Fortesque666, Reaper Eternal, Minimac, Marie Poise, Yaush, Dylan1946, EmausBot, Racerx11, GoingBatty, Carbosi, Thecheesykid, ZéroBot, Cogiati, Jjspinorfield1, Suslindisambiguator, Quondum, Maschen, Zueignung, Davidaedwards, RockMagnetist, Lom Konkreta, ClueBot NG, Gilderien, Iloveandrea, Vacation9, Heyheyheyhohoho, Fortune432, The ubik, Zak.estrada, Widr, Helpful Pixie Bot, Guy vandegrift, Evanescent7, Ykentluo, Martin.uecker, Walterpfeifer, Pfeiferwalter, Klilidiplomus, W.D., CarrieVS, Khazar2, Momo1381, Dexbot, Cerabot~enwiki, Garuda0001, AHusain314, Thepalerider2012, A.entropy, Mark viking, Faizan, Aj7s6, संजीव कुमार, Lemnaminor, BerFinelli, Axel.P.Hedstrom, Kclongstocking, Mutley1989, I art a troler, Liquidityinsta, Prokaryotes, DemonThuum, Dingdong2680, Asherkirschbaum, Monkbot, Gjbayes, Thedarkcheese, BradNorton1979, UareNumber6, Teelaskeletor, YeOldeGentleman, Mret81, KasparBot, CAPTAIN RAJU and Anonymous: 302

- **Position and momentum space** *Source:* https://en.wikipedia.org/wiki/Position_and_momentum_space?oldid=696996171 *Contributors:* Dr Greg, Biggerj1, Addbot, Legobot, Yobot, Hhhippo, Maschen, Wbm1058, Proksimulo and Anonymous: 6

- **Pilot wave** *Source:* https://en.wikipedia.org/wiki/Pilot_wave?oldid=696045538 *Contributors:* William M. Connolley, Clementi, Lumidek, Chris Howard, Foobaz, Mytildebang, Gpvos, Linas, MONGO, Jonnabuz, Rjwilmsi, Nemesis75, Conscious, Bhny, Salsb, Gareth Jones, Simxp, Profero, KasugaHuang, Chymicus, JorisvS, Pogsquog, Dr Smith, Dubrillion1, Cydebot, Krauss, Headbomb, Phoe, Widefox, Felix116, JamesBWatson, R'n'B, NinjaLore, Barraki, Myrikhan, Felixaldonso, XLinkBot, Addbot, Lightbot, Luckas-bot, Yobot, AnomieBOT, Edguy99, Deadly Nut, FrescoBot, Zicovich, Difu Wu, EmausBot, John of Reading, Racerx11, Spatrick99, RobertCurrey, Mikhail Ryazanov, Gilderien, Kasirbot, Ben morphett, Bibcode Bot, Horn.imh, Harizotoh9, Plutoniumjesus, Mogism, Lepixie2013surleweb, FizykLJF, Acadfandom and Anonymous: 25

- **Hidden variable theory** *Source:* https://en.wikipedia.org/wiki/Hidden_variable_theory?oldid=696151220 *Contributors:* Roadrunner, Michael Hardy, Ryan Cable, J-Wiki, Jonathan Chang, Charles Matthews, Patrick0Moran, Trevor mendham, Goethean, Blainster, Lethe, Cardmagic, Mckaysalisbury, Fuzzy, ConradPino, Lumidek, BeatePaland, Thorwald, Chris Howard, Freakofnurture, Arensb, Caroline Thompson, Guettarda, Gunboat Diplomat, Diego Moya, Plumbago, BRW, Count Iblis, Deathphoenix, Falcorian, Linas, RHaworth, Benhocking, Cannibal-Smith, Marudubshinki, Zoz, Rjwilmsi, Smithfarm, The wub, Bubbleboys, Chobot, DVdm, YurikBot, Hornplease, Rsrikanth05, Anomalocaris, Hwasungmars, StuRat, Curpsbot-unicodify, Profero, Sardanaphalus, SmackBot, Fulldecent, GaeusOctavius, Jordansmith, Jprg1966, Frap, Chlewbot, DJIndica, JorisvS, Antonielly, Iridescent, Chetvorno, Vyznev Xnebara, Vnvnfls, Dragon's Blood, Peterdjones, Michael C Price, PKT, Headbomb, JustAGal, MichaelMaggs, Widefox, Christopher Cooper, PChalmer, David Eppstein, Foober, Cpiral, Inquam, Tarotcards, Sheliak, VolkovBot, Pierre-Alain Gouanvic, Jim E. Black, Phe-bot, SimonTrew, Felixaldonso, Myrvin, Orangedolphin, Phrygiankleinbottle, Quarsaw, Elsweyn, Agge1000, Wildspell, Keithbowden, El bot de la dieta, Chippy87, TimothyRias, Fastily, Duncan, Addbot, Luckas-bot, Yobot, Alexander hendrix, Jacobs, Ciphers, Materialscientist, Citation bot, Deadly Nut, NOrbeck, Dvtausk, Chjoaygame, Paine Ellsworth, WQUlrich, Jonesey95, RedBot, Thecurran91, Full-date unlinking bot, RjwilmsiBot, EmausBot, Zeeber78, Manicjedi, A2soup, אבן ראשה, Suslindisambiguator, Staszek Lem, Loggerjack, Tommaso Ferrara, ChuispastonBot, ClueBot NG, Vincent Russo, Bibcode Bot, BG19bot, Prephysics, Mrt3366, Makecat-bot, Kleinerkai, Sethur2, Orly.alter, Y-S.Ko and Anonymous: 72

- **Quantum potential** *Source:* https://en.wikipedia.org/wiki/Quantum_potential?oldid=693843633 *Contributors:* Michael Hardy, Phil Boswell, Chris Howard, Rjwilmsi, Colonies Chris, Md2perpe, Cydebot, Michael C Price, Repliedthemockturtle, Addbot, Yobot, AnomieBOT, LilHelpa, Xqbot, HRoestBot, John of Reading, Helpful Pixie Bot, Addihockey10 (automated), BG19bot, Malyszkz, CarrieVS, Dexbot, Absinthia Stacy, Paragrin and Anonymous: 13

- **Interference (wave propagation)** *Source:* https://en.wikipedia.org/wiki/Interference_(wave_propagation)?oldid=696399933 *Contributors:* Sodium, Bryan Derksen, Andre Engels, William Avery, DrBob, Waveguy, Heron, FlorianMarquardt, Michael Hardy, Mac, Theresa knott, Qwert, Mxn, Reddi, Zoicon5, Radiojon, Patrick0Moran, Furrykef, Hyacinth, Omegatron, Sandman~enwiki, Cdnc, Sverdrup, Giftlite, Edcolins, Toytoy, Karol Langner, H Padleckas, PSzalapski, Icairns, Tsemii, Sonett72, Deglr6328, Danh, Rich Farmbrough, Guanabot, Vsmith, Solkoll~enwiki, Bobo192, Army1987, Interiot, Atlant, Snowolf, KingTT, Wtshymanski, Falcorian, Oleg Alexandrov, Nuno Tavares, Gerd Breitenbach, Jftsang, Cleonis, Pol098, Burgher, MONGO, Isnow, SDC, Waldir, Prashanthns, Pfalstad, Mandarax, Rnt20, Graham87, BD2412, Saperaud~enwiki, Pdelong, Tangotango, Jmcc150, Mathbot, Margosbot~enwiki, JdforresterBot, Gurch, J S Lundeen, Srleffler, Chobot, Krishnavedala, Cookie4869~enwiki, Hellbus, Shell Kinney, Jonathan Webley, BOT-Superzerocool, Enormousdude, Ethan Mitchell, Profero, Mejor Los Indios, KasugaHuang, SmackBot, Royalguard11, Skizzik, Kaiserb, Bluebot, MalafayaBot, Colonies Chris, OrphanBot, Kukini, DJIndica, SashatoBot, Harryboyles, Pflatau, Silvarbullet1, Rainwarrior, Rubikfreak, Ginkgo100, BranStark, Iridescent, RekishiEJ, Chetvorno, JohnTechnologist, Jordan Brown, RenamedUser2, Chasingsol, Michael C Price, Quibik, Christian75, Thijs!bot, Prarxol, JAnDbot, CosineKitty, Akuyume, Magioladitis, VoABot II, Dekimasu, MartinBot, Akurn, Rpetrenko, Mange01, Rgoodermote, Maurice Carbonaro, Rp:cs, LordAnubisBOT, Sheliak, VolkovBot, JohnBlackburne, TXiKiBoT, Rei-bot, Una Smith, Trevorcox, Mouse is back, Anoko moonlight, EmxBot, SieBot, Erguvan7~enwiki, YonaBot, BotMultichill, Nestea Zen, Gerakibot, Mbz1, LoopyEditor, Eurion, Mewasul, Aly89, BrokenBinary, Mygerardromance, ClueBot, Avenged Eightfold, Unica111, Neverquick, DragonBot, Xasthom, Brews ohare, SchreiberBike, Thingg, Aitias, Etabackman, XLinkBot, Rob-bob7-0, Addbot, Some jerk on the Internet, Fgnievinski, Epzcaw, KaiKemmann, Tide rolls, Yobot, The Earwig, QuantumCyclops, Jalal0, AnomieBOT, 4e to 4e, Materialscientist, TheAMmollusc, Corrigendas, Ahmed1994, JKaver18, Skarl the Drummer, Master27~enwiki, Omnipaedista, Chongkian, Anla001, Lookang, Steve Quinn, Weetoddid, BenzolBot, Wjh31, HamburgerRadio, Aldy, HRoestBot, LiborX, Cnwilliams, Ismathsadhir, Sarsene, TjBot, Garuh knight, EmausBot, WikitanvirBot, Jmencisom, BigJoe Physics, HCPotter,

Wstraub, Ishanarora11 11, ClueBot NG, Ulflund, Natural Philo, Lolo Sambinho, Nupsgc, Dalcde, Helpful Pixie Bot, Plnsaisabarish, Ashshydv, SIMMINARULA, Supersam654, Mark Arsten, Aranea Mortem, Anbu121, Fizped~enwiki, Qocheedy daiin, BrightStarSky, Dexbot, Final321, Qwertz987654321, Siw1939, Student2602, Evilcomedian and Anonymous: 191

- **Unified field theory** *Source:* https://en.wikipedia.org/wiki/Unified_field_theory?oldid=697873107 *Contributors:* M~enwiki, Michael Hardy, Ahoerstemeier, William M. Connolley, Charles Matthews, Reddi, Dysprosia, Pakaran, SJRubenstein, Josh Cherry, Academic Challenger, Rursus, Dbenbenn, JamesMLane, Xerxes314, StargateX1, Gzornenplatz, Karol Langner, Gscshoyru, Vitaleyes, Svdb, Caroline Thompson, Brianhe, H0riz0n, Cfailde, DPFJr, Pjacobi, Paul August, Dmr2, ESkog, Lentando~enwiki, BenjBot, Oldsoul, Etimbo, Noren, Bobo192, Ablathanalba, Mordemur, John Vandenberg, I9Q79oL78KiL0QTFHgyc, Nsaa, Hdeasy, Count Iblis, Falcorian, Blaze Labs Research, Simetrical, Linas, Mindmatrix, Decrease789, Savantnavas, Mpatel, Knuckles, Marudubshinki, BD2412, Nightscream, Macumba, R.e.b., Protez, Srleffler, Chobot, Moocha, GangofOne, DVdm, YurikBot, RadioFan2 (usurped), Rsrikanth05, NawlinWiki, Trovatore, Syrthiss, Steve G~enwiki, KasugaHuang, SmackBot, McGeddon, Gilliam, Chris the speller, Bluebot, Jjalexand, Crazy8s, Jeysaba, Silly rabbit, Timneu22, Redattore, Colonies Chris, QFT, InnocentMind, Xyzzyplugh, Jgwacker, Corby, RolandR, Marcus Brute, Sadi Carnot, Vina-iwbot~enwiki, Lambiam, Nishkid64, ArglebargleIV, Titus III, John, Gobonobo, Kevlarmry, Ckatz, Slakr, Rainwarrior, Trounce, Twas Now, Bridg, Courcelles, Shedsan, Prof.Maque, JForget, Will314159, Friendly Neighbour, NickW557, Gregbard, Peripitus, Gogo Dodo, Michael C Price, Qwyrxian, Roger Anderton, Headbomb, Marek69, Twcjr, KrakatoaKatie, AntiVandalBot, Emeraldcityserendipity, Tim Shuba, Verticordia~enwiki, Alphachimpbot, AndreasWittenstein, Blaine Steinert, Yill577, Wasell, VoABot II, Tobogganoggin, Arrowcatcher, DAGwyn, Catgut, Web-Crawling Stickler, Ours18, Skylights76, Jke310, Stephenchou0722, R'n'B, J.delanoy, Captain panda, Trusilver, Eliz81, Dogstar11, CardinalDan, VolkovBot, Butwhatdoiknow, Fennmeister, Alphanon, Cheffoxx, Betanon, ARUNKUMAR P.R, Deanlsinclair, PaddyLeahy, Gaelen S., Travisbmoore, Gammanon, Zharradan.angelfire, Anakin101, De728631, ClueBot, NossB, EhJJ, Bhushan foryou, Versus22, Qwfp, Pandanator75, Arthur chos, Addbot, Cxz111, Bobtron5000, Bte99, MrOllie, Deepthought137, Favonian, AtheWeatherman, Whitematter, Scientryst, MuZemike, Yobot, TaBOT-zerem, AnomieBOT, 9258fahsflkh917fas, Kanat Abildinov, Materialscientist, JohnnyB256, Addihockey10, Shirik, Mathonius, Pereant antiburchius, Natural Cut, Shadowjams, Lovelylilian, A. di M., Paine Ellsworth, Ottokar~enwiki, Knowandgive, RandomStringOfCharacters, Grandunifier, Tennant uk, Miracle Pen, Afteread, RA0808, Tommy2010, Wikipelli, Hhhippo, Susfele, AvicAWB, S.Lenane, DanielBurnstein, Barendjacobus, Davidaedwards, Mkh025, Dudge1983, ClueBot NG, StevenPower, Theopolisme, MerllwBot, Mophedd, Orphadeus, Peter Donald Rodgers, Joe0x7F, Dilaton, Astralbound, LynnetteA11, GabeIglesia, Jamesmcmahon0, NfrHtp, Titusfox, KEVIN123456789, Froglich, BuilderE, Michelle1881, Balbinder1706, Stowcalj, KasparBot, Chemistry1111, Nigel Hands, Dr Peter Donald Rodgers, FloRGX420S and Anonymous: 197

- **Many-worlds interpretation** *Source:* https://en.wikipedia.org/wiki/Many-worlds_interpretation?oldid=695228091 *Contributors:* AxelBoldt, Derek Ross, LC~enwiki, CYD, Zundark, Timo Honkasalo, The Anome, WillWare, Eclecticology, Josh Grosse, Darius Bacon, Nate Silva, Roadrunner, Maury Markowitz, FvdP, Stevertigo, Michael Hardy, JakeVortex, Oliver Pereira, Gabbe, Ixfd64, Alfio, William M. Connolley, Aarchiba, Evercat, HPA, Richj, Charles Matthews, Timwi, Terse, Grendelkhan, Phys, Fairandbalanced, Bevo, Andy Fugard, Nnh, BenRG, Fredrik, Vespristiano, Altenmann, Arkuat, Rholton, Bkell, ElBenevolente, Carnildo, Pablo-flores, Giftlite, Gtrmp, Barbara Shack, Wolfkeeper, Lethe, Bfinn, Wwoods, Ebonmuse, Ryanaxp, Wmahan, Mmm~enwiki, ChicXulub, Quadell, Piotrus, MadIce, Khaosworks, CSTAR, Togo~enwiki, Latitude0116, Sam Hocevar, Lumidek, Jcorgan, Eyv, Robin klein, Acsenray, Chris Howard, D6, Freakofnurture, Rich Farmbrough, Guanabot, Leibniz, Florian Blaschke, Dbachmann, Pavel Vozenilek, Nchaimov, Ben Standeven, El C, Haxwell, Kotuku33, SamRushing, Ehaque, Danski14, Plumbago, SlimVirgin, Batmanand, Hdeasy, Schaefer, Fourthords, Count Iblis, Pauli133, DV8 2XL, Kromozone, Dan100, Falcorian, Firsfron, Woohookitty, Linas, Dandv, JonBirge, Thruston, SDC, Joke137, Pfalstad, Marudubshinki, Mandarax, QuaestorXVII, Graham87, Grammarbot, Rjwilmsi, Ckoenigsberg, Eyu100, Aero66, HappyCamper, Theodork, Wragge, Arnero, ZoneSeek, Diza, Saswann, Chobot, DVdm, Korg, McGinnis, YurikBot, NTBot~enwiki, RussBot, Fabartus, Bhny, Gaius Cornelius, Vincej, Bboyneko, Thane, Draeco, David R. Ingham, Joncolvin, Nti2005, Schlafly, Brian Olsen, Kxjan, Inhighspeed, Pnrj, Emersoni, Zwobot, Kkmurray, Light current, Geoffrey.landis, Radioflux, Nixer, Ilmari Karonen, NeilN, Teo64x, SmackBot, Rex the first, JohnSankey, C.Fred, Jrockley, Dave Kielpinski, MalafayaBot, Silly rabbit, Imaginaryoctopus, Salmar, Shantrika, Calbaer, DenisDiderot, Savidan, DavidBoden, Ryan Roos, Hunter2005, DJIndica, Lambiam, Ser Amantio di Nicolao, Dr. Sunglasses, Mgiganteus1, Ckatz, Santa Sangre, Dicklyon, Hypnosifl, Eridani, Stephen B Streater, HisSpaceResearch, Dreftymac, Joseph Solis in Australia, Brian Wowk, Courcelles, Laplace's Demon, Bstepp99, Mustbe, 8754865, JForget, Will314159, CmdrObot, Vyznev Xnebara, AshLin, Denis MacEoin, Rgonsalv, Myasuda, Gregbard, Dragon's Blood, Cydebot, Peterbyrne, Peterdjones, Michael C Price, Alexnye, Geewee, Michael D. Wolok, Mbell, Headbomb, Parsiferon, Cj67, Stannered, Cyclonenim, AntiVandalBot, Mukake, Czj, MER-C, CosineKitty, Txomin, Magioladitis, SHCarter, Skew-t, Brusegadi, Torchiest, Duendeverde, Plexos, Sm8900, R'n'B, J.delanoy, Maurice Carbonaro, Kevin aylward, 5Q5, Coastal593, Shawn in Montreal, Gill110951, Tarotcards, Eli the Barrow-boy~enwiki, Sarge009, Phatius McBluff, Ajfweb, Inwind, Sheliak, Neuromath, VolkovBot, Jdcaust, Eve Hall, Xnquist, Aymatth2, Don4of4, TBond, Wiae, Ar-wiki, Sapphic, Spinningspark, Hrimpurstala, AlleborgoBot, Quantum Person, Yintan, Wing gundam, Likebox, Flyer22 Reborn, JohnSawyer, JohnnyMrNinja, Longi 93, AussieScribe, VanishedUser sdu9aya9fs787sads, Martarius, ClueBot, EoGuy, General Epitaph, EMC125, Djr32, Tianasez, Estirabot, Vanhoabui, Sun Creator, Resuna, SchreiberBike, Carriearchdale, Thingg, Ctkohl, DumZiBoT, BarretB, XLinkBot, Sanchoquixote, Kbdankbot, Addbot, Miskaton, Gregz08, Zahd, Tanhabot, Thirteenangrymen, Mac Dreamstate, Proxima Centauri, Ld100, Lightbot, Taketa, زرشک, Zorrobot, Whitneyz32, Legobot, MichelCPrice, Yobot, Pink!Teen, AnomieBOT, Unara, Citation bot, DirlBot, LilHelpa, Measles, Omnipaedista, Targeran, WaysToEscape, FrescoBot, ایلیا2010, Machine Elf 1735, MorphismOfDoom, Citation bot 1, I dream of horses, Jonesey95, Chatfecter, Grok42, Jusses2, Fredkinfollower, SkyMachine, Dc987, Duke159, Lhollo, GregWooledge, Jamesabloom, George Richard Leeming, Rekcana, Feelingsman22, Slightsmile, Italia2006, ZéroBot, Cogiati, Gahr gardner, Quantumavik, RUBEN TESOLIN, H3llBot, Quondum, Quantholic, L Kensington, Senjuto, Surajt88, Ihardlythinkso, Kartasto, Teapeat, Rememberway, ClueBot NG, Nobody60, Incompetence, Kazzie1995, Kasirbot, Helpful Pixie Bot, Bibcode Bot, BG19bot, Boriaj, Stelpa, FiveColourMap, AIMW32, Huntingg, Eqb1987, Willempramschot, Brendan.Oz, Harizotoh9, Dule1101, Modalizer, Khazar2, JYBot, Mogism, Corn cheese, Reatlas, Epicgenius, Jamesmcmahon0, A Certain Lack of Grandeur, Thevideodrome, Arfæst Ealdwrítere, Bardoligneo, Stamptrader, Fixture, Ismael755, 2PeterElls, Monkbot, Kalipsos, MazeHatter, Jonkirstenhof, CV9933, Azealia911, Fixing.your.problems., Karhu360 and Anonymous: 271

- **Dirac delta function** *Source:* https://en.wikipedia.org/wiki/Dirac_delta_function?oldid=692063852 *Contributors:* AxelBoldt, Zundark, The Anome, Tarquin, RAE, Ap, XJaM, Camembert, Patrick, Chas zzz brown, Michael Hardy, David Martland, Kidburla, TakuyaMurata, Looxix~Stevenj, Cyan, AugPi, Charles Matthews, Jitse Niesen, Fibonacci, Omegatron, Donarreiskoffer, Robbot, Benwing, Sverdrup, Henrygb, Lzur,Tobias Bergemann, Tosha, Centrx, Giftlite, Smjg, Mikez, BenFrantzDale, Lethe, MathKnight, Jason Quinn, Mboverload, Delta G, Mark-

23.16.2 Images

- **File:Einstein_patentoffice.jpg** *Source:* https://upload.wikimedia.org/wikipedia/commons/a/a0/Einstein_patentoffice.jpg *License:* Public domain *Contributors:* Transferred from en.wikipedia; transferred to Commons by User:Guerillero using CommonsHelper. *Original artist:* Lucien Chavan [#cite_note-author-1 [1]] (1868 - 1942), a friend of Einstein's when he was living in Berne.

- **File:Erwin_Schrodinger2.jpg** *Source:* https://upload.wikimedia.org/wikipedia/commons/9/99/Erwin_Schrodinger2.jpg *License:* Public domain *Contributors:* [1] [2] *Original artist:* ?

- **File:ExperimentCouder-Young.png** *Source:* https://upload.wikimedia.org/wikipedia/commons/9/90/ExperimentCouder-Young.png *License:* CC BY-SA 3.0 *Contributors:* Own work *Original artist:* Krauss

- **File:Finitepot.png** *Source:* https://upload.wikimedia.org/wikipedia/commons/b/b2/Finitepot.png *License:* CC-BY-SA-3.0 *Contributors:* Transferred from en.wikipedia *Original artist:* Original uploader was Bamse at en.wikipedia

- **File:Gold_leaf_electroscope_diagram.svg** *Source:* https://upload.wikimedia.org/wikipedia/commons/8/8a/Gold_leaf_electroscope_diagram.svg *License:* Public domain *Contributors:*

- Gold_leaf_electroscope_diagram.jpg *Original artist:* Gold_leaf_electroscope_diagram.jpg: Luke FM (talk)

- **File:Guassian_Dispersion.gif** *Source:* https://upload.wikimedia.org/wikipedia/commons/5/56/Guassian_Dispersion.gif *License:* CC0 *Contributors:* This mathematical image was created with Mathematica *Original artist:* Teply

- **File:HAtomOrbitals.png** *Source:* https://upload.wikimedia.org/wikipedia/commons/c/cf/HAtomOrbitals.png *License:* CC-BY-SA-3.0 *Contributors:* ? *Original artist:* ?

- **File:Heinrich_Rudolf_Hertz.jpg** *Source:* https://upload.wikimedia.org/wikipedia/commons/5/50/Heinrich_Rudolf_Hertz.jpg *License:* Public domain *Contributors:* http://wellcomeimages.org/indexplus/imageM0014750.html *Original artist:* Robert Krewaldt

- **File:Heisenberg_gamma_ray_microscope.svg** *Source:* https://upload.wikimedia.org/wikipedia/commons/b/bc/Heisenberg_gamma_ray_.svg *License:* CC-BY-SA-3.0 *Contributors:* Wikimedia commons *Original artist:* parri

- **File:Heisenbergbohr.jpg** *Source:* https://upload.wikimedia.org/wikipedia/commons/1/1a/Heisenbergbohr.jpg *License:* Public domain *Contributors:* http://www.fnal.gov/pub/inquiring/timeline/images/heisenbergbohr.jpg shown on http://www.fnal.gov/pub/inquiring/timeline/05.html *Original artist:* Fermilab, U.S. Department of Energy

- **File:Hilbert.jpg** *Source:* https://upload.wikimedia.org/wikipedia/commons/7/79/Hilbert.jpg *License:* Public domain *Contributors:* ? *Original artist:* ?

- **File:Hugh-Everett.jpg** *Source:* https://upload.wikimedia.org/wikipedia/en/c/cf/Hugh-Everett.jpg *License:* Fair use *Contributors:*
  http://ucispace.lib.uci.edu/handle/10575/1060
  http://sites.uci.edu/ucisca/2011/09/15/hugh-everett-iii-and-quantum-physics/ *Original artist:* ?

- **File:Hydrogen_Density_Plots.png** *Source:* https://upload.wikimedia.org/wikipedia/commons/e/e7/Hydrogen_Density_Plots.png *License:* Public domain *Contributors:* the English language Wikipedia (log). *Original artist:* PoorLeno (talk)

- **File:Infinite_potential_well.svg** *Source:* https://upload.wikimedia.org/wikipedia/commons/2/27/Infinite_potential_well.svg *License:* Public domain *Contributors:* Created by bdesham in Inkscape. *Original artist:* Benjamin D. Esham (bdesham)

- **File:Interference_colours_in_soap_film_1.jpg** *Source:* https://upload.wikimedia.org/wikipedia/commons/2/20/Interference_colours_in_soap_film_1.jpg *License:* CC BY-SA 3.0 *Contributors:* Own work *Original artist:* Natural Philo

- **File:Interference_of_a_quantum_particle_with_itself.gif** *Source:* https://upload.wikimedia.org/wikipedia/commons/7/7d/Interference_of_a_quantum_particle_with_itself.gif *License:* CC0 *Contributors:* Own work *Original artist:* Thierry Dugnolle

- **File:Interference_of_plane_waves_3.svg** *Source:* https://upload.wikimedia.org/wikipedia/commons/b/b3/Interference_of_plane_waves_3.svg *License:* CC0 *Contributors:* Own work *Original artist:* Epzcaw

- **File:Interference_of_two_waves.svg** *Source:* https://upload.wikimedia.org/wikipedia/commons/0/0f/Interference_of_two_waves.svg *License:* CC BY-SA 3.0 *Contributors:* Vecorized from File:Interference of two waves.png *Original artist:*

- original version: Haade;

- **File:Interferences_plane_waves.jpg** *Source:* https://upload.wikimedia.org/wikipedia/commons/d/d8/Interferences_plane_waves.jpg *License:* Public domain *Contributors:* No machine-readable source provided. Own work assumed (based on copyright claims). *Original artist:* No machine-readable author provided. Fffred~commonswiki assumed (based on copyright claims).

- **File:Many-worlds-bloch.png** *Source:* https://upload.wikimedia.org/wikipedia/en/d/d6/Many-worlds-bloch.png *License:* Cc-by-sa-3.0 *Contributors:* ? *Original artist:* ?

- **File:Many-worlds.svg** *Source:* https://upload.wikimedia.org/wikipedia/commons/8/82/Many-worlds.svg *License:* CC-BY-SA-3.0 *Contributors:* en:Image:Many-worlds.png *Original artist:* Traced by User:Stannered

- **File:Max_Born.jpg** *Source:* https://upload.wikimedia.org/wikipedia/commons/f/f7/Max_Born.jpg *License:* Public domain *Contributors:* ? *Original artist:* ?

- **File:Max_Planck_(1858-1947).jpg** *Source:* https://upload.wikimedia.org/wikipedia/commons/a/a7/Max_Planck_%281858-1947%29.jpg *License:* Public domain *Contributors:* http://www.sil.si.edu/digitalcollections/hst/scientific-identity/CF/display_results.cfm?alpha_sort=p *Original artist:* Unknown<a href='//www.wikidata.org/wiki/Q4233718' title='wikidata:Q4233718'><img alt='wikidata:Q4233718' src='https://upload.wikimedia.org/wikipedia/commons/thumb/f/ff/Wikidata-logo.svg/20px-Wikidata-logo.svg.png' width='20' height='11' srcset='https://upload.wikimedia.org/wikipedia/commons/thumb/f/ff/Wikidata-logo.svg/30px-Wikidata-logo.svg.png 1.5x, https://upload.wikimedia.org/wikipedia/commons/thumb/f/ff/Wikidata-logo.svg/40px-Wikidata-logo.svg.png 2x' data-file-width='1050' data-file-height='590' /></a>

- **File:Millikan.jpg** *Source:* https://upload.wikimedia.org/wikipedia/commons/2/2f/Millikan.jpg *License:* Public domain *Contributors:* http://nobelprize.org/nobel_prizes/physics/laureates/1923/millikan-bio.html *Original artist:* Nobel foundation

23.16.3   Content license

www.ingramcontent.com/pod-product-compliance
Lightning Source LLC
Chambersburg PA
CBHW06013921050526
45169CB00025B/2041